SOILS and GROUNDWATER POLLUTION and REMEDIATION

Asia, Africa, and Oceania

Edited by

P.M. Huang
I.K. Iskandar

Coeditors

M. Chino, T.B. Goh, P.H. Hsu,
D.W. Oscarson, and L.M. Shuman

Library of Congress Cataloging-in-Publication Data

Soils and groundwater pollution and remediation : Asia, Africa, and Oceania [sic] /
edited by P.M. Huang and Iskandar K. Iskandar.
p. cm.
Includes bibliographical references and index.
ISBN 1-56670-452-9
1. Soil pollution—Asia. 2. Groundwater—Pollution—Asia. 3. Soil remediation—Asia.
4. Groundwater—Purification—Asia. 5. Soil pollution—Oceania. 6. Groundwater—
Pollution—Oceania. 7. Soil pollution—Africa. I. Huang, P.M. (Pan Ming), II. Iskandar,
I.K. (Iskandar Karam), 1938–
TD878.4.A78S65 1999
363.739′6′095—dc 98-8491
CIP

International Standard Book Number 1-556670-452-9
Library of Congress Card Number 98-8491
Printed in the United States of America 1 2 3 4 5 6 7 8 9 0
Printed on acid-free paper

Table of Contents

Preface vii
About the Editors ix
Affiliation of Editors, Coeditors, and Contributors xi

1 **Environmental Impacts of Heavy Metals in Agroecosystems and Amelioration Strategies in Oceania** 1
K.G. TILLER, M.J. McLAUGHLIN, AND A.H.C. ROBERTS
Introduction 1
Background Concentrations of Heavy Metals in Australian and New Zealand Soils 1
Sources of Metals in Agroecosystems 3
Environmental Impact 10
Amelioration Strategies 17
Regulatory Control 28
Summary and Conclusions 35
References 35

2 **Environmental Concerns of Pesticides in Soil and Groundwater and Management Strategies in Oceania** 42
B.K.G. THENG, R.S. KOOKANA, AND A. RAHMAN
Introduction 42
Usage 43
Behavior and Fate in Soil 50
Pesticide Residues in the Environment 59
Management and Remediation Options 68
Summary and Conclusions 70
Acknowledgments 70
References 71

3 **Metal Pollution of Soil and Groundwater and Remediation Strategies in Japan** 80
M. CHINO
Introduction 80
Regulations and Control 81
Metals in Soil, Groundwater, Fertilizers, and Sewage Sludges of Japan 84
Remediation Strategies 87
Daily Intake of Heavy Metals by Japanese 90
Application of Sewage Sludge and Other Biosolids to Land and Heavy Metal Accumulation 91
References 95

4 **Impact of Radionuclides on Soil, Groundwater, and Crops and Radionuclide Cleanup in Japan** 96
S. UCHIDA AND Y. OHMOMO
Introduction 96

Meteorological Data in Japan 97
Fallout Radionuclides (^{137}Cs, ^{90}Sr, Pu, etc.) Deposited on Soil 99
Migration of Radionuclides in Soil and Groundwater 101
Transfer of Radionuclides to Crops 108
Decontamination of Radionuclides 118
Summary and Conclusions 121
Acknowledgments 122
References 122

5 **Effect of Anthropogenic Organic Compounds on the Quality of Soil and Groundwater and Remediation Strategies in Japan** 126
O. NAKASUGI AND T. HIRATA
Introduction 126
Nationwide Survey Results 127
Site Investigation for Volatile Organochlorines 132
Remediation Technologies for Subsurface Pollution 141
Summary and Conclusions 147
References 148

6 **Heavy Metal Pollution in Soils and Its Remedial Measures and Restoration in Mainland China** 150
H.K. WANG
Introduction 150
Background Levels of Trace Elements in Soils of China 151
Site Research of Soil Pollution in China 151
Heavy Metal Pollutants and Their Influencing Factors 154
Management Practices for the Mitigation of Heavy Metal Pollution in Soils of China 156
Conclusions 163
References 163

7 **Environmental Impacts of Metal and Other Inorganics on Soil and Groundwater in China** 167
G.X. XING AND H.M. CHEN
Introduction 167
Background Values of Elements in Soils of China 168
Soil Loading Capacity for Heavy Metals 171
Forms of Heavy Metals in Soils 179
Heavy Metal Pollution of Soils in China 186
Effect of the Amount of Dissolved Heavy Metals and Enrichment of Other Inorganics in Groundwater on Ecological Environment 193
Summary and Conclusions 196
References 197

8 **Environmental Problems in Soil and Groundwater Induced by Acid Rain and Management Strategies in China** 201
G. L. JI, J. H. WANG, AND X. N. ZHANG
Introduction 201
Acid Rain 201

Acidity and Acidification of Soils 205
Acidification of Groundwater 219
Strategies for Controlling the Acidification of the Soil Environment 221
Summary and Conclusions 222
Acknowledgments 222
References 222

9 Dynamics, Fate, and Toxicity of Pesticides in Soil and Groundwater and Remediation Strategies in Mainland China 225
D. J. CAI AND Z. L. ZHU
Introduction 225
History of Pesticide Production and Usage in China 226
Dynamics, Fate, and Toxicity of Pesticides in Soil 228
Groundwater Contamination by Pesticides in China 244
Remediation Strategies for Soil and Groundwater Contamination by Pesticides in China 246
Summary and Conclusions 247
Acknowledgments 248
References 248
Appendix: Chemical Names of Pesticides Mentioned in Chapter 9 251

10 Perspectives of Environmental Pollution in Densely Populated Areas: The Case of Hong Kong 254
M.S. YANG AND M.H. WONG
Introduction 254
Major Sources of Environmental Pollutants and Control Strategies 256
Effects of Pollution on Groundwater 263
Control and Remediation of Degraded Soil and Land 265
Future Perspectives 267
Summary and Conclusions 268
Acknowledgments 268
References 268

11 Environmental Impacts and Management Strategies of Trace Metals in Soil and Groundwater in the Republic of Korea 270
J.E. YANG, Y.K. KIM, J.H. KIM, AND Y.H. PARK
Introduction 270
Present Status and Environmental Impacts of Trace Metals in Korean Soils ..271
Management Strategies for Trace Metals in Korean Soils 281
Trace Metals in Groundwater 284
Management Strategies of Trace Metals in Groundwater 285
Summary and Conclusions 286
References 287

12 Transport, Residues, and Toxicological Problems of Agrochemicals in Agroecosystems and a Remediation Plan in the Republic of Korea 290
K.S. LEE AND B.H. SONG
Introduction 290
Changes of the Patterns in Pesticide Use 291

Pesticide Residues in Agricultural Environments 294
Evaluation of the Effects of Pesticide Residues on Agricultural Environment....300
Protection Strategies for Agricultural Ecosystems 312
Summary and Conclusions 322
Acknowledgments 323
References 323
Appendix: List of Pesticides Mentioned in Chapter 12 328

13 Toxic Metals and Agrochemicals in Soils in Malaysia: Current Problems and Mitigation Plans 330
Y.M. KHANIF, I.C. FAUZIAH, AND J. SHAMSHUDDIN
Introduction 330
Fertilizer and Agrochemical Usage 331
Ecological Problems of Toxic Metals and Agrochemicals 335
Mitigation Plans 342
Summary and Conclusions 344
References 344

14 Status of Cadmium, Lead, and Selenium in the Soils of Selected African Countries and Perspectives of Their Effects on Human and Environmental Health 347
B. WAIYAKI
Introduction 347
Soil Characteristics in Five Selected African Countries (Ethiopia, Ghana, Malawi, Sierra Leone, and Tanzania) 348
Status of Cadmium, Lead, and Selenium in the Soils of Selected African Countries 349
Status of Cadmium 351
Status of Lead 354
Status of Selenium 358
Sources, Forms, and Distribution of Cadmium and Human Exposure to It ..362
Environmental Aspects of Cadmium 364
Sources, Forms, and Distribution of Lead and Human Exposure to It 364
Environmental Aspects of Lead 367
Sources, Forms, and Distribution of Selenium and Human Exposure to It ...367
Summary and Conclusions/Recommendations 369
Acknowledgments 372
References 373

Preface

A large proportion of the world's arable lands fall in soil orders with variable charge, and the vast proportion of the world's population resides in these areas. With the high population density in Asia, threats to the ecosystem by intensifying agriculture to produce the food and fiber for the growing population through application of chemical fertilizers, manures, herbicides, insecticides, and the disposal of wastes should increase with time. Environmental pollution is, thus, of increasing concern in these areas. The potential for restoring ecosystem health through remediation in these areas is enormous. Therefore, we need to devote much more attention to the understanding and management of land in these densely populated areas in Asia. The information on the perspectives of soil and groundwater pollution and remediation in Oceania and Africa is also essential for the sake of comparative studies on soil environmental quality and the development of preventive and remedial land resource management strategies.

This book, composed of 14 chapters, covers the information on metals, radionuclides, other inorganics, pesticides, and other anthropogenic organic compounds in soil environments in Asia, Oceania, and Africa. The distribution, transformations, and dynamics of these environmental pollutants are discussed. Acid-rain-induced environmental problems are also dealt with. Besides agricultural intensification, the impact of industrialization on soil quality is discussed. Therefore, the book addresses the current status and future prospects on soil and groundwater pollution in Asia, Oceania, and Africa and the remediation strategies for years to come.

The contributors are well-known scientists in the respective regions. Many of them are internationally renowned. The book should be useful to researchers, educators, consultants, personnel in government regulatory agencies, and students in environmental and agricultural sciences.

We are grateful to the authors who have contributed chapters to this book and to the external referees and coeditors who have provided invaluable critical inputs to maintain the quality of this publication. Gratitude is extended to the University of Saskatchewan and the U.S. Army Cold Regions Research and Engineering Laboratory, which have provided the funding to facilitate the publication of this book. Sincere appreciation is also extended to Yun Yin Huang, as well as E.A. Wright and Donna Harp of the U.S. Army Cold Regions Research and Engineering Laboratory, for their invaluable editorial assistance during the preparation of this publication.

P.M. Huang
I.K. Iskandar

About the Editors

P.M. Huang received his Ph.D. degree in soil science at the University of Wisconsin, Madison, in 1966. He is currently Professor of Soil Science at the University of Saskatchewan, Saskatoon, Canada. His research work has significantly advanced knowledge of the nature and surface reactivity of mineral colloids and organomineral complexes of soils and sediments and their role in the dynamics, transformations, and fate of plant nutrients, toxic metals, and xenobiotics in terrestrial and aquatic environments. His research findings, embodied in over 200 refereed scientific publications, including research papers, book chapters, and 10 books, are fundamental to the development of sound strategies for managing land and water resources.

He has developed and taught courses in soil physical chemistry and mineralogy, soil analytical chemistry, and ecological toxicology. He has successfully trained and inspired M.Sc. and Ph.D. students and postdoctoral fellows, and received visiting scientists worldwide. He has served on numerous national and international scientific and academic committees. He also has served as a member of many editorial boards such as the *Soil Science Society of America Journal, Geoderma, Chemosphere,* and *Advances in Environmental Science*. He served as a titular member of the Commission of Fundamental Environmental Chemistry of the International Union of Pure and Applied Chemistry and is the founding and current Chairman of the Working Group MO "Interactions of Soil Minerals with Organic Components and Microorganisms" of the International Union of Soil Sciences. He received the Distinguished Researcher Award from the University of Saskatchewan. He is a fellow of the Canadian Society of Soil Science, the Soil Science Society of America, the American Society of Agronomy, and the American Association for the Advancement of Science.

I.K. Iskandar received his Ph.D. degree in soil science and water chemistry at the University of Wisconsin, Madison, in 1972. He is currently a Research Physical Scientist at the Cold Regions Research and Engineering Laboratory (CRREL) and a Distinguished Visiting Research Professor at the University of Massachusetts, Lowell, Massachusetts. During his tenure at CRREL, he developed a major research program on land treatment of municipal wastewater. He successfully coordinated the program for eight years and supervised the research on transformation and transport of nitrogen, phosphorus, and heavy metals. In the early 1980s, his research efforts were focused on the fate and transformation of toxic chemicals in soils, development of nondestructive methods for site assessments, and development and evaluation of in-situ remediation alternatives. Dr. Iskandar was the first to propose the use of a frozen ground barrier for containment of toxic waste.

Dr. Iskandar has edited or coedited 10 books, written more than 20 chapters for books, published more than 100 technical and reference papers and reports, presented more than 55 invited lectures, seminars, and symposia, and made 45 other presentations.

Dr. Iskandar has organized and coorganized many national and international workshops and symposia. He has received many awards including the Army Science Conference, 1979; CRREL Research and Development Award, 1988; and several exceptional performance awards at the U.S. Army Cold Regions Research and Engineering Laboratory. He is a fellow of the Soil Science Society of America, and a member of the American Society of Agronomy and the International Union of Soil Science.

CONTRIBUTORS

Affiliation of Editors

P.M. Huang
Department of Soil Science
University of Saskatchewan
51 Campus Drive
Saskatoon, SK S7N 5A8
Canada

I.K. Iskandar
The U.S. Army Cold Regions Research and Engineering Laboratory
72 Lyme Road
Hanover, NH 03755-1290
U.S.A.

Affiliation of Coeditors

M. Chino
Department of Applied Biological Chemistry
The University of Tokyo
Bunkyo-ku
Tokyo 113
Japan

T.B. Goh
Department of Soil Science
University of Manitoba
Winnipeg, MB R3T 2N2
Canada

P.H. Hsu
Department of Environmental Science
Rutgers University
The State University of New Jersey
New Brunswick, New Jersey 08903
U.S.A.

D.W. Oscarson
Atomic Energy of Canada Limited Research Company
Whiteshell Nuclear Research Establishment
Pinawa, MB R0E 1L0
Canada

L.M. Shuman
Department of Agronomy
University of Georgia
Griffin, GA 30223-1797
U.S.A.

Contributors

D.J. Cai
Nanjing Research Institute of Environmental Sciences
National Environmental Protection Agency of China
8 Jiangwangmiao Street
P.O. Box 4202
Nanjing 210042
China

H.M. Chen
Institute of Soil Science
Academia Sinica
Nanjing 210008
China

M. Chino
Department of Applied Biological Chemistry
The University of Tokyo
Bunkyo-ku
Tokyo 113
Japan

I.C. Fauziah
Department of Soil Science
Universiti Putra Malaysia
43400 Serdang
Selangor, Malaysia

T. Hirata
Department of Environmental Systems
Wakayama University
930 Sakaedani
Wakayama, Wakayama 640-8510
Japan

G. L. Ji
Institute of Soil Science
Academia Sinica
Nanjing 210008
China

Y.M. Khanif
Department of Soil Science
Universiti Putra Malaysia
43400 Serdang
Selangor, Malaysia

J.H. Kim
Division of Environmental and Biological Engineering
Kangwon National University
Chunchon 200-701
Republic of Korea

Y.K. Kim
Division of Environmental and Biological Engineering
Kangwon National University
Chunchon 200-701
Republic of Korea

R.S. Kookana
CSIRO Land and Water
Private Mail Bag 2
Glen Osmond, S.A. 5064
Australia

K.S. Lee
Department of Agricultural Chemistry
College of Agriculture
Chungnam National University
Taejon 405-764
Republic of Korea

M.J. McLaughlin
CSIRO Land and Water
Private Mail Bag 2
Glen Osmond, S.A. 5064
Australia

O. Nakasugi
The National Institute for Environmental Studies
16-2 Onogawa
Tsukuba
Ibaraki 305-0053
Japan

Y. Ohmomo
Institute for Environmental Sciences
1-7 Ienomae, Obuchi
Rokkasho-mura
Aomori 039-3212
Japan

Y.H. Park
Korea Environmental Institute
1049-1 Sadang-dong
Dongjak-gu
Seoul 156-090
Republic of Korea

A. Rahman
Ag Research
Ruakura Agricultural Research Centre
Private Bag 3123
Hamilton, New Zealand

A.H.C. Roberts
Ag Research
Ruakura Agricultural Research Centre
Private Bag 3123
Hamilton, New Zealand

J. Shamshuddin
Department of Soil Science
Universiti Putra Malaysia
43400 Serdang
Selangor, Malaysia

B.H. Song
Pesticide Safety Department
Agricultural Science and Technology Institute
Suwon 441-707
Republic of Korea

B.K.G. Theng
Manaaki Whenua-Landcare Research
Private Bag 11052
Palmerston North, New Zealand

K.G. Tiller (deceased)
CSIRO Land and Water
Private Mail Bag 2,
Glen Osmond, S.A. 5064
Australia

S. Uchida
Environmental and Toxicological Sciences
Research Group
National Institute of Radiological Sciences
4-9-1 Anagawa, Inage-ku
Chiba-shi, Chiba 263-8555
Japan

B. Waiyaki (deceased)
Soils and Agricultural Unit
United Nations Environment Programme
Nairobi, Kenya

H.K. Wang
Department of Agricultural Ecology and
Environmental Sciences
China Agricultural University
Beijing 100094
China

J. H. Wang
Institute of Soil Science
Academia Sinica
Nanjing 210008
China

M.H. Wong
Institute for Natural Resources and Waste
Management, and Department of
Biology
Hong Kong Baptist University
Kowloon Tong, Hong Kong
China

G.X. Xing
Institute of Soil Science
Academia Sinica
Nanjing 210008
China

J.E. Yang
Division of Biological Environment
College of Agriculture and Life Sciences
Kangwon National University
Chunchon 200-701
Republic of Korea

M.S. Yang
Institute for Natural Resources and Waste
Management, and Department of
Biology
Hong Kong Baptist University
Kowloon Tong, Hong Kong
China

X. N. Zhang
Institute of Soil Science
Academia Sinica
Nanjing 210008
China

Z.L. Zhu
Nanjing Research Institute of
Environmental Sciences
National Environmental Protection
Agency of China
8 Jiangwangmiao Street
P.O. Box 4202
Nanjing 210042
China

CHAPTER 1

ENVIRONMENTAL IMPACTS OF HEAVY METALS IN AGROECOSYSTEMS AND AMELIORATION STRATEGIES IN OCEANIA

K.G. Tiller, M.J. McLaughlin, and A.H.C. Roberts

INTRODUCTION

The agroecosystems of Australia and New Zealand encompass an enormous range of both soil types and climatic conditions. The main feature that they have in common is a relative freedom from influences of urban and industrial pollution. Soil and associated crop contamination has primarily resulted from agricultural practices arising from fertilizers, agricultural sprays, and organic wastes. Atmospheric contamination arising from industrial emissions has only influenced relatively small areas near a few regional centers. This review will discuss background concentrations of heavy metals in Australian and New Zealand soils and crops, sources of contamination, their environmental impact, some amelioration strategies especially in relation to farm management practices, and regulatory controls. Data on heavy metals in groundwater are few and will not be discussed. Those available are of questionable quality because of the likelihood of contamination from access tubes and bore casings. Information on a range of heavy metals will be presented; most research has been carried out on cadmium (Cd) in particular, as well as lead (Pb), arsenic (As), copper (Cu), and zinc (Zn), with less work still on nickel (Ni), mercury (Hg), and chromium (Cr).

BACKGROUND CONCENTRATIONS OF HEAVY METALS IN AUSTRALIAN AND NEW ZEALAND SOILS

Initially we will outline our attitude on the topic of background values. Background values are often used to determine if a soil has become enriched with an undesirable substance; such soils with values exceeding the local normal background values are then deemed "contaminated." If the soil becomes so contaminated that it has an observed environmental impact, it is deemed "polluted." As discussed by Tiller (1992), there is a continuum of soil concentrations from the natural or pristine state through to highly polluted soils. The level at which a soil is deemed polluted therefore depends on the contaminant

Table 1.1. Background Concentrations (mg kg^{-1}) of Heavy Metals in New Zealand Soil Types (Data for the 0 to 7.5-cm layer)[a]

		Soil Type[b]								
Element	**Nat./Past.**	**YBL**	**YBP**	**YBE**	**YGE**	**GLEY**	**PEAT**	**BGL**	**ALLUV**	**MEAN**
As	Native	3.9	7.0	4.2	2.1	5.3	8.4	5.3	3.0	4.3
	Pasture	7.4	7.8	3.0	2.4	5.1	15.1	5.4	3.6	4.9
Cd	Native	0.23	0.31	0.16	0.13	0.24	0.22	0.19	0.13	0.18
	Pasture	0.70	0.75	0.22	0.12	0.42	0.69	0.49	0.16	0.44
Cu	Native	26.9	7.2	14.6	8.0	17.9	20.4	19.4	11.5	17.0
	Pasture	32.1	10.9	12.3	8.9	18.2	27.0	25.5	17.3	17.7
Pb	Native	11.6	6.8	13.7	12.5	17.6	16.9	18.0	15.0	13.3
	Pasture	19.3	7.4	7.6	9.3	11.6	10.0	19.1	10.9	11.7
Zn	Native	77	37	65	54	92	51	56	72	65
	Pasture	97	47	55	61	66	43	82	78	68

[a]Source: Roberts et al. (1996).
[b] Soil types (Taylor and Pohlen, 1970): YBL = yellow-brown loam; YBP = yellow-brown pumice soil; YBE = yellow-brown earth ; YGE = yellow-grey earth ; GLEY = gleys; PEAT = peats; BGL = brown granular loam; ALLUV = alluvial soils. Sites sampled = 398.

of interest, its bioavailability and toxicity, and the pathways through which health or environmental risk are exercised.

In the view of most health authorities where urban and industrial pollution are the main concern, "background" values for metals in soil are often assessed in relation to rural soils. These normal background values will therefore include soils with mildly elevated metal concentrations due to normal agricultural fertilization practices. To assess the impact of farm management practice, metal concentrations in native (uncultivated) soils must be compared to their farmed or pastoral equivalents. For the purposes of this review, therefore, background values will be referred to in terms of the range of rural soils without distinction between native (i.e., unfertilized) and agricultural soils. These are the "normal" reference levels, which assume some contamination associated with everyday activities such as normal fertilization practice and the global smear of atmospheric pollution. Background data for New Zealand soils are presented on the basis of total metal concentrations in key soil types for both native soils and those under developed pastures (Table 1.1).

Of special note are the elevated concentrations of Cd in the pastoral soils, presumably due to inputs from fertilizers. This will be discussed later. The lower concentrations of Pb in some pastoral soil types may result from dilution by cultivation of Pb accumulated at the surface of soils in their native state. Since New Zealand pasture soils are rarely cultivated, there is tendency for metals, especially Cd, which is added in atmospheric accessions, dung, and fertilizers, to decrease with depth. This trend is more marked for Cd than for other metals (Roberts et al., 1994). The range of concentrations of some potentially toxic elements in Australia-wide agricultural surface soils (Table 1.2) and in pasture soils near Adelaide (Table 1.3) have been collated from published (Tiller, 1992; Merry et al., 1983; Merry and Tiller, 1991) and unpublished data from the investigations in the laboratories of the CSIRO Land and Water (formerly Division of Soils) over several decades.

The highest "background" concentrations of Co, Cu, Mo, Ni, and Pb found in Australian rural soils (Tiller, 1991) often exceed those quoted for Dutch soils (Moen et al., 1988), the latter often being used as a benchmark by many countries. The reverse is true for As. These differences reflect the great range of soil parent materials and the kind and

Table 1.2. Ranges of Background Concentrations of Some Potentially Toxic Elements in Australian Rural Surface Soils [a]

Element	Approx. No. of Samples	Range of Values (mg kg^{-1})	Dutch A Value[b]
Co	250	<2–170	20
Cu[c]	500	1–190	50
Mn	800	4–13000	NR
Zn	500	<2–200	200
Mo	300	<1–20	10
Ni	200	2–400	50
Cr	180	5–110	100
Pb[c]	160	2–160	50
Cd	180	<1	1
B[a]	50	10–75	NR
As[c]	26	<1–8	20

[a]B. Cartwright (unpublished data).
[b]Upper limit of background values in Netherlands (Moen et al., 1988), NR denotes no value reported.
[c]Excludes orchard soils treated with Pb, As, and Cu.

Table 1.3. Total Concentrations of Some Heavy Metals in Surface Pasture Soils (0–10 cm) of the Mt. Lofty Ranges, South Australia (mg kg^{-1}): Range, Median, and Mean Values (mg kg^{-1}) [a]

Element	Range[b]	Mean	Median
As	<1–8	3.9	NR
Cd	<1	NR	NR
Cr	5–110	27	25
Mn	32–5200	254	148
Cu	1–59	15	12
Pb	2–160	21	18
Ni	2–28	9.8	8.0
Zn	4–200	27	18

[a]Data are from unpublished reports of CSIRO Division of Soils, Adelaide.
[b]n = 185, except for Cu and Pb, n = 157; and As, n = 15. NR= no value reported.

duration of soil-forming processes existing in continental Australia, compared to the Netherlands, as well as differences in contaminants associated with more intensive agricultural activity in the Netherlands. Concentrations of As, Cd, Cu, Pb, and Zn in South Australian and New Zealand soils agree closely.

SOURCES OF METALS IN AGROECOSYSTEMS

Inputs of heavy metals into soils globally (Nriagu and Pacyna, 1988) have been assessed at up to 38 thousand tonnes of Cd and about one million tonnes of Pb per year, with major contributions from atmospheric fallout, fly ash, and urban refuse, and minor contributions, globally, from sewage sludge and fertilizers (Table 1.4).

This global perspective of increasing accumulation of heavy metals should stimulate our efforts to control emissions and disposal, and to rehabilitate contaminated land. In practice, the minor sources of contamination on a global scale may be the most important environmentally because of their regional or local impacts. Sewage sludge and mineral wastes are added to land relatively near their sources for economic reasons; fertilizer usage is controlled by suitability of land for agriculture and pressures for food production.

As far as natural atmospheric sources are concerned, much of the atmospheric emis-

Table 1.4. Additions of Lead, Cadmium, and Mercury to Soils [a]

Source	Lead	Cadmium	Mercury
		kt yr^{-1}	
Agricultural and food wastes	1.5–2.7	0–3.0	0–1.5
Animal wastes, manure	3.2–20	0.2–1.2	0–0.2
Logging and other wood wastes	6.6–8.2	0–2.2	0–2.2
Urban refuse	18–62	0.9–7.5	0–0.26
Municipal sewage sludge	2.8– 9.7	0.02–0.34	0.01–0.8
Miscellaneous organic wastes, and excreta	0.02–1.6	0–0.01	–
Metal manufacturing wastes	4.1–11	0–0.08	0–0.08
Coal ash	45–242	1.5–13	0.4–4.8
Fertilizer	0.4–2.3	0.03–0.25	–
Peat (agricultural and fuel use)	0.4–2.6	0–0.11	0–0.02
Commercial product waste	195–390	0.8–1.6	0.6–0.8
Atmospheric fallout	202–262	2.2–8.4	0.6–4.3
TOTAL	479–1113	5.6–38	1.6–15

[a]After Nriagu and Pacyna, 1988.

sions are ascribed to ash from forest and bush fires and from direct losses from vegetation, e.g., from leaf exudates. The main anthropogenic sources of emission for heavy metals are from smelters and urban sources such as coal and refuse combustion.

Even though atmospheric emissions can result in a global smear of contamination by fine particles with sufficiently long residence times, land near emission sites will receive the highest rates of deposition because of the well-recognized decreasing rate of deposition with distance.

Accessions to Soils from Urban/Industrial Sources

Cadmium

Inputs of Cd and other heavy metals from atmospheric sources can only be expected within the depositional zones of significant sources of emission, whether individual factories, such as smelters, or the emissions from major city areas. These latter sources mainly affect the suburban areas and probably have negligible effect on the rural hinterland of urban industrial areas. Lower-level contamination in rural areas of Australia have rarely been investigated in detail, except around Port Pirie, site of a major lead-zinc smelter for over 100 years in South Australia. Studies in our laboratory reviewed by Tiller (1989) showed that at least 3400 km^2 of land were contaminated by heavy metals to some extent (Fig. 1.1), and that pastures, sheep meat products, and cereal grain had heavy metal contents appreciably above background levels up to 20 km downwind from the smelter.

A potential environmentally hazardous situation with respect to food quality was prevented by the alkaline pH of the surrounding farm soils. Total accessions of Cd to the region surrounding the smelter were estimated at 300 tonnes accumulated over about 100 years, with only 10% within a radius of 10 km of the smelter.

Atmospheric deposition of Cd emitted from the Adelaide metropolitan area onto land about 20 km downwind from the urban fringe was much less than from the Port Pirie smelter and estimated from rainfall collection to be about 3 g ha^{-1} y^{-1} (Merry and Tiller, 1991). However, beyond 40 km downwind from the Adelaide metropolitan boundary, deposition of urban contaminants was indistinguishable from background. Measured Cd concentrations in rainfall in this study area were in the range of <0.1 to 0.4 µg L^{-1} (Merry and Tiller, 1991), with Cd concentrations in stream waters of about 0.05 µg L^{-1} (Ford-

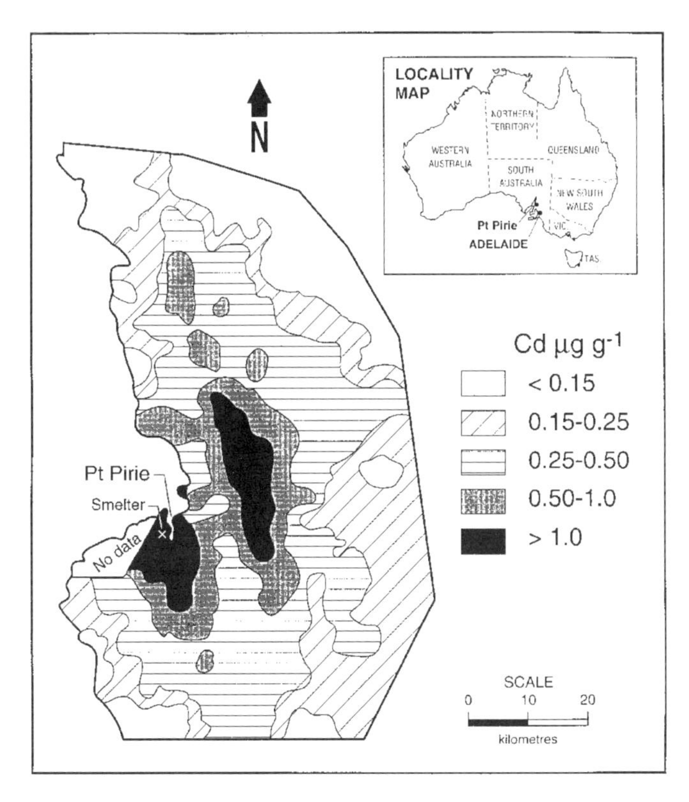

Figure 1.1. Atmospheric deposition of cadmium on soils in the region around the Port Pirie lead-zinc smelter (EDTA-extractable Cd in $\mu g\ g^{-1}$). Source: Cartwright et al. (1977).

ham, 1978). The atmospheric input of Cd in agricultural areas near Adelaide (about 3 g $ha^{-1}\ y^{-1}$) accords with values of 1–5 g $ha^{-1}\ y^{-1}$ quoted for European rural areas (Hovmand, 1981; Tjell et al., 1981; Gunnarsson, 1983; Hutton and Symon, 1986; Jones et al., 1987), where urban and industrial atmospheric pollution is more widespread and affects a large proportion of the agricultural land in many countries.

We assume that only a very small fraction of Australian agricultural land outside the hinterland of major cities and known industrial emitters receives any net gain in Cd from atmospheric sources. The main sources of Cd deposited on Australian land are summarized in Table 1.5.

Lead

Although emissions of Pb from bush fires may be a significant natural source in some regions, the main global source of Pb, namely emissions from combustion of petrol in motor vehicles, will probably also dominate in Australia and New Zealand. The second major global source, from primary metal production, will only be important in Australia near our several major smelters. All sources of Pb deposited on Australian agricultural land are summarized in Table 1.6. Unfortunately, detailed statistics are not available to provide accurate estimates of these identified sources.

Some earlier sources, such as agricultural Pb arsenate sprays, have now been eliminated, whereas others such as smelter and automotive emissions have been appreciably reduced in recent years. Agricultural areas near major cities will still receive some Pb from automotive emissions and metal industries emitting Pb into the atmosphere. Studies of a pastoral agroecosystem near Adelaide showed that Pb concentrations in surface soils had measurably increased up to about 50 km from the city boundary (Tiller et al., 1987).

Table 1.5. Main Sources of Cadmium Deposited on Australian Agricultural Land

Source	Dispersal	Comments
Fertilizer (phosphates)	Broadscale	Over 6000 t Cd added to Australian soils this century, now 40 t y^{-1} (McLaughlin et al., 1996).
Lime	Broadscale in some regions	Restricted to acidic soils and contributes only negligible Cd.
Sewage sludge	Localized	Usage is increasing; estimated potential sludge-Cd loading is 2–4 ty^{-1}. (Tiller et al., 1997).
Urban/industrial	Mainly restricted to hinterland of urban areas	
Smelter emissions	Locally and up to 50 km downwind	The Port Pirie smelter spread at least 300 t over 100 years on 3400 ha of surrounding agricultural land (Cartwright et al., 1977); averaging about 5 g ha^{-1} y^{-1} at 20 km from smelter), and 15 g ha^{-1} y^{-1} at 10 km.
Other urban/industrial atmospheric emissions, e.g., power stations, incinerators	Near source and downwind	Deposition at 20 km from city of Adelaide was estimated at about 3 g $ha^{-1}y^{-1}$ (Merry and Tiller, 1991).

Table 1.6. Main Sources of Lead Deposited on Australian Agricultural Land

Source	Dispersal	Comments
Fertilizer (phosphates)	Broadscale	Total accession about 3000 t Pb; assumes 20 mg Pb kg^{-1} fertilizer; averages about 2 g Pb ha^{-1} y^{-1} (McLaughlin et al., 1996).
Lime	Broadscale in some regions	Restricted to acidic soils; occasional dressings may result in an average of about 2 to 15 g Pb $ha^{-1}y^{-1}$.
Agricultural sprays	Localized	Discontinued in late 1960s: sprays contaminated some orchard soils to ~300 mg kg^{-1} soil; average rates were equivalent to about 5–10 g ha^{-1} y^{-1} during use (Merry et al., 1983).
Sewage sludge	Localized	Usage will increase. Potential addition of Pb to soil from sludge is about 70 t Pb y^{-1}, assumes 220,000 dry t used with 300 mg kg^{-1} sludge. EC limit values of 15 kg Pb ha^{-1} y^{-1} unlikely to be exceeded.
Urban/industrial	Mainly restricted to hinterland of urban areas	
Smelter emissions	Locally and up to 50 km downwind	The Port Pirie smelter spread at least 30,000 t Pb over 100 years on agricultural land downwind; equivalent to about 0.5 kg Pb ha^{-1} y^{-1} at 20 km from smelter (calculated from Cartwright et al., 1977).
Petrol	Near main roads and up to 50 km downwind from metropolitan areas	Australia-wide about 5400 t Pb y^{-1} are emitted from vehicles. Dispersed about 50 km downwind from Adelaide. Accession rate in 1980 was 50 g ha^{-1} y^{-1} at 20 km E of Adelaide city limits (Tiller et al., 1987).

Identification of the Pb isotopes constituting surface soil Pb confirmed that the Pb was primarily of automotive origin (Gulson et al., 1981). Atmospheric fallout of automotive Pb varied little from year to year but was seasonally variable, with highest deposition rates in winter. Calculation of an environmental Pb budget suggested that only about 3% of the Pb burned in the petrol in the metropolitan area was dispersed beyond the immediate highway zone and deposited on the land surface within 50 km of the city. Furthermore, a significant proportion of the total petrol Pb burned could not be accounted for and was presumed to have dispersed beyond the study area and to have contributed to continental and global pollution.

Sources such as sewage sludge will become more important in some areas near major centers because of recent decisions to cease marine disposal from metropolitan areas. However, sewage sludge application will appreciably affect Pb levels of a small proportion of agricultural land.

Traditional inorganic N and P fertilizers will have a much lesser impact on Pb content of agricultural soils, compared to Cd, and a negligible impact on crops. Accessions from all sources, whether atmospheric, fertilizer, etc., will continue to decrease in response to stricter regulatory controls, but of course the impact of past Pb accessions to land remains.

Other Metals

The dispersal of other metals such as As and Zn in the smelter fume emissions around Pt. Pirie was similar to those of Cd and Pb. Detailed studies of other areas and of other elements are not available for Australia and New Zealand.

Sewage Sludge

Sewage sludges need to be considered as a source of potentially toxic elements because they are likely to be increasingly used as fertilizer materials in both Australia and New Zealand. The heavy metal content of the processed sewage becomes trapped in the solid phase of sewage sludge during treatment; the liquid effluents are relatively metal free. Concern about toxic heavy metals and associated tightening of regulations controlling the discharge of industrial wastes into sewers is resulting in a gradual decrease in metal concentrations of sewage worldwide and in our region. Metal concentrations in sewage sludge in many industrialized countries ranged up to several thousand mg kg^{-1} sludge (dry weight) but median concentrations were only several hundred (Tiller, 1989); median concentrations of Cd ranged from about 10–30 mg kg^{-1}. Values for several metals in 20 Australian sewage sludges (deVries, 1983) are summarized in Table 1.7. Cadmium concentrations ranged at that time from 2 to 285 mg kg^{-1} dry sludge with mean and median concentrations of 41 and 26, respectively. In the last decades, metal concentrations from many sewage treatment plants have decreased dramatically. Cadmium in sludge from Adelaide's main sewage works, for example, is only about one-tenth of the values found in the late 1970s (Adelaide Engineering and Water Supply Dept., 1994).

Sewage sludge, until recently, has not been applied in appreciable amounts or to significant areas of Australian and New Zealand agricultural land. This situation will soon change in Australia because of recent government decisions to ban discharge of sewage, treated or otherwise, into the ocean and to restrict its incineration. Tiller et al. (1997) have recently estimated the total potential sludge-Cd loading to Australian agricultural soils to be about 2–4 t y^{-1}. The corresponding loading for Pb is about 150–300 t y^{-1} assuming Pb concentrations of 200–400 mg Pb kg^{-1} dry sludge. These loading rates will probably decrease in response to further reduction in industrial discharges to the sewers as

Table 1.7. Some Metal Concentrations in Australian Sewage Sludges (mg kg^{-1} Dry Weight) [a]

Metal	Maximum	Minimum	Median	Mean
Cadmium	285	2.0	26	41
Copper	2480	251	671	856
Manganese	645	114	242	266
Nickel	318	20	60	88
Lead	1980	55	424	562
Zinc	5510	241	1930	2070

[a] Based on 20 sewage treatment plants from all mainland states (after de Vries, 1983).

pressures for beneficial use of sewage sludge increase. This estimate of potential loading of Australian soils with Cd from sewage sludge is clearly much less than current loading from fertilizers (40 t y^{-1}, McLaughlin et al., 1996), but the impact could be much greater because of its use closer to sources of production. The situation is the reverse for Pb where the total annual loading of Australian soils is only about 20 t y^{-1} (McLaughlin et al., 1996).

The application of large amounts of treated sewage sludge to land should stimulate investigations to provide a better basis for either developing locally valid guidelines or for adapting foreign guidelines to Australian soil and climatic conditions. Much of the published local research experience on sewage sludge application to farmland is restricted to the work of de Vries and Tiller in the late 1970s near Adelaide, South Australia (e.g., de Vries and Tiller, 1978) and investigations of Ross et al. (1992) near Sydney, New South Wales, but some new research is currently under way in New South Wales and in Adelaide, South Australia.

Fertilizers

In Australia, most soils on the ancient, highly weathered land surface are of low phosphorus (P) status, and are associated in the virgin state with widespread P deficiency in crops that need regular applications of phosphatic fertilizers. The main exceptions are the vertisol-type soils on basalt, for example, in southern Queensland. Walkley (1940) identified the high concentrations of Cd in Australian superphosphates, but the environmental consequences for soil and food quality were highlighted much later by the investigations of Dr. C. Williams and colleagues at the CSIRO, Canberra (e.g., Williams and David, 1973). In both Australia and New Zealand, the main sources of rock phosphate used for manufacture of phosphatic fertilizers were, until recently, from the oceanic sedimentary, guano-based deposits with Cd concentrations ranging from 42–99 mg kg^{-1}. These rock sources had higher Cd concentrations than rocks found in other countries. Increasingly, rock phosphate and manufactured phosphate fertilizers are being imported from other overseas sources. A list of the Cd concentrations in rock phosphates was recently compiled by McLaughlin et al. (1996), and is shown in Table 1.8. The highest Cd-containing deposits appear to be in the Pacific islands, and western United States.

Rayment et al. (1989) compared a small range of low and high analysis fertilizers manufactured in Queensland and found a similar Cd:P ratio (413±40 mg Cd kg^{-1} P) in both low and high analysis formulations. However, Cd concentrations in fertilizers currently manufactured in Australia and New Zealand are likely to be much lower than this, due to manufacturers switching to lower Cd sources of phosphate rock. Zarcinas and Nable

Table 1.8. Cadmium and Phosphorus Concentrations of Some Phosphate Rocks [a]

Phosphate Rock	Cd ($mg\ kg^{-1}$)	P (%)	Cd ($mg\ Cd\ kg^{-1}\ P$)	Reference
USSR (Kola)	0.2	17.2	1	Singh (1991)
South Africa (Phalaborwa)	4	17.2	23	Williams (1974)
Chatham Rise phosphorite	2	8.9	23	Syers et al. (1986)
China (Yunan)	5	14.4	35	Bramley (1990)
Jordan	6	14.8	27	Bramley (1990)
Australia (Duchess)	7	13.9	50	Williams (1974)
Mexico	8	14.0	57	Syers et al. (1986)
Egypt (Quseir)	8	12.7	61	McLaughlin (unpub. data)
Makatea	10	13.0	77	Syers et al. (1986)
Peru (Sechura)	11	13.1	84	Syers et al. (1986)
Israel (Arad)	12	14.1	85	Syers et al. (1986)
Tunisia (Gafsa)	38	13.4	108	Syers et al. (1986)
Israel (Zin)	32	14.0	228	Bramley (1990)
Morocco (Boucraa)	38	15.7	240	McLaughlin (unpub. data)
Christmas Island	43	15.3	275	David et al. (1978)
North Carolina	47	15.1	311	McLaughlin (unpub. data)
Togo	51	16.0	320	McLaughlin (unpub. data)
Banaba (Ocean Island)	99	17.6	563	Williams (1974)
Nauru	100	15.6	641	Syers et al. (1986)
Western USA	60-340	NR[b]	NR	Auer (1977)

[a] After Mclaughlin et al., 1996.
[b] NR = not reported.

(1992) have noted a significant reduction in the Cd:P ratio of fertilizers manufactured in South Australia after 1989.

McLaughlin et al. (1996) recently calculated that the current total annual loading by Cd of Australian soils due to fertilizers is about 40 t y^{-1}; the equivalent value for Pb is 20 t y^{-1}. They estimated that over 6000 tonnes of Cd have been added to Australian soils through use of phosphatic fertilizers in the last 90 years. Of the current annual import of Cd to Australia from fertilizer or fertilizer raw materials, it is estimated that of the 40 tonnes, one-third is imported in finished fertilizers and the remainder as phosphate rock. From knowledge of the total amount of Cd applied to Australian soils and the total area fertilized (ABARE, 1993), an average Cd application rate of 2.5 g Cd $ha^{-1}\ y^{-1}$ can be calculated. This figure can be compared to fertilizer Cd input values (mean 1985-1987) quoted for the U.K. of 4.3 g $ha^{-1}\ y^{-1}$ (Hutton and Symon, 1986), 3.5 to 4.3 g $ha^{-1}\ y^{-1}$ in Germany (Sauerbeck, 1982; Kloke et al., 1984), 3.0 g $ha^{-1}\ y^{-1}$ in Denmark (Hovmand, 1981), 0.3 to 1.2 g Cd ha^{-1} in the United States (Mortvedt, 1987), and an average for EEC countries of 2.5 g $ha^{-1}\ y^{-1}$ (Biberacher and Shah, 1990).

Cadmium concentrations in nitrogenous and potassic fertilizers, agricultural limes, and natural gypsum materials are generally low and of no consequence in terms of soil contamination. Where materials are by-products of industrial processes, e.g., phosphogypsum, fly ashes, etc., then significant contamination may occur through these soil amendments.

The major difference between Cd inputs in Australia/New Zealand and European countries therefore is that fertilizers provide the bulk of the Cd inputs to agricultural soils, whereas in Europe contributions from atmospheric inputs and sewage sludge or waste material sources may be appreciable. Many reports quote approximately equal inputs of Cd from atmospheric and fertilizer sources, a situation rarely applicable in Australia and New Zealand.

Other Sources

There are several other sources of heavy metal contamination of agricultural land. Fungicidal sprays, such as Cu oxychloride, were formerly used on orchards, vineyards, and vegetable crops for about 100 years. Lead arsenate was used on orchards for a similar length of time to control chewing insects, especially codling moth. Commercial use of Pb arsenate was phased out in the 1970s. Around the same time, large areas of orchards were removed from production in response to market forces and converted to other crops or pastures. A survey of orchards and former orchards showed concentrations (mg kg^{-1}) of up to 320, 550, and 115, for Cu, Pb, and As, respectively (Merry et al., 1983). The main hazard from the contamination of horticultural soils by agricultural sprays is not to crops being treated but rather to the new uses of the soil following orchard clearance.

The use of As to control the growth of an undesirable weed can adversely affect the following crop; the control of one hazard can lead to another. The eradication of *Harrisia* cactus in north Queensland several decades ago by As pentoxide is a good example (Cuddihy and Fergus, 1974). About 650 tonnes of As were applied at variable rates over 24,300 ha. Arsenic concentrations reached 300 mg kg^{-1} and caused uneven, patchy crop failure. Because of the threat to the cattle industry of tick infestations, a widespread network of cattle dip sites was established in northern New South Wales and southern Queensland. Arsenic compounds were used as tickicides at over 1,600 sites until the mid-1950s when DDT was substituted (until 1980). Residues of these chemicals are now still of great concern because of possible transfer through the food chain (e.g., pasture-animal), hazard to children in housing, which is now expanding into areas formally occupied by dip sites, and transfer to ground and surface waters. Concentrations of As and DDT greatly exceeding regulatory guidelines have been found (sometimes >100 times). Arsenical compounds have also been widely used for general weed control, as insecticides and herbicides in cotton and banana production and as desiccants to facilitate harvesting of cotton.

Summary for Cadmium and Lead

The main sources of Cd and Pb are summarized in Tables 1.5 and 1.6. For Pb the important sources only have quite local impacts, primarily due to smelter emissions, car exhausts, and reuse of sewage sludges. Applications of soil amendments such as fertilizers in broadscale agriculture are of minor importance. For Cd, smelter and other industrial emissions and sewage sludge application to land can have significant local impact, yet have negligible influence on broadscale agriculture. Phosphate fertilizers are the only important source of Cd in Australian and New Zealand agroecosystems.

ENVIRONMENTAL IMPACT

Soil Quality

There are only limited data available on the impact of atmospheric accessions of heavy metals onto agricultural land. Calculations based on the work of Cartwright et al. (1977) in the region of the lead-zinc smelter at Pt. Pirie (summarized in Tables 1.5 and 1.6) showed that the average rate of Cd accumulation in local soils over 100 years was about 0.004 mg kg^{-1} ha^{-1} y^{-1} at 20 km from the smelter and 0.01 mg kg^{-1} ha^{-1} y^{-1} at 10 km. The corresponding values for Pb were 0.4 and 1.0 mg kg^{-1} ha^{-1} y^{-1}, respectively. Current

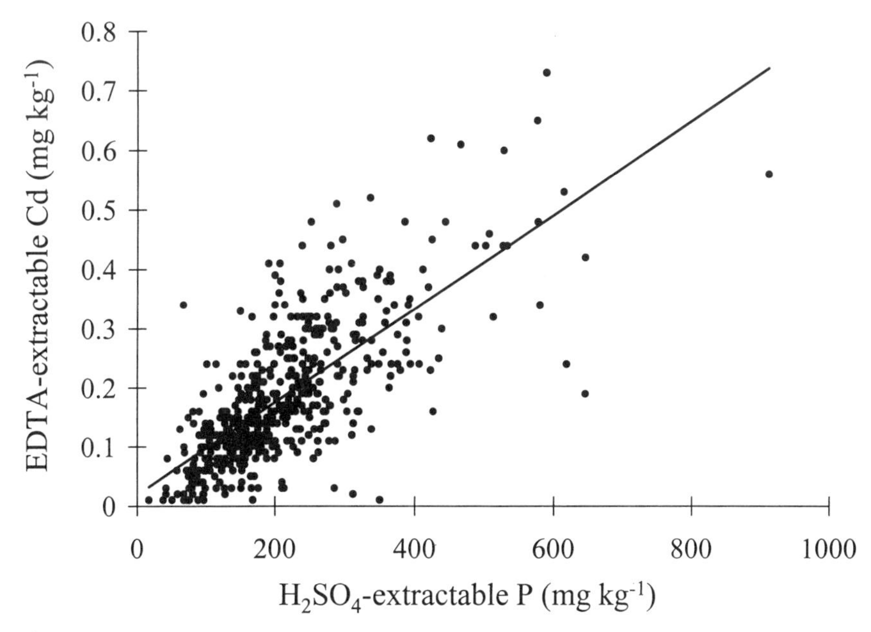

Figure 1.2. Relation between cadmium and phosphorus in soils of a pastoral agroecosystem east of Adelaide, South Australia. The fitted line is y = 0.008X + 0.0183: $R^2 = 0.53$, *P<0.001. Source: Merry (1988).*

rates of deposition are very much less because of improved smelter design and enforced emission controls.

The increasing use of sewage sludge near major cities will result in increased concentrations of many heavy metals on a restricted area of soils. The actual environmental effect of these applications will be subject to regulatory controls that are currently being developed in Australian states and in New Zealand. The New South Wales government has established interim limits that are similar to regulations of the European Community (see Table 1.17 later), but place greater emphasis on concentrations of contaminating heavy metals in soils. Increases in Cd concentrations in soil due to phosphate fertilization have been estimated in different Australian reports, with measured increases in Cd concentrations in surface soils of up to 0.14 and 0.18 mg kg^{-1} soil for cereal and pasture soils, respectively, near Canberra (Williams and David, 1973), and about 0.1 mg kg^{-1} for pasture soils near Adelaide (Merry and Tiller, 1991). As phosphate fertilization is the major source of Cd in Australian and New Zealand soils, there is usually a good relationship (Figure 1.2) between soil Cd concentrations and soil P concentrations (Merry, 1988).

An issue of greater importance than the total concentration is the concentration of phyto- and bio- available heavy metals, as controlled by soil and environmental factors. The CSIRO Land and Water Soil Contamination Group has used an exhaustive EDTA extraction (Clayton and Tiller, 1979) as a first approximation of the total potentially available toxic metal. Under many conditions this procedure measures the amount of metal added to the soil. The concentrations of EDTA-extractable Cd in soil commonly reported in Australian studies (usually <0.5 mg kg^{-1}) may appear low in comparison to European or American soils, but constitute a major proportion of the potentially available forms of Cd in soils. Provision of reliable estimates of both the total quantity of toxic metal and the soluble species in the equilibrium soil solution remains an important challenge to soil environmental scientists.

Quality of Agricultural Products

There are only a few data to demonstrate the environmental impact of low levels of contamination, such as from fertilizers, on the quality of agricultural produce. The link between fertilizer use, adding Cd to soils, and increases in soil Cd concentrations has now been well established in many studies both internationally (Mulla et al., 1980; Mortvedt et al., 1981; Andersson and Hahlin, 1981; Smilde and van Luit, 1983; Rothbaum et al., 1986; Andersson and Siman, 1991; Nicholson et al., 1994) and in Australia and New Zealand (Williams and David, 1973, 1976; Merry, 1988; Roberts et al., 1994).

However, while the link between fertilizer Cd and crop Cd concentrations can be demonstrated in greenhouse experiments (Mulla et al., 1980; Williams and David, 1973; Reuss et al., 1978), under field conditions the link is often less clear (Mulla et al., 1980; Andersson and Hahlin, 1981; Mortvedt, 1987; Andersson and Siman, 1991; Nicholson et al., 1994).

The effect of addition of fertilizer Cd to soil may or may not demonstrably increase crop Cd concentrations under field conditions due to spatial variability of field experiments (Andersson and Hahlin, 1981); seasonal variation in crop yield and rooting patterns (Jaakkola et al., 1977); growth dilution effects reducing Cd concentrations in plants due to responses to P (Mortvedt et al., 1981; Smilde and van Luit, 1983); low rates of Cd addition in relation to a high background concentration in soil which obscures any effect of Cd added (Vetter and Früchtenicht, 1982); and very strong retention of added Cd so that phytoavailability is unaffected by Cd additions to soil (Reuss et al., 1978; Mulla et al., 1980; Mortvedt et al., 1981).

In contrast, the impact of point source metal pollution on crop quality is easier to demonstrate (e.g., Merry et al., 1981), as metal addition rates to soil are often higher and a range of pollutant loadings can be studied at a single point in time by studying crop metal concentrations in relation to distance from the point source. This approach overcomes seasonal variation errors.

Impacts of both point source and diffuse sources of metal pollution on the quality of Australian agricultural produce are checked by a biennial survey of national retail markets (see Table 1.9).

The generally low concentrations of heavy metals in Australian agricultural produce reflects the low level of soil contamination, primarily because of the low areal intensity of industrialization and urbanization. Details of environmental impact of metal pollution on major agricultural systems in Australia and New Zealand are outlined below.

Wheat

Concentrations of Cd and Pb in Australian and New Zealand cereals have been measured in a number of detailed surveys (see Table 1.10). It is clear that overall Cd and Pb concentrations in both Australian and New Zealand wheat are very low and easily meet food regulations. Values are invariably lowest in the eastern Australian states, but some regions of concern in the context of Australia's clean food policy occur in South and Western Australia.

In Australia the highest Cd and Pb values occur on farms close to the Port Pirie lead-zinc smelter and these have been the topic of investigation (Merry et al., 1981). With the exception of Cd in whole wheat tops, Cd, Pb, and Zn in both wheat grain and whole tops grown in a smelter-contaminated area were found to be highly correlated ($p<0.001$)

Table 1.9. Concentrations of Cadmium and Lead in Some Foods Analyzed in the 1992 Australian Market Basket Survey (Stenhouse, 1994) in mg kg^{-1} Fresh Weight

	Cadmium			Lead		
Food	**Average[a]**	**Maximum**	**Minimum**	**Average**	**Maximum**	**Minimum**
Beans, green	0.003	tr[b]	nd[c]	0.010	nd	nd
Beef, minced	0.007	0.038	nd	0.023	tr	nd
Bread, wholemeal	0.008	0.036	tr	0.013	0.050	nd
Carrots	0.029	0.050	0.008	0.012	tr	nd
Lettuce	0.014	0.110	tr	0.014	tr	tr
Liver, lamb	0.163	0.740	0.014	0.076	0.280	nd
Livers, chicken	0.008	0.019	nd	0.013	tr	nd
Milk, full cream	0.003	nd	nd	0.014	0.070	nd
Potatoes	0.033	0.068	0.013	0.012	tr	nd
Tomatoes	0.007	0.170	nd	0.011	tr	nd

[a] No. of samples ranged from 24 to 32.
[b] tr (trace) represents < 0.004 mg kg^{-1}.
[c] nd (not detected) <0.002 mg kg^{-1}.

Table 1.10. Concentrations of Cadmium and Lead in Australian and New Zealand Cereals (mg kg^{-1} FW)

			Cadmium		Lead	
Region	**No. of Samples**	**Cereal**	**Mean**	**Max.**	**Mean**	**Max.**
Victoria[a]	12	Barley	0.006	0.018	0.123	0.218
S. Australia[a]	12	Barley	0.009	0.016	0.116	0.149
W. Australia[a]	16	Barley	0.006	0.011	0.046	0.120
Queensland[b]	267	Wheat	0.006	0.042	0.027	0.244
S. Australia[c]	1557	Barley	0.010	0.073	0.045	1.76
S. Australia[c]	1749	Wheat	0.015	0.256	no data	no data
W. Australia[d]	20	Wheat	0.049	0.081	no data	no data
New Zealand[e]	70	Wheat	0.070	0.190	no data	no data

Sources: [a]Best (1989); [b]Best (1991); [c]Richards et al. unpublished data; [d]Oliver et al. (1995b); [e]Roberts (unpublished data, 1998).

(Merry et al., 1981) with EDTA-extractable (Clayton and Tiller, 1979) concentrations in the soil. Associated work (Tiller et al., 1975) involving washing procedures showed that concentrations in the grain reflected uptake only by the roots; surficial contamination was not a factor.

Similar correlations are difficult to achieve in normal agricultural soils not subject to point source pollution. Studies of contamination of pastures on the same smelter-affected area (Merry and Tiller, 1978) also showed strong correlations between metal concentrations in pasture dry matter and metals extracted by EDTA. In this situation the relationship for Pb was due to adherence of particulate Pb on leaf surfaces; for Cd the correlation was due to uptake by the roots. These investigations showed that low levels of contamination can have an impact on the quality of agricultural products but that the impact can be difficult to predict.

In terms of diffuse sources of metal pollution (i.e., fertilizers) there are no long-term field studies in Australia that could be used to investigate the long-term impact of fertilizer Cd on crop quality. There is some evidence of the link between fertilizer use and crop Cd concentrations in the data of Williams and David (1976). They demonstrated on one

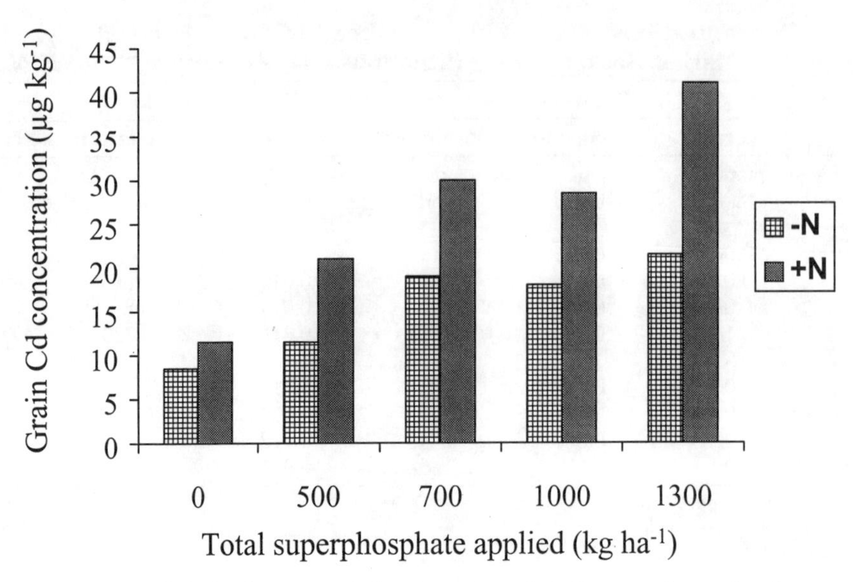

Figure 1.3. Relation between cadmium in grain and application of superphosphate. Source: Williams and David (1976).

site that Cd concentrations in wheat grain increased with increasing rate of fertilizer applied (Figure 1.3).

The only other example of a clear-cut relationship of metals in the soil and in agricultural crops was found near the smelter at Port Pirie. Wheat grain produced within about 15 km from the smelter over a farming area of 50–100 km^2 had Cd concentrations in excess of the Australian regulatory food limit. The potential hazard of this situation must, however, be viewed both in relation to the conservative food limit in force then, and the appreciable dilution of the contaminated grain with uncontaminated grain at the silo on delivery.

Potatoes

In horticultural soils in Australia, Sparrow et al. (1993a) found that increasing the rate of applications of banded triple superphosphate (TSP), up to 240 kg P ha^{-1}, increased concentrations of Cd in the flesh of potato tubers, although the TSP used in their work was highly Cd contaminated (151 mg kg^{-1} or 720 mg Cd kg^{-1} P).

In subsequent Australian work on potatoes, Sparrow et al. (1993b) compared increasing rates of double superphosphate (DSP) containing 15 mg Cd kg^{-1} (90 mg Cd kg^{-1} P) to similar treatments using DSP containing 90 mg Cd kg^{-1} (530 mg Cd kg^{-1} P). Cadmium concentration in the fertilizer had little effect on tuber Cd concentrations and the main effect observed was an increase in crop Cd concentrations due to P fertilization, independent of fertilizer Cd concentration. This result confirms earlier findings of Williams and David (1977) that addition to soil of P, without Cd, can significantly increase Cd concentrations in plants due to increases in root proliferation, allowing the plants to access more Cd in the soil.

A major survey of Cd in potatoes and their corresponding soils in South Australia (McLaughlin et al., 1994b) showed an appreciable range of Cd concentrations in both tubers (0.01 to 0.23 mg kg^{-1}) and soils, but no correlation between them (see Amelioration Strategies section). The high degree of phytoavailability and the obscuring of the rela-

Table 1.11. Cadmium Concentrations in Australian and New Zealand Potato Tubers in Comparison to International Data [a]

		Cd Concentration in Fresh Potatoes (mg kg^{-1} FW)				
Country	**No. Sites**	**Median**	**Mean**	**Min.**	**Max**	**Reference**
Australia[b]	429	0.037	0.046	0.004	0.232	McLaughlin (unpubl. data)
Aust. – WA	116	0.030	0.035	0.005	0.120	Adelaide Eng. & Wat. Sup. Dept. (1994)
New Zealand	20	–	0.020	0.010	0.100	Roberts (unpub. data)
U.K.	20	–	<0.100	<0.010	0.060	U.K. Ministry of Agriculture (1983)
U.S.	297	0.028	0.031	0.002	0.182	Wolnik et al. (1983)
Netherlands	94	0.030	0.030	0.010	0.080	Wiersma et al. (1986)
Spain	8	–	0.013	0.008	0.017	Zurera et al. (1987)
Poland 1986	70	0.006	0.010	0.001	0.067	Zawadzka et al. (1990)
Poland 1987	94	0.014	0.013	0.001	0.080	Zawadzka et al. (1990)
Poland 1988	90	0.010	0.018	0.001	0.090	Zawadzka et al. (1990)
Switzerland	101	0.013	0.015	0.003	0.044	Andrey et al. (1988)
Germany	133	0.030	0.047	<0.013	0.200	Weigert et al. (1984)
Sweden	62	–	0.016	0.005	0.055	Jorhem et al. (1984)

[a]After McLaughlin et al., 1997b.
[b]All states.

tionship with soil Cd are both due to the influence of chloride in soil solution (McLaughlin et al., 1994).

In a subsequent extended survey in Australia (McLaughlin et al., 1997b), Cd concentrations in potato tubers varied by almost two orders of magnitude across sites, but the range of Cd concentrations found were similar to those reported in surveys from other countries (Table 1.11). Analysis of a recent survey of potato crops in South Auckland, New Zealand, showed elevated soil Cd levels (relative to native sites) and elevated tuber Cd concentrations similar to those in the Australian survey.

Animal Issues

Pastures

Merry and Tiller (1991) in studies of trace metals in soils and pastures near Adelaide found that mean concentrations of Cd in four seasonal collections of subterranean clover (*Trifolium subterraneum*) and African capeweed (*Arctotheta calendula*) over three years were 0.30 to 0.34 and 1.57 mg kg^{-1}, respectively. Capeweed is a member of the daisy family (*Compositae*) known to accumulate some toxic metals (Matthews and Thornton, 1982). Equivalent results for Pb were 1.01 to 1.38, and 2.03 mg kg^{-1}, respectively. The Cd concentrations in clover were correlated ($p<0.001$) with Cd extracted from the soil by both EDTA and DTPA in three out of four seasonal pasture collections made. The principal source of Cd in soils was phosphatic fertilizers. Lead concentrations in clover also correlated well ($p<0.001$) with Pb extracted by the same extractants, but only in two of the four collections. In contrast to Cd, Pb was mainly derived from atmospheric accessions of automobile exhaust emissions that were transferred from the city of Adelaide (Tiller et al., 1987). Both Cd and Pb in soils also correlated well ($p<0.001$) with their concentrations in capeweed.

Similar work on a long-term fertilization trial in Victoria by Tiller et al. (1997) con-

Table 1.12. Mean Cadmium, Lead, Copper, Zinc and Arsenic Concentrations (mg kg^{-1}) in the Legume Component of New Zealand Improved Pastures on a Range of Soil Types [a]

Soil Type[b]	Cadmium	Lead[c]	Copper	Zinc	Arsenic
YBL	0.07	0.45	11.3	107	0.09
YBP	0.22	0.11	10.6	95	0.07
YBE	0.12	0.15	12.9	106	0.11
YGE	0.09	0.18	11.8	118	0.11
Gley	0.07	0.41	10.7	80	0.09
Peat	0.09	0.20	10.2	82	0.11
BGL	0.05	0.24	11.9	108	0.13
ALLUV	0.10	0.22	11.2	115	0.12

[a]Source: Roberts et al. (1994) and unpublished data.
[b]Soil types were described under Table 1.1.
[c]Lead concentrations were at or below detection limits.

Table 1.13. Mean Concentrations of Cadmium, Lead, Copper, Zinc, and Arsenic in the Grass, Legume, and Weed Components of New Zealand Pastures Averaged Over All Soil Types [a]

Herbage	Pasture	Cadmium	Lead	Copper	Zinc	Arsenic
Grass	Native	0.08	0.29	9.7	83	0.13
	Improved	0.10	0.29	10.3	90	0.12
Legume	Native	0.07	0.29	11.1	109	0.10
	Improved	0.06	0.21	12.0	105	0.10
Weeds	Nature	0.14	0.51	14.0	143	0.08
	Improved	0.28	0.26	13.0	128	0.12

[a]Source: Roberts et al. (1994) and unpublished data.

firmed that there was a significant correlation between fertilizer P applied and Cd concentrations in dry-standing herbage. Cadmium concentrations in weed species were raised to a greater extent by fertilizer addition than Cd concentrations in grasses or clover. Cadmium concentrations in sheep feces were also closely related (r^2 = 0.77) to fertilizer P applied.

More detailed and extensive information from New Zealand research (Roberts et al., 1994 and unpublished data) is summarized in Tables 1.12 and 1.13 for pasture legumes on several soil types and for three classes of herbage averaged over all soil types.

Differences in metal concentrations of pasture plants were noted between soil types, as might be expected, but differences between different kinds of herbage did not generally vary much on any particular soil type except for Cd, Zn, and Pb, which showed higher concentrations in weeds than in grasses or legume species. On some soil types, As and Pb concentrations in grasses were higher in native than in introduced species. Despite higher Cd concentrations in soils under improved pastures, Cd concentrations in legumes were identical on native compared with improved pastures on most soil types. Cadmium concentrations in grasses were significantly greater (p<0.05) on improved pastures, but the difference was small in magnitude. Cadmium concentrations in grasses did correlate with soil Cd in each soil type. Only weed species, some of which have a greater ability to absorb soil Cd, had appreciably higher Cd concentrations on improved pasture soils compared to native soils of lower soil Cd status. Cadmium-accumulating weeds like African capeweed, which strongly influences the Cd status of some southern pastures in Australia, are not an important factor in New Zealand pastures because of better rainfall and soil fertility conditions. The legume component of improved pastures in New Zealand principally consists of white clover (*Trifolium repens L.*) varieties, and the data in Table 1.12 would reflect heavy metal concentrations for this species. While other workers have indicated that other legume species, such as subterranean clover (*Trifolium subterraneum L.*), accumulate heavy metals, especially Cd (Merry, 1988), there was no suggestion of this for Cd or the other heavy metals in the New Zealand study.

In the case of the pastoral ecosystem the positive relation between metals in plants and soils is only clear for weed species. As noted previously, relationships between metals in soils and plant metal concentrations are usually better defined where point source pollution has occurred (Merry et al., 1981).

Elevated concentrations of Cd in pasture species pose no appreciable hazard to human health, except for a possible increase in Cd levels in offal products of sheep grazing on capeweed-dominant pastures.

Animal Products

National maximum permissible concentrations (MPCs) of Cd in animal products in Australia and New Zealand have recently been reviewed by the newly formed Australia and New Zealand Food Authority (ANZFA), and are currently 0.2, 2.5, and 1.25 mg Cd kg^{-1} fresh weight for muscle tissue, kidney, and liver, respectively (Table 1.19 later). In New Zealand the domestic MPC for offal and meat products was previously 1.0 mg Cd kg^{-1}, but will be revised under the new merged ANZFA standards. *Codex Alimentarius*, the body in the Food and Agriculture Organization (FAO) responsible for standardizing food quality regulations for traded commodities, is considering a limit of 2.0 mg Cd kg^{-1}, while the European Community requires that imported sheep offal should have less than 1.0 mg Cd kg^{-1}. Langlands (1988) indicated that only 1% of the muscle, 2% of liver samples, and 8% of kidney samples exceeded the Australian limits, with some regions having a higher proportion of violations than others. Cadmium levels in muscle rarely exceeded the MPC. In New Zealand about 25% of ovine and 17 % of bovine kidneys with a higher proportion for dairy cattle, exceeded the old New Zealand MPC of 1.0 mg Cd kg^{-1} (Roberts et al., 1994). The responsible on-farm factors are likely to be due to the interactions of fertilizer practice with soil and climatic conditions. In the interim, while farm practices are being modified where possible to reduce uptake of Cd, offal from animals in the most affected regions is being segregated according to animal age because Cd levels increase with maturity.

AMELIORATION STRATEGIES

It is well recognized that prevention of contamination of agricultural land should have the highest priority of all strategies to protect soil, crop, and water quality and ultimately human and animal health. To this end Australia and New Zealand have instituted, or are in the process of establishing, stricter controls on industrial atmospheric emissions, on water used for agricultural purposes, e.g., for irrigation and livestock purposes and on solid agricultural amendments, e.g., sewage sludge and fertilizers (see Regulatory Control section). However, where soil contamination has occurred, strategies are needed to control transfer of contaminants from soil through the food chain to animals or humans.

Fertilizer Macronutrients and Fertilizer Quality

As the links between fertilizer use and crop quality are often hard to establish under field conditions (discussed in the Environmental Impact section), it is therefore often difficult to determine if changing fertilizer type affects Cd uptake by field crops. Many workers have demonstrated the impact of P and N fertilizer type and quality on crop Cd concentrations in glasshouse experiments (Williams and David, 1976, 1977; Street et al., 1978; Reuss et al., 1978; Singh et al., 1988; Eriksson 1990; Willaert and Verloo, 1992; He and Singh, 1993). Unfortunately, some of these studies used soils spiked with inorganic Cd salts (Street et al., 1978; Singh et al., 1988; Eriksson, 1990). This practice is likely to lead to erroneous conclusions in interpretation of experimental results, due to reduced binding of Cd with soil surfaces (Tiller, 1989) and the likelihood of cation exchange reactions unrepresentative of reactions in normal agricultural soils. Concentrations of Cd added to soils in these studies ranged from 1 mg kg^{-1} (Eriksson, 1990) to 50 mg kg^{-1} (Singh et al., 1988), while most agricultural soils have Cd concentrations less than 0.5 mg kg^{-1}.

The other studies which used soils amended with commercial fertilizers reached a number of conclusions with regard to fertilizer addition and fertilizer quality:

1. Increasing Cd concentration in the fertilizer increases Cd concentrations in the plant (Williams and David, 1976, 1977; Reuss et al., 1978; Singh, 1990; He and Singh, 1993)
2. Except under acute P deficiency, adding P fertilizer to soil increases plant Cd concentrations due to enhancement of root proliferation (Williams and David, 1977);
3. Phosphorus fertilizers which produce alkaline reaction products, e.g., diammonium phosphate, tend to reduce the plant availability of fertilizer Cd compared to fertilizers having acidic reaction products, e.g., single superphosphate, monoammonium phosphate, or NPK mixes containing ammonium sulfate (Reuss et al., 1978; Eriksson, 1990; Willaert and Verloo, 1992)
4. Addition of fertilizers under greenhouse conditions may increase Cd concentrations in plants due to ion exchange reactions of cations in the fertilizer displacing Cd into solution and ionic strength effects reducing Cd sorption (Andersson, 1976; Eriksson, 1990).

Under field conditions there are less conclusive results. For commercial P fertilizers it is often difficult to separate effects of P from effects of Cd in the fertilizer. Sparrow et al. (1993b) showed that for potato production, addition of P increases tuber Cd concentrations at constant rates of Cd addition. McLaughlin et al. (1993) could find no effect of changing P fertilizer type (mono- and diammonium phosphate, single superphosphate and reactive phosphate rock) or changing fertilizer Cd concentration on Cd concentrations in potato tubers.

In Australia there has been little work on the interaction between N fertilization and Cd uptake by crops. Williams and David (1976) found that addition of ammonium nitrate to a soil significantly increased wheat grain Cd concentrations, but were unable to suggest possible mechanisms to explain this result. In general terms, farmer decisions can result in gradual changes in organic matter, pH, and possibly redox conditions, and thus affect heavy metal availability to crops at some later stage. Gradually declining pH values over several decades, for example as occur on soils associated with permanent leguminous pastures in Australia (Helyar, 1976) or possibly resulting from some crop rotations (Oliver et al., 1993), could lead imperceptibly to additional hazards if the soils were contaminated with heavy metals such as Cd.

There is conflicting evidence for the effect of changing K fertilizer type on crop Cd concentrations. Sparrow et al. (1994) found that in four out of six sites studied, potato crops fertilized with potassium sulfate instead of potassium chloride had lower tuber Cd concentrations. In a separate study McLaughlin et al. (1993) could find no effect of changing K fertilizer type on tuber Cd concentrations. Possible reasons for these conflicting results are that the rates of K applied by Sparrow et al. (1994) were higher (160–350 kg K ha^{-1}) than those used by McLaughlin et al. (1993) (150 kg K ha^{-1}), and irrigation water quality was poorer in the studies of McLaughlin et al. (1994a). Chloride is known to readily complex Cd in solution (Hahne and Kroonjte, 1973) and soil salinity has been demonstrated to markedly increase Cd uptake by field crops (McLaughlin et al., 1994a). Impacts of changing K fertilizer type on crop Cd concentrations will therefore be limited to crops heavily fertilized with K and grown on nonsaline soils or with good quality irrigation water.

In Australia and New Zealand, where rates of fertilizer application are generally low in extensive agriculture, it is unlikely that changing fertilizer type will affect crop Cd concentrations in the short term. Even in horticultural situations where fertilizer additions are greater, it is difficult to demonstrate significant changes in crop Cd concentrations due to fertilizer management. Compared to greenhouse conditions, under field conditions it is likely that different rooting patterns and conditions in the soil solution induced through rainfall and irrigation may minimize effects of fertilizer management on crop Cd concentrations in any one season. In the long term, however, it is prudent to use low Cd phosphatic fertilizers and reverse any soil acidification developing through the use of ammonium-based nitrogenous fertilizers and legume-containing crop rotations, as ignoring these management options will eventually lead to enhanced Cd concentrations in crops in the future.

Micronutrient Interactions

The interactions between Cd and Zn in the soil-plant system have been long recognized but are not clear-cut. Effects may be additive (Chaney and Hornick, 1978; Page et al., 1981), antagonistic (Haghiri, 1974) or nonexistent (McLean, 1976), even positive for plant foliage but negative for grain (Chaney and Hornick, 1978). These complex observations probably result from differences in test crop, soil type, soil factors, and actual and relative levels of Cd and Zn used. Negative interaction was the most frequently observed effect.

Cadmium/zinc interactions affecting cereal quality under farm-relevant conditions on Australian soils contaminated by residual Cd originating from fertilization were studied by Oliver et al. (1994). The total concentrations of Cd in the soils studied were less than 1 mg kg^{-1}, and Zn additions were < 20 kg ha^{-1}. Results were based on nine field experiments established in widely separated areas and varied soil types of South Australia characterized as being marginally Zn deficient. At most sites, grain Cd concentrations were decreased appreciably by small applications of Zn to the soil (<10 kg Zn ha^{-1}), especially where the grain was harvested in the year of Zn application (Figure 1.4). None of the decreases in Cd concentrations in grain at four of these sites could be attributed to growth dilution but in the other two, growth dilution partially contributed to the effect of Zn in decreasing Cd concentration in grain.

The mechanisms for this marked decrease in grain Cd by small applications of Zn are not known. Direct competition between Zn and Cd for root uptake may be possible (Cataldo et al., 1983; Robson and Pitman, 1983). Loss of root membrane integrity under conditions of nutrient deficiency, facilitating uptake of Cd, and other ions by mass flow (Welch et al., 1982; Cakmak and Marschner, 1988) may be involved. Alternatively, Cd uptake may also be enhanced by mobilization of soil-bound Cd by phytosiderophores released by wheat roots in response to nutrient deficiency (Crowley et al., 1987; Chaney, 1988). Lack of interaction between grain Cd and foliar-applied Zn (Oliver et al., unpublished data) would support the view that root function and translocation are the key determinants. Higher application rates of Zn (to 320 kg ha^{-1}) on Cd in wheat grown on a Zn-sufficient site within the Zn-deficient area discussed above (Figure 1.5) resulted in

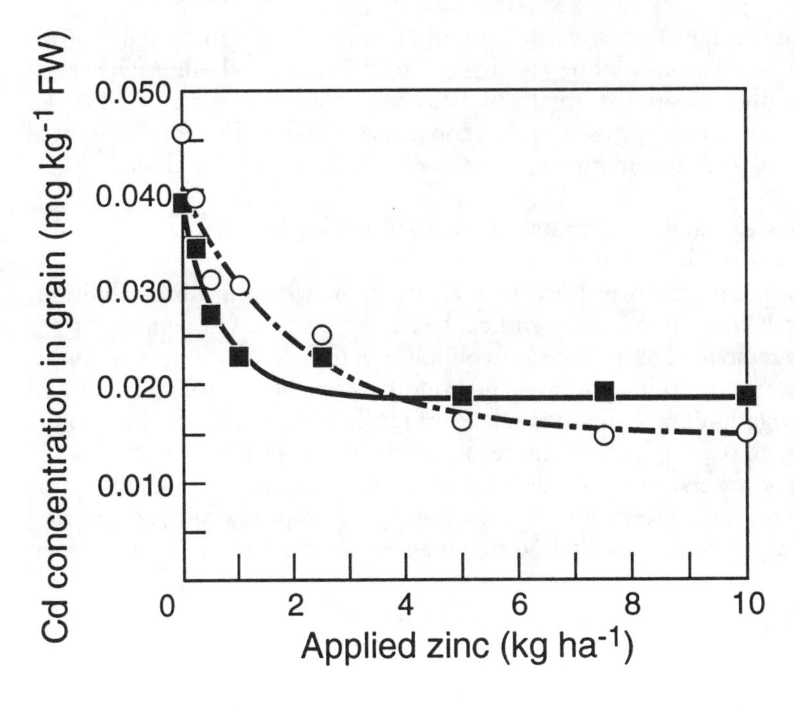

Figure 1.4. Effect of low application rates of zinc on concentrations of cadmium in wheat grain grown on two marginally zinc deficient South Australian soils in different regions. Source: Oliver et al. (1994).

about 20% decrease in Cd concentrations at 40 kg ha $^{-1}$ and about 30% at 320 kg ha^{-1} in relation to the control (Oliver et al., unpublished data). This approach would not provide a practical solution to the reduction of Cd concentrations in grain.

Cadmium–zinc interactions were also investigated for potato crops and pastures. Most published work on this topic had previously been restricted to greenhouse studies. Field experiments by McLaughlin et al. (1996) using Zn applications of up to 100 kg ha^{-1} showed a negative interaction between Cd concentration in potato tubers and Zn at four

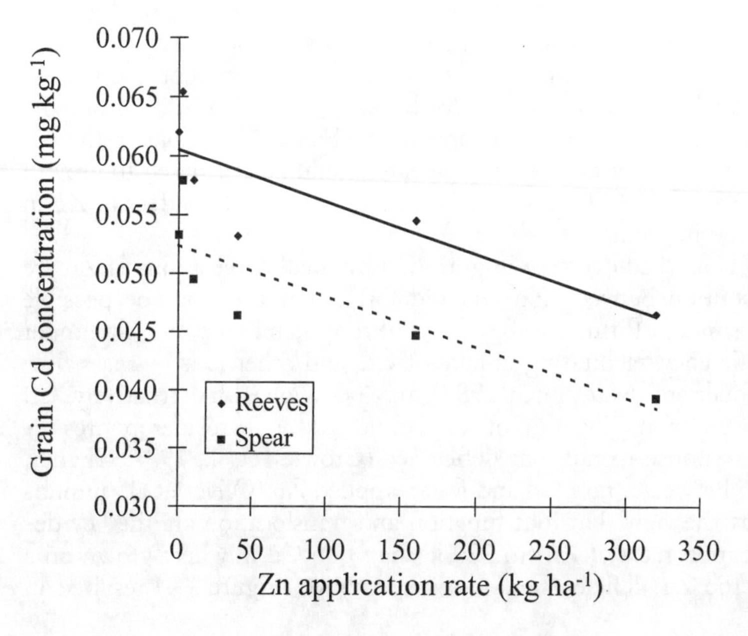

Figure 1.5. Effect of high application rates of zinc on concentrations of cadmium in grain of two varieties of wheat grown on a zinc sufficient South Australian soil. The fitted lines are $y = -0.0004X + 0.0605$: $R^2 = 0.72$, $P<0.001$ (variety Reeves); $y = 0.0004X + 0.0524$: $R^2 = 0.73$, $P<0.001$ (variety Spear). Source: Oliver and Tiller (unpublished data).

of the field sites studied. Reduction in Cd levels in tubers varied from 0 to 50%. Application of Zn up to 64 kg ha^{-1} (Tiller et al., 1997) did not significantly affect Cd concentrations in subterranean clover.

From a practical agronomic point of view, Cd concentrations in cereals and possibly other crops grown on marginally Zn-deficient land have good prospects of being ameliorated by the low applications of Zn required to alleviate Zn deficiency. Despite Zn deficiency usually not occurring in commercial potato crops, amelioration of undesirable Cd levels in tubers using Zn has met with some success (McLaughlin et al., 1996).

Crop Rotations and Some Other Cultural Practices

Cereal crops are often grown in sequence with other crops such as legumes, e.g., peas, beans, lupins, and clover- or medic-based pastures. Such rotations have been a common management practice in Australia and New Zealand to minimize root disease, control weeds, and increase nitrogen (N) status. Associated soil changes may include improvement in soil structure and soil fertility through increased organic matter, but negative impacts such as some soil acidification are likely in poorly buffered soils.

The effects of various crop rotations and other agronomic practices on Cd in grain were investigated (Oliver et al., 1993) at field trial sites in South Australia. Experiments included rotations with continuous wheat, subterranean clover-based pasture, and various other legumes. The experimental design also involved three tillage practices and two fertilizer regimes, one equivalent to local recommended practice and the other at double those rates. Some legume crops in the rotation, and especially lupins, increased Cd concentrations in wheat grain but it is not yet possible to fully explain these results. The association of soil acidification with legume growth under southern Australian growing conditions (Helyar, 1976) could partly explain the results. However, the slow rate of soil acidification by legumes and N fertilization (normally much less than 0.05 pH unit y^{-1}), during the five years growth of lupins in rotation would not be expected to have had an impact on the average soil pH of the cultivated layer. Localized rhizosphere pH effects are possible, especially below the plow layer. Lupins are also known to produce exudates (Nambiour, 1976; Warembourg and Billes, 1979) such as citric acid (Gardiner, 1981), which may combine with Cd and enhance its mobility and availability. Our ongoing soil chemical research in CSIRO Land and Water, Adelaide is investigating the mechanisms by which lupins, in particular, can increase the grain Cd concentrations of a following wheat crop.

The effect of different tillage practices on Cd concentration in grain was also studied in one of the rotation trials discussed above. Cadmium concentrations of direct drilled wheat in 1989 were about one-third higher than for either conventional or reduced tillage. This was explained by a restriction of root growth under direct drilling compared to conventional cultivation as found by other researchers (Hamblin and Tennant, 1979; Gates et al., 1981) on similar soil types. This would maximize in some seasons the uptake of both nutrients as well as Cd from the surface soil horizon where the highest soil Cd concentrations are located.

Different methods of treating wheat stubble were also investigated because of their possible effect on soil organic matter status and thus indirectly on Cd uptake by a later wheat crop. Treatments including burning, incorporation into the cultivated soil, and retention of straw on the soil surface had no significant effect on grain Cd concentration in the two years tested.

Soil pH

During recent decades it has been almost axiomatic in most published work and in discussions that soil pH has the key role in controlling the mobility and availability of most trace metals to crops. The general applicability and the economic benefits of liming practices for particular farming systems, especially in relation to the uptake of heavy metals, need to be critically assessed by field experiments. Few such investigations have been reported. We assessed these liming effects on wheat, potatoes, and pastures in a series of field experiments and by survey approaches in Victoria, New South Wales, and South Australia.

Wheat

In investigations of Oliver et al. (1995a) in Victoria and New South Wales, locally grown varieties of wheat, barley, and triticale were grown for three successive years under identical experimental conditions on soils adjusted in a pH($CaCl2$) range of 4.0 to 6.0.

Notwithstanding the similarity of the soils at the different sites, Cd concentrations in wheat and barley grain and the response slopes of the grain Cd vs. soil pH varied appreciably between sites and growing seasons. Examples of field results are shown in Figure 1.6. Greater and unexplained variance of the barley data compared to wheat prevented evaluation of differences in the dose-effect relationship of barley at some sites.

Although grain yields did not vary appreciably from site to site or from season to season at any site, Cd concentrations in grain at any site often varied appreciably from season to season, sometimes five-fold. The response of grain Cd concentration to soil pH change sometimes had similar slopes in different seasons but differed in Cd concentration. Grain Cd may clearly respond to soil pH in one growing season but not at all in another (Fig. 1.6). Growing seasons in which grain Cd concentrations did not respond to pH change provided grain with the lowest grain concentrations.

These observations are possibly explained by the wheat root distribution in relation to soil moisture status as determined by seasonal climatic conditions, especially rainfall distribution and frequency. Both fertilizer, especially P with associated Cd, and lime were re-

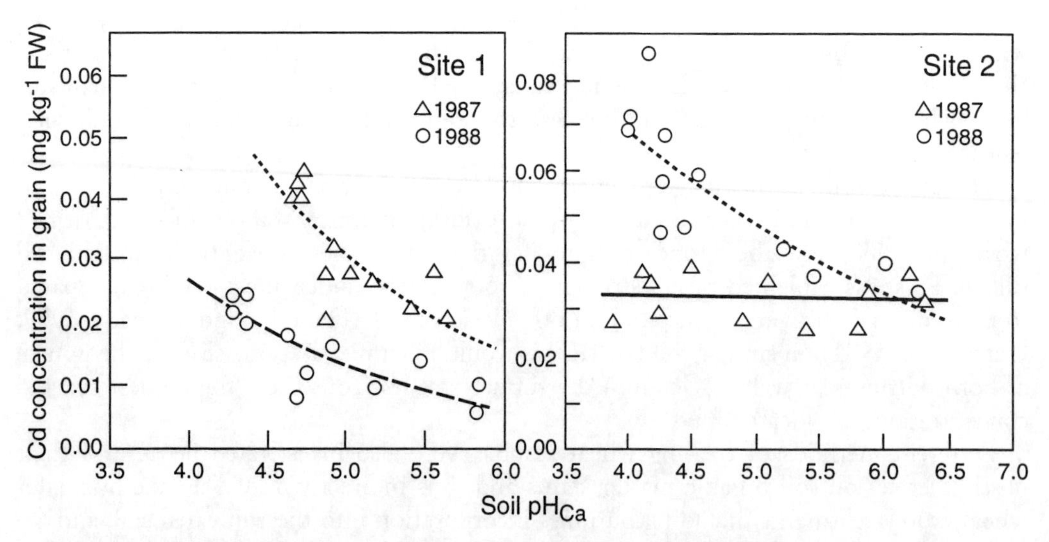

Figure 1.6. Effect of surface soil pH on the concentration of cadmium in field-grown wheat grain in two seasons and at two sites in New South Wales. Source: Oliver et al. (1995a).

stricted to the shallow cultivated layer, yet root distribution will be determined by soil moisture distribution with depth. If adequate moisture is maintained in the plow layer, cereal roots will remain in this Cd- and nutrient-enriched layer. This is also the layer with variable pH as determined by liming and thus restriction of the roots to this layer will not only maximize the Cd uptake, but also the influence of surface soil pH. In seasons with less regular rainfall, cereal roots will be more active in layers below the plow layer which are lower in Cd and have a more uniform soil pH. This will result in both a decrease in Cd uptake by cereals and a reduced or lack of pH response.

From these results it is clearly futile to attempt prediction of dose-response effects for added lime or added Cd in the field from laboratory or controlled environment experiments. Also, field experiments in one season may well not indicate likely outcomes in another growing season. Although liming should improve yields on acidic soils when economically justified, amelioration of Cd concentrations in wheat grain cannot be assured by this practice.

Potatoes

Both field and greenhouse liming experiments have been carried out on potato crops in various regions of southern Australia. Sparrow et al. (1993b) examined high rates of lime (up to 40 t ha^{-1}) application at three sites on heavily (pH)-buffered oxisols in Tasmania and found no effect of lime application at two sites and a significant increase in tuber Cd at the third. Soil pH was increased by 1 pH unit at most in these experiments. In a more comprehensive series of greenhouse and field trials (Maier et al., 1997), liming was found to significantly reduce tuber Cd concentrations in five out of six greenhouse trials, but the only significant effect under field conditions was an increase in tuber Cd at two of nine trials (Fig. 1.7). These experiments were carried out on mostly sandy soils having initial pH (water) values varying from 4.7 to 5.6, with final pH values after liming varying from 5.9 to 7.7.

We postulate that the lime-induced increase in tuber Cd concentrations may be due to

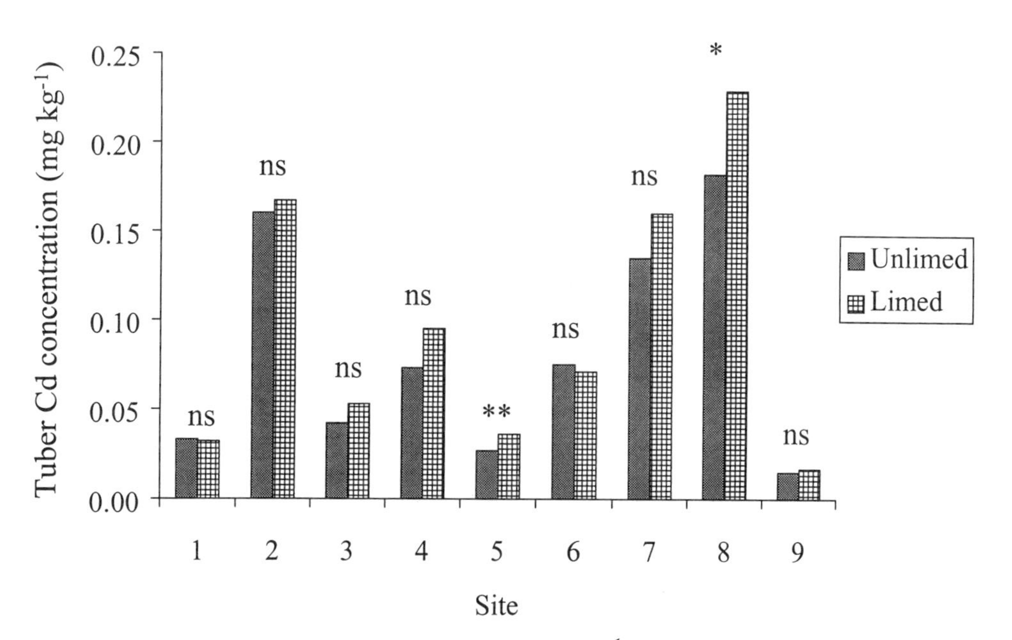

Figure 1.7. Effect of incorporated lime at rates up to 15 t ha^{-1} on Cd concentrations in potato tubers at nine field sites in South Australia. Source: Maier et al. (1997).

Ca^{2+} ions released from the lime displacing Cd^{2+} ions into solution, which on these soils may dominate any effects of pH on increasing Cd retention by soil surfaces. This hypothesis is currently under investigation.

Soil Salinity

It has been known for some time that the Cl ion can complex some trace metals, particularly Cd, Hg, and Zn, thus rendering them less prone to adsorption reactions in soils (Garcia-Miragaya and Page, 1986) and increasing mobility (Doner, 1978). Bingham et al. (1983) suggested from results of greenhouse experiments that Cl increases the phytoavailability of soil Cd. However, they related this to increased concentrations of the free Cd^{2+} ion in solution, an artifact of their experimental technique using soils dosed with Cd salts. In this situation Na^+, added as the counter ion to Cl^-, displaced Cd^{2+} from soil surfaces as much of the Cd would have been held only weakly by the soil, a situation not applicable in normal agricultural soils. More recently, McLaughlin et al. (1994a) demonstrated that soil salinity indeed increased phytoavailability of Cd for crops under field conditions. In this study, regional variation in Cd concentrations in commercial potato crops was related to water-extractable Cl concentrations in soil. Cadmium concentrations in tubers were also found to be significantly and inversely related (Fig. 1.8) to EDTA-extractable soil Zn concentration and to the potato variety.

Investigation of Cd concentrations in soil solution revealed that Cl significantly increased total concentrations of Cd (McLaughlin et al., 1997a), but that concentrations of Cd^{2+} ion were not high under high Cl conditions (Fig. 1.9). This would be expected, given the strong buffering of Cd^{2+} activity in most soils.

McLaughlin et al. (1994a) postulated that the Cl effect was due to the alleviation of a diffusion limitation to Cd transport to the root surface. However, subsequent work in our laboratories investigating the effect of Cl on Cd uptake indicate that $CdCl_n^{2-n}$ complexes may be taken up by plant roots, so that increased uptake under high Cl conditions may be explained by the high concentrations of $CdCl_n^{2-n}$ in solution (Smolders and McLaughlin, 1996).

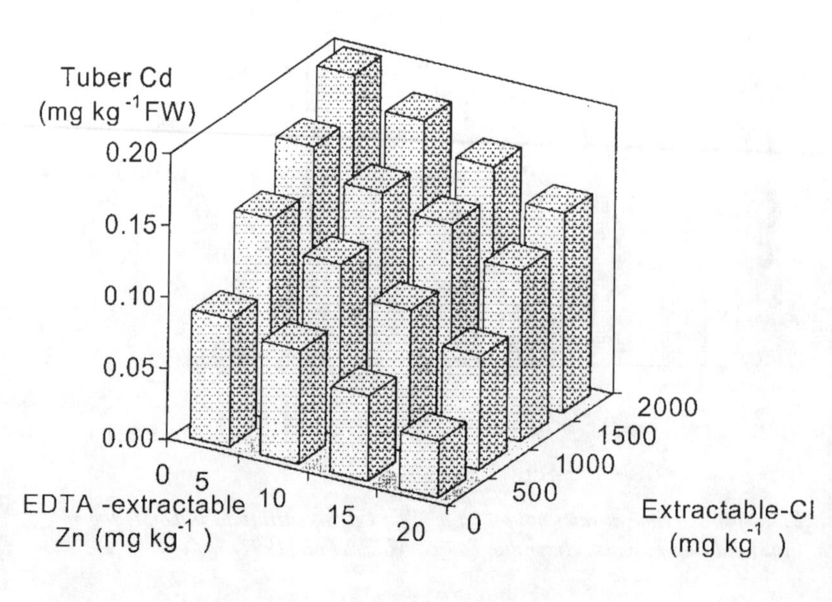

Figure 1.8. Three-dimensional relationship among tuber Cd concentration, water extractable Cl, and EDTA-extractable Zn in topsoils (0–150 mm). Source: McLaughlin et al. (1994b).

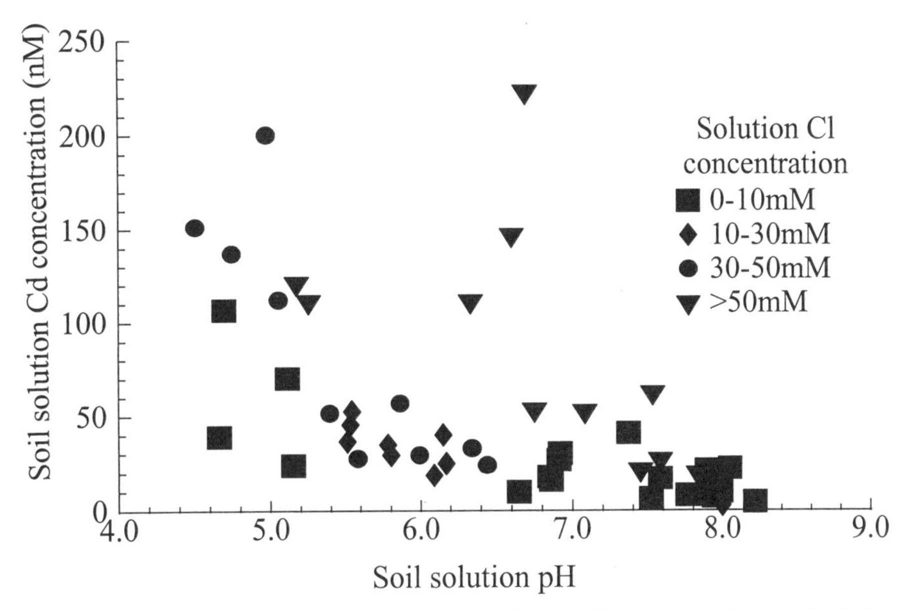

Figure 1.9. Relationship between pH, Cl concentration, and Cd concentration in soil solution from 50 soils in South Australia. The multivariate relationship between factors is log Cdss = 7.19 - 0.64pH + 0.01[Cl]: R^2 = 0.73, P<0.001. Source: McLaughlin et al. (1997a).

Choice of Crop Species and Variety

Crops have been bred to overcome many of the problems encountered in normal agronomic practice associated with increasing yield potential such as resistance to disease, insect attack, nutrient deficiency and excess, and more recently the limitation of the uptake of undesirable soil contaminants such as Cd. In Australia the genetic impact on the uptake of contaminants by crops has been restricted to problems of elevated Cd concentrations in key food commodities, and has been limited to the evaluation of varieties in existing plant breeding programs. Since these varieties were not chosen on the basis of ability to absorb soil Cd, there is a real risk that inadvertent selection for Cd may occur. Under Australian conditions, results are only available for Cd in wheat and potatoes.

Wheat

Cadmium concentrations were examined in wheat grain from national and state wheat variety trials (Oliver et al., 1995b). In the national trials about 15 varieties were sampled from 13 sites in two different years. Additional sites and varieties were studied in Western Australia. The sites covered a very large range of soil and environmental conditions. Although significant varietal effects were found, variation due to site effects were much more important. A low proportion of the sites produced wheat grain with Cd in excess of the superseded MPC of 0.05 mg kg^{-1}. Mean Cd concentrations of several varieties at a few sites approximated 0.1 mg kg^{-1}, but about 50% of the sites had grain with Cd concentrations <0.02 mg kg^{-1}, which were too low to assess varietal effects and were of no concern in any case.

The average range in Cd concentrations between the lowest and the highest absorbing wheat varieties at any site was two- to three-fold but ranged up to 19-fold at one site. Cadmium concentrations in any particular variety ranged to 20-fold over the 14 national trial sites, thus indicating the great importance of soil/site factors. Two varieties that had the highest Cd concentrations at a number of sites for several years had similar pedigrees, while a third variety that also tended to accumulate Cd had a different pedigree. Varieties which had low Cd concentrations also had similar pedigrees. This work suggests that in the breeding process, selection for some desirable characteristic have inadvertently also favored absorption of Cd by wheat plants.

Potatoes

The effect of potato variety on the uptake of Cd by tubers has been assessed at 12 sites with a broad range of soil and environmental conditions. Mean Cd concentrations of the 14 most commonly used varieties averaged over all sites varied from 0.03 to 0.05 mg kg^{-1}, thus not exceeding the previous Australian MPC (0.05 mg kg^{-1}), but at most sites there was a significant difference between varieties. The main commercial varieties demonstrated a two-fold difference between the lowest and highest in terms of Cd accumulation. Statistical analysis showed that no variety had consistently high or low Cd concentrations in all soil environments. Although differences between varieties were significant, soil/site factors, such as salinity of irrigation water (McLaughlin et al., 1994b), played a dominant role in determining the Cd concentrations of commercial potato varieties (see Environmental Impact section). This importance of site factors may explain the contradictory reports in the literature on effect of potato varieties in relation to Cd (e.g., Isermann et al., 1983). Notwithstanding the impact of soil salinity, the large difference in tuber Cd concentrations between varieties at some sites encourages the view that selection for varieties resistant to Cd accumulation may be achieved by breeding.

Pastures

Pastures are an important pathway of Cd to humans through grazing animals. Elevated concentrations of Cd in some animal offal products in some parts of Australia and New Zealand have stimulated research into the soil, plant, and animal factors involved. Soil, fertilizer, and grazing management are all accepted as affecting the amount of Cd intake by animals through their effect on soil and pasture composition and relative amounts of each ingested. Animals consume significant quantities of soil (Healy, 1973), either from mud splash or worm casts in humid areas, or as dust on leaves in dry areas. Animals foraging at ground level in overgrazed pastures, perhaps seeking pasture seed reserves may consume soil directly. However, research suggests that soil ingestion is likely to contribute only 8–11% of the total annual intake of Cd by grazing sheep in New Zealand (Longhurst et al., 1994). Broadcast fertilizer particles adhering to plant leaves may also be ingested if not washed off by rain before grazing; this pathway can be regulated by withholding grazing animals from freshly fertilized pastures.

The transfer of Cd through the pasture pathway by grazing animals to human food can also be controlled in several ways by farm management.

The important role of phosphatic fertilizer in supplying most of the Cd of pasture soils is well recognized (Williams, 1974). Merry (1992) studied the soil-plant relationships of Cd in field trials with sheep stocking rate (animals ha^{-1}) and phosphate fertilizer application as treatments. The concentrations of Cd in the dominant pasture plants responded linearly with applied P, especially capeweed (*Arctotheca calendula*). Sheep feces and surface soils also had Cd with concentrations that were closely related to added P and therefore added Cd. Similar results have been found in New Zealand (Longhurst et al., 1994). Amounts of Cd added in superphosphate were largely accounted for in the 0–10 cm soil layer. Cadmium concentrations were highest in the upper 0–1 cm layer which is most prone to ingestion by sheep. In other words, a farm practice involving high rates of Cd-containing fertilizer will result in high soil and pasture Cd levels. Investigations of the effect of adding noneconomic rates of Zn as $ZnSO_4$, up to 64 kg Zn ha^{-1}, to pastures ranging from deficiency to sufficiency in Zn did not appreciably decrease Cd concentrations in subterranean clover, as was found for wheat and potato in some areas.

The varying abilities of different plant species to absorb trace metals from soils also provides a practical means of limiting Cd through the pasture-animal pathway. Further field experiments by Merry (1992) confirmed this for pasture species of importance to the grazing industry in southern Australia. African capeweed contained 10 and 40 times, respectively, the Cd concentrations in subterranean clover and ryegrass (*Lolium perenne*). Merry and Tiller (1991) had earlier shown in a pasture survey of over a hundred sites during 1972–1974 that capeweed Cd levels exceeded clover by five times. Members of the daisy family (*Compositae*) are known to be metal accumulators (Matthews and Thornton, 1982). Manipulation of species available to grazing animals by farm management can clearly have a major impact on Cd intake by animals.

The practicalities of controlling Cd levels in pastures by liming was also investigated at a series of field trials on acidic pasture soils in the higher rainfall areas in South Australia (Merry, 1992). Significant negative responses of Cd levels to liming were found (Fig. 1.10) especially in subterranean clover, but not sufficient to appreciably decrease Cd intake of grazing animals when the expense of liming just to beneficiate Cd levels is considered.

When the total budget of Cd in the pasture system is considered, the uptake of Cd by the animals from ingestion of soil and fertilizers is potentially significant in the absence of Cd-accumulating weeds. Clearly in Australia, a significant pasture component of Cd-accumulating weeds should appreciably affect the exposure of Cd to grazing animals. Quantitative animal studies have not been undertaken, but high concentrations of Cd in animal offal occur in areas of pastures that are often dominated by African capeweed.

In New Zealand, a breeding ewe weaning one lamb per year requires a feed intake of at least 550 kg dry matter. Even at relatively low average concentrations of 0.3 mg kg^{-1} of mixed pasture containing less than 10% of weeds, the annual intake of Cd by the sheep is about 135–153 mg Cd $sheep^{-1}$ (Longhurst et al., 1994). This represents a situation in

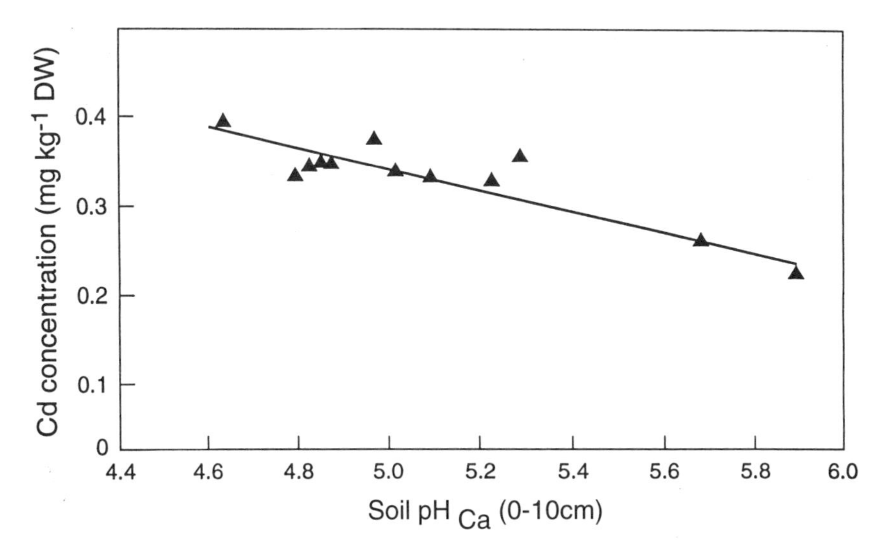

Figure 1.10. Effect of surface soil pH on the cadmium concentrations in subterranean clover (Trifolium subterraneum). *Source: Merry (1992).*

which both weeds and ingested soil contribute only a relatively small proportion of the total dietary intake of Cd, affecting the accumulation of Cd in offal of grazing animals. Clearly, permanent pasture represents the main source of dietary Cd to New Zealand sheep.

The management approach required to overcome any perceived problems of animal product quality involves minimizing future Cd inputs through fertilizer choice and practice, and adopting grazing and fertilizer strategies to maintain pasture compositions that are both nutritionally adequate and provide a low Cd intake. Another marketing strategy already in place is to restrict the age of animals from which offal products are taken for human consumption, since Cd in the animal accumulates with its age.

REGULATORY CONTROL

Soil

The control and remediation of contaminated soil sites in Australia is managed by state environmental and health agencies and is largely determined by the requirements of strongly contaminated urban sites. Regulations vary between different state authorities but the general approach is that followed internationally in many countries; that is, to define a soil limit which triggers an investigation. The thoroughness of the investigation will depend on the particular situation, especially land use. These limits have been usually based on the Dutch soil standards, which were set in the mid-1980s (Moen et al., 1988), and which have been recently revised (van den Berg, 1994). Some investigation limits in Australia corresponded to the presumed upper limit of the range of background levels, others to the old Dutch B levels that were usually 2–3 times higher.

National guidelines for limits to initiate some investigation of soil contamination are under continuing review by the NHMRC and the Australian and New Zealand Environment and Conservation Council (ANZECC). Health investigation level guidelines for nonagricultural soils have recently been revised in Australia (Table 1.14), based on various exposure scenarios (Imray and Langley, 1996). These criteria are applied in relation to the health risk associated with various pathways for exposure by humans to the soil contaminant, e.g., soil ingestion, dermal contact, inhalation, etc.

Environmentally-based regulations for contaminants have yet to be set in Australia and New Zealand. Some suggested soil quality guidelines for environmental impact were proposed by NHMRC/ANZECC (1992). Table 1.15 summarizes some soil background values and several investigation levels for some elements of concern.

Recently, New South Wales has provisionally adopted a set of "phytotoxicity investigation levels" (PILs - Table 1.16) based on the potential for contaminants to adversely affect plant growth on a sandy loam soil at pH 6–8 (NSWEPA, pers. comm., 1997).

These values are to be used fairly flexibly at this stage since they are not related to Australian field experimental experience, but are compiled from reviews of international phytotoxicity data sets. These may not be related to soil and climatic conditions that are relevant to Australia. Some of the suggested environmental soil quality guidelines to trigger site investigation fall well within the range of natural background values for which no evidence of environmental threat exists (Tiller, 1992). These environmental guide values are for general guidance only with a need to exercise professional judgment, until more data are available with which to set limits relevant to Australia.

The NHMRC/ANZECC and other guidelines are established primarily for contaminated sites affected by industrial on-site and atmospheric accessions. There are no soil contamination guidelines for Australian or New Zealand agricultural land. Normal

Table 1.14. Health-Based Investigation Levels in Australia for Heavy Metals [a]

	Exposure Scenarios[b]					
	A	B	C	D	E	F
Element	mg kg^{-1}					
As	100			400	200	500
Cd	20			80	40	100
Cr(III)[c,d]	12%			48%	24%	60%
Cr(VI)	100			400	200	500
Cu	1000			4000	2000	5000
Hg (inorg.)	15			60	30	75
Hg (methyl)	10			40	20	50
Ni	600			2400	60[d]	3000
Pb	300			1200	600	1500
Zn	7000			2.8%	1.4%	3.5%

[a]Imray and Langley, 1996.
[b]Exposure scenarios:
A = "Standard" residential;
B = Residential with vegetable garden + poultry (site specific, no guideline value);
C = Residential with vegetable garden (site specific, no guideline value);
D = Residential, little exposed soil;
E = Recreational areas/parks;
F = Commercial/industrial.
b = Need to ensure valency state by site history/analysis/knowledge of environmental behavior.
c = Soil discoloration may occur at these concentrations.
d = Need to ensure form of substance by history/analysis/knowledge of environmental behavior.

agricultural practice usually does not include use of high metal-containing sewage sludges and fertilizers mixed with polluted industrial by-products.

The New Zealand Government recently passed legislation which in defining sustainable management of natural resources has a requirement for "avoiding, remedying or mitigating any adverse effects of activities on the environment" (Resource Management Act, 1991). Clearly, this applies to heavy metal concentrations in soils. However, the regional authorities charged with regulating the management of the resource under the auspices of the Resource Management Act have yet to release heavy metal level guidelines for rural and urban soils. In lieu of any other information, these authorities may opt for those limits already defined by the N.Z. Department of Health for the application of sewage effluent and sludge on land (see Table 1.21 later). The quality of sewage sludge and fertilizers that may be used in agriculture is discussed below. If the NHMRC/ANZECC and Dutch investigation levels are sound, and they may well not be, at least for Australian conditions, then addition of fertilizers containing Cd and F would seem to pose the greatest risk of exceeding these limits.

Fertilizer Quality

Until recently Australia had no regulations for fertilizer quality, apart from voluntary controls adopted by the local manufacturing industry. This situation is changing rapidly and the various states have introduced legislation to control Cd concentrations in fertilizers. Limits vary from state to state and there are currently moves to harmonize the various state regulations (Table 1.17).

Lead concentrations are also being regulated; e.g., in Tasmania Pb in fertilizers must be less than 500 mg kg^{-1} and in soil conditioners (e.g., lime, gypsum), less than 10 mg

Table 1.15. Background Metal Concentrations (mg kg^{-1}) and Investigation Levels for Soils in Various Countries

	Background Values			Investigation Levels			
	Australia (Tiller, 1991)[a]	Australia (NHMRC/ ANZECC 1992)	U.S. (Holmgren, et al. 1993)[b]	Dutch A Value[c]	NHMRC/AN ZECC (1992)	Dutch[d] Action Values	E.C. Limit for Sludged Soil[e]
Zinc	<2–200 (27)	2–180	<3–264 (56.5)	200	200	720	150–300
Cadmium	<1	0.04–2	<0.01–2.0 (0.26)	1	3	12	1–3
Lead	2–160 (21)	<2–200	<1–135 (12.3)	50	300	530	50–300
Copper	1–190 (15)	1–190	<0.06–495 (29.6)	50	60	190	50–140
Mercury	–	0.001–0.1	–	0.5	1	10	1–1.5
Nickel	2–400 (9.8)	2–400	0.7–26.9 (23.9)	50	60	210	30–75
Chromium	5–110 (27)	0.5–110	–	100	50	380	–
Selenium	–	–	–	–	–	–	–
Boron	10–75	1–75	–	–	–	–	–
Arsenic	<1–6 (3.9)	0.2–30	–	20	20	55	–
Molybdenum	<1–20	<1–20	–	10		200	–
Fluorine	3002	–	–	200		–	–
Manganese	4–13000 (254)	4–13000	–	–	500	–	–

[a] Mean values for Mt. Lofty Ranges near Adelaide in brackets.
[b] Mean U.S. values shown in brackets.
[c] Moen et al.(1988), now superseded by Intervention or Dutch Action Values (van den Berg, 1994).
[d] van den Berg (1994).
[e] CEC (1986).

Table 1.16. Phytotoxicity-Based Investigation Levels (PILS)[a]

Element	Proposed PIL mg kg^{-1}
As	20
Cd	3
Cr	400(III), 10(VI)
Cu	100
Pb	600
Ni	60
Zn	200

[a] NSWEPA, Pers. Comm., 1997.

kg^{-1}. Other states in Australia are in the process of drafting similar legislation, although again, MPCs may vary from state to state.

New Zealand has no regulations regarding permitted concentrations of heavy metals in fertilizers, although N.Z. fertilizer manufacturers are currently adopting voluntary controls as in Australia.

Table 1.17. Regulations Controlling Cd in Fertilizers in Australia

	Queensland[a]		Victoria[b]		Tasmania[c]		NSW[d]		WA[e]	
Type of Fertilizer/ Soil Amendment	mg Cd/kg Product	mg Cd/ kg of P	mg Cd/kg Product	mg Cd/ kg of P	mg Cd/kg Product	mg Cd/ kg of P	mg Cd/kg Product	mg Cd/ kg of P	mg Cd/ kg Product	mg Cd/kg of P
Non-P fertilizer, other than phosphogypsum, or trace element fertilizer	10									
Non-P fertilizer, other than trace element fertilizer					10		10			
Non-P fertilizer			10						80	
P fertilizers (contains at least 2% P)		350		350		350				
P fertilizers (not defined)								350		500
Phosphogypsum	15									
Trace elements	50				80		50			

Sources:
[a] Agricultural Standards Regulation 1997 (Subordinate Legislation 1997, No 277 of Agricultural Standards Act 1994). [Sections 20 & 21, effective 1 October 1997], Government of Queensland.
[b] Agricultural and Veterinary Chemicals Act 1992; Agricultural and Veterinary Chemicals (Fertilizers) Regulations 1995. Regulation 35 (1a & b), Government of Victoria.
[c] Statutory Rules 1993, No. 200; Fertilizer Regulations 1993. Regulation 30 (2 a, b & c), Government of Tasmania.
[d] Fertilizers Act 1985; Fertilizers Regulation 1997 [effective 1 September 1997]. Regulation 6 (Schedule 1, Part 1, Sub-part 2) and 9. Government of New South Wales.
[e] Gov. Gazette, WA , 31 December 1992, Reg. 2 amended.

Table 1.18. European Community Limit Values for the Use of Sewage Sludge on Agricultural Land (after CEC, 1986)[a]

Metal	Concentration in Soil mg kg^{-1}	Concentration in Dry Sewage Sludge (mg kg^{-1})	Annual Application Rate (kg ha^{-1} y^{-1})[b]
Cadmium	1–3	20–40	0.15
Copper	50–140	1000–1750	12
Nickel	30–75	300–400	3
Lead	50–300	750–1200	15
Zinc	150–300	2500–4000	30
Mercury	1–1.5	16–25	0.1

[a]Assumes soil pH in range 6–7.
[b]Based on 10–year average.

Table 1.19. U.S. Environmental Protection Agency Limit Values for the Use of Sewage Sludge on Land [a]

Pollutant	Limit Conc. for Sludge[b] (mg kg^{-1} dry wt)	Cumulative Loading Limit: Area Basis kg ha^{-1}	Cumulative Loading Limit: Soil Basis[c] mg kg^{-1} soil	Limit Conc. for "Safe Sludge"[d] (mg kg^{-1} d.wt)	Annual Loading Limit (kg ha^{-1} y^{-1})
Arsenic	75	41	31	41	2.0
Cadmium	85	39	29	39	1.9
Chromium	3000	3000	2260	1200	150
Copper	4300	1500	1130	1500	75
Lead	840	300	226	300	15
Mercury	57	17	13	17	0.85
Molybdenum	75	18	14	18	0.90
Nickel	420	420	316	420	21
Selenium	100	100	75	36	5.0
Zinc	7500	2800	2100	2800	140

[a] USEPA, 1993.
[b] Absolute limit for beneficial use.
[c] Soil limit based on 10 cm plow depth × bulk density = 1.3 t m^{-3}; values would be correspondingly lower for deeper plow layers.
[d] Unrestricted.

Application of Sewage Sludge to Land

The European Community has established regulations (CEC, 1986) for the disposal of sludge on land. This is achieved by restricting the quality of the sludge being applied, to the amount of pollutant metal that can be added over a prescribed period, and by defining a MPC for heavy metals on agricultural soils with pH in the range 6–7 (Table 1.18).

The U.S. Environmental Protection Agency (US EPA, 1993) has revised rules and regulations, commonly referred to as Part 503, for the disposal of sewage sludge on land (Table 1.19).

These rules were based on a major research effort in the USA during recent decades which was interpreted in terms of the many pathways which affect human, animal, and environmental health. The USEPA rules were developed in terms of ceiling concentrations (limit values), above which sludge cannot be used on land and a cumulative pollutant loading, which effectively defines a soil limit when soil background values are included. Pollutant limit concentrations are also defined for a "clean sludge" which may be used without restriction.

Regulations for sludge application to land in Australia are currently under review both nationally as well as by some individual states. Although some research is currently underway in New South Wales and some research on sewage sludge has been published (e.g., de Vries and Tiller, 1978; Ross et al., 1992), environmental authorities are under pressure to develop guidelines without a sound basis of locally relevant results. Authorities are accordingly obliged to depend heavily on overseas regulations which were developed under different climatic, soil, and political conditions.

New South Wales and South Australia have recently (NSWEPA, 1994; SAEPA, 1996) published provisional guidelines for the application of sewage sludge on land which provide limitations relating to sludge quality and land use. Cumulative loadings are controlled by measured concentrations of the prescribed contaminants in the soil. Although the general approach is similar to that of the USEPA, the quality of sludge for unrestricted use and the total permitted loading of contaminants is much more restrictive, especially for Cd which is the heavy metal of greatest concern to agriculture. This is justified because the research of CSIRO Land and Water in Adelaide (Tiller et al., 1997; McLaughlin et al., 1996) has shown that unacceptably high concentrations of Cd in agricultural produce can be produced on the widespread saline soils of Australia, even on soils with less than 0.5 mg Cd kg^{-1}. The classes of land use and the limiting concentrations of contaminants permitted by the New South Wales regulations are summarized in Table 1.20. The 1992 New Zealand guidelines for maximum permissible applications of heavy metals to land and maximum concentrations in soil are shown in Table 1.21.

Table 1.20. New South Wales Interim Code of Practice for Use and Disposal of Biosolids [a]

	Concentration ($mg\ kg^{-1}$)		
	Max. Allowable in Soil		
Contaminant	**For Food Production**	**Not for Food Production**	**Sludge for Unrestricted Use**
Arsenic	20	20	20
Cadmium	3	11	3
Chromium (total)	100	500	100
Copper	100	750	100
Lead	150	150	150
Mercury	1	9	1
Nickel	60	145	60
Selenium	5	14	5
Zinc	200	1400	200

[a]NSWEPA, 1994.

Table 1.21. New Zealand Guidelines for the Application to Land [a]

Element	Max. Annual Application ($kg\ ha^{-1}$)	Max. Concentration in Soil ($mg\ kg^{-1}$)
Arsenic	–	10
Cadmium	0.2	3
Chromium	15	600
Copper	12	140
Lead	15	300
Mercury	0.1	1
Nickel	3.0	35
Zinc	30	300

[a]NZ Dept. of Health, 1992.

Table 1.22. Maximum Permitted Concentrations (MPC) for Metals in Some Foodstuffs[a]

Metal	Food	MPC (mg kg^{-1} FW)
Cadmium	Chocolate	0.5
	Kidney	2.5
	Liver	1.25
	Leafy vegetable	0.1
	Meat	0.05
	Peanuts	0.05
	Rice	0.1
	Root and tuber vegetables	0.1
	Wheat	0.1
Lead	Beverages	0.2
	Bran and wheat germ	2.5
	Mollusks	2.5
	Vegetables	2.0
	All other foods	1.5
Mercury	Fish, crustaceans, and mollusks	0.5
	All other foods	0.03

[a]From ANZFA (NFA, 1997).

Food Quality

The main regulatory control of the contamination of agricultural land is exercised through the food regulations, which for Cd in Australia was, until recently, the strictest in the world. The MPCs for heavy metals and other potential contaminants are set by ANZFA, and represent the lowest achievable concentrations and levels which ensure that the provisional tolerable weekly intake of healthy food (PTWI) is not exceeded. A recent review of the Cd regulations resulted in alterations to the classes of food regulated, and downward or upward revision of the MPCs for various classes of foods. The new Australian and New Zealand MPC values for some metals in foods are shown in Table 1.22.

Metal intakes regarded acceptable in terms of health risk to humans have been assessed by the Joint FAO/World Health Organization Expert Committee on Food Additives (JECFA), who have recommended PTWI values for Cd, Pb, and Hg. The intake of many contaminants is assessed biennially in Australia in a broadly based market basket survey, the Australian Market Basket Survey (AMBS) (Stenhouse, 1994). The PTWI values for these metals are 7, 50, and 3 μg kg^{-1} body weight per week, respectively. The PTWI value for Pb absorbed by children is 25 μg kg^{-1} body weight. The estimated intakes of Cd, Pb, and Hg by Australian adults (through average dietary food consumption) as a percentage of the PTWI were about 27%, 20%, and 5% respectively (Stenhouse, 1994). Intakes of Cd and Pb by children, especially very young children, were proportionately higher. Their Cd intake most closely approaches the PTWI and is of greatest concern. The 1992 and 1994 AMBS especially highlighted the incidence of Cd concentrations in potatoes and potato products, which exceeded the MPC. The MPC for Pb was not exceeded in any case. The majority of the Hg in the average Australian diet is contributed by seafood. The average and maximum concentrations of some metals in some representative foods as prepared for consumption were shown in Table 1.9.

SUMMARY AND CONCLUSIONS

Soils of Australian and New Zealand agroecosystems are characterized by low concentrations of heavy metals. Some limited areas of more contaminated soils exist but are primarily in the vicinity of some major urban centers and metal industries such as smelters. These restricted areas of contaminated soils do not pose any perceived health risk in terms of the food chain because of either crop selection or avoidance, product dilution, and monitoring. The main source of soil contamination results from widespread fertilization by phosphatic fertilizers. Stricter controls on industrial emissions in recent years and tighter guidelines for food quality, especially for Cd and for agricultural amendments in general, will result in even slower rates of addition of heavy metals to soils in future. However, the low concentrations of contaminants, especially Cd, resulting from fertilization for nearly a century, are still of appreciable concern under certain soil and agronomic conditions in Australia. In Australia, soils are generally sandier, more acidic in the topsoil (through pedogenic leaching processes and low use of lime) and are often Zn deficient or saline compared to many agricultural soils in United States, Europe, or even New Zealand. These conditions all enhance the transfer of Cd from soil to plant and hence despite low Cd concentrations in soil, significant concentrations of Cd have been found in food crops in certain areas of Australia. For example, research has shown that Cd concentrations in soil of even less than 0.5 mg kg^{-1} can produce agricultural produce with unacceptable concentrations of cadmium (0.25 mg kg^{-1} FW in potatoes), particularly where soils are saline. Thus the critical soil limits for Cd are much lower than commonly accepted (1.0 to 3.0 mg kg^{-1}) and this stresses the need to carry out research under environmentally relevant conditions.

REFERENCES

ABARE. *Commodity Statistical Bulletin.* Australian Bureau of Agricultural and Resource Economics. Australian Government Printing Service, Canberra, 1993.

Adelaide Engineering and Water Supply Department, *An Assessment of Metropolitan Adelaide's Sewage Sludges for Agriculture.* August, 1994., Adelaide, South Australia, 1994.

Andersson, A. On the Influence of Manure and Fertilizers on the Distribution and Amounts of Plant-Available Cd in Soils. *Swedish J. of Agric. REs.* 6, pp. 27–36, 1976.

Andersson, A. Mercury in Soils, in *The Biogeochemistry of Mercury in the Environment,* Nriagu, J.O. Ed., Elsevier, Amsterdam, The Netherlands, 1979, pp. 79–112.

Andersson, A. and M. Hahlin, Cadmium Effects from Phosphorus Fertilization in Field Experiments. *Swedish J. of Agric. Res.* 11, pp. 3–10, 1981.

Andersson, A. and G. Siman, Levels of Cadmium and Some Other Trace Elements in Soils and Crops as Influenced by Lime and Fertilizer Level. *Acta Agriculturae Scandinavia* 41, pp. 3–11, 1991.

Andrey D., T. Rihs, and E. Wirz, Monitoring-Program "Schwermetalle in Lebensmitteln" II. Blei, Cadmium, Zink und Kupfer in Schweizer Kartoffeln. *Mitteilungen aus der Gebiete der Lebensmitteluntersuchung und Hygiene* 79, pp. 327–338, 1988.

Auer, C. *Cadmium in Phosphate Fertilizer Production.* USEPA Midwest Research Instruction Report, October 1977. United States Environmental Protection Agency, Cincinnati, USA, 1977.

Best, E.K. *Screening Barley for Cadmium and Lead Content at Selected Depots within Australia.* Final Report DAQ-26B. Grains Research and Development Corporation, Canberra, Australia, 1989.

Best, E. K. *Cadmium and Lead in Queensland Wheat. Final Report of Project Q90457 W.* Grains Research and Development Corporation, Canberra, ACT, Australia, 1991.

Biberacher, G. and K.D. Shah, Fertilizer Manufacture and EEC Environmental Activities, in *Pro-*

ceedings IFA Technical Conference, Venice, Italy, 2–4 October 1990. pp. 1–10. International Fertilizer Industry Association, Paris, 1990.

Bingham, F.T., J.E. Strong, and G. Sposito. Influence of Chloride Salinity on Cadmium Uptake by Swiss Chard. Soil Science. 135, pp. 160–165, 1983.

Bramley, R.G.V. Cadmium in New Zealand Agriculture. *New Zealand J. Agric. Res.* 33, pp. 505–519, 1990.

Cakmak, I. and H. Marschner. Increase in Membrane Permeability and Exudation in Roots of Zinc Deficient Plants. *J.: Plant Physiol,* 132, pp. 356–361, 1988.

Cartwright, B., Unpublished data.

Cartwright, B., R.H. Merry, and K.G. Tiller. Heavy Metal Contamination of Soils Around a Lead Smelter at Port Pirie, South Australia. *Aust. J. Soil Res.* 15, pp. 69–81, 1977.

Cataldo D.A., T.R. Garland, and R.E. Wildung. Cadmium Uptake Kinetics in Intact Soybean Plants. *Plant Physiol.* 73, pp. 844–848, 1983.

CEC. Council Directive on the Use of Sewage Study in Agriculture. *Commission of the European Communities European Report,* L181, 1986, pp. 6–12.

Chaney, R.L. Metal Speciation and Interactions Among Elements Affect Trace Element Transfer in Agricultural and Environmental Food Chains, in *Metal Speciation, Theory, Analyses and Application,* Kramer J.R. and Allen, H.E. Eds., Lewis Publishers, Chelsea, MI, 1988.

Chaney, R.L. and S.B. Hornick. Accumulation and Effects of Cadmium on Crops, in *Proceedings 1st International Cadmium Conference, San Francisco.* Cadmium Metal Bulletin, London, 1978.

Chemistry Centre (WA)/West Australian Department of Agriculture/Health Department of Western Australia, *Cadmium in Western Australian Potatoes.* Perth, 1992.

Clayton, P.M. and K.G. Tiller. A Chemical Method for the Determination of the Heavy Metal Content of Soils in Environmental Studies. *Division of Soils Technical Paper* No. 41, CSIRO, Australia, 1979.

Crowley, D.E., C.P.P. Reid, and P.T. Szaniszlo. Microbial Siderophores as Iron Sources for Plants, in *Iron Transport in Microbes, Plants and Animals,* Weinheim, D. van der Helm and Neilands, J.B., Eds., Winkelman VCR Verlags, 1987, pp. 375–386.

Cuddihy,W.L. and I.F. Fergus. Arsenic Toxicity in the Collinsville Basin, Queensland. *Proceedings of the Australian Soil Science Conference, Melbourne, February, 1974,* Australian Soil Science Society, Melbourne, Australia, 1974.

David, D.J., A. Pinkerton, and C.H. Williams. Impurities in Australian Phosphate Fertilizers. *J. Aust. Inst. Agric. Sci.* 44, pp. 132–135, 1978.

deVries, M.P.C. Investigations of Twenty Australian Sewage Sludges—Their Evaluation by Means of Chemical Analyses. *Fert. Res.* 4, pp. 75–87, 1983.

de Vries, M.P.C. and K.G. Tiller. Sewage Sludge as a Soil Amendment, with Special Reference to Cadmium, Cu, Mn, Ni, Pb and Zinc—Comparison of Results from Experiments Conducted Inside and Outside a Glasshouse. *Environ. Pollut.* 16, pp. 231–240, 1978.

Doner, H.E. Chloride as a Factor in Mobilities of Ni(II), Cu(II) and Cd(II) in Soil. *Soil Sci. Soc. Am. J.* 42, pp. 882–885, 1978.

Eriksson, J.E. A Field Study of the Factors Influencing Cd Levels in Soils and in Grain of Oats and Winter Wheat. *Water, Air Soil Pollut.* 53, pp. 69–81, 1990.

Fordham, A.W. Correlation of Matrix Effects in the Analysis of Stream Waters by Flameless Atomic Absorption Spectrophotometry. *J. Geochem. Exploration,* 10, pp. 41–51, 1978.

Garcia-Miragaya, Cardenas, J., and A.L. Page, Surface Loading Effect on Cd and Zn Sorption by Kaolinite and Montmorillonite from Low Concentration Solutions. *Water Air Soil Pollut.,* 27, pp. 181–190, 1986.

Gardiner, W.K. The Soil-Root Interface of *Lupinus albus* L. and Its Significance in the Uptake of Manganese, Iron and Phosphate. Ph.D. thesis, University of Melbourne, Australia. 1981.

Gates, C.T., D.B. Jones, W.J. Muller, and J.S. Hicks. The Interaction of Nutrients and Tillage Methods on Wheat and Weed Development. *Aust. J. Agric. Res.* 32, pp. 227–241, 1981.

Gulson, B.L., K.G. Tiller, K.J. Mizon, and R.H. Merry. Use of Lead Isotopes in Soils to Identify the Source of Lead Contamination Near Adelaide, South Australia. *Environ. Sci. Technol.* 15, pp. 691–696, 1981.

Gunnarsson, D. Heavy Metals in Fertilizers. Do They Cause Environmental and Health Problems? *Fert. Agric.* 85, pp. 27–42, 1983.
Haghiri, F. Plant Uptake of Cadmium as Influenced by Cation Exchange Capacity, Organic Matter, Zinc and Soil Temperature. *J. Environ. Qual.* 3, pp.180–182, 1974.
Hahne, H.C.H. and W. Kroontje. The Simultaneous Effect of ph and Chloride Concentration Upon Mercury (II) as a Pollutant. *Soil Sci. Soc. Am. Proc.* 37, p. 838, 1973.
Hamblin, A.P. and D. Tennant, Interactions Between Soil Type and Tillage Level in a Dry Land Situation. *Aust. J. Soil Res.* 17, pp. 177–189, 1979.
He, Q.B. and B.R. Singh. Plant Availability of Cadmium in Soils. II. Factors Related to the Extractability and Plant Uptake of Cadmium in Cultivated Soils. *Acta Agriculturae Scandinavia* 43, pp. 143–150, 1993.
Healy, W.B. Nutritional Aspects of Soil Ingestion by Grazing Animals, in *Chemistry and Biochemistry of Herbage. Volume 1,* Butler, G.W. and Bailey, R.W., Eds., Academic Press, London, 1973, pp. 567–588.
Helyar, K.R. Nitrogen Cycling and Soil Acidification. *J. Aust. Inst. Agric. Sci.* 2, pp. 17–21, 1976.
Holmgren, G.G.S., M.W. Meyer, R.L. Chaney, and R.B. Daniels. Cadmium, Lead, Zinc, Copper, and Nickel in Agricultural Soils of the United States of America. *J. Environ. Qual.* 22, pp. 335–348, 1993.
Hovmand, M.F. Cirkulation af Bly, Cadmium, Kobber, Zink og Nickl I Dansk Landbrug, in *Slammets Jordbrugsandvendels, Vol. II Fokusering,* Polyteknisk Forlag, Lyngby, Denmark, 1981, pp. 85–118.
Hutton, M. and C.J. Symon. The Quantities of Cadmium, Lead, Mercury and Arsenic Entering the U. K. Environment from Human Activities. *Sci. Total Environ.* 57, pp. 129–150, 1986.
Imray, P. and A. Langley. *Health-Based Soil Investigation Levels.* National Environmental Health Forum Monographs Soil Series No. 1., 1996.
Isermann, V.K., P. Karch, and J.A. Schmidt. Cd-Gehalt des erntegutes verscheidener Sorten mehrerer Kulturpflanzen bei Anbau auf Stark mit Cadmium belastetem, neutralem Lehmboden. *Landwirtschaft Forschung* 36, pp 283–294, 1983.
Jaakkola, A. Effect of Fertilizers, Lime and Cadmium Added to Soil on the Cadmium Content of Spring Wheat. *J. Sci. Agric. Soc. Finland* 49, pp. 406–414, 1977.
Jones, K.C., C.J. Symon, and A.E. Johnston. Retrospective Analysis of an Archived Soil Collection. II. Cadmium. *Sci. Total Environ.* 67, pp. 75–89, 1987.
Jorhem, L., P. Mattsson, and S. Slorach. Lead, Cadmium, Zinc and Certain Other Metals in Foods on the Swedish Market. *Varfoda* 36, pp. 135–168, 1984.
Kloke, A., D.R. Sauerbeck, and H. Vetter, The Contamination of Plants and Soils with Heavy Metals and the Transport of Metals in Terrestrial Food Chains, in *Changing Metal Cycles and Human Health,* Nriagu, J. O., Ed., Springer-Verlag, Berlin, 1984, pp. 131–141.
Langlands, J.P. Cadmium Status of Grazing Ruminants in Australia, in *Cadmium Accumulations in Australian Agriculture: National Symposium, Canberra 1–2 March 1988,* Simpson, J. and Curnow, W.C., Eds., Australian Government Publishing Service, Canberra, ACT, Australia, 1988, pp. 113–130.
Longhurst, R.D., A.H.C. Roberts, M.W. Brown, and B. Carlson. Cadmium Cycling in Sheep Grazed Hill Country Pastures, in *Occasional Report No. 7,* Currie, L.D. and Loganathan, P., Eds., Fertilizer and Lime Research Centre, Palmerston North, New Zealand, 1994, pp. 297–302.
Maier, N.A., M.J. McLaughlin, M. Heap, M. Butt, M.K. Smart, and C.M.J. Williams. Effect of Current Season Applications of Calcitic Lime on pH, Yield and Cadmium Concentration of Potato (*Solanum tuberosum* L.) Tubers. *Nutrient Cycling Agroecosyst.* 47, pp. 1–12, 1997.
Matthews, H. and I. Thornton. Seasonal and Species Variation in the Content of Cadmium and Associated Metals in Pasture Plants at Shipham. *Plant and Soil* 66, pp. 181–193, 1982.
McLaughlin, M.J., I.R. Fillery, and A.R. Till. Operation of the Phosphorus, Sulphur and Nitrogen Cycles, in *Australia's Renewable Resources: Sustainability and Global Change,* Gifford R.M. and Barson, M.M., Eds., Bureau of Rural Resources Proceedings Number 14, Canberra, ACT, Australia, 1992, pp. 67–116.

McLaughlin, M.J., N. Maier, C.M.J. Williams, K.G. Tiller, and M.K. Smart. Cadmium Accumulation in Potato Tubers–Occurrence and Management, in *Proceedings 7th National Potato Research Workshop, Ulverstone, May 1993,* Fennell, J., Ed., Tasmanian Department of Primary Industry, Launceston, Australia, 1993, pp. 208–213.

McLaughlin, M.J. and K.G. Tiller. Chloro-Complexation of Cadmium in Soil Solutions of Saline/Sodic Soils Increases Phyto-Availability of Cadmium. *Transactions 15th World Congress of Soil Science* Vol. 3b, 1994a, pp. 195–196.

McLaughlin, M.J., K.G. Tiller, T.A. Beech, and M.K. Smart. Soil Salinity Causes Elevated Cadmium Concentrations in Field-Grown Potato Tubers. *J. Environ. Qual.* 34, pp. 1013–1018, 1994b.

McLaughlin, M.J., K.G. Tiller, R. Naidu, and D.P. Stevens. Review: The Behaviour and Environmental Impact of Contaminants in Fertilizers. *Aust. J. Soil Res.* 34, pp. 1–54, 1996.

McLaughlin, M.J., K.G. Tiller, and M.K. Smart. Speciation of Cadmium in Soil Solutions of Saline/Sodic Soils and Relationship with Cadmium Concentrations in Potato Tubers. *Aust. J. Soil Res.* 35, pp. 1–16, 1997a.

McLaughlin, M.J., N.A. Maier, G.E. Rayment, L.A. Sparrow, G. Berg, A. McKay, P. Milham, R.H. Merry, and M.K. Smart. Cadmium in Australian Potato Tubers and Soils. *J. Environ. Qual.* 26, pp. 1644–1649, 1997b.

McLean, A.J. Cadmium in Different Plant Species and Its Availability as Influenced by Organic Matter and Additions of Lime, P, Cadmium and Zinc. *Can. J. Soil Sci.* 56, pp. 129–138, 1976.

Merry, R.H. Investigations on Cadmium in South Australia: Rainfall, Soils, Cereals, Pastures and Soil-Plant Interactions, in *Cadmium Accumulations in Australian Agriculture: National Symposium, Canberra 1-2 March 1988,* Simpson, J. and Curnow, W.C., Eds., Australian Government Publishing Service, Canberra, ACT, Australia, 1988, pp. 62–79.

Merry, R.H. *Effects of Farm Management Practices on Cadmium Uptake by Pasture Plants,* Final Report CS137, Meat Research Corporation, Canberra, ACT, Australia, 1992.

Merry, R.H. and K.G. Tiller. The Contamination of Pasture by a Lead Smelter in a Semi-Arid Environment. *Aust. J. Exptl. Agric. Anim. Husb.* 18, pp. 89–96, 1978.

Merry, R.H. and K.G. Tiller. Distribution of Cadmium and Lead in an Agricultural Region Near Adelaide, South Australia. *Water, Air Soil Pollut.* 57–58, pp. 171–180, 1991.

Merry, R.H., K.G.Tiller, and A.M. Alston. Accumulation of Copper, Lead and Arsenic in Australian Orchard Soils. *Aust. J. Soil Res.* 21, pp. 549–61, 1983.

Merry, R.H., K.G. Tiller, M.P.C. de Vries, and B. Cartwright. Contamination of Wheat Crops around a Lead-Zinc Smelter. *Environ. Pollut. (Series B)* 2, pp. 37–48, 1981.

Moen, J.E.T., J.P. Cornet, and C.W.A. Evers. Soil Protection and Remedial Actions: Criteria for Decision Making and Standardisation of Requirements, in *Contaminated Soil '88',* Assink, J.W. and van den Brink, W.J., Eds., Kluwer Academic Publishers, Dordrecht, the Netherlands, 1988, pp. 1495–1503.

Mortvedt, J.J. Cadmium Levels in Soils and Plants from Some Long-Term Soil Fertility Experiments in the United States of America. *J. Environ. Qual.* 16, pp. 137–142, 1987.

Mortvedt, J.J., D.A. Mays, and G. Osborn. Uptake by Wheat of Cadmium and Other Heavy Metal Contaminants in Phosphate Fertilizers. *J. Environ. Qual.* 10, pp. 193–197, 1981.

Mulla, D.J., A.L. Page, and T.J. Ganje. Cadmium Accumulations and Bioavailability in Soils from Long-Term Phosphorus Fertilization. *J. Environ. Qual.* 9, pp. 408–412, 1980.

Nambiar, E.S. Genetic Differences in the Copper Nutrition of Cereals. II. Genotypic Differences in Response to Copper in Relation to Copper, Nitrogen and Other Mineral Contents of Plants. *Aust. J. Agric. Res.* 27, pp. 465–77, 1976.

NFA. *Australian Food Standards Code,* March 1993, National Food Authority, Australian Government Publishing Service, Canberra, ACT, Australia, 1994.

NFA. *Australian Food Standards Code, Amendment 35,* August 1997, National Food Authority. Government Publishing Service, Canberra, ACT, Australia, 1997.

NHMRC/ANZECC. *Australian and New Zealand Guidelines for the Assessment and Management of Contaminated Sites.* Report January 1992. Australian and New Zealand Environment and

Conservation Council and National Health and Medical Research Council, Canberra, ACT, Australia, 1992.
Nicholson, F.A., K.C. Jones, and A.E. Johnson. Effect of Phosphate Fertilizers and Atmospheric Deposition on Long-Term Changes in the Cadmium Content of Soils and Crops. *Environ. Sci. Technol.* 28, pp. 2170–2175, 1994.
Nriagu, J.O. and J.M. Pacyna. Quantitative Assessment of Worldwide Contamination of Air, Water and Soils by Trace Metals. *Nature.* 333, pp. 134–139, 1988.
NSWEPA. *Interim Code of Practice for Use and Disposal of Biosolid Products.* Draft-June 1994. New South Wales Environment Protection Authority, Sydney, NSW, Australia, 1994.
NSWEPA. Personal communication, 1997.
Oliver, D.P., J.E. Schulz, K.G. Tiller, and R.H. Merry. The Effects of Crop Rotations and Tillage Practices on Cadmium Concentration in Wheat Grain. *Aust. J. Soil Res.* 44, pp. 1221–1234, 1993.
Oliver, D.P., R. Hannam, K.G. Tiller, N.S. Wilhelm, R.H. Merry, and G.D. Cozens. The Effects of Zinc Fertilization on Cadmium Concentration in Wheat Grain. *J. Environ. Qual.* 23, pp. 705–711, 1994.
Oliver D.P., K.G. Tiller, M.K. Conyers, W.J. Slattery, R.H. Merry, and A.M. Alston. The Effects of pH on Cd Concentration in Wheat Grain Grown in South-Eastern Australia, in *Plant Soil Interactions at Low PH: Principles and Management,* Date, R.A., Grundon, N.J., Rayment, G.E., and Probert, M.E. Eds., Kluwer Academic Publishers: Dordrecht, The Netherlands, 1995a, pp. 791–795.
Oliver, D.P., G.W. Gartrell, K.G. Tiller, R. Correll, G.D. Cozens, and B.L. Youngberg. Differential Responses of Australian Varieties to Cadmium Concentration in Wheat Grain. *Aust. J. Agric. Res.* 45, pp. 873–886, 1995b.
Oliver, D.P. Unpublished data.
Page, A.L., F.T. Bingham, and A.C. Chang. Cadmium, in *Effect of Heavy Metal Pollution of Plants,* Lepp, N.W., Ed., Applied Science, London, England, 1981, pp. 77–109.
Rayment, G.E., E.K. Best, and D.J. Hamilton. Cadmium in Fertilisers and Soil Amendments. *Proceedings RACI Chemistry International (1st Environmental Chemistry Division Conference), Brisbane, Queensland, August 1989*, Royal Australian Chemical Institute, Brisbane, Queensland, Australia, 1989.
Reuss, J.O., H.L. Dooley, and W. Griffis, Uptake of Cadmium from Phosphate Fertilizers by Peas, Radishes and Lettuce. *J. Environ. Qual.* 7, pp. 128–133, 1978.
Roberts, A.H.C., R.D. Longhurst, and M.W. Brown, Cadmium Status of Soils, Plants and Grazing Animals in New Zealand. *N. Z. J. Agric. Res.* 37, pp. 119–129, 1994.
Roberts A.H.C. Unpublished data, 1998.
Roberts, A.H.C., K.C. Cameron, N.S., Bolan, H.K. Ellis, and S. Hunt. Contaminants and the Soil Environment in New Zealand, in *Contaminants and the Soil Environment in the Australasia-Pacific Region,* Naidu, R., Kookana, R.S., Oliver, D.P., Rogers, S. and McLaughlin, M.J., Eds., Kluwer Academic Publishers, Dordrecht, The Netherlands, 1996, pp. 579–628.
Robson, A.D. and M.G. Pitman. Interactions Between Nutrients in Higher Plants, in *Inorganic Plant Nutrition Vol. 15A*, Lauchli A. and Bieleski, R.L. Eds., Springer Verlag, New York, 1983.
Ross, A.D., R.A. Lawrie, M.S. Whatmuff, J.P. Keneally and A.S. Awad. *Guidelines for the Use of Sewage Sludge on Agricultural Lands.* New South Wales Department of Agriculture and Fisheries, Sydney, Australia, 1992.
Rothbaum, H.P., R.L. Gogirel, A.E. Johnston, and G.E.G. Mattingley. Cadmium Accumulation in Soils from Long-Continued Applications of Superphosphate. *J. Soil Sci.* 37, pp. 99–107, 1986.
SAEPA. *South Australian Biosolids Guidelines for the Safe Handling, Reuse and Disposal of Biosolids.* South Australian Environmental Protection Agency, Department of Environment and Natural Resources, Adelaide, South Australia, 1996.
Sauerbeck, D. Zur Bedeutung des Cadmiums in Phosphatdungenmitteln. *Landbauforschung Volkenrode* 32, pp. 192–197, 1982.

Schroeder, H.A. and J.J. Balassa. Cadmium: Uptake by Vegetables from Superphosphate and Soil. *Science* 140, pp. 819–820, 1963.

Singh, B.R. Unwanted Components of Commercial Fertilizers and Their Agricultural Effects. *Proc. Fert. Soc.*, London, December 1991, 1991, pp. 2–28.

Singh, J.P., B.Singh, and S.P.S. Karwasa. Yield and Uptake Response of Lettuce to Cadmium as Influenced by Nitrogen Application. *Fert. Res.* 18, pp. 49–56, 1988.

Smilde, K.W. and B. van Luit. The Effect of Phosphate Fertilizer Cadmium on Cadmium in Soils and Crops. *Institute Bodemvruchtbaarheid, Rapport* 6–83, 1983.

Smolders, E. and M.J. McLaughlin. Effect of Cl on Cd Uptake by Swiss Chard in Nutrient Solutions. *Plant Soil* 179, pp. 57–64, 1996.

Sparrow, L.A., A.A. Salardini and A.C. Bishop. Field Studies of Cadmium in Potatoes (*Solanum tuberosum* L.). II. Response of cv. Russet Burbank and Kennebec to Two Double Superphosphates of Different Cadmium Concentrations. *Aust. J. Agric. Res.* 44, pp. 855–861, 1993a.

Sparrow, L.A., A.A. Salardini, and A.C. Bishop. Field Studies of Cadmium in Potatoes (*Solanum Tuberosum* L.). I. Effects of Lime and Phosphorus on cv. Russet Burbank. *Aust. J. Agric. Res.* 44, pp. 845–853, 1993b.

Sparrow, L.A., A.A. Salardini, and J. Johnsone. Field Studies of Cadmium in Potatoes (*Solanum tuberosum L.)*. III. Response of cv. Russet Burbank to Sources of Banded Potassium. *Aust. J. Agric. Res.* 45, pp. 243–249, 1994.

Stenhouse, F. *The 1992 Australian Market Basket Survey*. National Food Authority. Australian Government Printing Service, Canberra, ACT, Australia, 1994.

Street, J.J., B.R. Sabey, and W.L. Lindsay. Influence of pH, Phosphorus, Cadmium, Sewage Sludge, and Incubation Time on the Solubility and Plant Uptake of Cadmium. *J. Environ. Qual.* 7, pp. 286–290, 1978.

Syers, J.K. A.D. Mackay, M.W. Brown, and L.D. Currie, Chemical and Physical Characteristics of Phosphate Rock Materials of Varying Reactivity. *J. Sci. Food Agric.* 37, pp. 1057–1064, 1986.

Taylor, N.H. and I.J. Pohlen. Soil Survey Method. *N. Z. Soil Bur. Bull.* 25, 1970.

Tiller, K.G. Heavy Metals and Their Environmental Significance, in *Advances in Soil Science Vol. 9*, Steward, B.A., Ed., Springer-Verlag, New York, 1989, pp. 113–142.

Tiller, K.G. Determining Background Levels, in *The Health Risk Assessment and Management of Contaminated Sites,* Saadi, O.E. and Langley, A., Eds., South Australian Health Commission, Adelaide, South Australia, 1991, pp. 98–101.

Tiller, K.G. Urban Soil Pollution in Australia. *Aust. J. Soil Res.* 30, pp. 937–957, 1992.

Tiller, K.G., B. Cartwright, M.P.C. de Vries, R.H. Merry, and L.R. Spouncer. Environmental Pollution of the Port Pirie region. 1. Accumulation of Metals in Wheat Grain and Vegetables Grown on the Coastal Plain. *CSIRO Division of Soils Divisional Report* No.6, 1975.

Tiller, K.G., L.H. Smith, R.H. Merry, and P.M. Clayton. The Dispersal of Automotive Lead from Metropolitan Adelaide into Adjacent Rural Areas. *Aust. J. Soil Res.* 25, pp. 155–166, 1987.

Tiller, K.G., D.P. Oliver, M.J. McLaughlin, R.H. Merry, and R. Naidu. Managing Cadmium Contamination of Agricultural Land, in *Remediation of Soils Contaminated by Metals,* Iskandar, I.K. and Adriano, D.C., Eds., Science Reviews, Northwood, England, 1997, pp. 225–255.

Tjell, J.C., J.A. Hansen, T.H. Christensen, and M.F. Hovmand. Prediction of Cadmium Concentrations in Danish Soils, in *The Second European Symposium on Characterization, Treatment and Use of Sewage Sludge, Vienna 20–24 October, 1980,* Hermite, P.L. and Ott, H., Eds., D. Reidel, London, England, 1981, pp. 650–664.

United Kingdom Ministry of Agriculture. *Survey of Cadmium in Food: First Supplementary Report*, Fisheries and Food Surveillance Paper No.12, HMSO, London, 1983.

USEPA. Standards for the Use or Disposal of Sewage Sludge; Final Rules (40 CFR Parts 257, 403 and 503). *Fed. Reg.* 58, pp. 9248–9415, 1993.

van den Berg, R. *Human Exposure to Soil Contamination: a Qualitative and Quantitative Analysis Towards Proposals for Human Toxicological Intervention Values.* Report No. 725201011, National Institute for Public Health and the Environment, Bilthoven, The Netherlands, 1994.

Vetter, von H. and K. Früchtenicht. Erhöht Phosphatdüngung die Cadmiumbelastung. DLG-Mitteilungen 17/1982, 1982.

Walkley, A. The Zinc Content of Some Australian Fertilizers. *Aust. J. Counc. Sci. Ind. Res. Aust.* 13, pp. 255–260, 1940.

Warembourg, F.R. and G. Billes. Estimating Carbon Transfers in the Plant Rhizosphere, in *Soil-Root Interface,* Harley, J.L. and Russell, R.S., Eds., Academic Press, London, England, 1979, pp. 183–196.

Weigert, P., J. Muller, H. Klein, K.P. Zufelde, and J. Hillebrand. ZEBS Hefte: Arsen, Blei, Cadmium und Quecksilber in und auf Lebensmitteln, ZEBS Hefte 1/1984, Federal Republic of Germany, 1984.

Welch, R.M., M.J. Webb, and F. Loneragan. Zinc in Membrane Function and its Role in Phosphorus Toxicity, in *Proceedings International Plant Nutrition Colloquium* Scaife, A., Ed., Warwick Publishers, London, England, 1982.

Wiersma, D., B.J. van Goor, and N.G. van der Veen. Cadmium, Lead, Mercury and Arsenic Concentrations in Crops and Corresponding Soils in the Netherlands. *J. Agric. Food Chem.* 34, pp. 1067–1074, 1986.

Willaert, G. and M. Verloo. Effects of Various Nitrogen Fertilizers on the Chemical and Biological Activity of Major and Trace Elements in a Cadmium Contaminated Soil. *Pedologie* 43, pp., 83–91, 1992.

Williams, C.H. Heavy Metals and Other Elements in Fertilisers—Environmental Considerations, in *Fertilisers and the Environment,* Leece, D.R., Ed., Australian Institute of Agricultural Science, Sydney, New South Wales, Australia, 1974, pp. 123–130.

Williams, C.H. and D.J. David. The Accumulation in Soil of Cadmium Residues from Phosphate Fertilizers and Their Effect on the Cadmium Content of Plants. *Soil Sci.* 121, pp. 86–93, 1976.

Williams, C.H. and D.J. David. Some Effects of the Distribution of Cadmium and Phosphate in the Root Zone on the Cadmium Content of Plants. *Aust. J. of Soil Res.* 15, pp. 59–68, 1977.

Williams, C.H. and D.J. David. The Effect of Superphosphate on the Cadmium Content of Soils and Plants. *Aust. J. Soil Res.* 11, pp. 43–56, 1973.

Wolnik, K.A., F.L. Fricke, S.G. Capar, G.L. Braude, M.W. Meyer, R.D. Satzger, and E. Bonnin. Elements in Major Raw Agricultural Crops in the United States. 1. Cadmium and Lead in Lettuce, Peanuts, Potatoes, Soybeans, Sweet Corn and Wheat. *J. Agric. Food Chem.* 31, pp. 1240–1244, 1983.

Zarcinas, B. and R.O. Nable. Boron and Other Impurities in South Australian Fertilizers and Soil Amendments. *CSIRO Division of Soils Divisional Report* 118, 1992.

Zawadzka, T., H. Mazur, M. Wojciechowska-Mazurek, K. Starska, E. Brulinska-Ostrowska, K. Cwiek, R. Uminska, and A. Bichniewicz. The Content of Metals in Vegetable from Various Regions of Poland in the Years 1986–1988. Part I. Lead, Cadmium and Mercury. *Roczniki Panstwowego Zakladu Higieny* 41(3–4), pp.111–131, 1990.

Zurera, G., B. Estrada, F. Rincon, and R. Pozo. Lead and Cadmium Contamination Levels in Edible Vegetables. *Bull. Environ. Contam. and Toxicol.* 38, pp. 805–812, 1987.

CHAPTER 2

ENVIRONMENTAL CONCERNS OF PESTICIDES IN SOIL AND GROUNDWATER AND MANAGEMENT STRATEGIES IN OCEANIA

B.K.G. Theng, R.S. Kookana, and A. Rahman

INTRODUCTION

As the title suggests, this chapter is an attempt at describing some environmental concerns and issues arising from, and relating to, pesticide usage in Oceania. The term "pesticides" is used here to denote organic chemicals that control, and act against, diseases and pests in a given ecosystem. By "Oceania" is meant Australia, New Zealand, and the Pacific Island countries. However, since little documented information is available for Melanesia, Micronesia, and Polynesia, the following account largely refers to the situation in Australia and New Zealand. Even in the case of Australasia the data are rather fragmentary.

Since the late 1940s the use of chemicals for pest control has been a cornerstone of land-based production systems. It is generally accepted that the introduction and use of pesticides in this connection have made a significant contribution to the worldwide increase in agricultural productivity. A classic example is the spectacular rise in global grain output over the last 30 years. This feat has largely been achieved through the intensified use of existing arable land, involving the massive application of pesticides as well as fertilizers (Brady, 1986; Theng, 1991). Similarly, the output of dairy produce, meat, and wool in Australia and New Zealand would have been much depressed, had pesticides not been used in their production. Indeed, fresh fruit and vegetables for export must be treated with a variety of pesticides in order to meet the quarantine and quality standards of importing countries. Pesticide usage has also reduced the labor requirements for crop production, and facilitated the introduction in many countries, including Australia, of minimum tillage cropping systems.

The economic benefit to agriculture of pesticide usage in Oceania is difficult to assess, however. In Australia alone the loss of revenue due to weed infestation of cropped land has been estimated at $3.3 billion (Combellack, 1989). The generally accepted return to farmers, based on U.S. practices, is between 3 and 5 dollars for every dollar spent on pesticides (Pimentel, 1981; LeBaron, 1990). For export horticulture in Australasia, the benefit/cost ratio may be even higher as pest control measures make up about 5–10% of total expenditure while preventing 30-90% of crop losses. Since 1970 when the use of DDT in New Zealand was banned, the loss of productivity from grass grub infestation of grasslands has been estimated at up to $100 million per year (Williams, 1988).

For all its benefits, however, pesticide usage has been perceived by the general public as having an adverse effect on human health and nontarget organisms. Recent surveys in Australia indicated that some 40% of the population identified chemicals as the primary environmental concern (Young, 1989). With respect to New Zealand, the issues and options relating to pesticide usage have been discussed in a report by MacIntyre et al. (1989). Because of their visibility, such practices as the aerial spraying of 2,4,5-T to control brush weeds, and the laying of bait containing 1080 (monofluoroacetate) for rabbit and opossum control, have attracted much public and media attention in New Zealand. On the other hand, with the possible exception of the well-publicized persistence of DDT residues in pasture soils (Orchard et al., 1991; Boul, 1995), long-term environmental problems of soil and water contamination by pesticides are largely invisible to the community.

Although the people and environment in Australia and New Zealand may well be exposed to a greater number and volume of pesticide residues than counterparts in Western Europe and North America (MacIntyre et al., 1989), the impact of pesticide usage on Australasian ecosystems is likely to remain hidden from the public eye, at least in the short term. Furthermore, these countries frequently have some "fallback" option if a given ecosystem, or part of it, has been damaged through human or natural activities (Morrison and Brodie, 1985). The same cannot be said about the Pacific, however, as more often than not the soils here are thin and porous, the land area is small, groundwater supplies are limited, and terrestrial biodiversity is low. Pacific ecosystems are therefore intrinsically fragile and vulnerable to environmental change (Watts, 1993). In addition, the impact of pesticides on the environment in the Pacific cannot simply be predicted from data obtained elsewhere (Mowbray and Hicks, 1989). Furthermore, many Pacific countries do not have the technical expertise and resources to assess the environmental fate of introduced pesticides.

Because of space limitations and the dearth of information, it has not been possible to give a comprehensive account of the subject matter. Rather, in what follows we focus attention on a few important issues and concerns. To this end, and for the sake of clarity and convenience, we have divided the chapter into a number of sections. The Usage section deals with the quantitative and legislative aspects. Behavior and Fate in Soil gives an overview of the behavior and fate of pesticides in soil. The occurrence and assessment of pesticide residues in the environment are described in the next section. Remediation options and strategies are considered in the following section, and this is followed by a summary and conclusions.

USAGE

General

Information on actual quantities of pesticides used in Australia is not routinely available. Data for products for sale at the factory gate indicate a marked increase in overall usage during the last 15 years, with herbicides making up 66% of the total sale (Fig. 2.1). The large growth in herbicide usage, partly ascribable to the increasing adoption of reduced tillage practices, has caused concern among many farmers about the long-term effects on soil health and productivity. Statistics on regional use are not available although a useful effort has recently been made to carry out an audit for the Condamine-Balonne-Culgoa

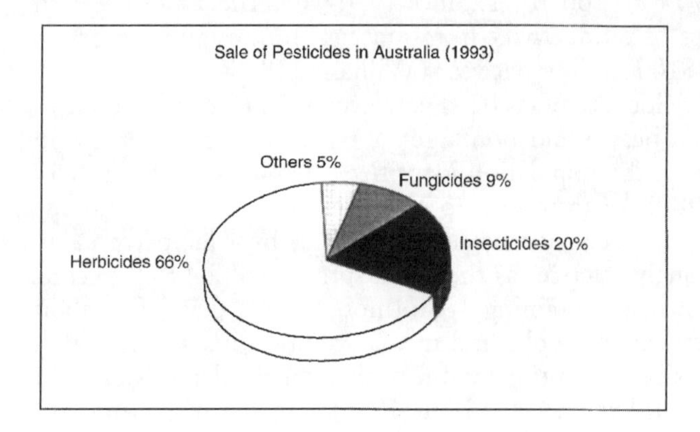

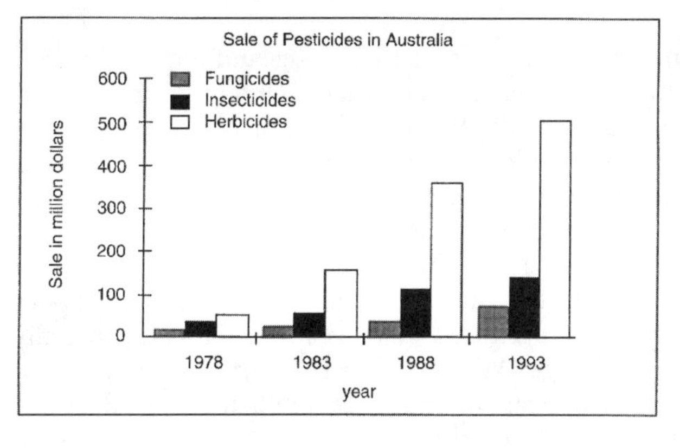

Figure 2.1. Top: percentages of major classes of pesticides sold at factory gate in Australia during 1993.
Bottom: value of pesticides sold at factory gate in Australia over the period 1978 to 1993.

River catchment in Queensland (Rayment and Simpson, 1993). The area involved is 163,000 km^2, comprising about 14% of the Murray-Darling River Basin catchment, and producing some 60% of dryland, and 11% of irrigated cotton in Australia. Some 144 different pesticides were identified to have registered usage in the catchment, of which herbicides (notably atrazine) and insecticides (notably endosulfan) made up 42 and 39%, respectively. Details of the audit are given in Table 2.1.

Table 2.2 gives a breakdown of pesticide sales in New Zealand for the years 1983, 1985, and 1987 (MacIntyre et al., 1989). The average annual value currently amounts to some $120 million, involving nearly 4000 tonnes of 270 different active ingredients in approximately 900 separate formulations (Wilcock, 1989; Wilcock and Close, 1990). The discrepancy between sale and usage may be as much as 20%. As in Australia, herbicides account for nearly 2/3 of total annual sales. Unlike the Australian situation, however, the sale of herbicides slightly declined from 1983 to 1987, while that of fungicides and insecticides increased by 80 and 25%, respectively. In addition, about 10 tonnes of organochlorine pesticides were sold annually during this period for grass grub control in pastures and for timber treatment. The North Island accounts for about 75% of the total amount of pesticides sold in New Zealand, and much of this is used by the horticultural sector.

By comparison with Australasia, pesticide usage in the Pacific is small. However, this is probably due more to financial constraints than concerns about the environment (Convard, 1993). Figures for total quantities by formulation and monetary value for the early

1980s are listed in Table 2.3 (Mowbray, 1988). More recent data indicate that approximately 286 active ingredients are available (Watts, 1993), and that current sale figures are much higher even after allowing for inflation. In 1990, for example, the Cook Islands imported pesticides to the value of about U.S.$ 200,000 (Samuel, 1991), representing a five-

Table 2.1. Pesticide Use Under Different Land Use Systems in the Condamine-Balonne-Culgoa Catchment, Queensland, Australia[a]

Land Use	Area (ha)	Insecticides (tonnes a.i. yr^{-1})	Herbicides (tonnes a.i. yr^{-1})	Major Insecticides[b]		Major Herbicides[b]	
Cotton							
Irrigated	26,000	148	222	Endosulfan	(110)	Fluometuron	(59)
Dryland	26,000	52		Endosulfan	(24)	Prometryn	
						diuron	(47)
							(37)
Wheat	450,000	5	127	Chlorpyrifos	(4.5)	2, 4-D	(58)
						triallate	(44)
Barley	273,000	7	128	Chlorpyrifos	(6.8)	Triallate	
						2, 4-D	(72)
							(58)
Sorghum	225,000	22	256	Endosulfan	(13.5)	Atrazine	(237)
Maize	NA	0.09	37	Chlorpyrifos	(0.09)	Atrazine	(27)
Sunflower	NA	3.4	9.3	Endosulfan	(3)	Trifluralin	(7.8)

[a]Rayment and Simpson, 1993.
[b]Figures in parentheses refer to usage in tonnes of active ingredients (a.i.) per year; NA = not available.

Table 2. 2. Pesticide Sales in New Zealand Expressed in Tonnes of Active Ingredients[a]

Pesticide Class	Year 1983	1985	1987
Ectoparasiticides	83.3	78.2	95.6
Fungicides	479.9	830.4	863.2
Herbicides	2467.9	2475.4	2245.0
Insecticides	358.6	437.1	448.3
Rodenticides	0.2	0.3	0.3
Total	3389.9	3821.4	3652.1

[a]MacIntyre et al., 1989.

Table 2.3. Pesticides Use in Different Countries of the South Pacific[a]

Country[b]	Year	Quantity by Formulation (tonnes or kiloliters)	Quantity by Active Ingredient (tonnes or kiloliters)	Quantity by Value (thousand $US)
Cook Islands	1981	na	na	40
Fiji	1981	505	na	2379
Guam	1981	8	na	39
Niue	1981	4	na	na
Papua New Guinea	1981	637	271	na
	1985	1300	na	3621
Solomon Islands	1981	na	na	na
Tonga	1982/83	20	7.2	83
Western Samoa	?	207	na	313

[a]Mowbray, 1988.
[b]In 1981 the South Pacific countries consumed only 0.1% of the total quantity of pesticides used in the world; na = not available at time of survey.

fold increase over the 1981 figure. However, pesticide importation into some Pacific Island countries has declined over the last 2–3 years. Thus, in the case of Vanuatu the total import of herbicides in 1994 has dropped by about 50% over the previous year (B. Tarilongi, pers. comm.). The proportions of the three major classes of pesticides sold and used vary greatly among Pacific Island countries (Mowbray, 1988). According to the Asian Development Bank (1987), insecticides make up nearly 76%, herbicides 14%, and fungicides 8% of overall sales. Of some concern is that the pesticides sold in the Pacific might be of low quality. Further, a number of pesticides that are banned, restricted, or unregistered in the country of origin, mainly Australia and New Zealand, have apparently found their way into the Pacific (Mowbray, 1988; Watts, 1993).

Usage by Sectors

In accord with global trends, agriculture is the largest single user of pesticides in Oceania. In Australia, herbicides are extensively used in "broadacre" agriculture; that is, the growing of wheat, barley, oats, lupins, sorghum, cotton, sunflower, sugarcane, and rice. Insecticides are largely used for locust control, and rodenticides against mice and rabbits, especially when their population reaches plague proportions. Horticulture is generally the most intensive user of pesticides because of the number of crops grown per year, together with repeated applications at high rates.

The dominant user of pesticides in New Zealand is pastoral agriculture (Table 2.4), although the combined consumption by trees, vines, market gardens, and vegetables is greater. Furthermore, from 1983 to 1987 the amount sold for horticultural use increased while usage in pastoral agriculture declined. This period coincided with the expansion of kiwifruit plantings, the gradual removal of subsidies on herbicides for weed control, and the cessation of supplementary minimum payments to sheep farmers. It is also interesting to note that organophosphates make up the largest proportion of insecticides used, while about half of the fungicides are dithiocarbamates.

In both Australia and New Zealand, forestry is a significant user of herbicides for the control of weeds in new and young plantations, and the maintenance of fire breaks. Fungicides are used to control pine needle blight on young trees, and prevent sap stain in peeled logs for export. On the other hand, insecticides are not, or only rarely, used by this sector.

The "other" (nonagricultural) users of pesticides, referred to in Table 2.4, are the industrial and urban sectors. In the former category, the timber industry is probably the largest single consumer. The urban sector uses herbicides to remove weeds which may damage

Table 2.4. Pesticides Usage by Different Sectors in New Zealand, Expressed in Tonnes of Active Ingredients[a]

	Year		
Sector	**1983**	**1985**	**1987**
Pastoral agriculture	1623.2	1332.0	1254.6
Trees and vines	463.2	694.4	883.9
Cereal and pea crops	500.8	691.6	589.6
Market gardens/vegetables	373.4	473.8	469.0
Forestry	202.4	304.2	168.3
Other	177.8	267.4	216.0
Household gardens	55.1	57.9	70.7
Total	3395.9	3821.4	3652.1

[a] MacIntyre et al., 1989.

road paving, choke water courses, or obstruct views of roads and railway lines. Pesticides are also used to protect parks, playing fields, ornamental plants, lawns, gardens, and household contents, and to assist in the conservation of native forests and endangered species.

Pesticide usage in the Oceania varies widely between countries. As in Australasia, pesticides here are used largely to maintain and improve the yield and quality of crops grown in plantations usually as monocultures (Watts, 1993). Examples are sugarcane in Fiji, cocoa, coconut, and oil palm in Papua New Guinea and the Solomons, rice in the Solomons, and coffee in New Caledonia, Papua New Guinea, and Vanuatu. Vegetable production on a commercial scale consumes an appreciable and increasing proportion of imported pesticides.

By contrast, subsistence agriculture uses very little, if any, chemicals. No reliable information on the breakdown of usage by the different sectors is available. A small proportion of pesticides is used for insect and rodent control. In malaria areas of the Solomons and Vanuatu, DDT and other organochlorines are still applied for mosquito control (Convard, 1993).

Legislative Aspects

Targeted Reductions

Many countries in Western Europe have introduced legislative programs for reducing pesticide usage but only in some cases have targeted reductions been achieved. Sweden and Denmark have notably achieved a 50% decrease in usage by 1990, while the Netherlands is aiming at a 60% reduction for arable farming by the turn of the century. However, as MacIntyre et al. (1989) have pointed out, the advantage of reducing total active ingredients is dubious because more toxic ingredients may be substituted, while more selective pest control measures may be overlooked. Further, such programs require substantial resources for their implementation and enforcement, and have more than an even chance of failure.

The issue of targeted reduction in pesticide use is also being closely examined in Australia (Evans and Rowland, 1993), but a strategy similar to that of some European countries has not been formulated. However, there is an undertaking between the Australian Consumer Association, the Apple and Pear Growers Association, the Rice Growers Cooperative, and other interest groups to seek ways of reducing pesticide usage (Australian Consumers Assoc. Conf., 1991). Apple growers, for example, aim at a 50% reduction by 1996 but this target will be difficult to achieve (Bower et al., 1993).

Registration

The registration of pesticides in Australia has traditionally been the responsibility of each state, and a national (federal) approach has been lacking until August 1993 when the National Registration Authority (NRA) was formed. The task of the NRA is to standardize the states' procedures, and provide a national focus for pesticide registration. For example, on the basis of recommendations by the National Health and Medical Research Council (NHMRC), and in consultation with the public, the NRA is currently examining the issue of deregistering aldrin/dieldrin for termite control.

In New Zealand there are at least seven government departments and other agencies that implement over 25 statutes and regulations concerned with pesticide usage. The principal statute is the Pesticide Act of 1979 which, through the Pesticides Board, controls the registration, sale, and use of pesticides for domestic, industrial, and public health purposes. Under the Pesticide Act any new product requires registration or application for an Experimental Use Permit. Development data, including the chemical and physical properties of the pesticide, its toxicology to humans, and its effects on environment and wildlife, together with field trials in New Zealand, must be provided to show that the product is safe to use and effective for the purposes claimed (MacIntyre et al., 1989). The thrust of the Pesticide Act and other related legislations is directed at ensuring that export produce (principally meat and dairy products) does not contain unacceptable concentrations of pesticide residues, and is free of pests. The procedures for export quality assurance are annually audited by the European Community and the U.S.A. Export fruit, vegetables, and timber must likewise comply with certain laws and regulations controlling the use of pesticides.

With respect to the Pacific, dependent territories of the U.S.A. (American Samoa, Guam, Palau) or France (New Caledonia, Wallis and Futuna Islands, French Polynesia) simply inherit the extensive legislation of their respective metropolitan countries. That is, any pesticide that is registered in the United States or France can, in theory, be registered in its territory. Since this is not always appropriate, some dependent territories (American Samoa, Guam, New Caledonia, French Polynesia) have developed their own legislation, imposing conditions that are often more stringent than those applicable in the corresponding metropolitan countries (Hicks and Mowbray, 1989). Most of the remaining Pacific Island countries have introduced some type of legislation which allows for either full or restricted registration of a pesticide. Apart from the length of time required in writing and approving the legislation, the majority of Pacific Island countries lack the infrastructure for its enforcement (Watts, 1993).

Guidelines for Public Health

In Australia the development of national guidelines and standards for public and occupational health rests with the NHMRC and the National Occupation Health and Safety Commission. To expedite the creation of uniform health and environmental guidelines, close linkages between NHMRC and the standing committee of the Australian and New Zealand Environmental and Conservation Council (ANZECC) have been established. Guideline development procedures in Australia involve some 16 steps from commissioned research, toxicological data assessment, and scientific consensus to extensive public consultation (Bentley, 1993). In relation to drinking water quality, for example, a "health value" and a "limit of determination" value have been specified. The former is based on 10% of the acceptable daily intake (ADI) of pesticides for an average adult weighing 70 kg and consuming an average of 2 L of water daily. The limit of determination is set on the basis that pesticides should not be present, and therefore reflects current analytical limits. Table 2.5 compares Australian water quality guidelines for commonly detected pesticides in waters (Cooper, 1994) with those set by Canada, the World Health Organization (WHO), and the U.S. Environmental Protection Agency (EPA) showing the wide range of values between countries and agencies.

In New Zealand new maximum allowable values (MAVs) for pesticides in drinking water, based on the 1993 WHO guidelines (Table 2.5), will shortly be issued by the Ministry of Health, taking effect from 1 January 1995 (P. Prendergast, pers. comm.). There will also be standards ("determinands") for a range of organic contaminants of health significance. Similarly, many food items have a maximum residue level (MRL) for pesticides. This has been set by the Foods Act 1981, and administered by the Department of Health. In the absence of an MRL, the limit is set at 0.1 mg kg^{-1} (ppm). The Toxic Substances Act 1979 is designed to protect public health and the environment as well as regulate the accidental spillage, importation, manufacture, sale, and transportation of toxic materials. There are also laws for maintaining safe working conditions at sites where pesticides are stored, formulated, or used. In essence, the laws and regulations relating to pes-

Table 2.5. Quality Guideline Values for Some Pesticides in Waters, Expressed in Micrograms Per Liter, as Prescribed by Various Agencies

Pesticide[a]	1987 NHMRC, Australia Drinking Water	1994 Draft NHMRC/ ARMCA NZ Drinking Water	1993 WHO[b] Drinking Water	1993 Canada Protection of Freshwater Life	1991 USEPA Office of Drinking Water Lifetime Health Advisory Levels
Endosulfan (I)	40	30	na	0.02	na
Profenofos (I)	0.6	0.3	na	na	na
Chlorpyrifos (I)	2	10	na	na	na
Atrazine (H)	na	20	2	2	3
Glyphosate (H)	200	1000	na	65	700
Diuron (H)	40	30	na	na	10
Fluometuron (H)	100	50	na	na	90
Metolachlor (H)	800	300	10	8	100
Prometryn (H)	na	na	na	na	na
Pendimethalin (H)	600	300	20	na	na
Trifluralin (H)	500	50	20	0.1	2

[a]I = insecticde; H = herbicide.
[b]The maximum allowable values for drinking water in New Zealand, prescribed by the Ministry of Health, are identical except for trifluralin which is set at 30 mg L^{-1}. NHMRC = National Health and Medical Research Council; ARMCANZ = Agricultural Resource Management Council of Australia and New Zealand; WHO = World Health Organization; USEPA = United States Environment Protection Agency.

ticide use in New Zealand are aimed at maintaining export market access. The protection of public health seems subordinate to the need of exporting agricultural and horticultural commodities (MacIntyre et al., 1989).

For many Pacific Island countries, the closest approach to developing regulations and guidelines for pesticide use has been the endorsement of the FAO (U.N. Food and Agriculture Org.) Code of Conduct on the Distribution and Use of Pesticides. Divided into 12 sections, the Code attempts at clarifying and defining the responsibilities of the parties that are associated with, and involved in, the development, distribution, and use of pesticides. Although the FAO code is voluntary, it has been very useful especially for those countries that do not have their own legislation. Data collected by Hicks and Mowbray (1989) indicated that, at best, only 55% of the countries complied with its provisions at the time of survey. Compliance is hindered by the reluctance of some countries to make the Code legally binding, and by the lack of resources for its enforcement (Watts, 1993).

National guidelines for contaminated soils in Australia are currently being developed by ANZECC/NHMRC (1992). The approach takes advantage of overseas research and experience, and incorporates defined criteria for initial assessment and site-specific data. No health investigation level guidelines for pesticides have so far been proposed. In the case of organic contaminants, guidelines for only polycyclic aromatic hydrocarbons (PAHs) and benzo(a)pyrene have been proposed. The environmental investigation guidelines will be based on threshold levels for (a) phytotoxicity and (b) contaminant uptake resulting in an unacceptable residue level. For pesticides, the interim environmental investigation levels have so far only been proposed for dieldrin at 0.2 mg kg^{-1}. Properties (farms) in Western Australia with an organochlorine residue > 0.1 mg kg^{-1} in soil are issued with a notice under the Agricultural Produce (Chemical Residues) Act 1983. If used for grazing, such properties can fall into one of three categories, (i) tested clear, (ii) provisionally clear, and (iii) restricted. The last category is applicable to properties producing marketable animals with an MRL of > 0.2 mg kg^{-1} for organochlorine in body fat, which essentially puts such farms under long-term quarantine.

In New Zealand there are guidelines for the land disposal of pesticides and general waste management. Standards for contaminated land with respect to pentachlorophenol (PCP) are currently under consideration. Two important regulations have recently been introduced that are relevant to pesticide use and pest management. These are the Resource Management Act 1991 and the Biosecurity Act 1993 (Blakeley, 1993). As the name suggests, the former is about sustainable management of natural and physical resources. The latter provides the legislative framework for the management of a wide range of organisms that are identified as pests, including unwanted animals, plants, microorganisms, and any entity capable of replicating itself or multiplying. The Biosecurity Act 1993 also consolidates and amends seven Acts that currently provide for agricultural security, and is closely aligned with the Resource Management Act 1991, particularly as it relates to national and regional pest management. Another piece of legislation that is being developed, and which will impact on the use of pesticides, is the Hazardous Substances and New Organisms Bill.

BEHAVIOR AND FATE IN SOIL

General

Soil is a major sink of pesticides, whether the compounds are directly (e.g., by incorporation) or indirectly (e.g., by aerial spraying) applied. A fraction of pesticides that enter soil is lost by leaching, photodecomposition, plant uptake, surface runoff, and volatilization. These processes combined may, under certain conditions, contribute appreciably to the total loss of a pesticide. In the majority of cases, however, the bulk of applied pesticides is sorbed by inorganic and organic soil constituents, chemically transformed, and microbially degraded or metabolized (Fig. 2.2). These processes of dissipation are more or less interdependent. Sorption, for example, may protect pesticides against microbial attack but also facilitate their abiotic transformation (Stevenson, 1982). By inhibiting desorption (release) into the soil solution, sorption also retards pesticide movement by leaching and runoff (Weed and Weber, 1974). Here we outline the pesticide-soil interaction, with particular emphasis on sorption, degradation, and transport processes, including a brief reference to volatilization. Where appropriate, existing gaps of knowledge and research needs will be indicated. For more details than could possibly be included in this review, we refer to the works written or edited by Goring and Hamaker (1972), Theng (1974), Hance (1980), Khan (1980), Sawhney and Brown (1989), Beck et al. (1993), and Nicholls (1993).

Sorption

Surface Interactions vs. Partition

Because of the complexity and heterogeneity of soils and the many variables involved, the interactions of pesticides with soils are difficult to predict under field conditions. Much of the data on pesticide sorption by soil (and sediment) refer to laboratory measurements using batch methods. The behavior of nonionic pesticides at soil surfaces has received particular attention. Although a broadly consistent picture is emerging from the mass of experimental data, a predictive model of universal applicability has yet to be developed.

It is generally accepted, however, that clays and organic matter (OM) are the principal

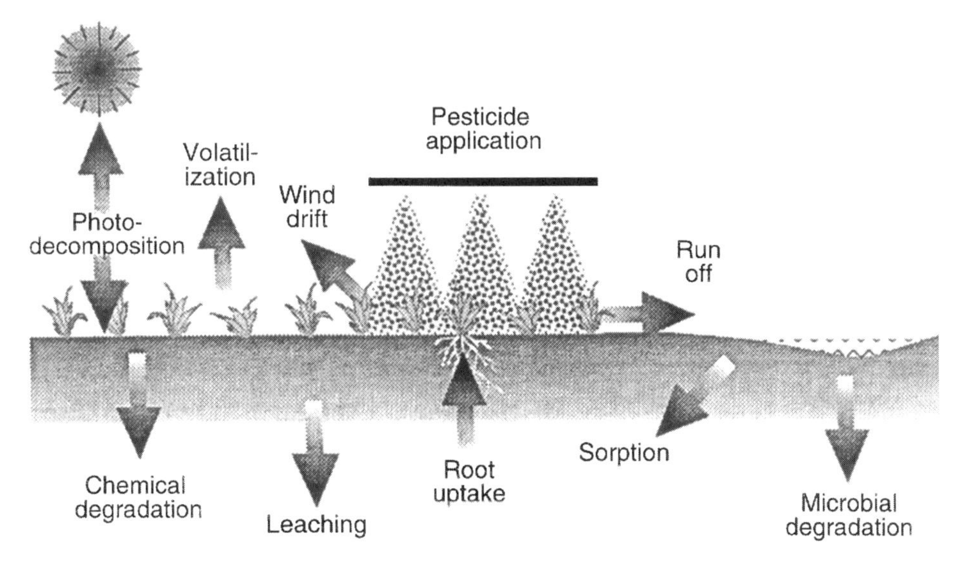

Figure 2.2. Pictorial diagram showing the fate of applied pesticides in the environment with reference to the various dissipation pathways.

constituents involved in pesticide sorption by soil (e.g., Koskinen and Harper, 1990). In highly weathered soils, the reactive mineral species may be identified more with the oxides and oxyhydroxides of iron and aluminium than with layer silicates, while in soils derived from volcanic ash, allophane and imogolite are important. Depending on the surface properties of the sorbents and the charge characteristics of the pesticides, different sorptive mechanisms are possible, of which ion exchange, cation-bridging, and ion-dipole interactions are important (Bailey and White, 1970; Mortland, 1970; Theng, 1974). In addition, ligand exchange, charge transfer, hydrogen-bonding, and van der Waals interactions may come into play. The sorption of nonionic pesticides by OM-rich topsoils is a special case in that "simple" partitioning of the compound between the aqueous solution and the OM phase often predominates over specific site interactions (Chiou et al., 1979).

On the other hand, the interactions of ionic or charged pesticides with soil and sediment are reasonably well understood. Positively charged species are principally sorbed by cation exchange, that is, by replacing the inorganic cations initially present on clay and OM surfaces. Being sorbed by electrostatic (Coulombic) attraction to the negatively charged sites on clay and OM surfaces, such cationic species as diquat and paraquat are strongly retained by soil. However, the energy of interaction is apparently higher for clays than for OM. Accordingly, the sorption of these compounds is largely controlled by both the content and type of clay as Kookana and Aylmore (1993) have demonstrated for some Western Australian soils. As might be expected, anionic pesticides interact only weakly with negatively charged surfaces of clay and OM. Sorption, in this instance, correlates better with OM than clay content, and increases with a decrease in ambient pH (Khan, 1980).

The correlation of sorption with OM content is very pronounced for nonionic pesticides. Indeed, their interactions with OM-rich topsoils and sediments can largely be rationalized in terms of partitioning into OM. The partition theory is supported by the linearity of the isotherms, the small enthalpy of sorption, and the lack of competition between solutes during sorption from binary mixtures (Chiou, 1989).

In order to compare sorption by different soils, the partition (also called sorption or distribution) coefficient (K_p), given by the ratio of the amount sorbed to that in the equilibrium solution, is normally expressed on an OM content basis

$$K_{om} = K_p / \% \text{ OM} \times 100 \quad (2.1)$$

Assuming that % OM = 1.724 × % organic carbon (OC), we can also write

$$K_{om} = K_{oc} / 1.724 \quad (2.2)$$

For some allophanic soils in New Zealand with a high OM content, Rahman and Matthews (1979) have suggested that K_p in Equation 2.1 may be replaced by a phytotoxicity or growth retardation index (GR_{50}), defined as the amount of pesticide required to reduce plant growth by 50%. This concept was based on the finding that the GR_{50} values for a number of pesticides (alachlor, atrazine, linuron, terbacil, trifluralin) were highly and positively correlated with soil OM content (Rahman, 1976).

Although partition into OM is sometimes referred to as "hydrophobic sorption" (Hassett and Banwart, 1989; Senesi, 1992), which implies a solute-surface interaction, the process is more akin to solute extraction from water into an organic solvent, such as 1-octanol. On this basis, the sorption of nonionic pesticides may be estimated from their octanol-water partition coefficient (K_{ow}). In studying the sorption of 25 nonionic organic compounds from aqueous solution by 17 Australian soils, Briggs (1981), for example, obtained the following relationship between K_{om} and K_{ow},

$$\log K_{om} = 0.52 \log K_{ow} + 0.69 \quad (2.3)$$

which was almost identical with what he found earlier for a number of British soils (Briggs, 1973). In investigating the sorption of atrazine by 26 Australian soils with different contents in clay and OM, Bowmer (1971) obtained a mean K_{om} of 68 which compared favorably with the value of 98 calculated from Equation 2.3. On the other hand, by substituting GR_{50} for K_p in Equation 2.1, Rahman and Matthews (1979) obtained K_{om} values of 2.8, 2.8, and 2.5 for atrazine in an allophanic soil from New Zealand containing 9.8, 15.5, and 20.6% OM, respectively. Similarly, the K_{om} values derived by Singh et al. (1990) for the sorption of fenamiphos and linuron by a number of Australian soils differed by an order of magnitude from that obtained using Equation 2.3.

Figure 2.3 shows the relationship between K_{ow} and K_{oc} for a large range and variety of soils worldwide. The regression lines are by no means parallel, suggesting that simple solute partitioning into OM, analogous to that between octanol and water, is not generally applicable (Gerstl, 1989). Nevertheless, the concept of partition uptake is a useful first approach to modeling pesticide sorption because only the OM content of the soil needs to be known to derive the K_p value of a given pesticide. Other indirect ways of obtaining sorption coefficients for modeling purposes have been discussed by Green and Karickhoff (1990).

The general applicability of the partition theory has also been questioned on the grounds that a number of nonionic compounds give rise to nonlinear isotherms (Khan, 1980; Weber and Miller, 1989), have high sorption enthalpies, and compete for sorption sites (Mingelgrin and Gerstl, 1983a; Pignatello, 1989). Further, for subsoils and soils low in OM, the close correlation of uptake with OM content no longer holds (Schwarzenbach and Westall, 1981) because clay and mineral surfaces now play a major part in sorp-

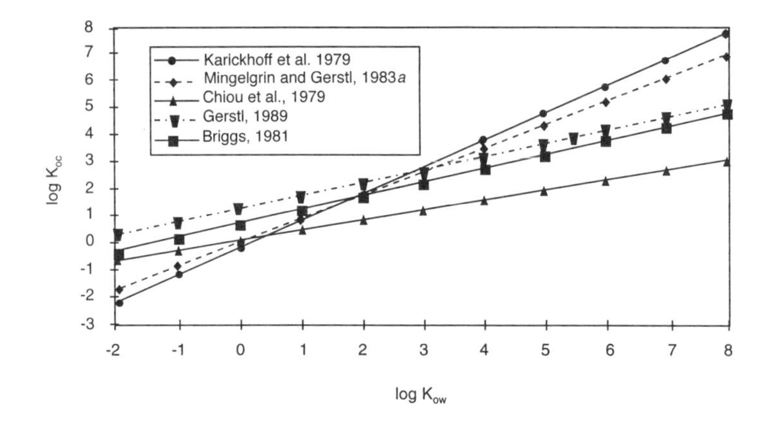

Figure 2.3. Some empirical relationships between the logarithm of the octanol/water partition coefficient (K_{ow}) and the logarithm of the sorption coefficient, expressed on the basis of organic carbon content (K_{oc}).

tion. Thus, although partition into OM is undoubtedly important, other mechanisms are apparently involved in the sorption of nonionic organic species by soils.

When nonlinear isotherms obtain, the data may be described in terms of the Langmuir, or more commonly, the Freundlich equation. The latter is written as

$$S = K_f C^{1/n} \quad (2.4)$$

where S is the mass of solute sorbed per unit weight of sorbent, C is the equilibrium solute concentration, and K_f and n are empirical constants. Singh et al. (1990) found that diquat sorption by some Australian soils followed the Langmuir equation, whereas the data for fenamiphos, linuron, and simazine fitted to the Freundlich equation. As in many studies elsewhere (Hamaker and Thompson, 1972), the value of $1/n$ in Equation (2.4) for these compounds was less than unity.

Nevertheless, it is often assumed that $1/n$=1, especially in modeling pesticide transport. The error introduced by this assumption will depend on the extent to which the isotherm deviates from nonlinearity, and on the value of C. However, a greater degree of refinement in describing sorption isotherms is perhaps unwarranted since kinetic and nonequilibrium processes, as we will see later, are often more important in determining pesticide movement in soil than is isotherm shape (Pignatello, 1989; Sparks, 1989; Kookana et al., 1992a; 1992b). A serious drawback in dealing with soils is to separate the relative contribution of OM (humus) and clay to sorption because both constituents are intimately associated, to form a clay-OM complex (Greenland, 1965; Theng and Tate, 1989). In an attempt to overcome this problem, Rebhun et al. (1992) measured the sorption of some nonionic compounds by a "synthetic" soil, composed of sand and (bentonite) clay coated with different amounts of humic acid (HA). The K_p values for sorption on HA were 8–20 times higher than those on the clay. For sorption by clay-HA complexes in the soil, K_p varied linearly with organic carbon content (f_{oc}) when this is greater than 0.5%. Below this level of f_{oc}, the mineral made a significant contribution to sorption although half of the sites on the bentonite were blocked by HA. Similarly, Walker and Crawford (1968) found that atrazine sorption by soils with an OM content of less than 8% was strongly correlated with the sum of clay and OM. But when the OM content of the soil exceeded this level, the clay component made little, if any, contribution to sorption.

The "cutoff" level might also be expected to depend on the type of clay and the nature of the associated OM. This is because clays differ widely in their capacity to sorb humic substances and pesticides, decreasing in the order imogolite > allophane >> smectite >

kaolinite (Calvet, 1980; Theng, 1986). Similarly, the sorption capacity of humic and fulvic acids is known to be different (Hayes et al., 1968; Khan, 1980). In line with these observations, Murphy et al. (1990) found that the magnitude of K_{oc} depended on both the type of the mineral substrate and the humic material coating it, with the most aromatic coating being the strongest sorbent. In a related study, Rutherford et al. (1992) found that K_{om} values increased as the (O+N)/C ratio; that is, the ratio of polar to nonpolar groups in the OM, decreased. Thus, the K_{om} value for a given nonionic compound would be greater in grassland soils where the OM is dominated by humic acid than that in forest soils of comparable OM content but which are relatively enriched in fulvic acid-type polymers. The wide range of K_{om} values for a given pesticide in different soils (Mingelgrin and Gerstl, 1983a) may therefore be ascribed, at least partly, to variations in the composition, and degree of humification, of OM in the soil.

Kinetics

Studies by batch methods, involving vigorous shaking or stirring of the reactants, have generally shown that equilibrium is attained instantaneously. However, both laboratory and field experiments indicate that sorption under flow conditions is almost invariably time-dependent (Murali and Aylmore, 1980; Brusseau et al., 1991; Kookana et al., 1992a). This effect may arise from structural limitations on solute diffusion into micropores within aggregates, and soil OM, as well as from sorptive interactions (Beck et al., 1993; Pignatello, 1993).

Using a flow technique similar to that developed by Sparks and coworkers (Sparks, 1989), Kookana et al. (1992b) found that the sorption of three pesticides by four Western Australian soils showed a rapid initial phase followed by a slower rate. Further, sorption is faster in a well-structured soil with a high OM content than in a dispersed soil containing little OM. In accord with the literature (Brusseau and Rao, 1989), these observations strongly suggest that sorption is not an instantaneous process, and to assume so would seriously underestimate the potential for the leaching and transport of pesticides to groundwater. The rate of desorption (release) from soil is likewise biphasic. Hysteresis (nonsingularity) in the sorption-desorption isotherms has been observed for a number of pesticides. As a result, the ratio of the desorption coefficient to that of sorption, referred to as the nonequilibrium index, commonly increases with the passage of time after application (Pignatello and Huang, 1991). Accordingly, the resistance to leaching and the persistence of many pesticides tend to increase with their time of residence in soil. Although nonequilibrium sorption and hysteresis are apparently real and significant, their effects are yet to be incorporated into models of pesticide transport.

Factors Affecting Sorption

The charge characteristics of many soils vary with the pH of the ambient solution (Theng, 1980). Soils with variable charge are usually rich in sesquioxides, short-range order minerals (allophane, imogolite, ferrihydrite), and humic substances. Soil (solution) pH is therefore an important factor affecting the sorption and desorption of pesticides, in particular of ionizable species. Since basic compounds (e.g., s-triazines) can acquire protons under acid conditions, both the extent and strength of their sorption would increase as ambient pH decreases (Weber et al., 1969; Singh et al., 1990). Conversely, acidic pesticides are proton donors, and behave essentially as anions when the ambient pH is one or more units above their pK_a.

Accordingly, the sorption by some Australian soils of simazine, chlorsulfuron, trisulfuron and, to a lesser extent, of linuron and fenamiphos decreased with increasing pH (Singh et al., 1990; Blacklow and Pheloung, 1992). These points are relevant to the management of soils in Australasia that have been acidified as a result of fertilization and cropping. The oft-practiced liming of such soils to alleviate the acidity problem would clearly affect the release and bioavailability of applied pesticides (Kookana, 1989).

Sorption is also influenced by the ionic strength of the soil solution. This is because inorganic cations compete directly with cationic and basic pesticides for sorption sites. Thus treatments that increase solution ionic strength would reduce the uptake, and promote the desorption of such compounds. By increasing the concentration of $CaCl_2$ in the soil solution from 0.005 to 0.05 M, Kookana and Aylmore (1993) observed a reduction in sorption capacity for diquat and paraquat ranging from 17% for a sandy soil to 40% for a clay soil from Western Australia. Similarly, some 60% of sorbed diquat could be released from a loamy sand soil by extraction with 0.05 M $CaCl_2$. By contrast, the sorption of linuron, simazine, and fenamiphos was little affected by variations in ionic strength (Kookana, 1989).

With uncharged polar organic compounds, water is a strong competitor for sorption sites on clay and mineral surfaces. This accounts for the observation that little, if any, sorption occurs from dilute solutions of such species until their molecules exceed a certain size or chain length (Theng, 1974). By the same token, water can displace polar compounds from mineral surfaces. By influencing the swelling and dispersion of smectite-rich soils, water also affects site accessibility. What is perhaps more important is the controlling effect that water has on the degradation and transformation of pesticides in soil.

Degradation and Transformation

Chemical

Organic compounds in soils and waters can be transformed by a number of chemical (abiotic) processes, the major ones being hydrolysis, oxidation-reduction, and photolysis (Valentine, 1986). Photolysis is an major dissipation pathway in aquatic environments. Its role in soil, however, is restricted to the top few centimeters of the profile (Miller et al., 1989) and losses by photodecomposition may only be significant during application (cf. Fig. 2.2). Likewise, oxidation is important in surface waters, and reduction in bottom sediments under anaerobic conditions (Triegel and Guo, 1994). However, the involvement of redox reactions in soil with respect to pesticide transformation is not well understood. Glass (1972) showed that the conversion of DDT (1,1,1-trichloro-2,2-bis(p-chlorophenyl) ethane to DDD, the 1,1-dichloro derivative, in anaerobic soils was correlated to the redox potential of the soil, and involved the formation of Fe^{3+}. This would indicate a free radical mechanism with Fe^{2+}, such as in Fe(II) porphyrin (Castro, 1964), serving as an electron donor. On the other hand, the abiotic transformation of pesticides in soil (and water) by hydrolysis is well documented (Khan, 1980; Wolfe et al., 1989).

As the name suggests, hydrolysis denotes the cleavage by water of an intramolecular bond, and the concomitant formation of a new bond involving the oxygen atom of water. The process may be facilitated by either acid or alkali, and is often promoted by the presence of OM (Stevenson, 1982). Clays can also enhance the hydrolytic conversion of many organic compounds, including pesticides (Yaron, 1978). A related reaction is the acid-catalyzed conversion of clay-sorbed pesticides (Lopez-Gonzalez and Valenzuela-Calahorro, 1970; Fusi et al., 1980). The high proton-donating (Brϕnsted) activity of clay surfaces derives from the dissociation of water associated with multivalent exchangeable cations (Mortland, 1970; Theng, 1982). As this type of acidity increases with a decrease in water content, protonation and transformation processes would be enhanced when the soil dries out.

Microbial

On the other hand, sorption by clays and OM can protect pesticides from biodegradation by making the molecules less accessible to microbes, and retarding their release into solution to below the biodegradable level (Scow, 1993). Nevertheless, microbial degradation is generally accepted as the principal pathway by which pesticides in topsoils are dissipated (Fig. 2.2). Once the compound leaches beyond the root zone, however, chemical pathways may become dominant. As might be expected, the rate of biodegradation is strongly influenced by temperature and water content (Hurle and Walker, 1980). However, OM content bears no simple relationship to degradation rate although microbial biomass increases with OM content (Theng et al., 1989). Rates of biodegradation, when modeled by a power equation, commonly follow greater than first-order kinetics. Other models have been proposed but no single one can adequately account for the fate of all organic species in so complex and heterogeneous an environment as soil (Alexander and Scow, 1989).

A pesticide that is quickly degraded to nontoxic metabolites has a much lower potential for polluting groundwater than a persistent one, although the former might be more mobile because it is weakly sorbed. Bowmer (1972), for example, found that in some Australian soils diuron had moved deeper down the profile after six annual applications than did simazine although less of the latter was sorbed. At the end of the experiment, diuron and simazine residues equaled 101 and 11% of annual application rate, respectively, indicating major differences in biodegradation propensity between the two herbicides.

Nor does microbial degradation necessarily reduce or remove the potential for environmental pollution because the metabolites produced can be as toxic as, or more so than, the parent compounds. The thio-oxidation of fenamiphos and several similar organophosphorus pesticides, for example, gives rise to their corresponding sulfoxide and sulfone analogues. Besides being at least as toxic, the metabolites are more polar, and hence potentially more mobile than the parent compounds. In an Australian soil fenamiphos was nearly completely transformed into fenamiphos sulfoxide within a week (Fig. 2.4). Since the latter was relatively persistent, it was the metabolite rather than the parent compound that provided the intended nematode control (Kookana et al., unpublished data). This parallels the behavior of DDT in aerobic New Zealand soils in that its breakdown product, DDE (1,1-dichloro-2,2-*bis*[p-chlorophenyl] ethylene) is more persistent, and becomes the primary residue unless there have been fresh applications of DDT (Boul, 1995).

The degradation of atrazine (Bowmer, 1973; Haigh and Ferris, 1991), sulfonylureas (Blacklow and Pheloung, 1992), and trifluralin (Jolly and Johnstone, 1994) in Australian

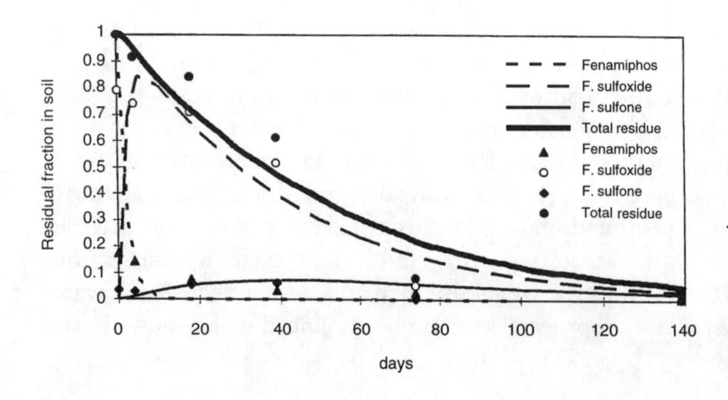

Figure 2.4. Degradation rates of fenamiphos and its two thio-oxidation products, fenamiphos sulfoxide and fenamiphos sulfone, in Bassendean sand (Western Australia) under laboratory conditions.

soils appeared to follow first-order kinetics. The same applies to the breakdown of metyalaxyl, linuron, metribuzin, and prometryn under laboratory conditions (Kookana et al., unpublished results). However, the degradation of atrazine and simazine in a sandy soil from Western Australia showed a rapid initial phase, followed by a slower first-order reaction that was pH-dependent (Walker, 1990). In allophane-rich soils from New Zealand, simazine degraded by first-order kinetics, and at application rates used for selective weed control was dissipated within 4–8 months (Rahman and Holland, 1985).

Long-term persistence has been reported for some pesticides in many Australian soils. For example, atrazine could have a half-life of more than 100 weeks under certain field conditions, and persist 13–20 times longer than in soils of Romania and the USA (Bowmer, 1991). Similarly, trifluralin has the potential for long-term persistence in droughty soils, although its activation energy for degradation is comparable to that observed overseas (Jolly and Johnstone, 1994). In the case of atrazine, the activation energy for degradation ranged from 12 to 25 kcal mol^{-1} as compared with a worldwide mean of 13 kcal mol^{-1} (Haigh and Ferris, 1991; Ferris and Haigh, 1993). In assessing the persistence of some organochlorines in the north coastal area of New South Wales, Harris (1987) found that heptachlor epoxide and dieldrin would probably have a half-life of less than 3 years, and dissipate faster under subtropical than temperate climates. In soils under cultivation and pasture, heptachlor residues declined by 20–30% in 8 months, and monthly cultivation slightly enhanced the dissipation rate.

The fate of sulfonylureas (chlorsulfuron, trisulfuron, metsulfuron-methyl) in Australian soils merits a separate description. Since their introduction in 1983 these compounds have gained rapid popularity with farmers in Australasia because of their effectiveness in controlling a wide spectrum of weeds at very low application rates (10-50 g ai ha^{-1}). In acidic soils these herbicides are largely degraded by chemical hydrolysis but in alkaline soils microbial decomposition is the preferred pathway. A large area of southern Australia, used for cereal production, is covered by alkaline soils. The pH increases with depth, sometimes reaching 10 (in water). Being weak acids, sulfonylureas exist as anionic (negatively charged) species under such pH conditions, and are not significantly sorbed by the soil. As a result, they move rapidly to subsurface layers where microbial activity is low and the pH is very alkaline, all of which are conducive to long persistence.

Intercropping with pasture/legumes is common practice in Australia as this counters pest problems associated with cereal monocultures and, at the same time, adds nitrogen to the soil through atmospheric N fixation by the legume. Since legumes are sensitive to sulfonylurea residues, their growth in alkaline soils can be adversely affected if recropping is carried out within a short period after herbicide application. However, in an allophanic soil from New Zealand (pH 5.7), and at application rates of up to 60 g ha^{-1}, the residues of a newly developed sulfonylurea herbicide (DPX-L5300) caused no significant reduction in the dry weight of lentils and subterranean clover planted 7 days after application. For an application rate of 120 g ha^{-1} a waiting period of 21 days was required before sensitive species could be planted (Rahman et al., 1988).

Volatilization

Volatilization can be a significant process of dissipation (Fig. 2.2). Indeed, for highly volatile compounds, losses by this means may exceed those through leaching and runoff (Taylor and Spencer, 1990). Irrespective of pesticide type, the rate of volatilization is high during the first few hours after application, and falls off rapidly as the soil surface becomes depleted. The rate then relates directly to the amount remaining in soil, and is controlled by the rate at which the compound moves from deeper layers to the surface (Glotfelty and Schomburg, 1989). Foliar and surface application would therefore enhance, whereas incorporation into the soil (e.g., by hoeing or irrigation) would retard, volatilization. In Australia gaseous losses of organochlorines from highly contaminated soils at cattle-dip sites may be important. The same may be true under dryland farming because as the soil begins to dry out, the residue of pesticides may diffuse or flow convectively ("wick effect") upward from the subsurface to the soil surface. As a general rule, volatilization is very low in dry soil because of strong sorption.

Transport

Many mathematical models have been developed for describing and predicting solute transport in soil under both saturated and unsaturated flow conditions. Process-based mechanistic models incorporating the classical convection-diffusion-dispersion equations are perhaps the most widely used in soil science (Brusseau and Rao, 1989; Jury and Ghodrati, 1989). Transport models may also be classified into screening, research, and management on the basis of their intended use (Addiscott and Wagenet, 1985).

Screening models evaluate the behavior of pesticides relative to each other under a standard set of soil and environmental conditions. By placing pesticides with similar behavior into groups, their relative potential for groundwater pollution may be assessed. Examples are the PESTicide ANalytical Solution (PESTAN) (Enfield et al., 1982), Behaviour Assessment Model (BAM) (Jury et al., 1983), Pesticide Mobility Index (PMI) (Rao et al., 1985), and Groundwater Ubiquity Score (GUS) (Gustafson, 1989). All these models are based on one-dimensional steady-state water flow using analytical solutions of the transport equation. In the PMI model, for instance, pesticide mobility is estimated through an attenuation factor, taking into account sorption (retardation) and degradation (half-life) processes. The factor ranges from 0 to 1; the larger the value, the higher the pollution potential. Using PMI interfaced with a geographical information system (GIS), Loague et al. (1990) have produced maps rating the qualitative potential for pesticide leaching to groundwater in Hawaii. The BAM model, modified to include OM decline with soil depth, was used by Kookana and Aylmore (1994) to assess the relative mobility of a large number of pesticides under irrigated horticulture in Australia. In environments favoring rapid leaching, nearly half of the pesticides were deemed to have the potential for leaching beyond a depth of 3 m. More details of this study are given in the next section.

Research models simulate dissipation processes to a higher degree of accuracy. Being more complex and data intensive, such models as the Pesticide Root Zone Model (PRZM) (Carsel et al., 1985) and the Leaching Estimation And CHemistry Model (LEACHM) (Hutson and Wagenet, 1992) are more valuable for understanding fundamental processes and their underlying mechanisms than for field applications.

The former model was used by Milne-Holme et al. (1991) to assess a regional risk of groundwater pollution by pesticides in New South Wales. Because local data are lacking, they used soil information from outside Australia, assuming that the generalized assessment would still be valid. Kookana et al. (unpublished data) used the LEACHM model in assessing the rate at which fenamiphos was transformed to its thioxidation products as well as its degradation rate in surface and subsurface layers of an Australian soil.

Management models are relatively less demanding of data acquisition and accuracy. A number of these, including CALculates Flow or CALF (Nicholls et al., 1982), Chemicals, Runoff, and Erosion from Agricultural Management Systems or CREAMS (Knisel, 1980), Productivity, Erosion, Runoff Functions to Evaluate Conservation Techniques or PERFECT (Littleboy et al., 1989), and the Walker-Barnes model (Walker and Barnes,

1981) have been used under Australian conditions with varying degrees of success. Thus for atrazine, CALF predicted longer persistence and less movement than was observed, possibly because bypass flow was not taken into account. The model has also been used to predict the persistence of trifluralin, diclofop-methyl and chlorsulfuron in some soils of Victoria (Haigh and Ferris, 1991; Ferris et al., 1994). Using the Walker-Barnes model, Bowmer (1991) found that concentrations of atrazine residues in two irrigated Australian soils were overestimated, possibly because the model did not account for losses by volatilization and anaerobic degradation. Similarly, Littleboy et al. (1989) have found deficiencies with CREAMS, a surface runoff model, but the similar PERFECT model was satisfactorily used in Queensland.

Soil solutions with an electrolyte concentration of < 0.01 M have traditionally been considered to be the major carrier of pesticides in soil. Compounds of low aqueous solubility are therefore not expected to leach through the profile. Recent studies have indicated, however, that suspended clay and organic colloids, and especially dissolved organic carbon (DOC), can substantially increase the solubility, and hence the mobility, of hydrophobic pesticides (Enfield and Yates, 1990; Mingelgrin and Gerstl, 1993b). This effect is very pronounced for compounds of high K_{ow}, such as DDT (Chiou et al., 1986). The concentration of DOC in soil solutions range from 10 to 1000 mg carbon/L, with a fulvic acid/humic acid ratio between 5:1 and 10:1 (Malcolm, 1993). Another important aspect of pesticide transport in field soils is that flow may occur through macropores, along ped faces and biopores, rather than through the soil matrix (Turco and Kladivko, 1994). Movement by preferential or bypass flow can have a dramatic effect on leaching depth and groundwater contamination. Macropores can also play a significant role in the gaseous transport of volatile pesticides.

PESTICIDE RESIDUES IN THE ENVIRONMENT

Soil, Food, Air, Sediments

Organochlorines (OCs) were extensively used in Australia to control agricultural and horticultural insect pests as well as parasites on cattle, sheep, pigs, and poultry. Although most OCs are now banned, their residues still occur in soil, and continue to cause problems of food and feed contamination. An outstanding example was the detection in mid-1987 of excessive OC residues in Australian export beef, following which thousands of soil samples from farms, deemed to be "at risk," were analyzed. Chlordane, DDT, dieldrin, and heptachlor were detected in 18.4, 39.6, 39.0, and 18.8 %, respectively, of 11,248 samples from Western Australia (EPAWA, 1989). Soils at cattle and sheep dip sites are massively contaminated with OCs and other pesticides. According to some estimates there are about 1700 cattle, and over 60,000 sheep, dip sites in northeastern New South Wales (NSW) alone. At the former sites the level of DDT contamination can be as high as 100 000 mg kg^{-1} soil, which is well above the threshold required to trigger investigation (Beard, 1993).

An analysis of soils under sugarcane from 69 farms in northern NSW showed an average dieldrin concentration of 0.12 mg kg^{-1} (Harris, 1987). In soils from banana plantations the value ranged from 0.06 to 3.2 mg kg^{-1}, depending on location. The NSW Environmental Protection Agency (NSWEPA) is currently evaluating the extent of soil contamination by OCs (and arsenic). A large proportion of contaminated land has either already been converted to residential use or is under great pressure for urban development.

BHC (β-benzenehexachloride), dieldrin, and heptachlor residues in beef cattle ap-

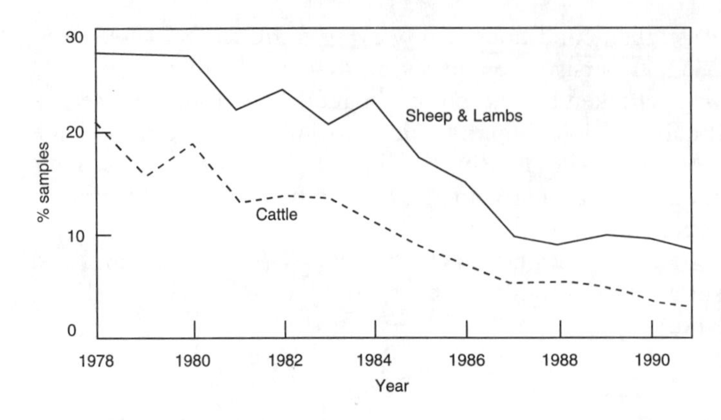

Figure 2.5. Long-term trends in levels of DDT residues for Australian cattle and sheep (adapted from Blackman, 1992).

peared to have come from grazing on land previously under horticulture and sugarcane in NSW (Wilson, 1987). A sensitivity study by Harradine and McDougall (1987) suggested that heptachlor residues in cattle fat could exceed the maximum allowable concentration of 0.2 mg kg^{-1} within a month of grazing treated land, and remained so for many years. Residue levels increased as pasture height decreased, indicating soil ingestion as the major means of pesticide intake by grazing animals. However, the proportion of animals with an MRL of 0.1 mg kg^{-1} in fat steadily declined from 1978 to 1991 (Blackman, 1992), while currently fewer than 1 in 5000 Australian cattle or sheep have residues above this value (Fig. 2.5).

In 1985–86 dairy milk was found to contain OC residues (King, 1987). A recent study by Stevens et al. (1993) on 128 breast-feeding women showed that dieldrin and heptachlor were taken in by 90 and 66% of their infants, respectively, in amounts greater than the ADI. On the other hand, only 5% of 604 vegetable samples surveyed in 1986-88 had OC residue concentrations above the MRL (Davies et al., 1993).

In New Zealand much DDT was used during the 1950s for controlling grass grub (*Costelytra zealandica* White) and porina caterpillar (*Wiseana* spp.) in pasture. Since then, its use was progressively restricted and finally banned in June 1970. Nevertheless, nearly 40% of lamb (meat) from Canterbury still contains more than 1 mg kg^{-1} of DDT residues (MacIntyre et al., 1989). Figure 2.6 shows that these residues, largely in the form of DDE, occur in the top 150 mm of the soil profile (Boul et al., 1994). When taken in by grazing animals through ingestion of contaminated topsoil (Harrison et al., 1970), DDT residues concentrate in the body and milk fat of sheep and cattle. As a result, the export of lamb to Europe may be restricted since the MRL for imported meat has been set at 1 mg kg^{-1} (ppm) by the European Community (Orchard et al., 1991).

With respect to the Pacific, DDT is banned in Fiji, chlordane in Fiji, Papua New Guinea, and Tonga, and dieldrin in Fiji and Tonga (Watts, 1993). DDT and other OCs are still heavily used in Vanuatu for mosquito control. Indeed, this source may contribute larger amounts of pesticides to the environment than does agriculture (Convard, 1993). The concentrations of some pesticide residues in Fiji soils are listed in Table 2.6. DDE levels here are three orders of magnitude lower than in many soils of Canterbury, New Zealand.

Little, if any, information is available on the presence and concentration of pesticides in the ambient air of Oceania. Some measurements on household air in Adelaide, South Australia (Gunn et al., 1992) and Sydney, New South Wales (Cantrell, 1993) indicated that median indoor levels of aldrin, chlordane, dieldrin, and heptachlor peaked after 7 days of application, and then declined gradually. After one year their concentrations de-

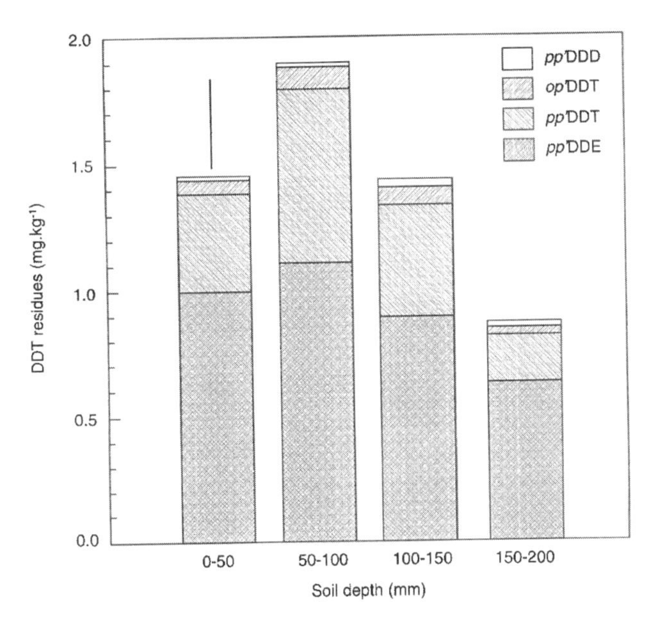

Figure 2.6. Concentrations of DDT and its principal degradation products, DDE and DDD, at different depths in a soil profile from Canterbury, New Zealand. Bar indicates LSD (P < 0.05) for total residues (Boul et al., 1994).

creased to 20-40% of peak levels. Di Marco (1993) has calculated that the dietary intake of OCs from ambient air by adults might approximate half of the total daily intake, which ranges from 0.1 to 0.28 ng kg^{-1} day^{-1} (depending on pesticide type), the other half coming from water.

The level of contamination by OCs of surficial sediments in and around Manukau Harbour, New Zealand, has been measured by Fox et al. (1988). Concentrations of DDT residues (DDT, DDE, DDD) ranged from 1.2 to 2.3 ng g^{-1}. These levels were comparable to those found in Galveston Bay, Texas, and San Francisco Bay, California, belying the notion that the marine environment of New Zealand is relatively unpolluted by pesticides. The total sediment load of chlordane was even higher, approximately twice that of DDT and its metabolites. Dieldrin was evenly distributed, ranging in concentration from 0.3 to 0.5 ng g^{-1}. The scarcity of amphipods at sites with the highest concentration of

Table 2.6. Concentration of Some Pesticide Residues in Fiji Soils[a]

Pesticide	Concentration (μg kg^{-1})
Aldrin	0.32
trans-Chlordane	0.27
o,p' - DDE	0.23
p,p' - DDE	0.38
Dieldrin	0.1
α-Endosulfan	0.1
Endrin	0.16
Hexachlorobenzene (HCB)	0.13
β-Benzenehexachloride (β-HCH)	0.44
α-Benzenehexachloride (α-HCH)	0.13
Heptachlor	1.4
Heptachlor epoxide	0.05
Methoxychlor	1.6

[a]Watts, 1993.

Table 2. 7. Concentrations (μg kg⁻¹) of Some Organochlorine Pesticides in Marine and River Sediments from Fiji, Tonga, and Vanuatu [a]

	Marine				River	
Pesticide	Suva Harbor Fiji	Lautoka Harbor, Fiji	Tonga	Vanuatu	Nadi, Fiji	Sigatoka, Fiji
Aldrin	0.25	nd [b]	nd	0.04	nd	0.24
cis-chlordane	0.12	nd	0.14	nd	0.11	nd
trans-chlordane	0.51	nd	2.22	nd	0.79	0.12
o,p'-DDD	4.1	nd	12.8	nd	1.7	0.56
p,p'-DDD	12.4	0.37	142.0	nd	6.3	1.4
o,p'-DDE	0.42	nd	0.95	nd	0.17	0.19
p,p'-DDE	4.3	0.17	17.1	0.06	4.3	0.42
p,p'-DDT	6.6	nd	854.0	0.31	1.5	2.8
Dieldrin	60.0	nd	nd	0.96	nd	nd
HCB	nd	nd	0.02	nd	nd	0.88
Heptachlor	0.57	nd	2.3	nd	nd	0.71
Heptachlor epoxide	0.65	nd	nd	0.61	0.13	0.03

[a]Watts, 1993.
[b]nd = not determined. Results of several samples have been combined; the value in the table was the highest found in the samples.

OCs (including polychlorinated biphenyls) suggests that these compounds accumulate in, and are toxic to, some marine species.

The available data for Pacific Island countries, collated by Watts (1993), indicate that OC levels in marine sediments are generally low except for DDT residues and dieldrin in Suva Harbour, Fiji (Table 2.7). Indeed, up to 142 and 854 ng g^{-1} of *p,p*′-DDD and *p,p*′-DDT were found in one sample from Tonga, taken near a residential area. This might be due to dumping, or heavy local use, of the insecticide (Morrison and Tulega, 1992).

Surface Waters

Because of the increased use of pesticides, concern is growing about the extent to which they may contaminate surface and groundwaters. In Australia as elsewhere there are numerous catchments close to areas of horticulture and intensive agriculture that are at risk of pollution by pesticides. Waters draining such catchments have recently been monitored. Runoff concentrations are strongly correlated with pesticide levels in the top 10 mm of the watershed soil, and total loss for each runoff event declines exponentially with time. Although the average loss by runoff (for atrazine) is less than 2% (Bowmer et al., 1994), it could be ecologically significant. On the other hand, soil loss in areas of high rainfall and intensive agriculture can be substantial. For sloping soils in the Darling Downs, Australia, the loss could be as high as 30–60 tonnes ha^{-1} yr^{-1}, whereas the figure for conventionally cultivated soils in the catchment is on the order of 10–15 tonnes ha^{-1} yr^{-1} (Kenway, 1993). By cutting back on soil loss, conservation tillage would also reduce the amount of pesticides that get into surface waters.

The NSW Department of Water Resources introduced an intensive pesticide monitoring system in 1990/91 under the Central and North Western Regional Water Quality Program, and in collaboration with water users of the Upper Darling Basin (Figure 2.7). In 1992/93 and 1993/94 seven and eight different pesticides, respectively, were detected in surface water samples (Cooper, 1994). Because of its wide usage in irrigated cotton

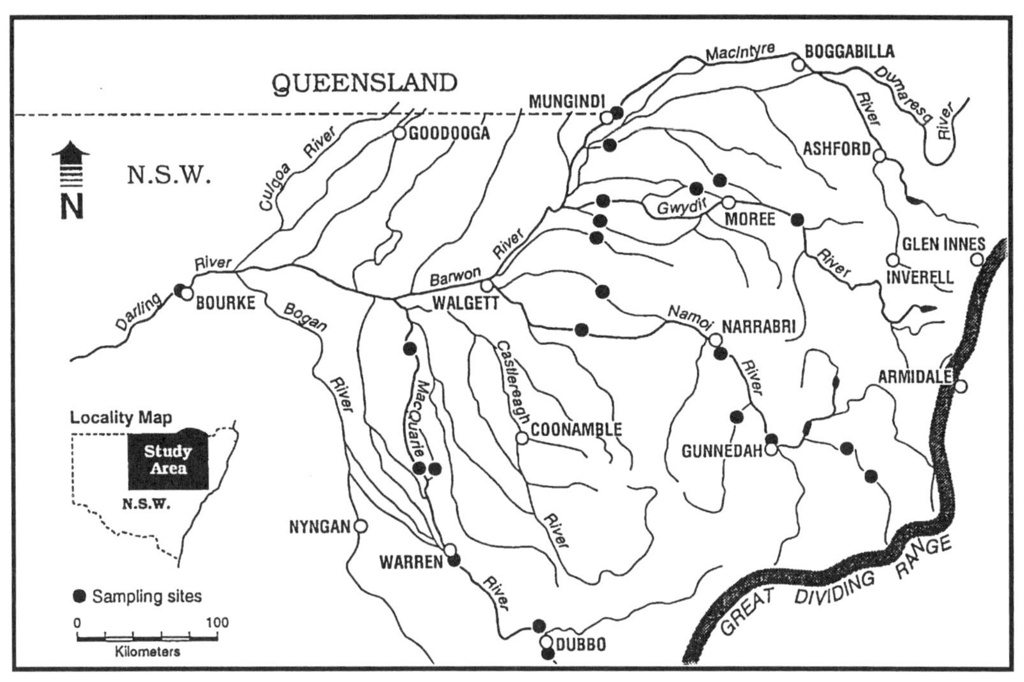

Figure 2.7. Study area covered under the Central and North Western Regional Water Quality Program of the New South Wales (NSW) Department of Water Resources (Cooper, 1994).

production, and to a lesser extent in oilseed crops, endosulfan is the most commonly detectable insecticide (Figure 2.8). The total concentration of endosulfan, however, has always been less than the NHMRC guideline for drinking water (40 μg L^{-1}) but often exceeds the ANZECC environmental guideline of 0.01 μg L^{-1}.

A large proportion of endosulfan and its principal degradation product, endosulfan sulfate, was associated with suspended sediments. The concentration in tailwater leaving cotton fields strongly depends on the timing of storm runoff after pesticide application. On this basis, endosulfan has been found in tailwaters several months after application (Cooper, 1994). The latter period is comparable with the half-life of 120-180 days that Ghadiri et al. (1993) measured in the laboratory.

The herbicides atrazine, diuron, prometryn, and fluometuron have been detected during the last three years of monitoring together with pendimethalin in 1992/93 and metalochlor and trifluralin in 1993/94. Atrazine, used for channel weed control, has been the most widely and frequently detected herbicide. However, the concentration of herbicide residues in surface waters from the region did not exceed the 1987 NHMRC drinking water guidelines (cf. Table 2.5). In Western Australia (WA) samples from farm dams, domestic rainwater tanks, and townwater supply dams in the eastern wheatbelt were analyzed in 1984/85 as a result of adverse publicity about herbicide usage in dryland agriculture. None of the samples, however, had detectable levels of diuron, 2,4-D, chlorsulfuron, trifluralin, triallate, paraquat, or glyphosate residues (EPAWA, 1989). In 1987 sediment and water samples from various estuaries and rivers were also analyzed for OCs and organophosphates (OPs). Except in one or two cases, residues of these insecticides in sediments were generally below detection level. Residues of OCs, but not of herbicides or

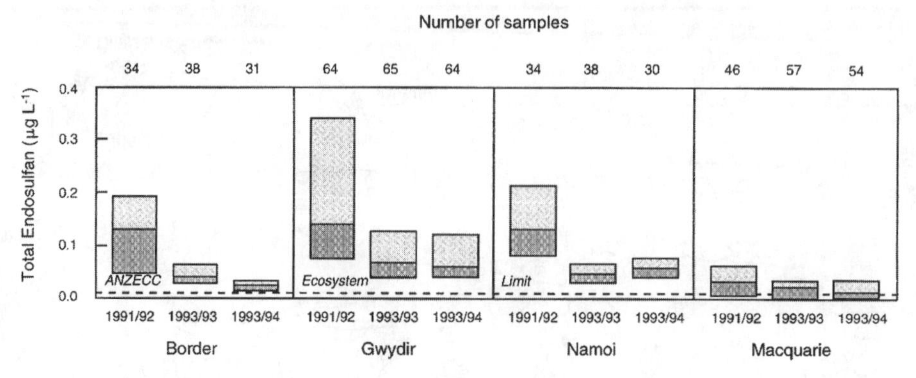

Figure 2.8. Concentration of total endosulfan in samples of surface waters collected in irrigated cotton-growing areas of New South Wales during March and November of the specified year. Each box represents the middle 50% of data collected in the season; the upper and lower boundaries in each box denote the 75th and 25th percentile values, respectively; the solid horizontal line is the median or 50th percentile value. The broken line indicates the ANZECC (environmental) limit (adapted from Cooper, 1994).

OPs, were detected in all water samples. In this respect, the Blackwood, Denmark, and Hay Rivers were the most contaminated (EPAWA, 1989). Earlier, Thurlow (1986) reported the presence of OCs in the Swan-Canning estuarine systems while Atkins (1982) detected residues of DDT, aldrin, dieldrin, and heptachlor in water samples from the Preston River. Although the level of dieldrin here was the highest among the rivers monitored, it had since fallen from 0.006 to 0.003 µg L^{-1} while the concentration of other OCs was comparable to that found in rivers elsewhere within the state (EPAWA, 1989). In surface water of a freshwater wetland in the Perth metropolitan region, Davis et al. (1990) measured concentrations of chlordane, dieldrin, and heptachlor in excess of the maximum permissible levels for the protection of aquatic life as recommended by USEPA. High levels of OCs were also found in the aquatic fauna, indicating that bioaccumulation was occurring.

In South Australia, pesticide residues in streams draining a horticultural catchment at Onkaparinga, supplying about 40% of Adelaide's water requirement, were monitored from July 1984 to January 1987 (Thoma, 1988). The survey showed that 83.5% of all water, and 100% of sediment samples tested contained pesticide residues, the concentration of which decreased in the following order: dachtal > propyzamide > DDT > endosulfan > chlorpyrifos > lindane > chlorthalonil. Dachtal was also the most frequently detected. On two occasions the concentration of DDT and chlorpyrifos residues in streams exceeded the MRL set by NHMRC. However, the quality of reservoir water was only marginally affected because of significant en route degradation and dilution.

In Tasmania an intensive sampling program was carried out from 1989 to 1992, involving 29 streams draining agricultural and forestry catchments (Davies et al., 1994). Of this total, 20 streams had detectable levels of several triazine residues. The median concentration for atrazine and simazine in 114 samples was 2.85 and 1.05 µg L^{-1}, respectively. The corresponding value for cyanazine, metribuzin, and propazine was less than 0.05 µg L^{-1} in all instances. The concentration of all herbicides, however, ranged over several orders of magnitude with a maximum of 53 mg L^{-1}. Streams draining (*Eucalyptus*) forests generally contained more pesticides than their agricultural counterparts, presumably because of differences in the (i) means of application (aerial spraying under forestry vs. ground spraying under agriculture), (ii) time of application (winter under forestry when the soil is moist vs. spring and early summer under agriculture), (iii) the nature of the compounds and their physicochemical interactions with soil (only atrazine and simazine of relatively low water solubility were used under forestry vs other triazines under agriculture). In addition, agricultural catchments have well-developed storage dams.

In assessing the fate of 2,4,5-T, applied by aerial spraying, in a steep hill country catchment under gorse and pasture in Waikato, New Zealand, Fox and Wilcock (1988) found that only 0.6% of the applied herbicide entered a nearby stream over 6 months, with 0.4% being washed out in the first storm. A mass balance, carried out 6 days after spraying, indicated that 80% of the applied amount was lost in spray drift.

Groundwaters

General

"Groundwater" refers to water that occupies and saturates pores and cracks in subsoils or rocks. If the saturated formation is sufficiently large in volume, porous, and can supply water to wells and springs, it is called an "aquifer." In 1988 the U.S. Environmental Protection Agency reported that the groundwaters of 26 states contained residues of 46 pesticides used in normal agriculture. Point source contamination of groundwaters has also been observed in many areas of the United States and Western Europe where pesticides (notably atrazine and alachlor) have been applied on a large scale (Hallberg, 1989; Leistra and Boesten, 1989). In the absence of preferential flow, the quantity that can leach through the vadose zone to groundwaters depends on the solubility, sorption characteristics (kinetics, hysteresis), and persistence (degradation, transformation) of the pesticide in question. Until more is known about the behavior and fate of pesticides in subsoils and aquifers, Nicholls (1993) has proposed that up to 0.1% of applied pesticides with a log K_{ow} value of less than 4, and a half-life in soil of more than 10 days, might be expected to reach shallow groundwaters.

Site Monitoring

Sporadic monitoring of pesticides in groundwaters has been carried out in various parts of Australia for some years. It is only recently, however, that a systematic approach has been made to assess groundwater quality on a national scale (Bauld et al., 1992). The available information, summarized below, indicates that although pesticides occur in many groundwaters, their concentrations are well below the MRL set by NHMRC.

In groundwaters of the Burdekin River delta in Northern Queensland, for example, the concentration of heptachlor and lindane residues did not exceed 3.2 and 4.8 ng L^{-1}, respectively (Brodie et al., 1984).

In Western Australia, Gerritse et al. (1988) analyzed for OC insecticides in 64 borehole samples taken along a transect through the Bassendean Sands. Aldrin, dieldrin, chlordane, and heptachlor (for termite control), and chlordane and heptachlor (against Argentinian ants) were applied at a rate of 50 and about 5 kg ha^{-1}, respectively. Despite a 10-fold difference in application rate, the concentration of OCs and their residues in the underlying groundwaters was below the NHMRC limit. Similarly, EPAWA (1989) reported that of some 130 borehole samples that have been analyzed since 1974, none contained pesticide residues in excess of the MRL. However, the number of samples having one or more pesticides at levels above the EPA criteria has risen significantly over recent years, a reflection of the increased use of pesticides in the agricultural and urban sectors. Effluent disposal on land or unlined ponds of agricultural chemicals has frequently led to contamination of groundwaters in their vicinity (Appleyard, 1993).

In New South Wales, Ang et al. (1989) analyzed for OCs, OPs, and other pesticides in drinking waters from five north coast shires where various crops and beef were produced. Despite the intensive usage of pesticides in this area, only 14 out of 110 samples from bores and springs contained pesticides at trace concentrations (0.05 to 0.5 μg L^{-1}).

Recent measurements by the NSW Department of Water Resources (Jiwan, pers. comm.) revealed the presence of atrazine in groundwaters at five sites in the Upper Namoi region. Of a total of 62 bores sampled between November 1992 and March 1993, only 8% contained atrazine. In some soil profiles atrazine was found at a depth of 20 m.

In Victoria, Bauld et al. (1992) sampled 10 observation wells in September 1990 and March 1991, analyzing their waters for more than 30 pesticides. Half of the samples contained atrazine and simazine at concentrations below the NHMRC draft guidelines. Apart from the occasional appearance of some degradation products of atrazine (deethylatrazine and deisopropylatrazine), no other pesticide species were detected.

Atrazine and simazine have also been found in groundwaters in South Australia (Stadter et al., 1992), and Bauld et al. (1992) reported the presence of atrazine and deethylatrazine in wells, located in irrigated vineyards.

In New Zealand, about 37% of the population depends wholly or partially on groundwater for drinking. Agriculture and industry also make extensive use of groundwaters. Because of the importance of groundwater and concerns about its quality, public water supplies have been monitored by the Ministry of Health at three-year intervals since 1987. Analysis of about 250 samples, carried out over the period April 1993 to June 1994, failed to detect any pesticides in groundwater (Close, 1994). A national survey of pesticides in groundwaters was initiated during 1990/91. To this end, 82 wells in shallow alluvial gravel, sand, basalt, and pumice aquifers, located in areas of high potential contamination, were sampled. Only nine wells, however, had detectable levels of pesticides although one contained 37 μg L^{-1} of atrazine, and another had 1.7 μg L^{-1} of procymidone (Close, 1993a). In an allophanic soil in Waikato, atrazine (from legitimate agricultural use) was detected at depths of 200–700 mm after 2 months of it being applied, and shortly afterward trace amounts appeared in the underlying groundwater (Rahman et al., 1994). Although pesticides have not previously been detected in groundwaters of Canterbury, V.R. Smith (1994) found simazine (0.01 to 1.09 μg L^{-1}) in samples from 38 wells in the Levels Plain and Temuka area. Atrazine and terbuthylazine were also detected at concentrations of 0.01 to 0.31 and 0.07 to 0.70 μg L^{-1} in samples from 18 and 5 wells, respectively. The groundwaters here are apparently more vulnerable to contamination than those of other areas of Canterbury because of the shallow depth of soil and water table. Further, recharge of the thin (< 12–m) unconfined gravel aquifer occurs mainly through irrigation, and pesticides are applied just before the irrigation season.

Among Pacific Island countries, Guam regularly monitors its groundwaters for pesticides but only in one area where golf courses are located (Watts, 1993). The presence of paraquat, lindane, and 2,4-D in Guam groundwaters has been reported (Morrison and Brodie, 1985). Because of lack of funds, very little monitoring is done in the other countries of the region. Trace concentrations (< 0.1 μg L^{-1}) of DDE and hexachlorobenzene were detected in some groundwater samples from Tonga (Watts, 1993).

Vulnerability Assessment

As mentioned earlier (Behavior and Fate in Soil section), a number of models have been used to assess the risk of groundwater pollution by pesticides on a catchment or regional scale in Australia and elsewhere. In addition to these, Aller et al. (1987) have proposed a system, called DRASTIC, which incorporates seven general variables, *viz.*, depth to water table, recharge rate of the aquifer, aquifer media, soil properties, topography (slope),

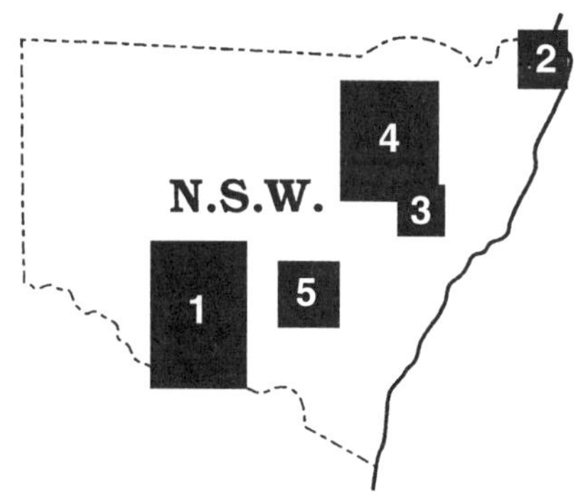

Figure 2.9. Map showing regions in New South Wales, Australia where vulnerability assessments of groundwater contamination by pesticides have been carried out. Each black box is drawn in proportion to the area of the specified regions (after Finlayson, 1989).

impact of the vadose zone, and hydraulic conductivity of the media. Each variable is assigned a score and a weighting, the products of which are added to yield an overall index of vulnerability to pollution. In Australia the DRASTIC system was tested by Barber et al. (1993) with respect to the Peel catchment in northern NSW. A vulnerability map was obtained using spatial data derived from Geographic Information System, and evaluated by sampling 130 bores for nitrate under four different land use categories. However, the system appeared to be of limited applicability due to the subjectivity of the assessment, the lack of data input, and the use of nitrate as a probe. Earlier, Finlayson (1989) assessed the vulnerability to pesticide contamination of groundwaters in five different groundwater regions of NSW, *viz.*, Murray Riverina, North Coast/Richmond River Valley, Upper Hunter, Namoi/Gwydir, and Lachlan (Fig. 2.9). A risk assessment analysis using the PRZM model suggests that the groundwater of the Richmond River Valley, which receives a high rainfall and is under intensive agriculture, is most vulnerable to contamination. Computer simulations, using 2,4-D and lindane, indicated that under a "worst-case scenario" (hydraulic class comprising deep sands with low runoff potential) one in 40 years would see approximately 70% of these pesticides leaching to a depth of 1.8 m.

In Western Australia, Kookana and Aylmore (1994) used the BAM model of Jury et al. (1987) to assess the vulnerability to pollution by a number of pesticides of groundwaters underlying a sandy soil in the Swan Coastal Plain. Hydrological parameters were estimated from local climatic and soil data, while sorption (K_{oc}) and persistence (half-life) values were obtained from the literature. Assuming a uniform distribution of OM content throughout the profile, the model predicts that 20 out of the 40 pesticides evaluated have the potential to leach below a depth of 3 m in excess of 0.1% of the applied amount. However, if the OM content was assumed to decrease exponentially with depth, 30 of the pesticides would reach groundwater and in much larger amounts. When the behavior of six pesticides was evaluated using site-specific values for K_{oc} and half-life, the concentration of some compounds leaching to groundwater was markedly different from what would be predicted on the basis of their corresponding literature values.

In areas of Victoria (Vic) and South Australia (SA) where cereals are grown under dryland farming conditions, pesticides would generally not be expected to leach much beyond the rooting zone. However, some modern ionic pesticides may be appreciably leached if sufficient water is available. A case in point are sulfonylureas, which behave as anions in alkaline soils and would move like chloride or nitrate down the profile. The potential for leaching is further increased by the persistence of these compounds at alkaline pH. At Walpeup (Vic) 55 mm of rain was sufficient to leach trisulfuron to a depth of 300–400 mm. Model simulations using chlorsulfuron indicated that leaching might be significant at Clare (SA), intermittent at Lameroo (SA), and negligible at Tamworth (NSW) and Minnipa (SA). At Clare, chlorsulfuron from annual applications was predicted to leach at a rate of 2 g ha^{-1} yr^{-1} (Ferris, 1993).

In New Zealand, Close (1993b) has developed an index of potential groundwater pollution based on the usage, mobility, and persistence of the pesticides involved, combining these ranking factors with the DRASTIC vulnerability score. Seventeen regions with a history of significant pesticide applications, and a well-characterized groundwater system, were selected. Of this number the Poverty Bay, Te Puke, and Motueka regions, used dominantly for growing kiwifruit and pipfruit, were assessed to have a high pollution potential.

MANAGEMENT AND REMEDIATION OPTIONS

Surface or runoff waters containing pesticides in dissolved form or attached to solids (soil and clay particles) can be filtered using buffer strips, which are strips of vegetation or grassed waterways draining fields. Riparian vegetation and wetlands, designed to delay and purify water coming to rivers and streams, may also be regarded as buffer zones.

Although buffer strips can protect streams from contamination by nutrients and pesticides and are commonly used in forestry, their effectiveness in Australia has not been closely examined (Barling and Moore, 1993). In measuring the level of pesticides in water draining from 20 *Eucalyptus nitens* plantations in Tasmania, Barton and Davies (1993) found a negative correlation between concentration on the day of spraying and buffer strip width. On this basis, a 30- and 50-m wide strip for atrazine and pyrethoids, respectively, would be required to minimize the short-term impact of these compounds on stream ecology. This accords with the finding that 30 m is the generally recommended width for stream buffers in forested areas (Clinnick, 1985). Soil conservation officers in Victoria, however, frequently use a slope factor in determining the width of buffer strips; that is, $W = 8 + 0.6S$ where W and S are the width (meters) and slope (%), respectively (Trimble and Sartz, 1957).

Several studies in New Zealand have supported the usefulness of buffer strips and riparian vegetation in reducing contaminants loading into streams. For example, C.M. Smith (1989) found that the concentrations of sediment, particulate phosphorus, and nitrate nitrogen in runoff water was significantly lower for retired pasture than for its grazed counterpart. Similarly, McColl (1978) noted that nutrient concentrations in runoff were inversely related to grass length.

Riparian wetlands and floodplains can also serve effectively as zones for sediment and nutrient deposition (Cooper et al., 1987). Similarly, storage dams and wetlands can act as sinks (Buser, 1990), providing time for pesticides to degrade or be removed by sorption to sediment particles. The work of Tuite (unpublished data) in NSW indicated that the total concentration of endosulfan in water could be halved in 17 days, mainly through transfer from the water column to sediment. Davies et al. (1994) suggested that well-developed storage dams between sections of streams draining a cropped catchment contributed to the relatively low frequency of stream contamination by triazines.

The abiotic remediation of lowly contaminated soils on a large volume is difficult and expensive. In Australia the only economically viable means is by encapsulation or landfill disposal (Davies et al., 1993) while in situ burial may be possible in rural areas. For soils contaminated with OCs in northern NSW, the common method of reducing pesticide levels is by dilution through deep plowing. In this way, Harris (1987) was able to achieve a 40% reduction in heptachlor levels. Management strategies for OC-contaminated land, used for grazing, have been developed by the Western Australian Department of Agriculture. These include fencing, planned grazing, and implementing alternative land-use systems (e.g., agroforestry, horticulture).

In New Zealand, Boul et al. (1994) showed that long-term irrigation and superphosphate application could significantly reduce the level of DDT residues in a contaminated soil. They suggested that this occurred through the enhanced degradation of *p,p'*-DDT via *p,p'*-DDD at anaerobic microsites. In line with this suggestion, Burke and Theng (1994) found that under laboratory conditions DDT was rapidly decomposed to DDD when the soil was flooded and amended with shredded grass. Since DDE is the principal residue in soil (cf. Fig. 2.6), this remediation option will only be practical where DDT makes up a significant proportion of total residues. Using 100–mm soil columns, Parfitt (1994) found that up to half of the total DDT residues in a contaminated soil could be leached with surfactants (Triton X-100, polypropylene glycolethoxylate). The finding, however, has not been tested under field conditions.

Other methods for remediating contaminated soils include solvent extraction, followed by base-catalyzed decomposition of the contaminant, electrokinetic soil processing, in-situ soil vitrification, and biological degradation (Beard, 1993; Chong, 1994). However, the applicability and commercial viability of these technologies have not been assessed.

Bioremediation is an attractive and promising option especially for soils contaminated at point source. James and Hogg (1994) reported that composting of a coarse silty soil from New Zealand, contaminated with OCs, led to a reduction of up to 91 and 9% in the concentration of DDT and dieldrin, respectively, over six weeks. For DDT a similar dissipation rate was observed under both aerobic and anoxic conditions. The level of DDD increased over the same period, suggesting partial biodegradation via dichlorobenzophenone. Composting was also effective in reducing the concentration of lindane and aldrin.

In this connection, the potential usefulness of methanotrophic bacteria for remediating contaminated soils in New Zealand has received attention. By producing a methane mono-oxygenase enzyme, these microbes are capable of dehalogenating and oxidizing a wide range of OC compounds. Enzyme production may be stimulated by adding a copper chelate (Schipper and Burke, 1994), or by injecting an organic waste into the subsoil (Thiele et al., 1994).

The establishment in soil of the appropriate degrading or assimilating microbes is an important aspect of the bioremediation of contaminated sites. Using particle size fractionation, Nishiyama et al. (1992; 1993) have been able to identify the microhabitat for growth, death, and survival in soil of *Sphingomonas paucimobilis*, a lindane-assimilating bacterium. Its survival is apparently related to association with soil aggregates of a certain size. The interaction of bacteria with soil particle surfaces together with the physiochemical basis of bacterial survival in soil have been discussed in a recent review by Theng and Orchard (1995).

Perhaps a better approach to reducing environmental pollution is to use fewer types, or smaller amounts of pesticides, in combination with biological means of pest control. In Australia, strategic grazing with sheep can reduce seed production by a number of weeds. Weeds may also be controlled by including certain crops in rotation (Combellack, 1992) while in areas under horticulture, and in home gardens, mulching is often effective. Sometimes a simple change in application time can markedly reduce pesticide consumption as in spraying against leaf rollers and codling moths in orchards of Hawke's Bay and Nelson, New Zealand (MacIntyre et al., 1989).

Biotechnological approaches to controlling diseases and pests of plants and animals require a large and long-term investment in basic research, the outcome of which is by no means assured or generally applicable (MacIntyre et al., 1989). They also pointed out that integrated pest management (IPM) solutions to crop pest problems must be developed and tested within each country, and cannot be bought "off-the-shelf" from outside, as can pesticides. This is because the animal and plant pests within their respective ecosystems are unique. Whatever appropriate solution may emerge, a reduction in pesticide usage, without compromising crop yield and quality, would be of enormous benefit to all countries in Oceania both financially and in terms of environmental and human health.

SUMMARY AND CONCLUSIONS

Pesticides have been applied in Oceania for a very long time to increase agricultural and horticultural productivity, but the economic benefit of their usage is difficult to quantify. The widespread and increasing usage of pesticides is of public concern although its impact on the environment of Australasia (Australia and New Zealand) is not always visible, at least over the short term. On the other hand, the ecosystem of most Pacific Island countries is intrinsically fragile and vulnerable to environmental change because the soils here are thin and porous, the land area is limited, and biodiversity is low.

Both Australia and New Zealand have enacted regulations and statutes for the registration, sale, and usage of pesticides, while many Pacific Island countries have endorsed the FAO Code of Conduct on the Distribution and Use of Pesticides. There are also guidelines for public health, and standards for (allowable) pesticide levels in food, drinking water, and soil. However, the appropriate legislations are by no means uniform or enforceable throughout Oceania.

As soil is a major sink of applied pesticides, the behavior and fate of these compounds in soil are discussed in terms of sorption, degradation/transformation, volatilization, and movement. In common with global trends, the sorption of nonionic pesticides is highly correlated with the organic matter (OM) content of the soil, indicating that such compounds are largely partitioned into OM. However, for subsoils and soils low in OM, this close relationship no longer holds because clay and mineral surfaces now play a major part in sorption. For ionic pesticides, sorption is dependent on the charge characteristics of the soil which, in turn, are determined by the pH and ionic strength of the ambient solution. Among the other pathways of dissipation, only the microbial degradation and transformation of certain pesticides (e.g., atrazine, simazine, sulfonylureas) are described at some length. As far as transport in soil is concerned, several models have been developed and used to assess the relative mobility, and potential for leaching, of different pesticides.

Pesticide residues in the environment of Oceania are described largely in terms of the occurrence of organochlorines in soil, food, and sediment. Detectable concentrations of several pesticides have also been measured in surface and groundwaters of Australia, New Zealand, and some Pacific Islands countries. In some areas of Australasia the vulnerability to pollution of groundwaters by some pesticides has been assessed using available transport (leaching) models.

Management options would include the construction of buffer strips, together with riparian vegetation and wetlands, to protect streams from contamination by pesticides. In some instances, the landfill disposal, and in-situ burial, of contaminated soils may be considered. Although bioremediation is an attractive and promising means of cleaning up soils, especially those that are contaminated at point source, its effectiveness has not been widely investigated. In the long term, however, the use of fewer types of pesticides, and in reduced amounts, in combination with biological pest control may well be the preferred management option for Oceania.

ACKNOWLEDGMENTS

We are grateful to numerous persons and organizations for copies of papers and reports, and in some cases, for unpublished data. Many colleagues assisted in finding information and references to the scattered literature. One of us (BKGT) acknowledges the Foundation of Research, Science and Technology (FRST) for financial support, and J. Williams and A. Habets for typing assistance. RSK wishes to thank Bruce Cooper, NSW Department of Land and Water Conservation (Sydney) for supplying unpublished data.

REFERENCES

Addiscott, T.M. and R.J. Wagenet. Concepts of Solute Leaching in Soils, A Review of Modelling Approaches. *J. Soil Sci.*, 36, pp. 411–424, 1985.

Alexander, M. and K.M. Scow. Kinetics of Biodegradation in Soil, in *Reactions and Movements of Organic Chemicals in Soils*, Sawhney, B.L. and Brown, K., Eds., Soil Science Society of America, Madison, WI, pp. 243–269, 1989.

Aller, L., T. Bennett, J. Lehr, R. Petty, and G. Hackett. *DRASTIC: A Standardized System for Evaluating Groundwater Pollution Using Hydrogeological Settings.* EPA/600/2-87/035, USEPA, Athens, GA, 1987.

Ang, C., K. Meleady, and L. Wallace. Pesticide Residues in Drinking Water in the North Coast Region of New South Wales, Australia, 1986–87. *Bull. Environ. Contam. Toxicol.*, 42, pp. 595–602, 1989.

ANZECC/NHMRC. (1992). *Australian Water Quality Guidelines for Marine and Fresh Water.* National Water Quality Management Strategy.

Appleyard, S.J. Resourcing the Investigation and Management of Groundwater Contamination in Western Australia: A Case Study of Contamination from Herbicide Manufacture near Perth. *AGSO J. Aust. Geol. Geophys.*, 14, pp. 177–182, 1993.

Asian Development Bank. *Handbook on the Use of Pesticides in the Asia-Pacific Region.* Manila, Philippines, 1987.

Atkins, R.P. (1982). *Organochlorine Pesticide Residues in the Preston River, Western Australia.* Report by the Waterways Commission, Perth, Australia.

Australian Consumer Association Conference. *Pesticides Charter. Towards a National Food Policy—Working Papers,* Australian Consumers Association Conference, Canberra, 1991, pp. 21–24.

Bailey, G.W. and J.L. White. Factors Influencing the Adsorption, Desorption, and Movement of Pesticides in Soils. *Residue Rev.*, 32, pp. 29–92, 1970.

Barber, C., L.E. Bates, R. Barron, and H. Allison. Assessment of the Relative Vulnerability of Groundwater to Pollution, A Review and Background Paper for the Conference Workshop on Vulnerability Assessment. *AGSO J. Aust. Geol. Geophys.*, 14, pp. 147–154, 1993.

Barling, R. and I.D. Moore. The Role of Buffer Strips in the Management of Water-way Pollution. *LWRRDC Occasional Paper No. 01/92*, 1993, pp. 1–45.

Barton, J.L. and P.E. Davies. Buffer Strips and Streamwater Contamination by Atrazine and Pyrethroids Aerially Applied to *Eucalyptus nitens* Plantations. *Aust. Forestry*, 56, pp. 201–210, 1993.

Bauld, J., W.R. Evans, and M.W. Sandstrom. Groundwater Quality Under Irrigated Agriculture, Murray Basin, Southeastern Australia, in *Groundwater and Environmen. Proceedings of International Workshop on Groundwater and Environment,* Fei Jin, Ed., Seismological Press, Beijing, pp. 447–457, 1992.

Beard, J. The Evaluation of DDT-Contaminated Land in New South Wales, in *Proceedings of the Second National Workshop on the Health Risk Assessment and Management of Contaminated Sites,* Langley, A. and van Alphen, M., Eds., South Australian Health Commission, Adelaide, pp. 119–122, 1993.

Beck, A.J., A.E. Johnston, and K.C. Jones. Movement of Nonionic Organic Chemicals in Agricultural Soils. *Crit. Rev. Environ. Sci. Technol.*, 23, pp. 219–248, 1993.

Bentley, K. Harmonising Health and Environment Criteria, Health Based Criteria, in *Proceedings of the Second National Workshop on the Health Risk Assessment and Management of Contaminated Sites,* Langley, A. and van Alphen, M., Eds., South Australian Health Commission, Adelaide, pp. 1–3, 1993.

Blacklow, W.M. and P.C. Pheloung. Sulfonylurea Herbicides Applied to Acidic Sandy Soils;

Movement, Persistence, and Activity Within the Growing Season. *Aust. J. Agric. Res.*, 43, pp. 1157–1168, 1992.
Blackman, N. DDT: Where Has It Gone?, in *Rural Resources Interface*, 1992, pp. 14–16.
Blakeley, R. Environmental Protection and Pest Management: A Conundrum, in *Plant Protection: Costs, Benefits and Trade Implications,* Suckling, D. M. and Popay, A. J., Eds., New Zealand Plant Protection Society, Rotorua, pp. 120–129, 1993.
Boul, H.L. DDT Residues in the Environment: A Review with a New Zealand Perspective. *New Zealand J. Agric. Res.*, 38, pp. 257–277, 1995.
Boul, H.L., M.L. Garnham, D. Hucker, D. Baird, and J. Aislabie. Influence of Agricultural Practices on the Levels of DDT and its Residues in Soil. *Environ. Sci. Technol.*, 28, pp. 1397–1402, 1994.
Bower, C.C., L.J. Penrose, and K. Dodds. A Practical Approach to Pesticide Reduction on Apple Crops Using Supervised Pest and Disease Control - Preliminary Results and Problems. *Plant Prot. Q.*, 8, pp. 57–62, 1993.
Bowmer, K.H. Atrazine Adsorption and Control of *Echinocola* in Australian Soils. *Proc. Weed Soc. NSW,* 4, pp. 12–20, 1971.
Bowmer, K.H. Aspects of the Long-term Use of Herbicides, Measurement of Residues of Diuron and Simazine in an Orchard Soil. *Aust. J. Expt. Agric. Anim. Husb.*, 12, pp. 535–543, 1972.
Bowmer, K.H. The Effect of Temperature and Soil Type on Atrazine Persistence, in *Proceedings of the Fourth Asian-Pacific Weed Science Society Conference,* Rotorua, New Zealand, 1973, pp. 242–246.
Bowmer, K.H. Atrazine Persistence and Toxicity in Two Irrigated Soils of Australia. *Aust. J. Soil Res.*, 29, pp. 339–350, 1991.
Bowmer, K.H., W. Korth, T. Thomas, and G. McCorkelle. River Pollution with Agricultural Chemicals (Pesticides and Fertilizers), in *The Murrumbidgee, Past and Present*, Roberts, J. and Oliver, R., Eds., CSIRO Division of Water Resources, Griffith Laboratory, 1994, pp. 7–19.
Brady, N.C. Soils and the World Food Supplies, in *Transactions of the 13th International Congress of Soil Science,* Volume 1. Plenary Papers, pp. 61–79, 1986.
Briggs, G.G. A Simple Relationship Between Soil Adsorption of Organic Chemicals and Their Octanol/Water Partition Coefficients, in *Proceedings of the 7th British Insecticide and Fungicide Conference,* Volume 1, pp. 83–88, 1973.
Briggs, G.G. Adsorption of Pesticides by Some Australian soils. *Aust. J. Soil Res.*, 19, pp. 61–68, 1981.
Brodie, J.E., W.S. Hicks, G.N. Richards, and F.C. Thomas. Residues Related to Agricultural Chemicals in the Groundwater of the Burdekin River Delta, North Queensland. *Environ. Pollut. Ser.* B8, pp. 187–215, 1984.
Brusseau, M.L. and P.S.C. Rao. Sorption Nonideality During Organic Contaminant Transport in Porous Media. *CRC Crit. Rev. Environ. Control,* 19, pp. 33–99, 1989.
Brusseau, M.L., R.E Jessup, and P.S.C. Rao. Nonequilibrium Sorption of Organic Chemicals: Elucidation of Rate-limiting Processes. *Environ. Sci. Technol.*, 25, pp. 134–142, 1991.
Burke, C. and B. Theng. *Decomposition of DDT, DDE and DDD in a Soil Amended With Organic Material, Under Anaerobic Conditions.* Abstract, lst AgResearch/Landcare Research Pesticide Residue Workshop, Lincoln, New Zealand, 1994.
Buser, H.R. Atrazine and Other s-Triazine Herbicides in Lakes and in Rain in Switzerland. *Environ. Sci. Technol.*, 24, pp. 1049–1058, 1990.
Calvet, R. Adsorption-Desorption Phenomena, in *Interactions Between Herbicides and the Soil,* Hance R.J., Ed., Academic Press, London, 1980, pp. 1–30.
Cantrell, P. "Exploratory air monitoring studies to determine the fate of domestic termiticides." Masters thesis, University of New South Wales, 1993.
Carsel, R.F., L.A. Mulkey, M.N. Lorber, and L.B. Baskin. The Pesticide Root Zone Model (PRZM): A Procedure for Evaluating Pesticide Leaching Threats to Groundwater. *Ecol. Model.*, 390, pp. 46–69, 1985.
Castro, C.E. The Rapid Oxidation of Iron(II) Porphyrins by Alkyl Halides. A Possible Mode of Intoxication of Organisms by Alkyl Halides. *J. Am. Chem. Soc.*, 86, pp. 2310–2311, 1964.

Cheng, H.H. (Ed.) *Pesticides in the Soil Environment: Processes, Impacts, and Modelling.* Soil Science Society of America, Madison, WI, 1990.

Chiou, C.T. Partition and Adsorption on Soil and Mobility of Organic Pollutants and Pesticides, in *Toxic Organic Chemicals in Porous Media,* Gerstl, Z., Chen, Y., Mingelgrin, U., and Yaron, B., Eds., Springer-Verlag, Berlin, 1989, pp. 163–175.

Chiou, C.T., R.L. Malcolm, T.I. Brinton, and D.E. Kile. Water Solubility Enhancement of Some Organic Pollutants and Pesticides by Dissolved Humic and Fulvic Acids. *Environ. Sci. Technol.,* 20, pp. 502–508, 1986.

Chiou, C.T., L.J. Peters, and V.H. Freed. A Physical Concept of Soil-Water Equilibria for Nonionic Organic Compounds. *Science,* 206, pp. 831–832, 1979.

Chong, R. *Non-biological Methods of Toxic Materials Disposal.* Abstract, 1st AgResearch/Landcare Research Pesticide Residue Workshop, Lincoln, New Zealand, 1994.

Clinnick, P.F. Buffer Strip Management in Forest Operations, a Review. *Aust. Forestry,* 48, pp. 34–45, 1985.

Close, M.E. Assessment of Pesticide Contamination of Groundwater in New Zealand. 2. Results of Groundwater Sampling. *New Zealand J. Mar. Freshwater Res.,* 27, pp. 267–273, 1993a.

Close, M.E. Assessment of Pesticide Contamination of Groundwater in New Zealand. 1. Ranking of Regions for Potential Contamination. *New Zealand J. Mar. Freshwater Res.* 27, pp. 257–266, 1993b.

Close, M.E. *Pesticides in New Zealand's Groundwater? Proceedings 1st AgResearch/Landcare Research Pesticide Residue Workshop,* Lincoln, New Zealand, 1994.

Combellack, J.H. The Importance of Weeds and the Advantages and Disadvantages of Herbicide Use. *Plant Prot. Q.,* 4, pp. 14–32, 1989.

Combellack, J.H. An Appraisal of Opportunities to Reduce Herbicide Use. *Plant Prot. Q.,* 7, pp. 66–69, 1992.

Convard, N. *Land-based Pollutants Inventory for the South Pacific Region.* South Pacific Regional Environmental Programme, Apia, Western Samoa, 1993.

Cooper, B. Central and North West Regions Water Quality Program. *1993/94 Report on Pesticide Monitoring.* NSW Dept. of Water Resources TS94.087, 1994.

Cooper, J.R., J.W. Gilliam, R.B. Daniels, and W.P. Robarge. Riparian Areas as Filters for Agricultural Sediments. *Soil Sci. Soc. Am. J.,* 51, pp. 416–420, 1987.

Davies, J., P. Dupen, and R. McFarland. Review of Pesticide-contaminated Land in New South Wales, in *Proceedings of the Second National Workshop on the Health Risk Assessment and Management of Contaminated Sites,* Langley, A. and van Alphen, M., Eds., South Australia Health Commission, Adelaide, pp. 177–186, 1993.

Davies, P.E., L.S.J. Cook, and J.L. Barton. Triazine Herbicide Contamination of Tasmania Streams; Sources, Concentrations and Effects on Biota. *Aust. J. Mar. Freshwater Res.,* 45, pp. 209–226, 1994.

Davis, J., S.A. Halse, G. Ebell, J. Blyth, M. Garland, and A.M. Pinder. *Persistence and Movement of Pesticides in Urban Wetland.* Proceedings of the 29th Congress of the Australian Society of Limnology, Jabiru, Northern Territory, 1990.

DiMarco, P. The Assessment and Management of Organochlorine Termiticides, in *Proceedings of the Second National Workshop on the Health Risk Assessment and Management of Contaminated Sites,* Langley, A. and van Alphen, M., Eds., South Australia Health Commission, Adelaide, pp. 153–175, 1993.

Enfield, C.G. and S.R. Yates. Organic Chemical Transport to Groundwater, in *Pesticides in the Soil Environment: Processes, Impacts and Modelling,* Cheng, H.H., Ed., Soil Science Society of America, Madison, WI, 1990, pp. 79–101.

Enfield, C.G., R.F. Carsel, S.Z. Cohen, T. Phan, and D.M. Walters. Approximate Pollutant Transport to Groundwater. *Groundwater,* 20, pp. 711–722, 1982.

EPAWA. *Monitoring Pesticides—A Review.* The Environment Protection Authority, Perth, Western Australia, Bulletin No. 407, 1989.

Evans, G. and P. Rowland. Improving the Efficiency of Pesticide Use in Sweden, Denmark and The Netherlands. *Agric. Sci.,* pp. 46–48, 1993.

Ferris, I.G. and B.M. Haigh. Herbicide Persistence and Movement in Australian Soils. Implications for Agriculture, in *Pesticide Interaction in Crop Production,* Altman, J., Ed., CRC Press, Boca Raton, FL, 1993, pp. 134–160.

Ferris, I.G. A Risk. Assessment of Sulfonylurea Herbicides Leaching to Groundwater. *AGSO J. Aust. Geol. Geophys,* 14, pp. 297–302, 1993.

Ferris, I.G., R. Fawcet, P. Stork, B.M. Haigh, N. Pederson, and A. Rovira. Sulfonylurea Behaviour in Sodic Soils, in *Proceedings of the National Conference on Sodic Soils, November 1992, Adelaide,* Naidu, R., Ed., CSIRO Press, Melbourne, 1994.

Finlayson, D. "A risk assessment of groundwater contamination by pesticide residues in selected aquifers of New South Wales," Masters thesis, University of New South Wales, 1989.

Fox, M.E., D.S. Roper, and S.F. Thrush. Organochlorine Contaminants in Surficial Sediments of Manukau Harbour, New Zealand. *Mar. Pollut. Bull.,* 13, pp. 217–218, 1988.

Fox, M.E. and R.J. Wilcock. Soil, Water, and Vegetation Residues of 2, 4, 5-T Applied to a New Zealand Hillside. *New Zealand J. Agric. Res.,* 31, pp. 347–357, 1988.

Fusi, P., G.G. Ristori, and A. Malquori. Montmorillonite-Asulam Interactions. I. Catalytic Decomposition of Asulam Adsorbed on H- and Al- clay. *Clay Miner.,* 15, pp. 147–155, 1980.

Gerritse, R.G., C. Barber, and J. Adeney. *The Effects of Urbanization on the Quality of Groundwater in Bassendean Sands.* CSIRO, Division of Water Resources, 1988.

Gerstl, Z. Predicting the Mobility and Availability of Toxic Organic Chemicals, in *Toxic Organic Chemicals in Porous Media,* Gerstl, Z., Chen, Y., Mingelgrin, U., and Yaron, B., Eds., Springer-Verlag, Berlin, 1989, pp. 151–162.

Ghadiri, H., C.W. Rose, and D. Connell. *Characteristics and Field-to-Stream Transport of Pesticides and Other Agricultural Chemicals.* Final project report. Land and Water Resources, Research and Development Corporation, Canberra, ACT, 1993.

Glass, B.L. Relation Between the Degradation of DDT and the Iron Redox System in Soils. *J. Agric. Food Chem.,* 20, pp. 324–327, 1972.

Glotfelty, D.E. and C.J. Schomburg. Volatilization of Pesticides from Soil, in *Reactions and Movement of Organic Chemicals in Soils,* Sawhney, B.L. and Brown, K., Eds., Soil Science Society of America, Madison, WI, 1989, pp. 181–207.

Goring, C.A.I. and J.W. Hamaker, Eds. *Organic Chemicals in the Soil Environment,* Volume 1. Marcel Dekker, New York, 1972.

Green, R.E. and S.W. Karickhoff. Sorption Estimates for Modelling. in *Pesticides in the Soil Environment: Processes, Impacts, and Modelling,* Cheng, H.H., Ed., Soil Science Society of America, Madison, WI, 1990, pp. 80–101.

Greenland, D.J. Interactions Between Clays and Organic Compounds in Soils. II. Adsorption of Organic Compounds and its Effect on Soil Properties. *Soils Fert.* 28, pp. 521–532, 1965.

Gunn, R.T., D.L. Pisaniello, C. Hann, and J. Crea. Organochlorine Pesticide Exposure and Uptake Following Treatment of Domestic Premises, Data from the First 6 Months of Follow-up. *Int. J. Environ. Health Res.,* 2, pp. 52–59, 1992.

Gustafson, D.I. Groundwater Ubiquity Score, a Simple Method for Assessing Pesticide Leachability. *Environ. Toxicol. Chem.,* 8, pp. 339–357, 1989.

Haigh, B.M. and I.G. Ferris. Predicting Herbicide Persistence using the CALF Model, in *Modeling the Fate of Chemicals in the Environment,* Moore, I.D., Ed., CRES, ANU, Canberra, 1991, pp. 50–60.

Hallberg, G.R. Pesticide Pollution of Groundwater in the Humid United States. *Agric. Ecosys. Environ.,* 26, pp. 299–367, 1989.

Hamaker, J.W. and J.M. Thompson. Adsorption, in *Organic Chemicals in the Soil Environment,* Volume 1, Goring, C.A.I. and Hamaker, J.W., Eds., Marcel Dekker, New York, 1972, pp. 49–143.

Hance, R.J., Ed. *Interactions between Herbicides and the Soil.* Academic Press, London, 1980.

Harradine, I.R. and K.W. McDougall. Residues in Cattle Grazed on Land Contaminated with Heptachlor, in *Pesticide Seminar, North Coast Agricultural Institute, Wollongbar*, Doyle, B.J. and McDougall, K.W., Eds., NSW Agriculture and Fisheries, 1987, pp. 70–75.

Harris, C.R. Management Strategies for Organochlorine Insecticides in Agricultural Soils, in *Pes-*

ticide Seminar, North Coast Agricultural Institute, Wollongbar, Doyle, B.J. and McDougall, K.W. Eds., NSW Agriculture and Fisheries, 1987, pp. 33–65.

Harrison, D.L., J.C.M. Mol, and W.B. Healy. DDT Residues in Sheep from the Ingestion of Soil. *New Zealand J. Agric. Res.,* 13, pp. 664–672, 1970.

Hassett, J.J. and W.L. Banwart. The Sorption of Nonpolar Organics by Soils and Sediments, in *Reactions and Movement of Organic Chemicals in Soils,* Sawhney, B.L. and Brown, K. Eds., Soil Science Society of America, Madison, WI, 1989, pp. 31–44.

Hayes, M.H.B., M. Stacey, and J.M. Thompson. Adsorption of s-Triazine Herbicides by Soil Organic Matter Preparations, in *Isotopes and Radiation in Soil Organic Matter Studies.* International Atomic Energy Agency, Vienna, 1968.

Hicks, J. and D. Mowbray. *Draft Report for Greenpeace on Pesticides in the Pacific.* University of Papua, New Guinea, 1989.

Hurle, K. and A. Walker. Persistence and its Prediction, in *Interactions between Herbicides and the Soil,* Hance, R.J., Ed., Academic Press, London, 1980, pp. 83–122.

Hutson, J.L. and R.J. Wagenet. *LEACHM: A Process Based Model of Water and Solute Movement, Transformation, Plant Uptake and Chemical Reactions in the Unsaturated Zone.* Version 3, New York State College of Agriculture and Life Sciences. Cornell University, Ithaca, New York, 1992.

James, T.K. and D. Hogg. *Biodegradation of DDT Using Composting.* Abstract, 1st AgResearch/Landcare Research Pesticide Residue Workshop, Lincoln, New Zealand, 1994.

Jolly, A. and P. Johnstone. Degradation of Trifluralin in Three Victorian Soils Under Field and Laboratory Conditions. *Aust. J. Expt. Agric.,* 34, pp. 57–65, 1994.

Jury, W.A., D.D. Focht, and W.J. Farmer. Evaluation of Pesticide Groundwater Pollution Potential from Standard Indices of Soil-chemical Adsorption and Biodegradation. *J. Environ. Qual.,* 16, pp. 422–428, 1987.

Jury, W.A. and M. Ghodrati. Overview of Organic Chemical Environmental Fate and Transport Modeling Approaches, in *Reactions and Movement of Organic Chemicals in Soils,* Sawney, B.L. and Brown, K., Eds., Soil Science Society of America, Madison, WI, 1989, pp. 271–304.

Jury, W.A., W.F. Spencer, and W.J. Farmer. Behaviour Assessment Model for Trace Organics in Soil. I. Model Description. *J. Environ. Qual.,* 12, pp. 558-564, 1983.

Karickhoff, S.W., D.S. Brown, and T.A. Scott. Sorption of Hydrophobic Pollutants on Natural Sediments. *Water Res.,* 13, pp. 241–248, 1979.

Kenway, S.J., Ed. *Water Quality Management in the Condamine-Balonne-Culgoa Catchment.* Condamine-Balonne Water Committee, Dalby, Queensland, Australia, 1993.

Khan, S.U. *Pesticides in the Soil Environment.* Elsevier, Amsterdam, 1980.

King, S. Pesticide Residues in Milk in the Hunter Valley, in *Pesticide Seminar, North Coast Agricultural Institute, Wollongbar*, Doyle, B.J. and McDougall, K.W., Eds., NSW Agriculture and Fisheries, 1987, pp. 65–70.

Knisel, W.G., Ed. *CREAMS, a Field-scale Model for Chemicals, Runoff, and Erosion from Agricultural Management Systems.* USDA-SEA Conservation Research Report 16. U.S. Government Print Office, Washington, D.C., 1980.

Kookana, R.S. "Equilibrium and kinetic aspects of sorption-desorption, and mobility of pesticides in soils," Ph.D. thesis, University of Western Australia, 1989.

Kookana, R.S. and L.A.G. Aylmore. Retention and Release of Diquat and Paraquat Herbicides in Soils. *Aust. J. Soil Res.,* 31, pp. 97–109, 1993.

Kookana, R.S. and L.A.G. Aylmore. Estimating the Pollution Potential of Pesticides to Ground Water. *Aust. J. Soil Res.,* 32, pp. 1141–1155, 1994.

Kookana, R.S., L.A.G. Aylmore, and R.G. Gerritse. Time-Dependent Sorption of Pesticides During Transport in Soils. *Soil Sci.,* 154, pp. 214–225, 1992a.

Kookana, R.S., C. Phang, and L.A.G. Aylmore. Transformation and Degradation of Fenamiphos Nematicide and Its Metabolites in Soils. *Aust. J. Soil Res.* 35, pp. 753–761, 1997.

Kookana, R.S., R.G. Gerritse, and L.A.G. Aylmore. A Method for Studying Nonequilibrium Sorption During Transport of Pesticides in Soil. *Soil Sci.,* 154, pp. 344–349, 1992b.

Koskinen, W.C. and S.S. Harper. The Retention Process: Mechanisms, in *Pesticides in the Soil En-*

vironment: Processes, Impacts and Modelling, Cheng, H.H., Ed., Soil Science Society of America, Madison, WI, 1990, pp. 51–78.

LeBaron, H.M. Weed Science in the 1990s: Will It Be Forward or in Reverse? *Weed Technol.,* 4, pp. 671–689, 1990.

Leistra, M. and J.J.T.I. Boesten. Pesticide Contamination of Groundwater in Western Europe. *Agric. Ecosys. Environ.,* 26, pp. 369–389, 1989.

Littleboy, M., D.M. Silburn, D.M. Freebairn, D.R. Woodruff, and G.L. Hammer. *PERFECT-a Computer Simulation Model of Productivity Erosion Runoff Functions to Evaluate Conservation Techniques,* Queensland Department of Primary Industries, Bulletin QB89005, 1989.

Loague, K.M., R.E. Green, T.W. Giambelluca, T.C. Liang, and R.S. Yost. Impact of Uncertainty in Soil, Climatic, and Chemical Information in a Pesticide Leaching Assessment. *J. Contam. Hydrol.,* 5, pp. 171–194, 1990.

Lopez-Gonzalez, J. and C. Valenzuela-Calahorro. Associated Decomposition of DDT to DDE in the Diffusion of DDT on Homoionic Clays. *J. Agric. Food Chem.,* 18, pp. 520–523, 1970.

MacIntyre, A., N. Allison and D. Penman. *Pesticides: Issues and Options for New Zealand.* Ministry for the Environment, Wellington, 1989.

Malcolm, R.L. Concentration and Composition of Dissolved Organic Carbon in Soils, Streams, and Groundwaters, in *Organic Substances in Soil and Water: Natural Constituents and Their Influences on Contaminant Behaviour,* Beck, A.J, Jones, K.C., Hayes, M.H.B., and Mingelgrin, U., Eds., Royal Society of Chemistry, Cambridge, 1993, pp. 19–30.

McColl, R.H.S. Chemical Runoff from Pasture: The Influence of Fertilizer and Riparian Zones. *New Zealand J. Mar. Freshwater Res.,* 12, pp. 371–380, 1978.

Miller, G.C., V.R. Herbert, and W.W. Miller. Effect of Sunlight on Organic Contaminants at the Atmosphere-Soil Interface, in *Reactions and Movement of Organic Chemicals in Soils,* Sawhney, B.L. and Brown, K., Eds., Soil Science Society of America, Madison, WI, 1989, pp. 99–110.

Milne-Holme, W.A., D. Finlayson, and X. Yu. Generalised Risk Assessment of Ground Water Contamination by Pesticides, An Expert System Approach to PRZM, in *Modelling the Fate of Chemicals in the Environment,* Moore, I.D., Ed., CRES, ANU, Canberra, 1991, pp. 40–49.

Mingelgrin, U. and Z. Gerstl. Reevaluation of Partitioning as a Mechanism of Nonionic Chemicals Adsorption in Soils. *J. Environ. Qual.,* 12, pp. 1–11, 1983a.

Mingelgrin, U. and Z. Gerstl. A Unified Approach to the Interaction of Small Molecules with Macrospecies, in *Organic Substances in Soil and Water: Natural Constituents and Their Influences on Contaminant Behaviour,* Beck, A.J., Jones, K.C., Hayes, M.H.B., and Mingelgrin, U., Eds., Royal Society of Chemistry, Cambridge, 1993b, pp. 102–127.

Morrison, R.J. and J. Brodie. *Pollution Problems in the South Pacific: Fertilizers, Biocides, Water Supplies and Urban Wastes.* Environment and Resources in the Pacific, UNEP Regional Seas Reports and Studies No. 69, 1985.

Morrison, R.J. and L.C. Tulega. *The South Pacific Regional Marine Pollution Assessment and Control Programme (SPREP POL).* Progress Report presented to the SREP Intergovernmental Meeting, September 1992, Apia, Western Samoa, 1992.

Mortland, M.M. Clay-Organic Complexes and Interactions. *Adv. Agron.,* 22, pp. 75–117, 1970.

Mowbray, D.L. *Pesticide Use in the South Pacific.* UNEP Regional Seas Reports and Studies No. 89 and SPREP Topic Review, No. 26, 1988.

Mowbray, D.L. and J. Hicks. Environmental Considerations in the Evaluation of Pesticides—Special Considerations for South Pacific Countries. *FAO/SPC Regional Workshop on Implementation of the International Code of Conduct on the Distribution and Use of Pesticides,* Noumea, New Caledonia, 1989.

Murali, V. and L.A.G. Aylmore. No-flow Equilibration and Adsorption Dynamics During Ionic Transport in Soils. *Nature,* 283, pp. 467–469, 1980.

Murphy, E.M., J.M. Zachara, and S.C. Smith. Influence of Mineral-bound Humic Substances on the Sorption of Hydrophobic Organic Compounds. *Environ. Sci. Technol.,* 24, pp. 1507–1516, 1990.

Nicholls, P.H., R.H. Bromilow, and T.M. Addiscott. Measured and Simulated Behaviour of Fluometuron, Aldoxycarb and Chloride Ions in a Structured Soil. *Pestic. Sci.*, 13, pp. 475–483, 1982.

Nicholls, P.H. Organic Contaminants in Soils and Groundwaters, in *Organic Contaminants in the Environment,* Jones, K.C., Ed., Elsevier Applied Science, London and New York, 1993, pp. 87–132.

Nishiyama, M., K. Senoo, and S. Matsumoto. Establishment of γ - 1, 2, 3, 4, 5, 6 - Hexachlorocyclohexane-assimilating Bacterium, *Sphingomonas paucimobilis* Strain SS86, in Soil. *Soil Biol. Biochem*., 25, pp. 769–774, 1993.

Nishiyama, M., K. Senoo, H. Wada, and S. Matsumoto. Identification of Soil Micro-habitats for Growth, Death and Survival of a Bacterium, γ - 1, 2, 3, 4, 5, 6-Hexachlorocyclohexane-assimilating *Sphingomonas paucimobilis*, by Fractionation of Soil. *FEMS Microbiol. Ecol.*, 101, pp. 145–150, 1992.

Orchard, V.A., B.K.G. Theng, P.L. Searle, C.M. Burke, R.L. Parfitt, D. Donnell, and S. Harper. *Interim Report on DDT Residues in Soil and Experiments to Enhance Their Degradation Rates.* DSIR Land Resources Technical Record No. 58, Department of Scientific and Industrial Research, Lower Hutt, N.Z., 1991.

Parfitt, R.L. *Moving DDT Through Soils with Surfactants.* Abstract, lst AgResearch/Landcare Research Pesticide Residue Workshop, Lincoln, New Zealand, 1994.

Pignatello, J.J. Sorption Dnamics of Organic Compounds in Soils and Sediments, in *Reactions and Movement of Organic Chemicals in Soils,* Sawhney, B.L. and Brown, K., Eds., Soil Science Society of America, Madison, WI, 1989, pp. 45–80.

Pignatello, J.J. Recent Advances in Sorption Kinetics, in *Organic Substances in Soils and Waters,* Beck, A.J., Jones, K.C., Hayes, M.H.B., and Mingelgrin, U., Eds., Royal Society of Chemistry, Cambridge, 1993, pp. 128–140.

Pignatello, J.J. and L.Q. Huang. Sorptive Reversibility of Atrazine and Metolachlor Residues in Field Soil Samples. *J. Environ. Qual.*, 22, pp. 222–230, 1991.

Pimentel, D., Ed. *Handbook on Pest Management in Agriculture.* CRC Press, Boca Raton, FL, 1981.

Prendergast, P., personal communication.

Rahman, A. Effect of Soil Organic Matter on the Phytotoxicity of Soil-applied Herbicides-Glasshouse Studies. *New Zealand J. Expt. Agric.*, 4, pp. 85–88, 1976.

Rahman, A. and P.T. Holland. Persistence and Mobility of Simazine in Some New Zealand Soils. *New Zealand J. Expt. Agric.*, 13, pp. 59–65, 1985.

Rahman, A. and L.J. Matthews. Effect of Soil Organic Matter on the Phytotoxicity of Thirteen s-Triazine Herbicides. *Weed Sci.*, 27, pp. 158–161, 1979.

Rahman, A., T.K. James, and D.R. Lauren. *Atrazine Residues in Soil and Groundwater in the Waikato.* Abstract, lst AgResearch/Landcare Research Pesticide Residue Workshop, Lincoln, New Zealand, 1994.

Rahman, A., T.K. James, and J. Mortimer. Persistence and Mobility of the Sulfonylurea Herbicide DPX-L5300 in a New Zealand Volcanic Soil. *Proc. Int. Congr. Crop Protection,* 53, (3b), pp. 1463–1470, 1988.

Rao, P.S.C., A.G. Hornsby, and R.E. Jessup. Indices for Ranking the Potential for Pesticide Contamination of Goundwater. *Proc. Soil Crop. Sci. Soc. Fla.*, 44, pp. 1–8, 1985.

Rayment, G.E. and B.W. Simpson. Pesticide Audit for the Condamine-Balonne-Culgoa Catchment, in *Water Quality Management in the Condamine-Balonne-Culgoa Catchment,* Kenway, S.J., Ed., Condamine-Balonne Water Committee, Dalby, Queensland, Australia, 1993, pp. 102–166.

Rebhun, M., R.Kalabo, L. Grossman, J. Manka, and Ch. Rav-Acha. Sorption of Organics on Clay and Synthetic Humic-Clay Complexes Simulating Aquifer Processes. *Water Res.*, 26, pp. 79–84, 1992.

Rutherford, D.W., C.T. Chiou, and D.E. Kile. Influence of Soil Organic Matter Composition on the Partition of Organic Compounds. *Environ. Sci. Technol.*, 26, pp. 336–340, 1992.

Samuel, N. *Country Paper—Cook Islands.* FAO Regional Workshop on the Implementation of Prior Informed Consent. Manila, Philippines, 1991.

Sawhney, B.L. and K. Brown, Eds. *Reactions and Movement of Organic Chemicals in Soils.* Soil Science Society of America, Madison, WI, 1989.

Schipper, L. and C. Burke. *Potential for Pollutant Bioremediation in the Wetland Plant Rhizosphere.* Abstract, lst AgResearch/Landcare Research Pesticide Residue Workshop, Lincoln, New Zealand, 1994.

Schwarzenbach, R.P. and J. Westall. Transport of Nonpolar Organic Compounds from Surface Water to Groundwater. Laboratory Sorption Studies. *Environ. Sci. Technol.,* 15, pp. 1360–1367, 1981.

Scow, K.M. Effect of Sorption-Desorption and Diffusion Processes on the Kinetics of Biodegradation of Organic Chemicals in Soil, in *Sorption and Degradation of Pesticides and Organic Chemicals in Soil,* Soil Science Society of America, Madison, WI, 1993, pp. 73–114.

Senesi, N. Binding Mechanisms of Pesticides to Soil Humic Substances. *Sci. Tot. Environ.,* 123/124, pp. 63–76, 1992.

Singh, R., R.G. Gerritse, and L.A.G. Aylmore. Adsorption-Desorption Behaviour of Selected Pesticides in Some Western Australian Soils. *Aust. J. Soil Res.,* 28, pp. 227–243, 1990.

Smith, C.M. Riparian Pasture Retirement Effects on Sediment, Phosphorus, and Nitrogen in Channelised Surface Runoff from Pastures, *New Zealand J. Mar. Freshwater Res.,* 23, pp. 139–146, 1989.

Smith, V.R. *Groundwater Contamination by Triazine Pesticides, Levels Plain and Temuka, South Canterbury, New Zealand.* Abstract, lst AgResearch/Landcare Research Pesticide Residue Workshop, Lincoln, New Zealand, 1994.

Sparks, D. L. *Kinetics of Soil Chemical Processes.* Academic Press, London, 1989.

Stadter, F., A. Emmett, and P. Dillon. *Occurrence of Atrazine in Groundwater in the Southeast of South Australia.* Centre for Groundwater Studies Report No. 45, 1992.

Stevens, M.F., F.E. Ebell, and P. Pasil-Savona. Organochlorine Pesticides in Western Australian Nursing Mothers. *Med. J. Aust.,* 158, pp. 238–241, 1993.

Stevenson, F.J. *Humus Chemistry: Genesis, Composition, Reactions.* Wiley-Interscience, New York, 1982.

Tarilongi, B., personal communication.

Taylor, A.W. and W.F. Spencer. Volatilization and Vapour Transport Processes, in *Pesticides in the Soil Environment: Processes, Impacts and Modelling,* Cheng, H.H., Ed., Soil Science Society of America, Madison, WI, 1990, pp. 213–269.

Theng, B.K.G. *The Chemistry of Clay-Organic Reactions.* Adam Hilger, London, 1974.

Theng, B.K.G., Ed. *Soils with Variable Charge.* New Zealand Society of Soil Science, Lower Hutt, 1980.

Theng, B.K.G. Clay-Activated Organic Reactions, in *International Clay Conference 1981*, van Olphen, H. and Veniale, F., Eds., Elsevier, Amsterdam, 1982, pp. 197–238.

Theng, B.K.G. Clay-Humic Interactions and Soil Aggregate Stability, in *Soil Structure and Aggregate Stability,* Rengasamy, P., Ed., Institute for Irrigation and Salinity Research, Tatura, Victoria, Australia, 1986, pp. 32–73.

Theng, B.K.G. Soil Science in the Tropics—the Next 75 Years. *Soil Sci.,* 151, pp. 76–90, 1991.

Theng, B.K.G. and V.A. Orchard. Interactions of Clays with Microorganisms and Bacterial Survival in Soil: A Physicochemical Perspective, in *Environmental Impact of Soil Component Interactions*, Volume II, Huang, P.M., Berthelin, J., Bollag, J.-M., McGill, W.B., and Page, A.L., Eds., Lewis Publishers, CRC Press, Boca Raton, FL, 1995, pp. 123–143.

Theng, B.K.G. and K.R. Tate. Interactions of Clays with Soil Organic Constituents. *Clay Res.,* 8, pp. 1–10, 1989.

Theng, B.K.G., K.R. Tate, and P. Sollins. Constituents of Organic Matter in Temperate and Tropical Soils, in *Dynamics of Soil Organic Matter in Tropical Ecosystems,* Coleman, D.C., Oades, J.M., and Uehara, G., Eds., University of Hawaii. Honolulu, NifTAL Project, 1989, pp. 5–32.

Thiele, J.H., V. Meduna, and P. Greenwood. *Soil Microcosms for Remediation Studies with Methanotrophic Bacteria.* Abstract, Ist AgResearch/Landcare Research Pesticide Residue Workshop, Lincoln, New Zealand, 1994.

Thoma, K. *Pilot Survey of Pesticide Residues in Streams Draining a Horticultural Catchment, Pic-*

cadilly Valley, South Australia. South Australian Department of Agriculture Technical Paper No. 131, 1988.
Thurlow, B.H. *Swan-Canning Estuarine System: Environment, Use, and the Future.* Report No. 9, Waterways Commission, Perth, Western Australia, 1986.
Triegel, E.K. and L. Guo. Overview of the Fate of Pesticides in the Environment, Water Balance; Runoff vs. Leaching, in *Mechanisms of Pesticide Movement into Ground Water,* Honeycutt, R.C. and Schabacker, D.J., Eds., Lewis Publishers, CRC Press, Boca Raton, FL, 1994, pp. 1–13.
Trimble, G.R. and R.S. Sartz. How Far from a Stream Should a Logging Road be Located? *J. Forestry,* 55, pp. 339–341, 1957.
Turco, R.F. and E.J. Kladivko. Studies on Pesticide Mobility: Laboratory vs. Field, in *Mechanisms of Pesticide Movement into Ground Water,* Honeycutt, R.C. and Schabacker, D.J., Eds., Lewis Publishers, CRC Press, Boca Raton, FL, 1994, pp. 63–80.
Valentine, R.L. Nonbiological Degradation in Soil, in *Vadose Zone Modeling of Organic Pollutants,* Hem, S.M. and Meloncon, S., Eds., Lewis Publishers, Chelsea, MI, 1986, pp. 223–243.
Walker, A. and A. Barnes. Simulation of Herbicide Persistence in Soil, a Revised Computer Model. *Pestic. Sci.,* 12, pp. 123–135, 1981.
Walker, A. and D.V. Crawford. The Role of Organic Matter in Adsorption of the Triazine Herbicides by Soil, in *Isotopes and Radiation in Soil Organic Matter Studies,* International Atomic Energy Agency, Vienna, 1968, pp. 91–108.
Walker, S.R. "Movement, persistence, and activity of simazine and atrazine in sandy soils; lupin x wheat regions of Western Australia," Ph.D. thesis, University of Western Australia, 1990.
Watts, M. *Poisons in Paradise - Pesticides in the Pacific.* Greenpeace, Auckland, 1993.
Weber, J.B. and C.T. Miller. Organic Chemical Movement Over and Through Soil, in *Reactions and Movement of Organic Chemicals in Soils,* Sawhney, B.L. and Brown, K., Eds., Soil Science Society of America, Madison, WI, 1989, pp. 305–334.
Weber, J.B., S.B. Weed, and T.M. Ward. Adsorption of s-Triazines by Soil Organic Matter. *Weed Sci.,* 17, pp. 417–421, 1969.
Weed, S.B. and J.B. Weber. Pesticide-Organic Matter Interactions, in *Pesticides in Soil and Water,* Guenzi, W.D., Ed., Soil Science Society of America, Madison, WI, 1974, pp. 39–66.
Wilcock, R.J. *Patterns of Pesticide Use in New Zealand. Part 1. North Island 1985-1988.* Water Quality Centre Publication 15, 1989.
Wilcock, R.J. and M.E. Close. *Patterns of Pesticide Use in New Zealand. Part 2. South Island 1986-1989.* Water Quality Centre Publication 16, 1990.
Williams, T. Beating Grass Grub with its Old Enemy. *AgriSearch,* 3(4), pp. 14–15, 1988.
Wilson, M. BHC and Dieldrin Residues in Beef Cattle Grazing Land Previously Used for Sugar Cane Production, in *Pesticide Seminar North Coast Agricultural Institute, Wollongbar,* Doyle, B.J., and McDougall, K.W., Eds., NSW Agriculture and Fisheries, 1987, pp. 76–77.
Wolfe, N.L., M.E. Metwally, and A.E. Moftah. Hydrolytic Transformations of Organic Chemicals in the Environment, in *Reactions and Movement of Organic Chemicals in Soils,* Sawhney, B.L. and Brown, K., Eds., Soil Science Society of America, Madison, WI, 1989, pp. 229–242.
Yaron, B. Some Aspects of Surface Interactions of Clays with Organophosphorus Pesticides. *Soil Sci.,* 125, pp. 210–216, 1978.
Young, K. Agriculture Chemicals and Food – The Implications. *Agric. Sci.,* pp. 45–48, 1989.

CHAPTER 3

METAL POLLUTION OF SOIL AND GROUNDWATER AND REMEDIATION STRATEGIES IN JAPAN

Mitsuo Chino

INTRODUCTION

In the 1960s, heavy metal pollution problems occurred in Japan, which caused Itai-itai disease and Minamata disease. There were several reasons for this pollution.

1. Because Japan is a mountainous country, most of its rivers are swift (about 70% of the total area of Japan is mountainous area). These rivers carried waste materials from many small mines, which are mostly located in the mountain area (Fig. 3.1) to the plains, causing contamination of the agricultural lands all over Japan (Fig. 3.1).
2. Large portions of the population live on the plains, which occupy only about 30% of the total area. Industries, houses, and agricultural lands are close together in these small plains. Thus, the wastes from various industries easily contaminated agricultural lands.
3. The rapid industrialization in the 1960s discharged large amounts of heavy-metal-containing waste.
4. Most Japanese eat rice every day, and about half of the arable land consists of paddy fields, which require irrigation water from rivers flowing from the mountains. The rivers carried the waste material from the mining areas to the paddy fields in the plains.

In 1970, Japan revised the Basic Law for Environmental Pollution Control at the 64th Extraordinary Diet Meeting and added soil pollution as one of the typical pollutions along with water and air pollution. According to this revision, the Agricultural Land Soil Pollution Prevention Law was established and several guidelines and measures were proposed to prevent metal pollution of agricultural land and to remediate polluted soil. This law and the guidelines were effective in preventing soil pollution, and now the problem of heavy-metal pollution from mining or related industries (smelting, refining, metal plating, etc.) is diminished. In the future, problems of heavy-metal pollution of soil may be related largely to the application of sewage sludge.

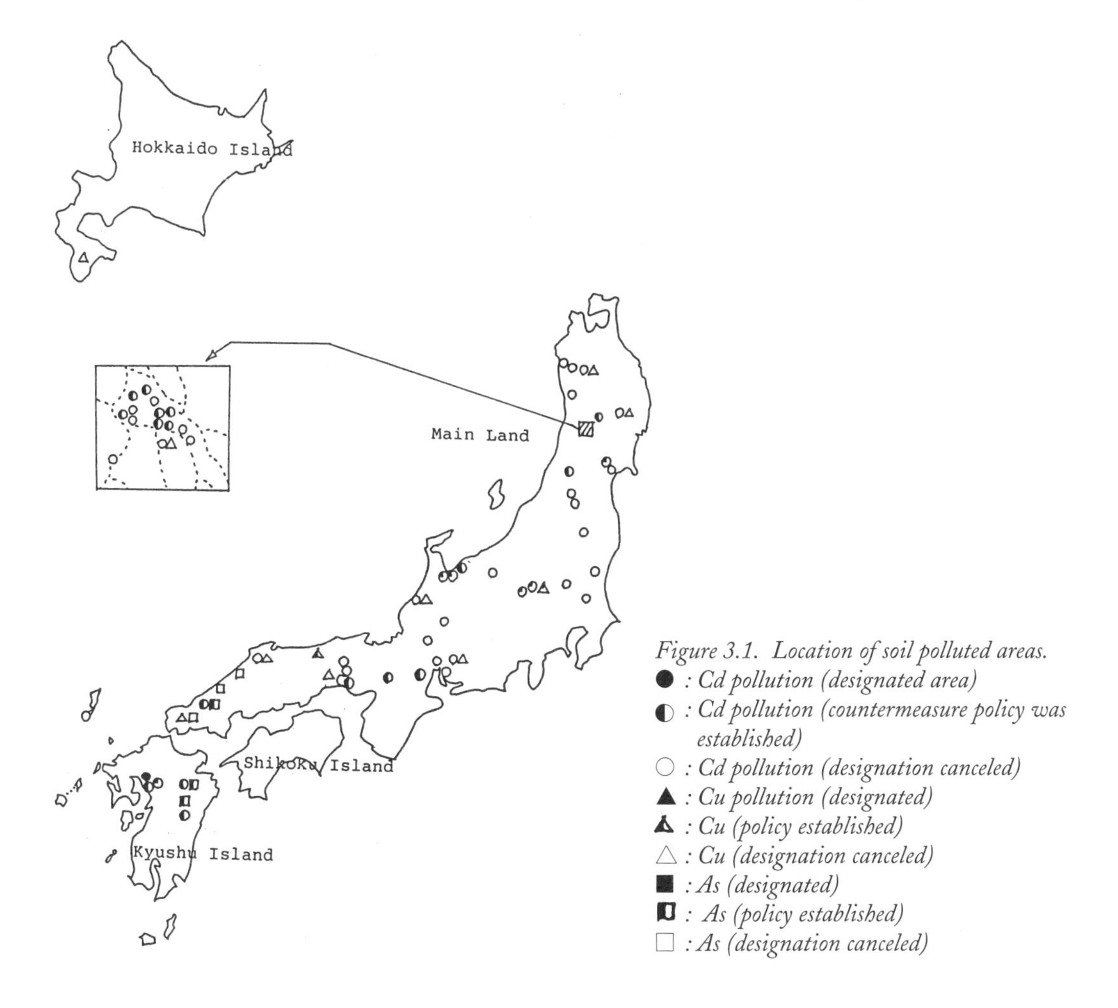

Figure 3.1. Location of soil polluted areas.
● : *Cd pollution (designated area)*
◐ : *Cd pollution (countermeasure policy was established)*
○ : *Cd pollution (designation canceled)*
▲ : *Cu pollution (designated)*
▲ : *Cu (policy established)*
△ : *Cu (designation canceled)*
■ : *As (designated)*
◧ : *As (policy established)*
□ : *As (designation canceled)*

REGULATIONS AND CONTROL

Soil

The Agricultural Land Soil Pollution Prevention Law designated Cd, Cu, and As as specific harmful substances and set their maximum allowable limit in soil. According to this law, agricultural land soil pollution for Cd, Cu, and As is defined as follows:

1. Cd-polluted areas are (a) those where rice grains containing ≥ 1 ppm of Cd are produced, and (b) those adjacent to area 1 where the Cd concentration of the soil is equal to or greater than that in area 1.
2. Cu-polluted areas are those where the content of 0.1 N HCl-soluble Cu of the soil is ≥ 125 ppm (for paddy field only).
3. As-polluted areas are those where the content of 1 N HCl-soluble As of the soil is ≥ 15 ppm (paddy fields only).

The prefectural governors are asked to carry out measurement surveys concerning the state of the pollution of agricultural land soil with Cd, As, and Cu. According to the results of the survey, the prefectural governors designate a region that is contaminated with the metals over the maximum allowable limits as an agricultural land soil pollution policy area after hearing the opinions of the prefectural council and the leaders of the cities, towns, and villages concerned. After the designation of the policy area and with approval of the Director-General of the Environmental Agency and the Minister of Agriculture, Forestry and Fisheries, the governors establish an agricultural land soil pollution policy project for prevention of further pollution and for remediating polluted areas. The cost of these projects is wholly or partially borne by the party responsible for the pollution, and the balance by the central government and prefectural government.

In 1991, new guidelines for soil pollution, Urban Soil Pollution Control Measures, were set for urban soil. The Agricultural Land Soil Pollution Prevention Law is also active for agricultural land. The new guidelines include Cd, Pb, Cr (VI), As, Hg and others like CN, organic phosphate compounds, and PCB. In the new guidelines, all the metals in soils are removed by the same extractant (water of pH 5.8 to 6.3 adjusted by HCl) and the critical values for Cd, Pb, Cr (VI), As, and Hg are 0.01, 0.1, 0.05, 0.05, and 0.0001 mg per liter of the extractant, respectively. Several local governments enacted the laws for disposal standards for soils contaminated with heavy metals, which required landowners who sell land or construct buildings on the land to check if the soil is polluted, and if so, the owners should take the necessary measures.

Chemical Fertilizers and Organic Waste

The guidelines for the sewage sludge applications to agricultural land soil, "Control Standards for the Prevention of Accumulation of Heavy Metals in Agricultural Land Soil," were proposed by the Environmental Agency in 1984. According to the guidelines, sewage sludge can be applied to soil if Zn concentration is <120 mg/kg soil. Zn is known to be quickly accumulated in sludge amended soil and can be a good indicator for controlling metal accumulation due to sludge application. About 95% of Japan's soils contain Zn below this value (120 mg/kg). Concentrations of Cd, As, and Hg in sewage sludges applied to soils should be less than 5, 50, and 2 mg/kg dry sludge, respectively.

The heavy metal concentrations in chemical fertilizer are generally low and they do not cause soil pollution. However, some phosphate rocks contain high levels of Cd (1–100 ppm), and the Japanese Ministry of Agriculture, Forestry and Fisheries set 20 ppm Cd for 10% P_2O_5 as the guideline for phosphate fertilizers.

Groundwater

The Water Pollution Control Law was established in 1970 in Japan and set the guidelines for polluted groundwater. This law requires Prefectural Governors to survey the quality of groundwater periodically and report the results annually. According to this law, clean groundwater should not contain Cd at levels > 0.01 mg/L, PB > 0.1 mg/L, Cr(VI) > 0.05 mg/L, As > 0.05 mg/L, and Hg > 0.005 mg/L.

Legal Standards for Cd in Japan

The Japanese legal standards for Cd in air, water, soil, food, and industrial wastes are summarized in Table 3.1.

Table 3.1. Japanese Legal Environment Standards for Cadmium

Medium	Material	Type of Standard	Value of Standard	Legislation	Date of Enforcement
Air	Ambient	Emission	≤ 1 mg/m^3	Air Pollution Control Law	1971
	Mining smoke	Emission	≤ 1 mg/m^3	Mining Safety Law	1972
	Working place	Quality for indoor air	≤ 0.05 $mg/^3$	Industrial Safety and Health Law	1975
Water	Drinking water	Quality	≤ 0.01 mg/L	(Administrative guidance) Water Supply Law	(1969) 1979
	Public water	Quality	≤ 0.01 mg/L	Basic Law for Environmental Pollution Control	1970
	Drainage	Effluent	≤ 0.1 mg/L	Water Pollution Contral Law	1971
	Mining drainage	Effluent	≤ 0.1 mg/L	Mining Safety Law	1972
	Public sewage	Effluent	≤ 0.1 mg/L	Sewerage Law	1976
Soil	Paddy field	Quality for rice	≤ 1.0 ppm	Agricultural Land Soil Pollution Prevention, etc., Law	1970
	Soil	Solution of analysis	≤ 0.01 mg/L	Basic Environment Law	1991
	Agricultural land	Quality for rice	< 1.0 ppm	Basic Environment Law	1991
Food	Unpolished rice	Quality	< 1.0 ppm	Food Sanitation Law	1970
	Polished rice	Quality	< 0.9 ppm	Food Sanitation Law	1970
Industrial Wastes	Sludge, slag, cinder, fly ash	Quality for landfilling	≤ 0.3 mg/L	Waste Disposal and Public Cleansing Law	1973
	Organic sludge, water-soluble inorganic sludge	Quality for ocean dumping ocean dumping	≤ 5 mg/L	Law for Prevention of Marine Pollution and Maritime Disasters	1973
	Slightly water-soluble inorganic sludge, slag, cinder, fly ash	Quality for ocean dumping ocean dumping	≤ 0.1 mg/L	Law for Prevention of Marine Pollution and Maritime Disasters	1973
	Acid-waste fluid, alkaline-waste fluid	Quality for ocean dumping	≤ 1 mg/L	Law for Prevention of Marine Pollution and Maritime Disasters	1973

Table 3.2. Ranges of Heavy Metal Concentrations of Japanese Land Soil

	Urban Soil			Agricultural Land Soil			Orchard Soil		
Metal	**Min.**	**Max.**	**Med.**	**Min.**	**Max.**	**Med.**	**Min.**	**Max.**	**Med.**
Hg	0.010	2.33	0.110	–	–	–	–	–	–
Cd	0.093	4.82	0.470	0.10	0.84	0.38	0.10	1.20	0.40
Pb	6.20	324	42.3	6.60	38.15	18.10	8.10	352.5	28.05
As	0.700	36.7	5.20	1.80	20.23	7.20	1.89	63.18	9.95
Cr(VI)	ND	ND	ND						

METALS IN SOIL, GROUNDWATER, FERTILIZERS AND SEWAGE SLUDGES OF JAPAN

Urban Soils

Concentration ranges of Hg, Cd, Pb, As, and Cr(VI) in Japanese urban soils, agricultural land soils, and orchard soils are given in Table 3.2. As shown in this table, urban soil contains more Cd and Pb than agricultural land and orchard soil. The sources of heavy metal contamination of urban soil are incinerators, cars, factories, and others. Heavy metal concentrations in the air of some Japanese cities are shown in Table 3.3 (Komai, 1981). The air of heavily industrialized cities like Tokyo, Osaka, and Kawasaki contains higher levels of Cd, Pb, Zn, and Ni than cities like Sapporo, Matsue, and Kurashiki.

The roadside soils generally contain higher levels of heavy metals than other soils (Table 3.4) (Komai, 1981). It is thought that gasoline and rubber tires are the main sources of contamination of roadside soil, although Pb is not used as an antiknocking agent in Japan. Komai (1981) investigated the heavy metal concentrations of soil in parks of Sakai City (heavily industrialized) and Kishiwada City (residential and less industrialized). Some of his data are shown in Table 3.5. The soils in the parks of Sakai city contained higher levels of Zn, Cu, and Mn than those in Kishiwada City.

The numbers of heavy metal pollution of urban soils are increasing annually and are reported to be 96 in 1987, 116 in 1988, 130 in 1989, 149 in 1990, 169 in 1991, and 177 in 1992 (Japanese Environment Agency Annual Report, 1992). The main sources of pollution are metal plating factories (12% of total cases) that caused Cd, Pb and Cr(VI), As, Hg, Cu, and Zn contamination; electric machine- and tools-producing factories (10%) that caused Cd, Pb, Cr(VI), and Zn contamination; natural science research institutes (4%) that caused Cd, Pb, As, Hg, and Zn contamination; and others, including cement, steel, general machine, gas industries, construction, etc. The main causes of the pollution were breaking of apparatus and machines by accidents (17 cases), dumping of the waste (17), mistreatment of the pollutants (12), and leakage from waste treatment apparatus or plants (9). Some countermeasures as mentioned below (Remediation Strategies section) to remediate the soil were conducted at 112 polluted sites among these 177 sites (Japanese Environment Agency Annual Report, 1992). The occurrence of metal pollution of urban soil often results in groundwater pollution. In 1992, five cases of Cd pollution of groundwater, one of Pb, one of As and 14 of Cr(VI) were reported to be caused by urban soil contamination (Japanese Environment Agency Annual Report, 1992).

Pollution of Agricultural Land Soil

Detailed surveys by prefectural governments of soils polluted with heavy metals revealed that the number of the areas that were designated as heavy metal polluted soil areas ac-

Table 3.3. Heavy Metal Concentrations in the Air of Some of Japanese Cities (1970)

City	Total Suspended Particles (μg/m^3)	Cd	Pb	Zn	Ni
Nonindustrialized Cities					
Sapporo	238	0.005	0.32	0.8	0.031
Matsue	91	0.002	0.06	0.3	0.019
Kurashiki	175	0.009	0.38	0.4	0.030
Industrialized Cities					
Tokyo	117	0.016	0.67	1.6	0.059
Kawasaki	301	0.024	1.46	1.2	0.149
Osaka	224	0.030	1.05	2.0	0.085

Table 3.4. Distances from Road and Heavy Metal Concentrations in Soils in Chiba Prefecture [a]

Investigation Site[b]	Distance from the Road[c] (m)	Metal Concentration (ppm)		
		Pb	Zn	Cd
Matsudo City	1	240	316	1.50
	30	57	365	1.23
	300	49	238	1.44
Kashiwa City A	1	132	235	2.82
	60	74	253	1.82
	200	31	111	1.25
Kashiwa City B	1	397	325	1.42
	30	49	130	1.21
	120	50	164	1.30

[a]Average values of Apr.-Nov., 1972. Samples were decomposed with $HNO_3^-HClO_4$.
[b]Town names were omitted.
[c]Route No. 6.

Table 3.5. Areal Mean Values of Heavy Metal Concentrations in Surface Soil of Parks in Sakai and Kishiwada

		Metal Concentration (ppm)		
		Zn	Cu	Mn
Sakai City	Range [a]	1,580–108	300–41	990–205
	Mean of total areas (I)	410	122	470
	Mean of area A	747	183	421
	Mean of area B	328	134	428
	Mean of area C	276	93	597
	Mean of area D	151	66	443
Kishiwada City	Range	560–55	120–31	540–188
	Mean of total areas (II)	177	65	332
	Mean of area A	233	80	293
	Mean of area B	168	72	354
	Mean of area C	179	57	342
	Mean of area D	103	46	327
Ratio of II/I (%)		43.2	53.3	70.6

[a] Determined by X-ray fluorescence analysis. Samples were taken at 0- to 3-cm depth.

cording to the guidelines of the Agricultural Land Soil Pollution Prevention Law was 63, and the total area was approximately 6150 ha, up to 1988. The number and total area of Cd-polluted soil areas were 54 and 6000 ha, those of Cu polluted soil areas were 13 and 1250 ha, and those of As-polluted soil areas were 7 and 160 ha up to 1988. Among these pollution designated areas, the areas where the soil was remediated were 4090 ha (3940 ha from Cd, 100 ha from Cu, and 160 ha from As) and 61 sites (52 cities for Cd, 13 sites for Cu, and 7 sites for As) up to 1988. No significant increase in the number and the area of polluted soils has been found since 1988.

Pollution of Groundwater

The 1992 survey for metal concentration of groundwater from about 2500 wells in Japan showed that 5 wells were contaminated with Cd, 5 with As, 14 with Cr(VI), and 3 with Hg. Maximum concentrations of the polluted groundwaters were 0.023 mg/L for Cd, 0.19 mg/L for As, and 0.003 mg/L for Hg. These well waters were not used for drinking. The sources of these pollutants were not clearly identified and some were suggested to be natural sources or urban soil contamination.

Heavy Metal Concentration of Chemical Fertilizers and Sewage Sludges

As some phosphate rocks contain high levels of Cd, various phosphate fertilizers produced in Japan were analyzed for Cd in 1971 (Table 3.6) (Shibuya et al., 1975). Maximum Cd concentration of the fertilizers was 28 ppm, which is higher than the guidelines of 20 ppm Cd for phosphate fertilizers. However, most phosphate fertilizers contain Cd in amounts far less than 20 ppm. Fused magnesium phosphate contains Cd in much smaller amounts than the other phosphate fertilizers such as superphosphate and concentrated superphosphate, which suggests Cd is volatilized in the process of fusion under high temperature. Copper and Zn concentrations of various chemical fertilizers contain high levels of Cu and Zn (Table 3.7). Fused magnesium phosphate contains rather high levels of these elements, which indicates that fusion does not volatilize Zn and Cu as much as Cd. Silicate fertilizer and calcium carbonate fertilizer also contain higher levels of Cu and Zn than ammonium sulfate, calcium nitrate, calcium cyanide, and potassium fertilizers. Magnesium sulfate contains high levels of Zn but low levels of Cu. These values of metal concentrations of fertilizers suggest that chemical fertilizers contribute small amounts of Cd, Zn, and Cu to agricultural fields and do not cause soil contamination as long as the fertilizers are not applied in excess.

Table 3.6. Cd Concentrations of Phosphate Fertilizers (Ministry of Agriculture and Forestry)

Fertilizers [a]	Number of Factories for Sample Collection	Number of Samples	Cd Concentration (ppm)		
			Maximum	Minimum	Average
SP	22	24	17	< 0.4	5
CSP	5	5	16	2	6
FMP	12	17	1	< 0.4	1
CF	46	102	28	1	7

[a]SP: superphosphate; CSP: concentrated superphosphate; FMP: fused magnesium phosphate; CF: compound fertilizer (or mixed fertilizer).

Table 3.7. Cu and Zn Concentration of Various Chemical Fertilizers

Fertilizer [a]	Cu (ppm)	Zn (ppm)	Fertilizer [a]	Cu (ppm)	Zn (ppm)
AS	0.2	3.0	K^2SO_4	ND	3.3
CaNit	2.8	2.6	KCI	1	0.9
$CaCN_2$	2.1	5.7	$MgSO_4$	1.3	87.0
SP	15.7	159.0	CF	13.0	92.7
CSP	32.5	269.5	SiF	13.0	44.0
FMP	14.2	133.7	$CaCO_3$	3.8	33.0

[a]AS: ammonium sulfate; CaNit: calcium nitrate; $CaCN_2$: calcium cyanamide; AP: superphosphate; CSP: concentrated superphosphate; FMP: fused magnesium phosphate; K_2SO_4, potassium sulfate; KCl: potassium chloride; $MgSO_4$, magnesium sulfate; CF: compound fertilizer; SiF: silicate fertilizer; $CaCO_3$, calcium carbonate.

Table 3.8. Heavy Metal Concentration of Various Sewage Sludges in Japan [a]

	pH	Cd	Hg	As	Cu	Zn	Pb	Ni	Cr
No. of measurements	48	72	72	69	81	59	34	33	32
Maximum value (ppm)	12.6	19.1	2.3	52.0	610.0	3000.0	121.4	110.0	470.0
Minimum value (ppm)	5.6	0.24	0.01	0.45	3.0	143.6	1.73	2.00	9.04
Average (ppm)	9.3	3.26	1.03	6.90	173.0	960.9	48.2	37.4	48.96

[a]Japanese Environment Agency.

Table 3.9. Mean Heavy Metal Concentrations in Four Kinds of Japanese Sewage Sludges (ppm)

Sludge [a]	Cd	Cu	Zn	Pb	Hg	N
A	10.2	652	2009	126.9	1.01	2.90
B	3.83	370	1455	70.1	1.26	1.24
C	2.17	233	1240	24.2	0.14	6.85
D	4.26	141	908	38.7	0.64	3.11

[a]A:Sewage sludge from high level of industry wastewater contamination, treated with inorganic coagulator;
B:Sewage sludge from low level of industry wastewater contamination, treated with organic coagulator;
C:Sewage sludge without industry wastewater contamination, treated with inorganic coagulator;
D:Sewage sludge from low level of industry wastewater contamination, treated with inorganic coagulator.

The Japanese Environment Agency and Japanese Ministry of Agriculture, Forestry and Fisheries investigated heavy metal concentrations of various kinds of sewage sludges in Japan. The results are shown in Table 3.8 (Japanese Environment Agency) and Table 3.9 (Japanese Ministry of Agriculture, Forestry and Fisheries). The sewage sludge A from industrial wastewater contains high levels of the heavy metals as shown in Table 3.9 and one derived from domestic wastewater (sewage sludge C in Table 3.9) contains low levels of heavy metals.

Another investigation on heavy metal concentrations of Japanese sewage sludges showed that the average concentrations of Hg, As, Cd, Pb, Zn, Cu, Ni, and Cr, were 0.87 (sample size 38, std. dev. 0.34), 4.88 (37, 3.47), 1.72 (38, 0.87), 39.3 (22, 27.2), 835 (68, 492), 250.6 (67, 154.2), 59.6 (16, 67.7), and 104.6 (21, 196.3), respectively.

REMEDIATION STRATEGIES

Various countermeasures to decrease Cd uptake by rice plants from polluted soils have been evaluated. Application of various soil amendments were sometimes effective to some

Table 3.10. Effect of the Application of Soil Amendment Materials on Cadmium Concentration of Rice Grains

Treatment	Soil pH	Cd ppm of Rice of Grains	Grain Yield (g/pot)
Control	5.7	0.48	32.4
$CaSiO_3$ 10 g/pot	7.2	0.11	27.2
$CaSiO_3$ 10 g/pot	7.5	0.09	24.9
Control	5.6	0.43	26.6
$CaSiO_3$ 10 g/pot	6.4	0.58	34.7
$CaSiO_3$ 10 g/pot	7.7	0.69	9.8
Control		0.45	4.3 t/ha
$CaSiO_3$, 5.5 t/ha		0.33	5.4 t/ha
Fused Mg-P 14.5 t/ha		0.25	6.0 t/ha
Control		1.21	67 g/pot
$Ca(OH)_2$ 20 g/pot		0.49	71 g/pot
Fused Mg-P 20 g/pot		0.64	66 g/pot
Compost 100 g/pot		0.94	65 g/pot
EDTA 1 g		0.74	68 g/pot
Alluvial soil			
Control	7.1	0.88	
Fused Mg-P 15 t/ha	7.5	0.52	
Fused Mg-P 30 t/ha	8.1	0.51	
Volcanic ash soil			
Control	7.6	0.16	
Fused Mg-P 15 t/ha	8.1	0.36	
Fused Mg-P 30 t/ha	8.2	0.24	
Control	5.7	0.78	58.3 g/pot
$CaCO_3$ 30 g/pot	6.8	0.47	56.0 g/pot
$CaSiO_3$ 33 g/pot	7.2	0.21	65.8 g/pot
Fused Mg-P 33 g/pot	6.9	0.54	59.9 g/pot

extent (Table 3.10) (Chino, 1981). Liming is expected to decrease Cd uptake by increasing soil pH. However, in paddy fields, soil pH is usually higher than 6.5 because of high concentration of soluble Fe_{2+}. Thus, liming of paddy fields decreased only 30–50% of Cd of rice grain. Application of phosphorous compounds to paddy fields was also as effective as liming (Table 3.10). Screening for rice varieties that absorb little Cd was investigated but no practical varieties were found (Table 3.11) (Chino, 1981). Removal of Cd from polluted soil by Cd accumulator plants was not successful. Leaching of soil Cd with HCl, $CaCl_2$ solution, and EDTA solution was found to be effective at the level of 30–90% reduction (Table 3.12) (Chino, 1981). Effectiveness of leaching was greater for a sandy soil than a volcanic ash soil. However, this may cause contamination of groundwater and surface water and it may leach nutrients from soil. Decrease in the metal concentration of the polluted soil by mixing unpolluted soil was not effective in decreasing Cd concentration of rice grains because the mixing of unpolluted soil often made the polluted soil oxidative and increased solubilities of Cd compounds (Table 3.13) (Chino, 1981). Growing rice plants under submerged conditions throughout the growing season decreased Cd

Table 3.11. Varietal Difference in Cadmium Concentration of Rice Grains Harvested from the Same Field

Period of Variety	Cd ppm of Maturation	Rice Grains
Sasanishiki	Early	0.38
Etsunan No. 77	Middle	0.38
Nihonbare	Late	0.29
Hohnenwase	Early	0.12
Koshijiwase	Early	0.16
Fuko No. 66	Early	0.18
Tomisakae	Middle	0.13
Koshihikari	Middle	0.08
Yomomasari	Middle	0.07
Yamabiko	Late	0.05
Nihonbare	Late	0.04
Norin No. 22	Late	0.06

Table 3.12. The Effect of Removing Cadmium from the Soil with HCl, $CaCl_2$, EDTA Solution by Leaching on Cadmium Concentration of Rice Grains

Treatment	Soil pH	Cd ppm of Grains	Yield
Fukui soil			
Control	6.2	0.33	3.5 t/ha
0.1 N HCl	6.0	0.20	2.8 t/ha
HCl+liming	6.7	0.06	3.5 t/ha
Fuchu soil			
Control	5.5	0.16	27.2 g/pot
0.1 N HCl	3.9	0.02	36.2 g/pot
0.1 N CH_3 COOH	5.2	0.02	30.1 g/pot
1 N $CaCl_2$	5.4	0.02	17.1 g/pot
Kurobe soil			
Control		0.28	34.2 g/pot
0.1 N HCl		0.33	9.5 g/pot
0.1 N CH_3 COOH		0.23	6.5 g/pot
Annaka soil			
Control	6.34	0.103	11.4 g/pot
EDTA 7 g/pot	6.33	0.073	34.6 g/pot
EDTA 14 g/pot	6.31	0.003	24.5 g/pot
Takasaki soil			
Control	5.47	0.432	29.5 g/pot
EDTA 7 g/pot	5.52	0.232	24.2 g/pot
EDTA 14 g/pot	5.45	0.215	24.1 g/pot

Table 3.13. Effect of Mixing of Unpolluted Soil with Polluted Soil on Cadmium Concentration of Rice Grains

Treatment	Cd ppm of grains	Yield
Ishikawas soil		
Control	0.57	23.1 g/pot
3 : 2 mixing	1.73	21.4 g/pot
1 : 1 mixing	2.16	18.7 g/pot
Fuchu soil		
Control	0.12	4.4 t/ha
4 : 1 mixing	0.10	4.4 t/ha
Kurobe soil		
Control	0.36	3.6 t/ha
4 : 1 mixing	0.40	3.3 t/ha

concentrations of rice grains by converting Cd^{2+} to insoluble CdS under the reductive conditions of the soil (Table 3.14) (Chino, 1981). The best and most reliable countermeasure was to place an unpolluted soil layer of 20–30 cm thickness over polluted soil (Table 3.15) (Chino, 1981).

The urban soils polluted with metals were also remediated in several ways according to the level of contamination. These were (1) soil covering and vegetation to prevent the dispersion of soil in the case of lowest contamination, (2) water insulation with setting of impermeable sheets or clay to prevent effects on groundwater in case of medium contamination, and (3) insulation by containing metal polluted soil in a concrete vessel to isolate the soil from the environment in the case of heavy contamination. In 1992, the soils of 59 polluted sites among 177 were recovered by the ways mentioned in (2) or (3), while 23 sites were by (1).

DAILY INTAKE OF HEAVY METALS BY JAPANESE

Ohmomo and Sumiya (1981) calculated daily intake of heavy metals through usual fare by Japanese (Table 3.16). The average daily intake is 49 μg for Cd, 12.5 mg for Zn, 1.5 mg for Cu, 176 μg for Pb, 329 μg for As, and 18 μg for Hg. Daily intake of heavy metals may be less than this estimation because heavy metals are removed in the cooking process. Main sources of heavy metal intake are rice and vegetables, while marine products are the major source of As and Hg. Among the vegetables, spinach and root vegetables are the main sources of metal intake.

We also estimated daily intake of heavy metals by analyzing metal contents of all the food supplied to workers three times a day at a dormitory of a factory for one week. The results suggest that the daily intake of heavy metals by a Japanese are 21–51 μg for Cd, 1.2 to 1.4 mg for Cu, 5.8 to 9.2 mg for Zn, and 178–457 μg for Ni. Analysis of the feces collected from 5 houses in the city of Yokosuka made it possible to estimate the daily intake of the heavy metals by inhabitants of these homes; the results are shown in Table 3.17. These values are about the same as those mentioned above. Similar values were obtained by Yamagata (1977) (Table 3.17).

According to the reports of the FAO/WHO Joint Committee, daily intake of Cd by a Japanese (between 1980 and 1985) was higher than that by people in the U.K., Australia,

Table 3.14. Effect of Flooding on Cadmium Concentration of Rice Grains

Treatment	Cd ppm of Grains	Yield
Toyama soil		
Flooding	0.30	16.3 g/pot
Nonflooding	2.25	10.1 g/pot
Takasaki soil		
Flooding	0.38	67.3 g/pot
Nonflooding	1.30	61.1 g/pot
Bandai soil		
Flooding	0.74	
Nonflooding	3.01	

Table 3.15. Effect of Placement of an Unpolluted Soil over a Polluted Soil at Various Depths

Depth of Unpolluted Soil Placed over Polluted Soil	Cd ppm of Rice Grains	Yield (t/ha)
0 cm (control)	1.27	5.4
5 cm	1.10	5.2
10 cm	0.80	5.7
20 cm	0.47	5.4
0 cm (control)	1.121	5.5
25 cm	0.080	5.5
30 cm	0.075	4.3
40 cm	0.080	5.5

and Hungary (Fig. 3.2). Post-1985 daily intake data for Cd (Table 3.18) shows that intake by Japanese is in the range of 20–38.7 μg, which is the highest among 12 countries.

The excretion of Cd by a Japanese was estimated by analyzing feces. The results are shown in Tables 3.19 and 3.20. Cd excretion through feces of people who live in the polluted areas was higher than that from those living in nonpolluted areas near the polluted areas (Table 3.19). The Cd excretion through feces decreased after soil remediation, which clearly indicates that soil remediation was effective in decreasing the Cd intake through food (Table 3.20).

APPLICATION OF SEWAGE SLUDGE AND OTHER BIOSOLIDS TO LAND AND HEAVY METAL ACCUMULATION

The number of wastewater treatment plants and the amount of sewage sludge increases every year. Much work on the application of sewage sludge to soil has been done in relation to heavy metal accumulation. Several guidelines have been established for the application of sewage sludges to land as described in the section on metals in Soil, Groundwater, Fertilizers, and Sewage Sludge. These guidelines exclude the use of sewage sludges produced from industrial wastewaters. Thus, the sewage sludges produced from domestic wastewaters are mainly used for land application because of their low metal concentra-

Table 3.16. Estimation of Heavy Metal Intake Through Common Foods

Category	Average Daily Intake (wet g/d/p) [a]	Calculated Daily Intake (μg/d/p)							
		Cd	Zn	Mn	Cu	Pb	As	Hg	Fe
Cereals	328.7								
Rice	234.5 as unpolished rice	26.0	4,800	6,100	660	73	—	—	2,600
Wheat	92.4 as soft flour	2.8	468	215	103	—	—	1.8	924
Others	1.7 as buckwheat flour	0.2	49	29	10	1	0.1	0.1	60
Potatoes	61.9 as potato	1.2	210	180	80	3	5.0	—	310
Sugars	14.0 as white sugar	— [b]	3	—	1	—	0.6	0.1	28
Confectioneries	27.1 as biscuits	4.3	180	—	43	7	—	—	410
Oils and fats	17.7 as margarine	—	12	1	1	—	3.0	0.4	35
Pulses	67.7								
Soybeans and soybean products	64.8 as soybean curds (*tofu*)	1.3	472	308	104	11	1.9	0.7	907
Others	2.9 as azuki beans	0.1	82	40	29	—	0.3	0.2	140
Vegetables	270.2								
Green and yellow vegetables	59.3 as spinach	6.5	690	240	100	9	3.6	1.2	2,000
Others	210.9 as cabbage	—	460	550	65	—	—	2.1	840
Fruits	180.9 as *Citrus unshin*	—	160	90	65	—	20	—	360
Seaweeds	5.0 as wakame seaweed	2.9	130	41	71	5	160	1.3	650
Seasonings and beverages	116.4 as soy sauce	2.3	910	982	29	28	2.3	—	—
Marine products	88.5 as Jack mackerel	0.9	449	14	77	9	122	8.0	620
Meat, poultry and whales	68.4 as port, fresh	—	2,100	6	61	30	6.8	1.4	1,200
Eggs	40.8 as hen's egg	0.4	816	16	35	—	3.3	1.2	1,100
Milk	100.4 as market milk	—	270	4	11	—	—	—	100
Milk products	6.4 as processed cheese	0.1	220	2	3	—	0.2	—	13
Total		49.0	12,481	8,814	1,548	176	329.1	18.4	12,297

[a] gram/day/person.
[b]— : Not calculated.

Table 3.17. Daily Intake of Heavy Metals by Japanese

Cd	Ni	Pb	Cr	Hg	Mn	Cu	Zn	Fe	
	μg · day⁻¹ person⁻¹				mg · day⁻¹ person⁻¹				References
30–60	—	230–320	700–900	45	2.1	0.78–2.54	13.7	—	Yamagata (1977) Food analysis
38	204	94	13	—	4.91	1.6	8.9	8.5	Mori et al. (1986) Yokosuka, feces
44	247	35	56	—	6.86	1.83	12.8	13.6	Mori et al. (1986) Hachinohe, feces
34	224	33	39	—	5.98	2.00	10.2	12.1	Mori et al. (1986) Hakodate, feces

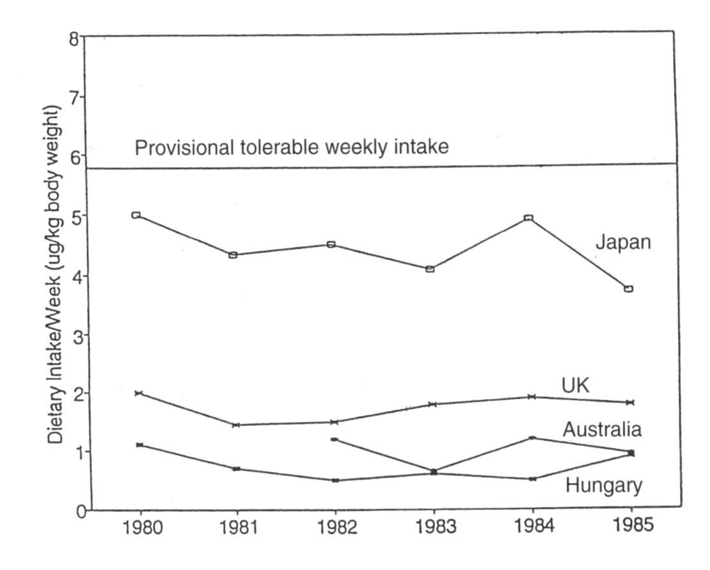

Figure 3.2. Dietary intake of cadmium in four countries, 1980–1985.

tions. At present the annual production of sewage sludge is about 2,000,000 m^3, 20% of which is estimated to be used as organic fertilizers. Effects of long-term application (17 years) of the two kinds of sewage sludge composts on the heavy metal concentration in soil has been investigated. The results showed that the concentration of Zn increased in surface soil to a depth of 10 cm soon after the start of sludge application, and a slight increase in Zn concentration in the subsoil (20–30 cm depth) was observed 10 years after the start of the application. Increases in the concentrations of Cu, Ni, and Mn in the surface soil were not detectable (Chino et al., 1992).

The main sources of the heavy metal contained in the sewage sludge produced from domestic wastewater are food and water. Average amounts of the metals discharged from a Japanese house in the city of Yokosuka have been estimated to be 192 $\mu g \cdot day^{-1} \cdot person^{-1}$ for Cd, 1661 $\mu g \cdot day^{-1} \cdot person^{-1}$ for Ni, 6753 $\mu g \cdot day^{-1} \cdot person^{-1}$ for Pb, 775 $\mu g \cdot day^{-1} \cdot person^{-1}$ for Cu, 8.19 $mg \cdot day^{-1} \cdot person^{-1}$ for Mn, 5.58 $mg \cdot day^{-1} \cdot person^{-1}$ for Cu, 44.3 $mg \cdot day^{-1} \cdot person^{-1}$ for Zn, and 111.1 $mg \cdot day^{-1} \cdot person^{-1}$ for Fe (Chino et al., 1991).

Table 3.18. Post-1985 Daily Intake of Cadmium

Country	Period	Method	μg Cd/day
USA	1985–1991	Total diet	
	16 year old F		9.2
	14–16 year old M		13.3
	25–30 year old F		9.5
	25–30 year old M		12.9
	60–65 year old F		8.8
	60–65 year old M		11.2
Canada	1987 (P)	Duplic. meal	13.8
Germany (Bavaria)	1992	Total diet	12.7
Germany	1986–1987	Duplic. meal	10.3
Netherlands	1984–1985	Market basket	23
	1984–1986	Duplic. meal	9
United Kingdom	1986	Total diet	17
	1987	Total diet	
	1988	Total diet	18 (maximum)
			11 (minimum)
Sweden (Stockholm)	1991 (P)	Duplic. meal	8.3
		Fecal excretion	9.4
Finland	1986	Total diet	10.4
		Market basket	12
	1987 (P)	Duplic. meal	8.8
		Total diet	9.3
Yugoslavia (Zagreb)	1991 (P)	Duplic. meal	8.3
		Fecal excretion	15
Japan (Yokohama)	1991 (P)	Duplic. meal	
		Fecal excretion	20
Japan	1985–1988	Market Basket	38.7
China (Beijing)	1991 (P)	Duplic. meal	7.1
		Fecal excretion	7.3
China	1985–1988	Market basket	13.3
Denmark	1983–1987	Total diet	19.5
Australia	1987	Market basket	22.6
	1990	Market basket	25.7
France	1983–1990	Duplic. portion	19

Table 3.19. Excretion of Cd by Feces from People Living in Polluted Areas

	Cd Concentration of Feces (μg/day/person)
Kosaka, a polluted town in Akita prefecture	177
Ikawa, a nonpolluted town near Kosaka	51
Iwahara, a polluted town in Nagasaki prefecture	255
Nonpolluted town near Iwahara	42

Table 3.20. Changes in Cd Excretion through Feces Before and After Soil Remediation of a Cd Polluted Area (Iwahara of Nagasaki Prefecture)

	Sampling Year	Cd Excretion through Feces (μg/day/person)
Before soil remediation	1969	213
Before soil remediation	1976	184
After soil remediation	1983	77

REFERENCES

Chino, M. The Assessment of Various Countermeasures Against Cd Pollution in Rice Grains in *Heavy Metal Pollution in Soils of Japan,* Yamane, I. and Kitagishi, K., Eds., Japan Scientific Societies Press, Tokyo, 1981, pp. 281–286.

Chino, M., K. Moriyama, H. Saito, and T. Mori. The Amount of Heavy Metals Derived from Domestic Sources in Japan. *Water Air Soil Pollut.*, 57–58, pp. 159–168, 1991.

Chino, M., S. Gota, K. Kumazawa, N. Owa, O. Yoshioka, N. Takechi, S. Inanaga, H. Inoo, D.L. Cai, and R.A. Youssef. Behavior of Zinc and Copper in Soil with Long-Term Application of Sewage Sludge. *Soil Sci. Plant Nutr.*, 38, pp. 159–167, 1992.

Japanese Environment Agency. 1992. Annual Report 1992 (in Japanese).

Komai, Y. Heavy Metal Pollution in Urban Soil, in *Heavy Metal Pollution in Soils of Japan*, Yamane, I. and Kitagishi, K., Eds., Japan Scientific Societies Press, Tokyo, 1981, pp. 193–218.

Ohmomo, Y. and M. Sumiya. Estimation of Heavy Metal Intake from Agricultural Products, in *Heavy Metal Pollution in Soils of Japan*, Yamane, I. and Kitagishi, K., Eds., Japan Scientific Societies Press, Tokyo 1981, pp. 235–244.

Shibuya, Y., F. Yamazoe, T. Ogata, and K. Nose. Kankyou Osen to Nougyou *(Environmental Pollution and Agriculture),* Hakuyusha, Tokyo (in Japanese), 1975, pp. 277–288.

Yamagata, N. Biryou Genso (*Trace Elements*), Sandyou-Tosho, Tokyo (in Japanese), 1977, pp. 123–277.

CHAPTER 4

IMPACT OF RADIONUCLIDES ON SOIL, GROUNDWATER, AND CROPS AND RADIONUCLIDE CLEANUP IN JAPAN

S. Uchida and Y. Ohmomo

INTRODUCTION

Historically, beginning in the 1950s, most man-made radionuclides released into the environment were through nuclear weapons testing. Among these fission products, ^{90}Sr and ^{137}Cs, with relatively long half-lives (28.8 and 30 years, respectively) and high fission yields, are important for internal dose assessment because both are incorporated and retained in the human body. Currently in Japan, many nuclear power plants are in operation and a commercial nuclear fuel reprocessing plant is under construction. Significant radionuclides released from nuclear facilities include ^{3}H, ^{131}I, ^{137}Cs, and ^{90}Sr. Long-lived radionuclides such as $^{239,240}Pu$, ^{129}I, and ^{99}Tc are also important, although the amounts released from the facilities are expected to be low.

In internal dose assessment, ingestion of contaminated crops and livestock products is well known as the most important pathway through which radionuclides are taken into the human body. Radiation dose estimation is usually made with the aid of mathematical models in which model parameters expressing the transfer of radionuclides from one environmental compartment to another are involved. These parameters are often described as the concentration ratio of a radionuclide between two compartments when the system is at equilibrium. For example, transfers of radionuclides from soil to crops, from grass to milk, and from grass to meat have been estimated by using transfer factor (*TF*), transfer coefficient to milk (F_m), and transfer coefficient to meat (F_f), respectively. A distribution coefficient, K_d, is used for predicting the mobility of a radionuclide in soil and groundwater.

This chapter describes work done in Japan on the transfer of radionuclides, particularly the transfer of fallout ^{90}Sr and ^{137}Cs, in the terrestrial environment. The most common transfer parameters obtained for Japan, such as *TF* (specifically *TF* to rice) and K_d are introduced. These are important in safety assessments before the nuclear facilities, including underground disposal facilities, can operate. Attention is also paid to the transfer of $^{239,240}Pu$ in soil. The levels and the *TF*s of the radionuclides are briefly discussed, based on studies carried out in the area contaminated by the plutonium-fueled atomic bomb exploded in Nagasaki in 1945.

METEOROLOGICAL DATA IN JAPAN

Meteorological Data

Meteorological data (National Astronomical Observatory, 1992) are important in understanding the behavior of radionuclides in terrestrial environments. Because radionuclides are absorbed with soil solution by plants through the roots, temperature affects the plants' transpiration. Runoff or percolation of water also affects radionuclides deposited on the ground; the amount of precipitation controls their migration rates.

Figure 4.1 shows monthly average temperatures in six cities in the north, central, and southern parts of Japan. Each value is an average for the 30-year period 1961 to 1990. The temperature in Sapporo City, the northernmost in the list, is below 0°C in winter, but higher than 20°C in mid-summer. The winter temperatures are not below 0°C in the other five cities.

Figure 4.2 shows annual precipitation in selected Japanese cities. These are averages for the same 30-year period. Most areas have an annual precipitation of much more than 1000 mm. The average value throughout Japan is about 1600 mm y^{-1}. The amount of precipitation in the north is lower than that in the south. Also, the amount of precipitation in areas facing the Sea of Japan (Akita and Niigata) is higher than that in areas facing the Pacific Ocean (Morioka and Sendai).

Water Budget in a Paddy Field

Realistic estimates of food-chain transfer of radionuclides require accurate knowledge of physical transfer media such as groundwater and irrigation water. In Japan and Southeast Asian countries, rice plants are cultivated in paddy fields under continuously submerged conditions. The characteristics of the water budget in Japanese paddy fields are described by Watanabe (1992). Briefly, under normal hydrological conditions, the water budget in a paddy field during the rice growing season consists of precipitation onto the field (R_W), irrigation water (Q_i) into the field, evapotranspiration (ET) from the plants and surface

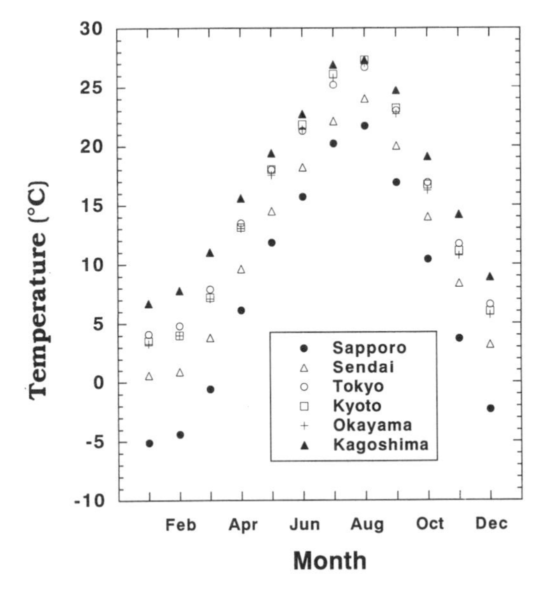

Figure 4.1. Monthly average temperature in six cities in northern (Sapporo and Sendai), central (Tokyo, Kyoto, and Okayama) and southern (Kagoshima) parts of Japan during 1961 to 1990 (National Astronomical Observatory, 1992).

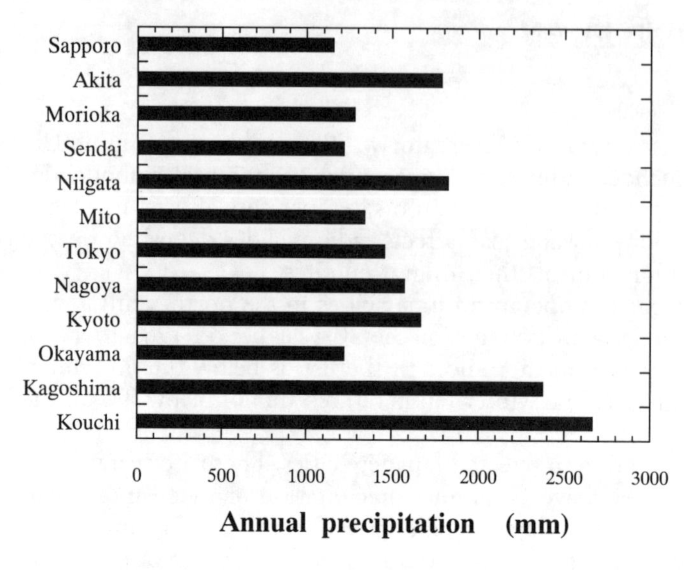

Figure 4.2. Average annual precipitation in selected Japanese cities during 1961 to 1990 (National Astronomical Observatory 1992).

water, and percolation (P_W) into the soil and surface runoff water (Q_0). A paddy field lot includes a land area where rice plants grow surrounded by embankments (bunds) to retain the irrigation water.

The water budget varies widely with field conditions and water management practices. The evapotranspiration rate changes with rice growing stages, weather conditions, etc., from a minimum of 0 to a maximum of 8 mm d^{-1}. Because percolation is related to soil type and soil structure, its rate also varies widely with a range of 5 to 10 mm d^{-1} in poorly drained soils to 15 to 30 mm d^{-1} in well-drained ones. The normal ranges of each factor in the water budget are shown in Figure 4.3. This gives a rough estimate of water volume per unit area for rainless days in summer under stable water management.

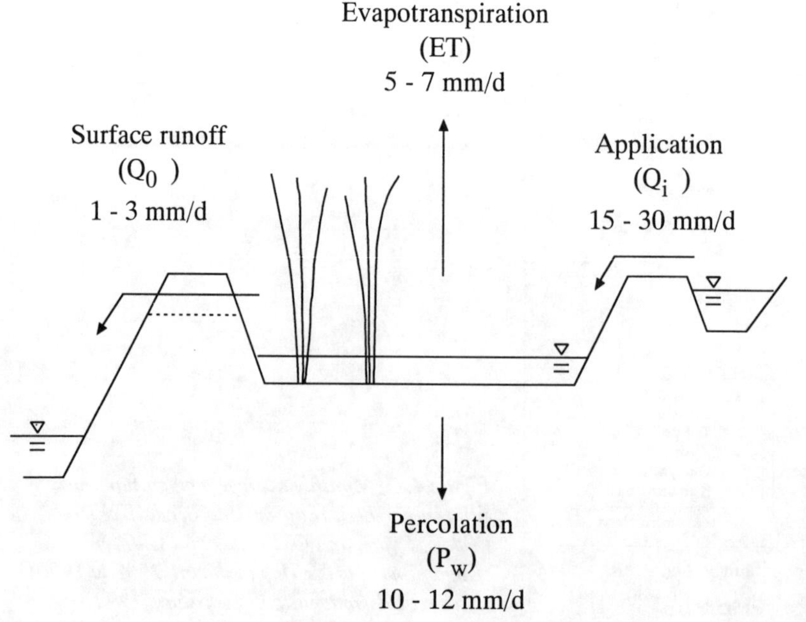

Figure 4.3. Diagram of water budget in a Japanese paddy field (Watanabe, 1992).

FALLOUT RADIONUCLIDES (^{137}CS, ^{90}SR, PU, ETC.) DEPOSITED ON SOIL

Fallout Radionuclides from Atmospheric Nuclear Weapon Tests

Since April 1957, monthly depositions in Japan of fallout radionuclides, derived from atmospheric nuclear weapon tests, have been studied at the Geochemical Research Division of the Meteorological Research Institute (MRI). The annual depositions of ^{90}Sr , ^{137}Cs, ^{239,240}Pu, and ^{238}Pu up to 1984 are listed in Table 4.1 (Hirose et al., 1987). From these data, the maximum annual depositions of ^{90}Sr, ^{137}Cs, and ^{239,240}Pu were recorded in 1963, following the large-scale atmospheric thermonuclear tests conducted during 1961–1962. The highest monthly deposition of radionuclides was observed in June 1963 at Tokyo, where ^{90}Sr and ^{137}Cs contributed 170 and 550 Bq m^{-2}, respectively (Katsuragi, 1983). It is well known that stratospheric fallout exhibits seasonal variation, with a maximum in spring or early summer and a minimum in autumn in the Northern Hemisphere (Miyake et al., 1962). After cessation of atmospheric tests by the U.S.A. and U.S.S.R. in 1962, the fallout of radionuclides in Japan decreased steadily for four years. During the period from 1967 to 1981, however, the annual deposition of each radionuclide showed

Table 4.1. Annual Deposition of ^{238}Pu, ^{239,240}Pu, ^{90}Sr and ^{137}Cs Measured by the Meteorological Research Institute (until the end of March 1980 in Tokyo and Thereafter in Tsukuba, about 60 km north of Tokyo)[a]

		^{239,240}Pu	^{238}Pu	^{90}Sr	^{137}Cs
	Year	Bq m^{-2}		kBq m^{-2}	
Up to	1958	7.5	0.32	0.53	1.5
	1959	3.6	0.19	0.30	0.95
	1960	1.6	0.063	0.089	0.23
	1961	1.4	0.059	0.078	0.27
	1962	4.1	0.24	0.30	0.81
	1963	7.4	0.27	0.71	1.9
	1964	6.8	0.20	0.32	0.60
	1965	4.5	0.18	0.16	0.39
	1966	2.7	0.048	0.067	0.19
	1967	0.78	0.12	0.030	0.081
	1968	0.93	0.093	0.048	0.10
	1969	0.44	0.11	0.044	0.081
	1970	0.22	0.070	0.052	0.10
	1971	0.48	0.026	0.041	0.085
	1972	0.19	0.019	0.022	0.044
	1973	0.11	0.0037	0.0074	0.015
	1974	0.19	0.0022	0.033	0.063
	1975	0.22	0.0037	0.019	0.037
	1976	0.037	0.00074	0.0074	0.0074
	1977	0.19	0.011	0.019	0.0030
	1978	0.26		0.022	0.033
	1979	0.15		0.0074	0.019
	1980	0.037		0.0037	0.0074
	1981	0.26		0.019	0.026
	1982	0.052		0.0026	0.0048
	1983	0.014		0.0013	0.0024
	1984	0.013		0.00056	0.0012
	Sum	43.99	2.03	2.91	7.52

[a]Hirose et al., 1987.

no decrease, mainly due to atmospheric thermonuclear tests carried out by the People's Republic of China. Beginning with the first Chinese test conducted in 1967, radioactive debris from these tests was responsible for most of the fallout in Japan. A peak in the fallout pattern was observed in 1981 and was attributed to the twenty-sixth Chinese nuclear test in October 1980. The fallout level thereafter decreased to the end of 1985. In 1985, the annual deposition of each radionuclide was the lowest since MRI started its observations in 1957. The cumulative amounts of ^{90}Sr, ^{137}Cs, and ^{239,240}Pu fallout at the MRI to the end of 1984 are estimated to be about 2.9 kBq m^{-2}, 7.5 kBq m^{-2}, and 44 Bq m^{-2}, respectively.

Although some studies have been made for ^{99}Tc in environmental samples, information is limited. This radionuclide, released from nuclear facilities, is important from a radiological viewpoint because it has a long half-life of 2.1 10^5 y and high thermal neutron fission yield (roughly 6%). Recently, Tagami and Uchida (1995) measured the concentration of ^{99}Tc in rain and dry fallout in Japan; the deposition was less than 0.4 mBq m^{-2} month^{-1}.

As described in the next section, most of the deposited ^{99}Tc remains in the surface soil layer (< 10 cm in depth). In order to clarify the behavior of Tc in the terrestrial environment, it is important to know its concentration in environmental samples, such as soils, crops, and rain, actually contaminated by nuclear weapon tests.

Fallout Radionuclides Due to the Chernobyl Accident

Radioactive fallout due to the Chernobyl accident on 26 April, 1986 was first detected on 3 May 1986 in Japan (Aoyama et al., 1986). The monthly depositions of ^{137}Cs, ^{134}Cs, and ^{90}Sr during the period from January 1986 to December 1988 are shown in Table 4.2. A markedly high monthly deposition was observed in May 1986, followed by a rapid decrease from June to November 1986. The monthly ^{137}Cs deposition in May 1986 was 131 Bq m^{-2}, the same as that in 1963. The apparent tropospheric residence time was calculated to be about 25 days. Thereafter, the monthly ^{137}Cs deposition pattern showed the same seasonal variation as observed previously. This suggests that most of the ^{137}Cs originated in stratospheric fallout. Its monthly deposition, derived from the Chernobyl accident, continued until the end of 1988.

Since May 1986, ^{134}Cs, which was not present in the debris from atmospheric nuclear weapon tests, has been detected in fallout samples together with the other short-lived radionuclides, ^{131}I and ^{103}Ru. The existence of these radionuclides is one of the distinctive features of the Chernobyl fallout (Aoyama et al., 1987). The monthly ^{134}Cs deposition was 67 Bq m^{-2} in May 1986. In June 1986, it was 1.2 Bq m^{-2}, which corresponded to 2% of that in May 1986 (Table 4.2). Thereafter, the monthly ^{134}Cs deposition dropped markedly from June to December 1986.

Monthly ^{90}Sr deposition showed similar temporal trends to those observed from 1957 to 1985. In May 1986, the ^{90}Sr deposition increased to 1.4 Bq m^{-2}, almost 60 times higher than the value for the previous month. In December 1986, the monthly deposition of ^{90}Sr decreased to 0.001 Bq m^{-2}, approximately two orders of magnitude lower than May. A subsequent spring peak in ^{90}Sr deposition was observed in April 1987, with a value of 0.021 Bq m^{-2}. This was also coincident with the peak in monthly deposition of ^{137}Cs (Table 4.2).

Before the Chernobyl accident, the activity ratio of ^{137}Cs to ^{90}Sr in monthly fallout ranged from 0.8 to 6.0. However, the ratio increased to 96 in May 1986 and ranged from 3 to 46 in the period from June 1986 to December 1987. This ratio is another characteristic feature of the Chernobyl fallout (Aoyama et al., 1991).

Table 4.2. Monthly Depositions of ^{137}Cs, ^{134}Cs, and ^{90}Sr Measured by the Meteorological Research Institute in Tsukuba from January 1986 to December 1988[a]

	Month	^{137}Cs (Bq m^{-2})	^{134}Cs (Bq m^{-2})	^{90}Sr (Bq m^{-2})
1986	Jan.	0.036	ND	0.021
	Feb.	0.083	ND	0.027
	Mar.	0.080	ND	0.094
	Apr.	0.097	ND	0.024
	May	131.0	67.1	1.235
	Jun.	2.51	1.24	0.164
	Jul.	0.783	0.358	0.015
	Aug.	0.369	0.167	0.018
	Sep.	0.087	0.029	0.011
	Oct.	0.106	0.079	0.016
	Nov.	0.071	0.033	0.019
	Dec.	0.078	0.020	0.010
1987	Jan.	0.117	0.034	0.013
	Feb.	0.137	0.042	0.017
	Mar.	0.105	0.041	0.014
	Apr.	0.158	0.058	0.021
	May	0.088	0.025	0.012
	Jun.	0.064	0.018	0.010
	Jul.	0.066	0.016	0.007
	Aug.	0.079	0.026	0.005
	Sep.	0.024	0.008	0.006
	Oct.	0.040	ND	0.006
	Nov.	0.041	ND	0.011
	Dec.	0.035	0.010	0.011
1988	Jan.	0.051	0.016	
	Feb.	0.053	0.007	
	Mar.	0.070	0.020	
	Apr.	0.088	0.012	
	May	0.076	0.006	
	Jun.	0.052	0.006	
	Jul.	0.028	ND	
	Aug.	0.016	ND	
	Sep.	0.008	ND	
	Oct.	0.027	ND	
	Nov.	0.025	ND	
	Dec.	0.074	0.014	

[a]Aoyama et al., 1991.

MIGRATION OF RADIONUCLIDES IN SOIL AND GROUNDWATER

Migration in Surface Soil

Vertical distributions of fallout radionuclides including ^{90}Sr and ^{137}Cs in soil have been studied by several workers. Morita et al. (1993) collected undisturbed core soil samples in Japan and determined the radioactivity of ^{137}Cs, ^{237}Np, $^{239,240}Pu$, and ^{99}Tc. The depth profiles of these radionuclides are illustrated in Figure 4.4. They show that more than 95% of each radionuclide is retained in the surface layer (upper 10 cm) which contains much organic material. They also measured concentrations of ^{99}Tc in the surface layers of uncultivated soils ranging from 0.081 to 0.31 Bq kg^{-1} dry soil. Tagami and Uchida (1993;

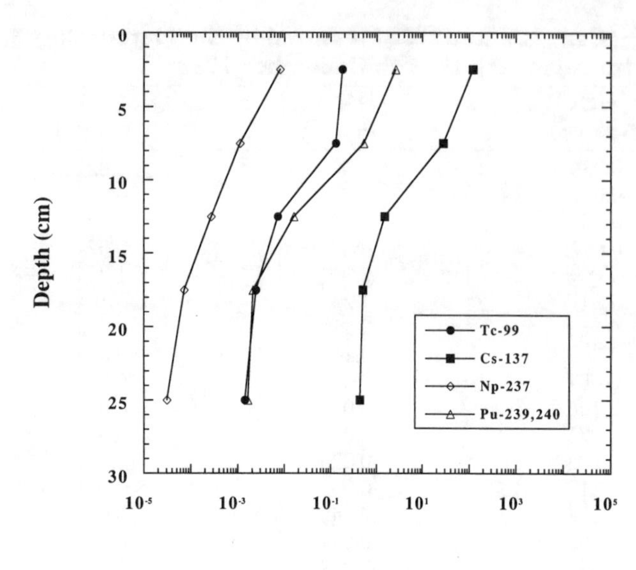

Figure 4.4. Soil profiles of ^{237}Np, ^{99}Tc, $^{239,240}Pu$, and ^{137}Cs (Morita et al., 1993).

1997) measured the concentration of ^{99}Tc in agricultural soils collected in Japan and noted that the activity ratio of ^{99}Tc /^{137}Cs in the soil was higher than the ratio of 1.4×10^{-4} in global fallout theoretically expected from their fission yields. These results seem to indicate that Tc might be retained in surface soil, which would be contrary to the hypothesis that ^{99}Tc would have high mobility in surface soil, because its chemical form is expected to be an anion, i.e., TcO_4^- in oxidizing environments (Brookins, 1988). It is therefore necessary to study the sorption mechanism on soil and organic soil components, as related to oxidation-reduction potential, and also to identify the role of microorganisms (Tagami and Uchida, 1996).

The plutonium-fueled "Fatman" bomb exploded 500 m above Urakami, Nagasaki City, on 9 August, 1945. Half an hour after the detonation, large amounts of fission products and unfissioned plutonium, in the form of fallout, reached the ground surface in Nishiyama, an eastern suburb of Nagasaki, 3 km from the hypocenter. Mahara and Miyahara (1984) examined the migration rate of Pu in soil. They found the concentration of ^{239,240}Pu in soil in Nishiyama was about 10 to 30 times higher than that in other areas of Nagasaki. A sampling spot was then determined for excavation of an uncultivated and undisturbed soil core so that a vertical distribution of ^{239,240}Pu in the subsurface environment after 36 years could be observed. A soil core with a diameter of 0.3 m and a length of 2.25 m was removed and was sliced from top to bottom into sections, 50 to 250 mm thick. The sections were analyzed for ^{239,240}Pu and ^{137}Cs.

Vertical distributions of ^{239,240}Pu and ^{137}Cs in Nishiyama are illustrated in Figure 4.5. Samples of up to 0.1 m in depth retained 90% of the total ^{239,240}Pu content and 96% of the total ^{137}Cs content. Comparison of the distribution of ^{239,240}Pu with that of ^{137}Cs in surface soil to a depth of 0.3 m showed that ^{239,240}Pu was slightly more mobile than Cs. These results are consistent with those of other investigations: most of the ^{239,240}Pu and ^{137}Cs deposited on the ground surface as a result of global fallout from weapons tests did not migrate below a depth of 0.2 m, and ^{239,240}Pu migrated faster than ^{137}Cs in soil (Mclendon, 1975; Harley, 1980).

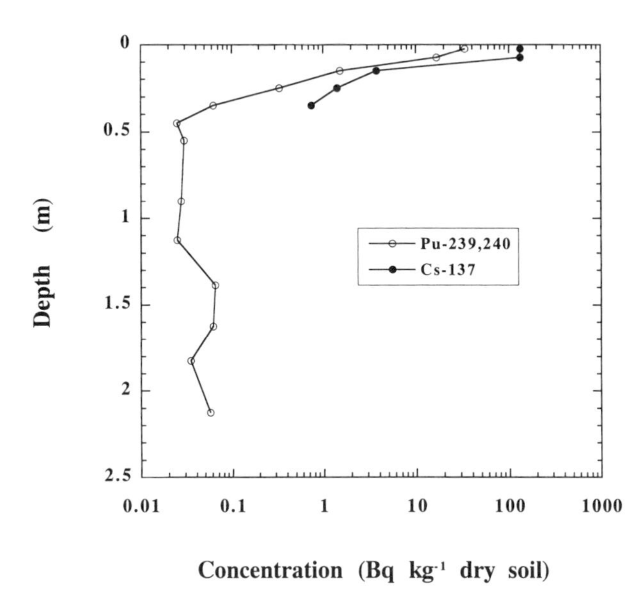

Figure 4.5. The distributions of $^{239,240}Pu$ and ^{137}Cs contents in the soil at Nishiyama, Nagasaki City, 36 years after the explosion of the atomic bomb in 1945 (Mahara and Miyahara, 1984).

Mahara and Miyahara (1984) calculated a mean migration rate of ^{239,240}Pu using the following one-dimensional diffusion equation,

$$\frac{\partial C}{\partial t} = \left(\frac{D}{R}\right) \times \frac{\partial^2 C}{\partial Z^2} - \left(\frac{U}{R}\right) \times \frac{\partial C}{\partial Z} \tag{4.1}$$

where C is the concentration of ^{239,240}Pu in soil solution, D is a diffusion coefficient, U is an average infiltrating rate of rainwater in soil, Z is soil depth, t is time, and R is a retardation factor, the ratio of the infiltration rate of rainwater to the migration rate of ^{239,240}Pu in soil. As most of the ^{239,240}Pu was retained in surface soil, initial and boundary conditions for the solution of Equation 4.1 were approximated as follows:

$$\begin{array}{lll} C = 0 & Z > 0 & t = 0 \\ C = C_o & Z = 0 & t > 0 \\ C = 0 & Z = \infty & t > 0 \end{array} \tag{4.2}$$

The analytical solution to Equation 4.1 under the above conditions is given by Lindstrom et al. (1967) as follows:

$$C = \frac{C_0}{2}\left\{ \mathrm{erfc}\left[\frac{Z - Vt}{(4D^* t)^{1/2}}\right] + \exp\left(\frac{VZ}{D^*}\right) erfc\left(\frac{Z + Vt}{(4D^* t)^{1/2}}\right)\right. \tag{4.3}$$

where $D^* = D/R$, the diffusion coefficient of ^{239,240}Pu, and $V = U/R$, the migration rate of ^{239,240}Pu in soil. Mahara and Miyahara (1984) estimated the ^{239,240}Pu retardation factor to be 2000 from comparison of theoretical curves with the measured ^{239,240}Pu content in soil. From this result, they deduced the mean ^{239,240}Pu migration rate to be 1.25 mm y^{-1}. About 97% of the total ^{239,240}Pu concentration was retained in a layer of up to 0.3 m in depth in the Nishiyama soil, but the remaining 3% had moved to a depth of 0.3 to 2.25

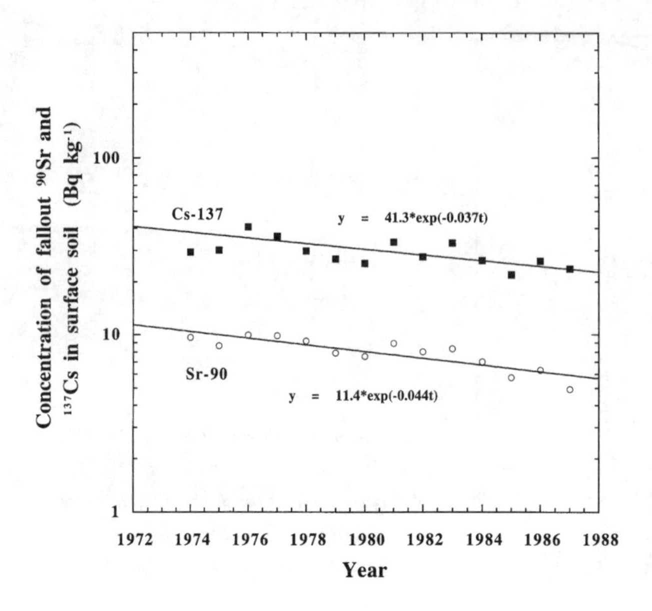

Figure 4.6. Concentrations of fallout ^{90}Sr and ^{137}Cs in uncultivated surface soil (0 to 5 cm depth) on nationwide basis (Uchida, 1990).

m, where the average concentration of the radionuclide was about 0.037 Bq kg^{-1} of dry soil. The ^{137}Cs content in the soil at the depths of 0.4 to 2.25 m was below the detection limit of gamma ray spectrometry of 0.37 Bq kg^{-1}.

Figure 4.6 shows the concentrations of fallout ^{90}Sr and ^{137}Cs in the uncultivated surface soil (0 to 5-cm depth) (Uchida, 1990). Each value is an average of soil samples collected at 31 locations in Japan by the Japan Chemical Analytical Center (National Institute of Radiological Sciences, 1975–1988). The effective removal constants from the surface soil, including the radiological decay constants, are estimated as 0.044 y^{-1} for ^{90}Sr and 0.037 y^{-1} for ^{137}Cs.

Concentrations of ^{90}Sr in farm soils are shown in Figure 4.7. During the periods with decreasing ^{90}Sr deposition rates, 1964 to 1980, the amounts of exchangeable ^{90}Sr in the paddy and upland soils decreased exponentially with time. The effective removal constant

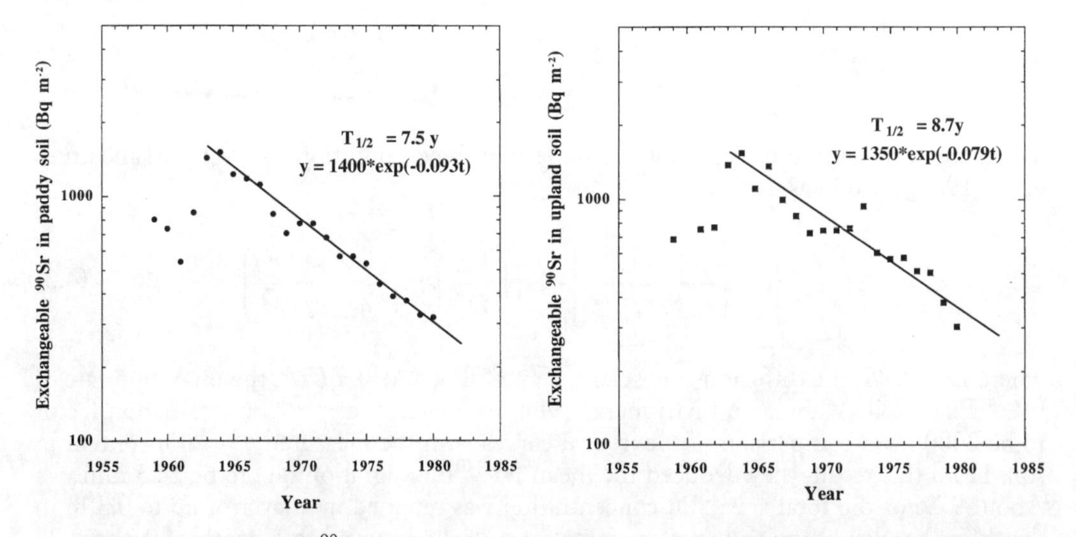

Figure 4.7. Exchangeable ^{90}Sr in plowed layers in paddy and upland soils (Kobayashi et al., 1984).

Table 4.3. Concentrations of ^{237}Np, $^{239,2240}Pu$, and ^{137}Cs in Paddy Soils Collected from Akita, Niigata, and Ishikawa Prefectures During 1959 to 1989[a,b]

Location	Year of Sampling	Sampling Depth (cm)	Dry Bulk Density (cm)	Content [Bq kg^{-1} dry]		
				^{237}Np	$^{239,240}Pu$	^{137}Cs
Akita	1959	12	0.83	0.82	0.36	29
	1963	12	0.71	2.02	0.73	73
	1967	13	0.81	2.26	0.73	68
	1971	13	0.81	1.84	0.66	56
	1975	13	0.81	1.87	0.70	51
	1979	13	0.81	1.60	0.62	43
	1984	16	0.81	1.66	0.63	38
	1989	13	0.81	1.57	0.62	35
Niigata	1959	13	0.95	0.72	0.48	40
	1963	12	0.83	2.64	1.14	109
	1967	12	0.91	4.55	1.41	150
	1971	15	0.88	2.96	1.11	103
	1975	14	0.88	1.81	0.89	78
	1979	14	0.88	2.22	0.86	72
	1984	14	0.88	1.58	0.58	39
	1989	12	0.88	1.66	0.66	43
Ishikawa	1959	13.5	1.07	0.78	0.33	27
	1963	16	1.05	2.22	0.55	67
	1967	16	1.05	2.61	0.65	76
	1971	15	1.04	2.11	0.58	60
	1975	15	1.04	2.27	0.74	68
	1979	15	1.03	1.76	0.85	50
	1984	16	1.03	1.51	0.57	44
	1989	14	0.70	0.39	0.20	12

[a]The ^{137}Cs value was corrected for decay to date of sampling.
[b]Yamamoto et al., 1994.

in the soils was estimated to be 0.093 y^{-1} for paddy soils and 0.079 y^{-1} for upland soils (Kobayashi et al., 1984). Atmospheric deposition of the radionuclides was not considered in the estimation because the annual increasing rates of ^{90}Sr deposition during the corresponding period were evaluated to be less than 2%. The constants of exchangeable ^{90}Sr for paddy and upland soils were a little larger than those for uncultivated soils. This may be due to elimination of plant-available ^{90}Sr by root uptake or transformation to unexchangeable forms in the soils.

Yamamoto et al. (1994) measured the global fallout of ^{237}Np, $^{239,240}Pu$, and ^{137}Cs in paddy soils collected periodically from 1959 to 1989 in Akita, Niigata, and Ishikawa prefectures (Table 4.3). Figure 4.8 shows the concentrations of these radionuclides as a function of time in paddy soils in Akita prefecture. The ^{237}Np contents ranged from 0.39 to 4.55 mBq kg^{-1} for three prefectures. The concentrations of $^{239,240}Pu$ ranged from 0.20 to 1.41 Bq kg^{-1} and ^{137}Cs from 12 to 150 Bq kg^{-1} on a dry soil basis. In each location, the highest values for the three radionuclides were shown in 1967 or 1963.

They also investigated residence times of ^{237}Np, $^{239,240}Pu$, and ^{137}Cs in paddy soil with a simple compartment model. The residence time of ^{237}Np in the soils was shorter than that of $^{239,240}Pu$ and ^{137}Cs, implying greater mobility of ^{237}Np. As shown in the section on Meteorological Data, paddy fields have greater infiltration, compared to upland or uncultivated fields. In spite of this, these radionuclides are retained tightly on sur-

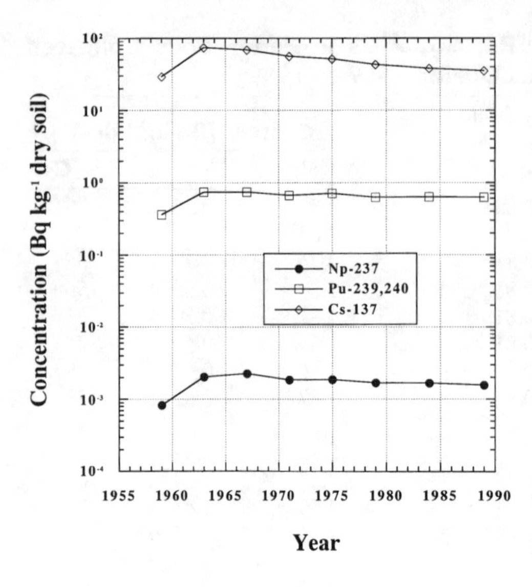

Figure 4.8. Concentrations of fallout ^{237}Np, ^{239,240}Pu, and ^{137}Cs in paddy soils in Akita City (Yamamoto et al., 1994).

face soils. However, plants cannot take up all of the radionuclides present in the surface soil layer. To predict the amount of the radionuclides transferred to crops, it is important to determine their plant-available form.

Migration in the Aerated Zone

The migration rate of ^{90}Sr in soil is higher than that of the other cationic radionuclides. Kamata (1984) obtained concentration profiles for ^{90}Sr in aerated zones in Japan. Figure 4.9 shows the profile of ^{90}Sr at Kibune, Yamagata prefecture. As seen in this figure, some peaks were observed in deep layers. This suggests that ^{90}Sr migrates into the groundwater along with precipitation from the surface soil. Morisawa et al. (1983) proposed a mathematical model to estimate the vertical migration of fallout ^{90}Sr in the aerated zone.

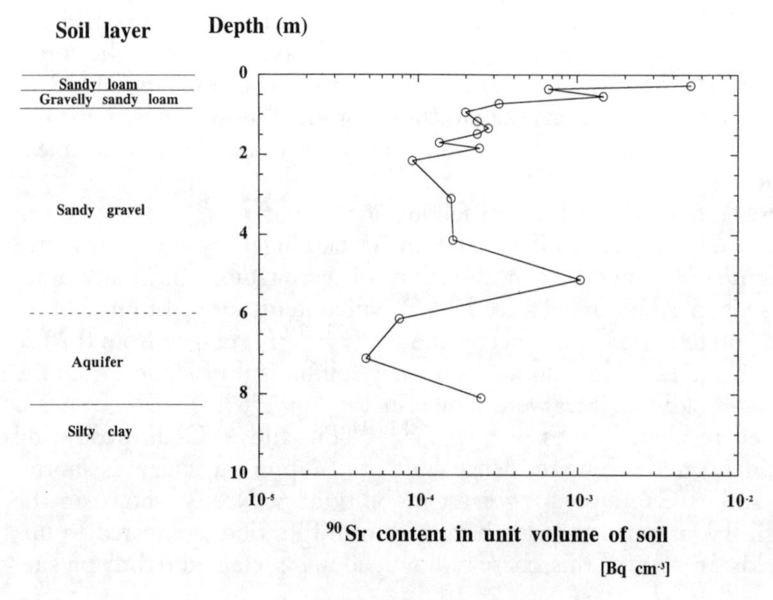

Figure 4.9. Depth profiles of fallout ^{90}Sr in uncultivated soil (Kamata, 1984).

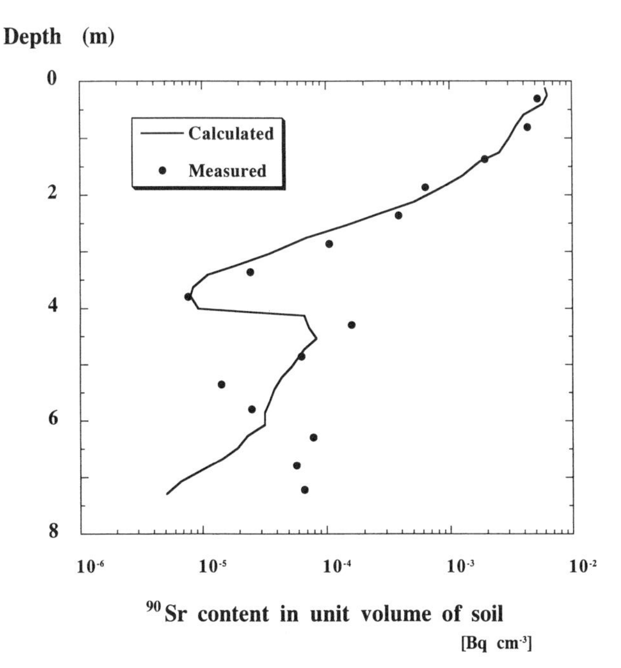

Figure 4.10. Comparison of calculated profile with measured one in Ibaraki, 1980 (Morisawa et al., 1983).

To examine the validity of the model, the calculated ^{90}Sr profiles for two locations, in Ibaraki and Yamagata prefectures, were compared with the measured ones. Figure 4.10 shows the calculated and measured ^{90}Sr profiles for the former location. The proposed model is applicable for estimating ^{90}Sr migration in the aerated zone. Among the parameters in the model, the distribution coefficient K_d is one of the most important ones for the calculated ^{90}Sr profile.

Distribution Coefficient K_d

Distribution coefficient K_d is often used to predict the behavior of a radionuclide in soil-water systems. This coefficient is defined as the concentration of a radionuclide per unit weight of soil divided by the concentration of a radionuclide per unit volume of solution at equilibrium. Table 4.4 shows the K_d values of radionuclides for soils obtained from Japan and other countries (Okabayashi and Uchida, 1990). The data were collected from publications, mainly from 1980 to 1989. The Japanese data were obtained mainly in laboratories, while the other data were obtained in the field. As shown in the table, there is little difference in K_d values between Japan and the other countries. The K_d values of these radionuclides on clay are higher than those on sand and loam, and the values depend not only on the size of the soil particles (surface area per unit weight), but also on the types of minerals in the soils. However, it is useful for prediction of radionuclide migration in soil to give K_d values for categorized soil types.

Yasuda et al. (1995) measured K_d values for five radionuclides (^{54}Mn, ^{60}Co, ^{65}Zn, ^{85}Sr, and ^{137}Cs) by a batch technique for paddy and upland soils collected throughout Japan. The values for each radionuclide showed a lognormal distribution except for ^{54}Mn (Fig. 4.11). When the correlations between the K_d values and soil properties were examined, the combinations giving the highest correlation for each radionuclide were the exchangeable Ca for Mn and Co; the water content for Zn; the cation exchange capacity (CEC) for Sr; and the exchangeable K_d for Cs. The Sr- K_d showed an adequate correlation with

Table 4.4. Distribution Coefficients (K_d: L kg^{-1}) for Various Soils Obtained from Japan and Other Countries[a]

	Sand	Loam	Clay
		Japan	
Co	$3.0 \times 10^0 - 1.0 \times 10^3$	$1.0 \times 10^1 - 3.3 \times 10^3$	$2.0 \times 10^1 - 1.2 \times 10^4$
Sr	$5.0 \times 10^{-1} - 6.8 \times 10^2$	$5.0 \times 10^{-1} - 2.7 \times 10^3$	$1.0 \times 10^1 - 9.0 \times 10^3$
I	4.0×10^1		
Cs	$3.0 \times 10^0 - 8.0 \times 10^3$	$1.0 \times 10^0 - 2.5 \times 10^3$	$5.0 \times 10^1 - 7.0 \times 10^4$
U	4.6×10^3		
Pu	$9.7 \times 10^2 - 1.9 \times 10^3$	$2.4 \times 10^3 - 4.6 \times 10^3$	$6.0 \times 10^2 - 7.0 \times 10^2$
Am	$1.0 \times 10^1 - 1.0 \times 10^3$	$6.0 \times 10^1 - 7.0 \times 10^4$	$1.6 \times 10^3 — 3.5 \times 10^4$
		Other Countries	
Co	$1.0 \times 10^2 - 4.8 \times 10^3$		$2.0 \times 10^0 - 1.2 \times 10^4$
Sr	$3.0 \times 10^{-1} - 1.7 \times 10^2$	$9.0 \times 10^{-1} - 1.5 \times 10^2$	$1.0 \times 10^0 - 1.0 \times 10^5$
I			$5.0 \times 10^{-1} - 1.5 \times 10^3$
Cs	$6.0 \times 10^0 - 3.8 \times 10^3$		$4.0 \times 10^{-1} - 1.1 \times 10^5$
U	$1.0 \times 10^{-1} - 2.2 \times 10^3$	$2.0 \times 10^0 - 7.3 \times 10^1$	$2.0 \times 10^0 - 6.5 \times 10^2$
Pu	$3.2 \times 10^1 - 1.8 \times 10^3$		$4.0 \times 10^1 - 4.0 \times 10^4$
Am	$7.5 \times 10^1 - 8.0 \times 10^2$		$4.0 \times 10^0 - 1.0 \times 10^7$

[a]Okabayashi and Uchida, 1990.

the value of the CEC divided by the supernatant electrical conductivity (EC) (Yasuda and Uchida, 1993).

TRANSFER OF RADIONUCLIDES TO CROPS

Critical Foods for Intake of Fallout Radionuclides

Ingestion is one of the important pathways for radionuclides to enter the human body. Eating habits, however, differ by countries. Figure 4.12 shows the relative contributions of 11 food groups to the total food intake (Ministry of Agriculture, Forestry and Fisheries, 1988). In European and North American countries, livestock products including

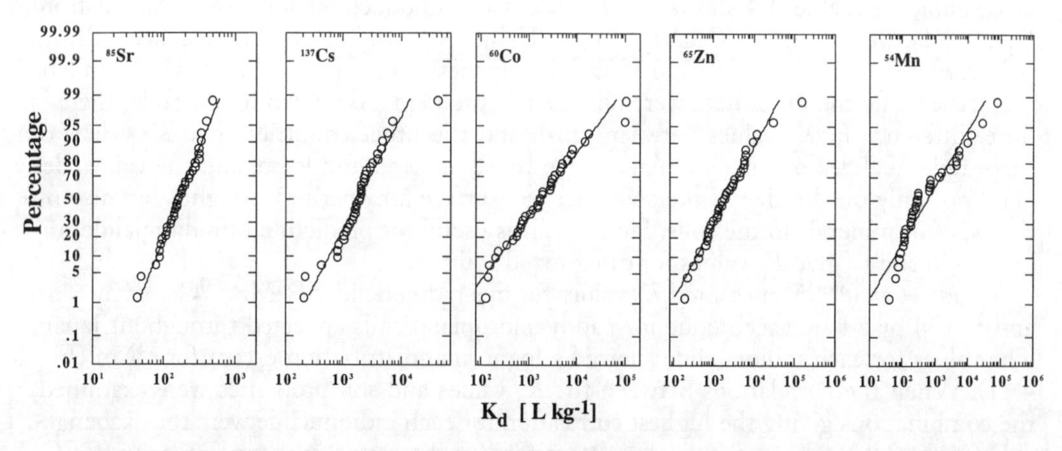

Figure 4.11. Probability distributions of K_ds for Sr, Cs, Co, Zn, and Mn, for paddy and upland soils throughout Japan (Yasuda et al., 1995).

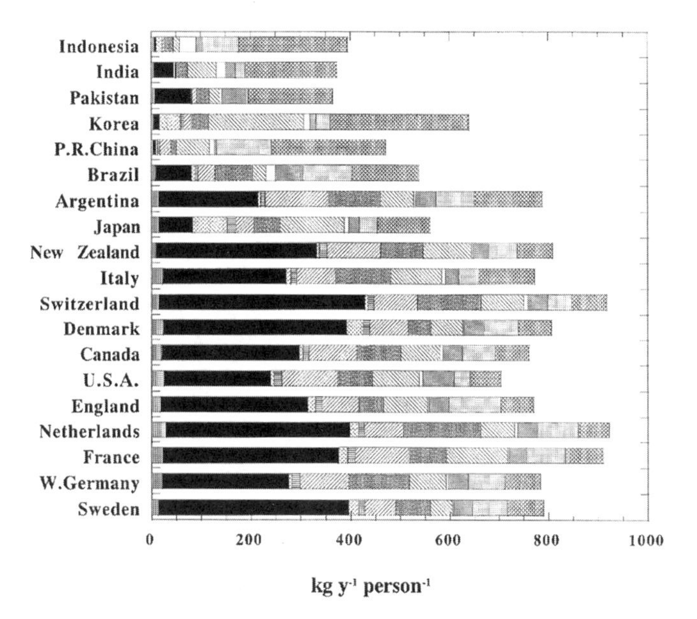

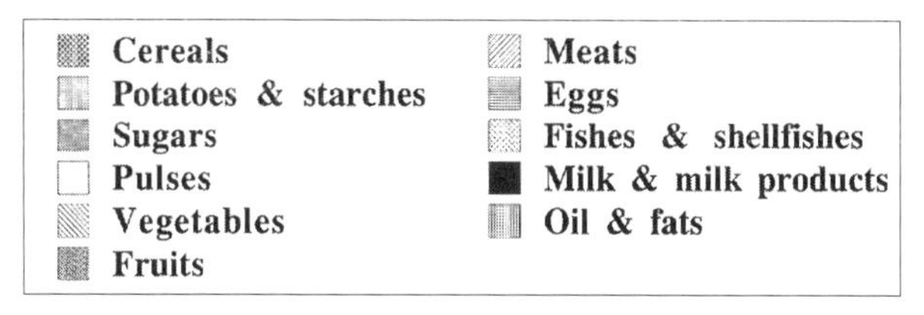

Figure 4.12. Contributions of 11 food groups to total food intake per year per capita in selected countries (Ministry of Agriculture, Forestry and Fisheries, 1988).

meat, eggs, and milk make a big contribution, whereas in Asian and South American countries, agricultural products including cereals and vegetables are the main contributors.

To find critical foods for intake of fallout ^{137}Cs and ^{90}Sr in Japan, radiochemical analysis of foods in nine categories (root vegetables, leafy vegetables, fish and shellfish, meat, eggs, milk and milk products, potatoes, beans, and cereals) was conducted by Ueda et al. (1974). The contribution of each food to the total intake of fallout ^{137}Cs and also ^{90}Sr by adults (25 to 46 years old) is shown in Figure 4.13. Percentage contributions given in the figure are averages during the period from 1966 to 1971 in three prefectures, Hokkaido, Niigata, and Kagoshima. The contribution from cereals for ^{137}Cs is the highest (about 40%), followed by beans, milk and milk products, leafy vegetables, and fish and shellfish. Among the cereals, consumption of rice is the highest in Japan. In the case of fallout ^{90}Sr, the highest contribution is from leafy vegetables, followed by beans and rice.

Hisamatsu and Takizawa (1990) measured ^{137}Cs ingestion rate for a variety of food types during 1983 to 1985. Food samples were purchased in Akita City using the market basket sampling method. The food samples were divided into 13 separate food categories: polished rice, other cereals, potatoes, confectionery/sugar, beans/seasoning, fruit, green vegetables, other vegetables, seaweeds, fish/shellfish, meat/poultry, eggs, and dairy products. Among the food categories, the largest contributor to ^{137}Cs intake was fish/shellfish, except in 1985, as shown in Table 4.5. The contribution by fish/shellfish was ap-

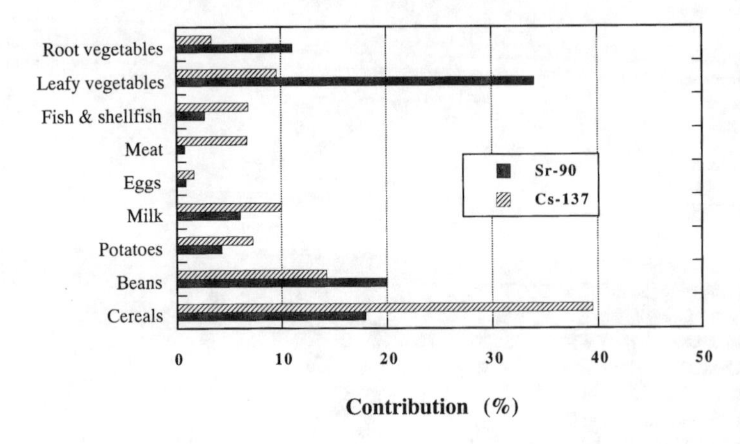

Figure 4.13. Relative contributions of nine food groups to the total intake of fallout ^{90}Sr and ^{137}Cs. Each value is a mean during the period from 1966 to 1971 in three prefectures (Ueda et al., 1974).

proximately 20% of total ^{137}Cs intake. The average of ^{137}Cs intake by consumption of rice, dairy products, and meat/poultry for 1983 to 1985 was about 13% of the total for each. The contributions by other food categories—cereals, potatoes, and vegetables (both green and other vegetables)—were roughly 10% each. It was noted that the relative contributions by each food category had not changed much from those in the 1960s.

Hisamatsu and Takizawa (1990) also obtained the post-Chernobyl ^{137}Cs ingestion rates from the same food categories in Akita City for 1986 to 1988. The total Chernobyl ^{137}Cs intake per capita in Akita City was estimated as about 60 Bq. The ^{137}Cs intake rate obtained during June to July 1986 was found to be 0.27 Bq day^{-1}, which is approximately three times higher than the preaccident rates during 1983 to 1985. The contribution of dairy products to the total ^{137}Cs intake was 42%. Green vegetables collected in Akita City from May to early June 1986 were heavily contaminated with ^{137}Cs from the accident (Hisamatsu et al., 1988), while ^{137}Cs in green vegetable samples collected during late June to July 1986 was notably low. The relative contributions of various food cate-

Table 4.5. ^{137}Cs Ingestion Rates Per Capita for the General Population in Akita City during 1983 to 1985[a]

	^{137}Cs Ingestion (Bq days^{-1} x 10^{-2})		
Sample	**1983**	**1984**	**1985**
Polished rice	0.82	0.88	2.3
Other cereals	1.9	0.47	0.63
Potatoes	—	0.94	0.88
Confectionery/sugar	0.46	0.20	0.03
Beans/seasoning	0.96	0.29	0.33
Fruit	0.48	0.25	0.03
Green vegetables	2.4[b]	0.07	0.24
Other vegetables	—	0.70	0.87
Seaweeds	—	0.07	0.37
Fish/shellfish	2.1	1.6	1.7
Meat/poultry	1.7	1.3	0.78
Eggs	0.09	0.09	0.10
Dairy products	1.3	1.4	1.1
Total	12.2	8.3	9.4

[a]Hisamatsu and Takizawa, 1990.
[b]Including other vegetables and potatoes.

gories to the total intake of ^{137}Cs from the accident differed from those from nuclear weapons tests. The ^{137}Cs intake rate from dairy products increased by a factor of 10 relative to the preaccident value.

The intake of fallout Pu from six or seven food groups collected in Japan during 1962 and 1983 to 1984 has also been reported (Hisamatsu et al., 1986). During 1962 the contribution from algae was approximately 60% of the total Pu ingested. That of beans/fruits-vegetables was second, followed by cereals, fish/shellfish, and meats/milk. During 1984, the contribution from algae was 74%. The second contributor was fish/shellfish (20%). The contribution of other food groups was minor. The contribution from whole marine products, i.e., algae and fish/shellfish, was approximately 70% of the total Pu intake in 1962, and more than 90% in 1983 to 1984. The higher Pu intake by Japanese than by people living in the United States is attributable to the greater consumption of marine food products by the former. From these results, it was concluded that marine products, especially algae, are the largest source of Pu intake in Japan.

Pathways of Radionuclides to Crops

There are two radionuclide pathways to crops: direct deposition and root uptake. To investigate the mechanism of fallout ^{90}Sr and ^{137}Cs contamination in rice, rice plants growing in the autumn of 1960 were analyzed for both radionuclides (Ichikawa et al., 1962). Half of the rice plants were covered with polyethylene sheets during the period from earshooting to harvest to eliminate floral absorption of the current fallout activity. Contents of ^{90}Sr and ^{137}Cs were compared between covered and uncovered plants. The ^{90}Sr contents in hulls, bran, and leaves of covered plants were 60 to 80% less than those of uncovered ones. These parts are more apt to absorb the fallout directly, whereas the polished rice is protected by the hulls. The ^{137}Cs contents of the covered and uncovered plants were almost the same.

The National Institute of Agro-Environmental Sciences investigated the contributions of direct and indirect transfers of the radionuclides to cereals by statistical analysis of the data of ^{90}Sr contents in brown rice and paddy soils measured since 1959 (Kobayashi et al., 1984). It was found that the ^{90}Sr content in brown rice during the periods of high ^{90}Sr deposition rates showed fairly good correlation with the ^{90}Sr deposition rate. Figure 4.14 shows the relationship of ^{90}Sr content between brown rice and polished rice under different conditions. A good correlation in ^{90}Sr content was observed between brown rice and polished rice under both conditions of high and low ^{90}Sr fallout rates in the nationwide survey. At a low deposition rate, much ^{90}Sr was found in polished rice, while there was less at a high deposition rate.

These results show that initially after a nuclear reactor accident the direct pathway, transfer of radionuclides from the atmosphere to crops, is very important. Subsequently, during a long-term contamination the root uptake is the main contamination pathway.

Transfer from the Atmosphere to Crops

Since radioiodine is released from nuclear facilities in gaseous forms into the environment, obtaining information on the transfer of the radionuclide onto crops is important to assess the radiation dose to the thyroid gland. The main chemical species of the gaseous iodine in the atmosphere are elemental iodine (I_2) and methyliodide (CH_3I) (Noguchi and Murata, 1988). Many workers have measured the deposition velocities of

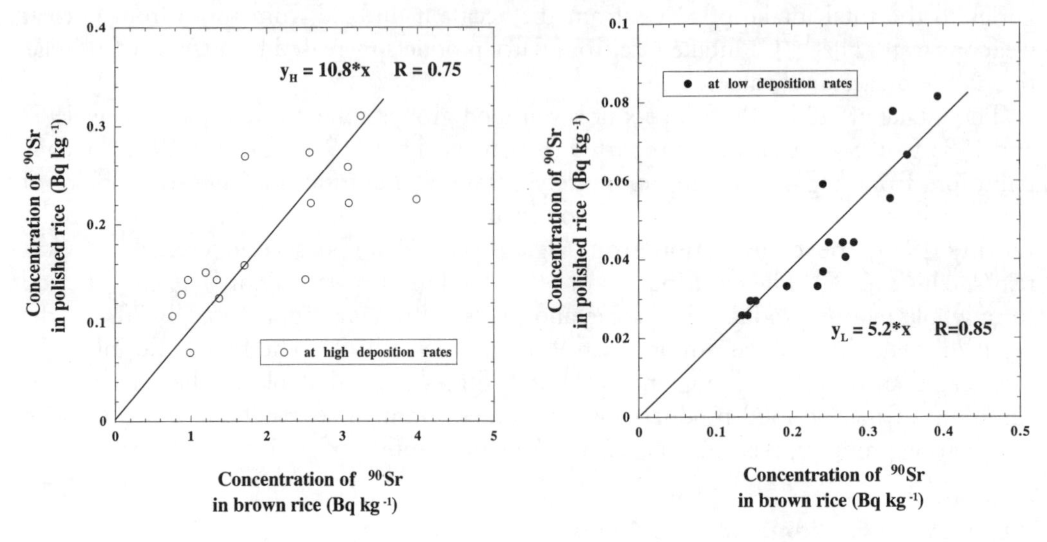

Figure 4.14. Relationship for fallout ^{90}Sr content between brown rice and polished rice during the high and low fallout ^{90}Sr deposition periods (Kobayashi et al., 1984).

these gaseous iodine species on the ground surface and on the crops, excluding cereals (Sehmel, 1980).

To obtain the deposition velocities of gaseous iodine on rice, exposure of rice plants to I_2 and CH_3I was carried out using an exposure chamber (Uchida et al., 1988a; Muramatsu et al., 1996). Experiments were also conducted with Chinese cabbage and spinach (Muramatsu et al., 1992).

Sumiya et al. (1987) observed that the translocation of iodine deposited on leaves and stalks of rice plants to rice grains both before and after the heading period was negligible. It was deduced that iodine found in brown rice and polished rice mainly originated from that deposited on the hull. Deposition velocity (V_g) has been widely used as a key parameter for estimating radioiodine concentration in crops. This parameter is defined as the iodine deposited on a unit area of ground per unit time divided by the average concentration of iodine in the atmosphere. It is applicable to pasture grass, which covers the ground almost completely. However, in the case of standing crops, it is better to use the mass-normalized deposition velocity (V_D) recommended by Hoffman (1976). The V_D is defined as the amount of iodine per unit weight of unhulled rice samples per unit time divided by the iodine concentration per unit volume of the air. The V_D value, however, varied according to the weight of unhulled rice exposed. For example, the V_D in the heading period was 10 times higher than that in the dough stage. The deposition of gaseous iodine was primarily related to the surface area of unhulled rice, which approximately corresponded to the grain number. Grain number normalized deposition velocity (VS) (Uchida et al., 1988a) was introduced instead of V_D and V_g. The V_S is defined as the amount of iodine deposited per unit number of unhulled rice samples per unit time divided by the iodine concentration per unit volume of the air.

Mean values of V_D and V_S of unhulled rice for I_2 were 0.15 $cm^3\ g^{-1}\ s^{-1}$ and 0.44 cm^{-3} (100 grains s)$^{-1}$ respectively (Uchida et al., 1988a). The values of V_D and V_S for CH_3I were 4.8 10^{-4} $cm^3\ g^{-1}\ s^{-1}$ and 9.0 10^{-4} cm^3 (100 grains s)$^{-1}$, respectively (Muramatsu et al., 1996). The V_D and V_S of I_2 were about 300 times higher than those of CH_3I. These results indicate that the amount of the radioiodine deposition on rice is highly influenced by the chemical forms of the radionuclide in the atmosphere. The lower deposition velocities for CH_3I compared to I_2 were also observed in other plants such as spinach (Nakamura and Ohmomo, 1980a; 1980b).

Loss of iodine on unhulled rice by biological and weathering processes, excluding washout, has also been examined (Uchida et al., 1991). It was observed that both species of iodine were retained on unhulled rice and little was removed.

As rice is eaten after threshing and usually after subsequent polishing, knowing the distribution of the deposited iodine in a rice grain is therefore necessary to estimate the decontamination by these procedures. The distribution ratios (on a grain basis) between unhulled rice: brown rice: polished rice were 1.0: 0.05: 0.02 for I_2 (Uchida et al., 1988a) and 1.0:0.49:0.38 for CH_3I (Muramatsu et al., 1996). About 95% of the elemental iodine remained in the hull and only a few percent remained in polished parts, whereas about 40% of the methyliodide was found in polished rice. This suggests that gaseous methyliodide is apt to enter the grains. Deposition characteristics of both species of iodine in Chinese cabbage were also examined. Elemental iodine remained on the outer leaves, specifically on green parts of the body, while more than 10% of the methyliodide was found in the inner leaves (Muramatsu et al., 1992).

Transfer from Soil to Crops

As mentioned before, in Japan rice is the most important crop for internal radiation dose assessment; therefore, the main focus in this section is the transfer of radionuclides from soil to rice. For estimating radionuclide transfer through this route, the transfer factor (*TF*) is commonly used. The *TF* is defined as the concentration of the radionuclide in a plant organ at harvest divided by the concentration of the radionuclide in dry soil.

TFs of ^{90}Sr, ^{137}Cs, $^{239,240}Pu$, and ^{99}Tc

The National Institute of Agro-Environmental Sciences has measured the concentration of fallout ^{90}Sr and ^{137}Cs in brown rice, wheat, and farm soil samples collected from various parts of Japan (Fig. 4.15). The concentrations of fallout ^{137}Cs in brown rice and paddy soils collected in 1988 are shown in Table 4.6 (Komamura et al., 1989). The *TFs* obtained from these data are also listed. The maximum *TF* value is 0.03, the minimum 0.0007, and the average 0.007. Table 4.7 shows the concentration of ^{137}Cs in husked wheat and upland soils, and *TFs*. The average *TF* value is 0.006. The range of the *TFs* of ^{137}Cs is within two orders of magnitude for brown rice and one order of magnitude for wheat. The *TFs* for brown rice based on the plant-available form of ^{90}Sr in paddy soils obtained in 1987 are shown in Table 4.8 (Komamura et al., 1988). The plant-available form of ^{90}Sr in the soil was extracted with 1 N ammonium acetate solution as exchangeable ^{90}Sr. The average *TF* of ^{90}Sr for brown rice is 0.034, the maximum *TF* value 0.075, and the minimum 0.011. As shown in this table, the variances of *TFs* are small.

Okajima et al. (1990) reported the *TFs* of $^{239,240}Pu$ in Nishiyama, Nagasaki City (cf. Migration section), where there was concentrated radioactive fallout with black rain following the atomic bombing. As this bomb was a Pu bomb, the fallout must have contained nonfissioned ^{239}Pu in addition to fission products. Cultivated soils and crops were collected in this area in 1982 and 1983. The *TFs* of $^{239,240}Pu$ and ^{137}Cs for six types of crops are shown in Table 4.9. Control samples were also collected from outside Nishiyama. The TFs for $^{239,240}Pu$ are 10^{-4} to 10^{-3}, being somewhat smaller in Nishiyama district than in the control district. For ^{137}Cs, the *TFs* are from 0.03 to 0.21 for Nishiyama and from 0.01 to 0.15 for the control. The *TF* of ^{137}Cs was 100 to 200 times higher than that of Pu obtained in Nishiyama. Variation of Pu uptake by plants can be

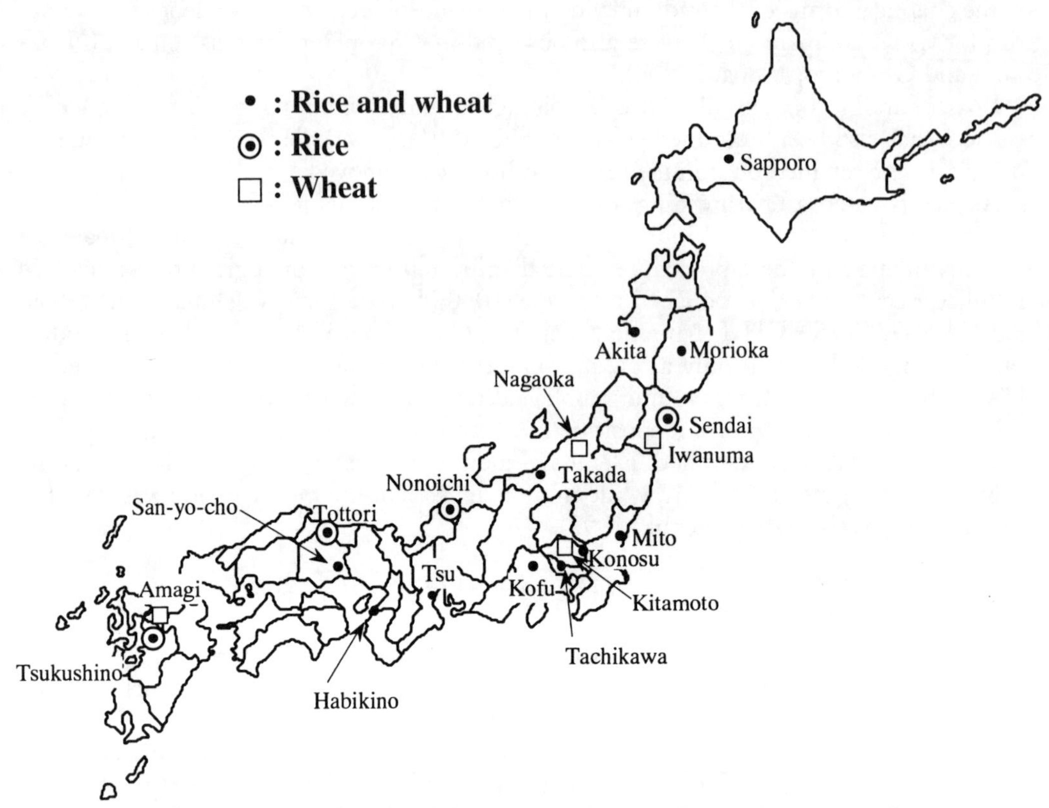

Figure 4.15. Sampling sites of rice and wheat in Japan (Kobayashi et al., 1984).

explained by differences in plant species, as described later, as well as edaphic parameters such as type of soil, climatic condition, and differences in the chemical forms of Pu initially added to the soil.

Yanagisawa et al. (1992) obtained the *TF*s of Tc from soil to rice and wheat plants by laboratory experiments. The factors for rice plants were 0.005 for the grains (hulled) and about 1.1 for the lower leaf blade. By contrast, higher *TF*s were found for wheat plants; 0.027 for the grains (hulled) and 230 for the lower leaf blade. The Tc concentrations in the grains of both plants were lower than those in the leaves. The level of Tc in the soil solution collected from the flooded soil decreased rapidly with time. In the case of wheat plants grown in a nonflooded soil, the decrease in Tc level in the soil solution was rather slow. It was suggested that the Tc tracer, added as TcO_4^- to the paddy soil, was readily transformed to insoluble forms under the reducing conditions in the flooded soil (Tagami and Uchida, 1996). The low *TF* observed for the rice plants could be explained by a fixation of Tc in the soil.

TFs of Iodine for Rice

From a radioecological point of view, ^{131}I ($T_{1/2}$ = 8.0 d) and ^{129}I ($T_{1/2}$ = 1.6 x 10^7 y) are two of the concerned radionuclides of iodine. For ^{129}I, rice root uptake may be more important, because this radionuclide has a long physical half-life and rice plants are cultivated under flooded conditions. Under flooded conditions, iodine is soluble in the soil solution and possibly taken up by crops through their roots (Tensho and Yeh, 1970).

The *TF*s of iodine from soil to rice were obtained by laboratory experiments using ^{129}I

Table 4.6. Concentrations of Fallout ^{137}Cs in Brown Rice and Paddy Soils in 1988 and Transfer Factors (*TFs*) [a]

Prefecture	Brown Rice Bq kg^{-1}	Paddy Soil Bq kg^{-1}	*TF* (–)
Hokkaido	0.24	8.2	0.030
Iwate	0.40	24.0	0.017
Miyagi	0.18	20.0	0.0092
Akita	0.26	29.8	0.0086
Niigata	0.079	44.2	0.0018
Ishikawa	0.026	5.3	0.0049
Ibaraki	0.039	14.5	0.0027
Tokyo	0.039	25.1	0.0016
Osaka	0.026	4.8	0.0054
Tottori	0.034	20.3	0.0017
Okayama	0.006	9.0	0.00067
Fukuoka	0.039	7.5	0.0052
Average	0.11	17.7	0.0073
Maximum	0.40	44.2	0.030
Minimum	0.006	4.8	0.00067

[a]Komamura et al., 1989.

Table 4.7. Concentrations of Fallout ^{137}Cs in Husked Wheat and Upland Soils in 1988 and Transfer Factors (*TFs*) [a]

Prefecture	Husked Wheat Bq kg^{-1}	Upland Soil Bq kg^{-1}	*TF* (–)
Hokkaido	0.13	11.4	0.012
Iwate	0.076	17.8	0.0043
Miyagi	0.051	10.4	0.0049
Niigata	0.059	28.8	0.0021
Ibaraki	0.075	5.2	0.014
Saitama	0.036	6.4	0.0056
Tokyo	0.034	13.1	0.0026
Okayama	0.023	9.5	0.0024
Average	0.061	12.8	0.0060
Maximum	0.13	28.8	0.014
Minimum	0.023	5.2	0.0021

[a]Komamura et al., 1989.

tracer (Muramatsu et al., 1989; Muramatsu et al., 1993). The *TF*s of each rice plant organ for Andosol and Gray lowland soil are shown in Table 4.10. The average *TF*s of radioiodine from soil to brown rice on Andosol and Gray lowland soil were 0.0064 and 0.0022, respectively. Because rice is usually consumed as polished rice in Japan, determining the concentration of radioiodine in polished rice is important. It was also found that ^{125}I was concentrated mainly in rice bran and the amount of radioiodine in rice depended on the degree of polishing. The weight ratio of brown rice to polished rice was usually about 1.0 : 0.9. At this ratio, about 30% of the radioiodine remained in the polished rice. The *TF*s of iodine from the soils to polished rice could be calculated as 30% of those of brown rice, i. e., about 0.002 for Andosol and 0.0007 for Gray lowland soil.

Iodine concentrations in the soil solutions during the experiments are shown in Figure

Table 4.8. Concentrations of Fallout ^{90}Sr in Brown Rice and Paddy Soils in 1987 and Plant Available Transfer Factors (*TFs*)[a]

Prefecture	Brown Rice Bq kg^{-1}	Paddy Soil[b] Bq kg^{-1}	*TF* (–)
Hokkaido	0.9	26	0.035
Akita	1.6	100	0.016
Niigata	5.5	139	0.040
Ishikawa	1.0	36	0.028
Tottori	1.0	41	0.024
Iwate	0.8	21	0.038
Miyagi	1.0	92	0.011
Ibaraki	0.9	29	0.031
lbkyo	0.3	10	0.030
Yamashina	0.3	4	0.075
Osaka	0.5	12	0.042
Okayama	0.4	10	0.040
Fukuoka	0.2	7	0.029
Average	1.1	41	0.034
Maximum	5.5	139	0.075
Minimum	0.2	4	0.011

[a]Komamura et al., 1988.
[b]Exchangeable^{90}Sr: extracted with 1N ammonium acetate solution.

Table 4.9. Transfer Factors (TFs) of $^{239,240}Pu$ and ^{137}Cs from Soil to Crops (on a Dry Weight Basis)[a]

	TF × 10^{-3}			
	^{137}Cs		$^{239,240}Pu$	
Sample	Nishiyama	Control	Nishiyama	Control
White potato	146	46	0.72	4.76
Sweet potato	130	10	1.40	3.67
Pumpkin	106	65	0.43	1.22
Taro	59	27	0.62	0.86
Radish	213	145	0.53	8.62
Leaf mustard	33	40	0.19	1.97

[a]Okajima et al., 1990.

4.16. The vertical axis shows "relative concentration of ^{125}I in the soil solution," defined as the amount of ^{125}I present in the soil solution in a pot divided by the amount of ^{125}I added to the pot. Most of the ^{125}I added to the soils was easily adsorbed by the soil solid phase, because the concentration of the radionuclide in the solutions was very low in the first growing stage. In spite of the initial high adsorption capacity of both soils, ^{125}I was gradually desorbed from the cultivated soils about one month after transplanting. The highest concentrations in the soil solution for both soils were observed around the flowering period, 42 days after transplanting. At this time, about 29% and 13% of the ^{125}I existed in the soil solutions of the flooded Andosol and Gray lowland soils, respectively. After these maximum values were achieved, the concentrations in the solutions decreased slowly. From these experiments, it seems that the iodine concentration in rice plants is influenced by that in the soil solution.

Takagi et al. (1985) analyzed stable I concentrations in rice and paddy soil in Japan and estimated the *TF*s of brown rice and polished rice. The results are shown in Table 4.11. The maximum I content in polished rice was 4 to 5 μg kg^{-1}, the minimum one is 2 μg

Table 4.10. Transfer Factors (*TFs*) for Iodine from Soil to Different Parts of Rice Plants (on a Dry Weight Basis) [a]

	Andosol			Gray Lowland Soil	
Organ	1-A	2-A	3-A	1-G	2-G
Brown rice	0.0044	0.0076	0.0073	0.0017	0.0027
Unhulled rice	0.056	0.045	0.10	0.024	0.024
Rachis	1.6	1.6	0.90	0.21	0.13
1st leaf blade	2.6	5.2	3.6	0.48	0.83
2nd leaf blade	3.8	5.1	9.3	0.87	1.7
Stem	3.4	2.3	1.3	0.36	0.39

[a]Muramatsu et al., 1989.

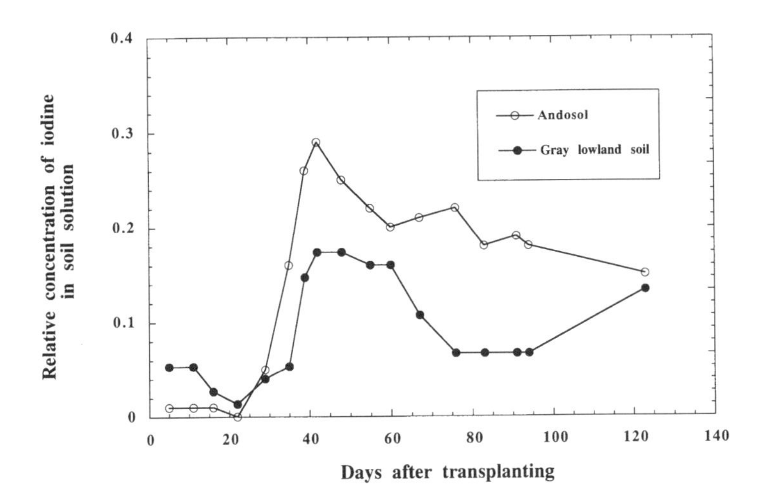

Figure 4.16. Relative concentration of iodine in soil solution after transplanting rice plants as a function of time (Muramatsu et al., 1993).

Table 4.11. Concentration of Stable Iodine in Brown Rice, Polished Rice, and Paddy Soils, Transfer Factors (TFs)[a]

	Rice (μg kg⁻¹)			Soil (mg kg⁻¹)		TF (–)		
Prefecture	1979 Polished	1980 Polished	1980 Brown	1979	1980	1979 Polished	1980 Polished	1980 Brown
Hokkaido	1.8	3.0	10.6	1.86	2.46	0.00097	0.0012	0.00431
Akita	2.2	4.6	12.2	1.58	1.29	0.0014	0.0036	0.0095
Niigata	2.3	3.7	10.4	1.15	1.4	0.0020	0.0026	0.0074
Ishikawa	1.6	—	—	0.6	0.61	0.0027	—	—
Tottori	2.2	4.6	5.9	0.98	1.27	0.0022	0.0036	0.0047
Iwate	2.1	3.6	9.4	10.2	9.47	0.00021	0.00038	0.00099
Miyagi	2.7	3.8	7.7	2.33	2.11	0.0012	0.0018	0.0037
Ibaraki	2.4	2.5	10.1	19.2	17.2	0.00013	0.00015	0.00059
Saitama	1.9	5.0	7.7	0.95	0.95	0.0020	0.0053	0.0081
Tokyo	2.5	4.2	4.7	1.29	1.35	0.0019	0.0031	0.0035
Yamanashi	3.2	2.9	4.1	0.41	0.39	0.0078	0.0074	0.011
Osaka	2.7	2.5	12.2	0.54	0.62	0.0050	0.0040	0.020
Okayama	4.0	4.7	4.9	0.61	0.81	0.0066	0.0058	0.0061
Fukuoka	2.9	3.8	6.9	0.86	0.61	0.0034	0.0062	0.011
Average	2.46	3.76	8.22	3.04	2.90	0.0027	0.0035	0.0069
Maximum	4.0	5.0	12.2	19.2	17.2	0.0078	0.0074	0.020
Minimum	1.6	2.5	4.1	0.41	0.39	0.00013	0.00015	0.00059

[a]Takagi et al., 1985.

kg^{-1}. For soil samples, the maximum value was 17 to 19 mg kg^{-1} in Ibaraki, which is close to the Pacific Ocean. As shown in this table, the *TF*s for polished rice ranged from 0.0001 to 0.008 and the average value was 0.003. The *TF* for brown rice was 0.007.

Yuita et al. (1988) collected rice and paddy soil samples from 40 locations throughout Japan, and measured the contents of stable I by neutron activation analysis. Yuita (1987) also carried out laboratory experiments using radioiodine. The average *TF* value from soil to polished rice was 0.0011 in field experiments and 0.0035 in the laboratory. Namiki et al. (1987) also collected 27 samples, pairs of polished rice and associated paddy soils, around Tokai-mura. They estimated that the *TF* of I to polished rice was 0.0015.

Although the species of rice plant, soil type, and experimental conditions were different, the values obtained in these studies are similar.

Differences in *TF*s by Crop Species

Compared with the *TF* of I in the edible part of plants (0.02) recommended by the IAEA (1982) and U. S. NRC (1977), the values obtained by Muramatsu et al. (1989; 1993) and others (Takagi et al., 1985; Yuita, 1987; Yuita et al., 1988; Namiki et al., 1987) for polished rice (about 0.002) are obviously lower, whereas those for leaves and stems (2 to 6) are much higher (Table 4.10).

Therefore, the values for I recommended by the IAEA and the U. S. NRC are not applicable to rice plants. This also suggests that different *TF*s must be applied for each plant species, or at least plant groups that are categorized by their edible parts.

Uchida and Okabayashi (1988b) proposed that *TF*s should be given by type of edible parts, i.e., rice, cereals other than rice, tubers, root vegetables, leafy vegetables, fruit and seeds (Table 4.12).

DECONTAMINATION OF RADIONUCLIDES

Removal of Radionuclides from Soil by Harvesting Plants

In some cases, elements taken up by roots are accumulated more in inedible than in edible parts. Attention, however, has scarcely been paid to transfer to the inedible parts. For example, the leaves and stem, which are inedible parts of cereals or fruits, could play an important role in lowering the amount of a radionuclide in the soil root zone by harvesting these crops.

Uchida and Muramatsu (1995) estimated the loss of ^{129}I from paddy soil by harvesting the edible and inedible parts of the rice plants. Radionuclides migrate with irrigation water. Particularly in paddy fields, much water infiltrates from the surface of the field into deep soil layers, as described in the section on Meteorology. However, for some elements, such as I and Tc, accurately estimating the concentrations of the elements in the soil solution is difficult, because the concentrations are affected by the plant growing stage (Muramatsu et al., 1993; Yanagisawa et al., 1992). Therefore, only the removal effect of ^{129}I from the paddy field by harvesting the plants was considered.

The transfer model proposed by Schwarz and Hoffman (1980) was used. The model was introduced to describe the removal of ^{129}I from the paddy fields by harvesting of rice plants only. It was assumed that no retranslocation of the radionuclide occurred, once it was taken up through the roots. Hence the transfer of the radionuclide to any organ was assumed to be independent of the others. Also, no loss of ^{129}I by other pathways was assumed; e.g., leaching from the soil and loss by surface water runoff or volatilization.

Table 4.12. Transfer Factors (*TF*s) of Radionuclides from Soil to Crops Which Were Categorized into Seven Groups Based on Edible Parts of the Crops[a]

	Rice	Cereals Other than Rice	Tubers	Root Vegetables	Leafy Vegetables	Fruit	Seeds
Mn	3×10^{-2}	5×10^{-3}–3×10^{-2}	3×10^{-3}	3×10^{-2}–4×10^{-1}	8×10^{-3}–2	1×10^{-3}	1×10^{-2}–9×10^{-1}
Fe				1×10^{-2}–3×10^{-2}	2×10^{-2}–1×10^{-1}		2×10^{-2}–1×10^{-1}
Co	4×10^{-3}	2×10^{-4}	1×10^{-3}	4×10^{-2}–3×10^{-1}	$4\times10^{-}$ –3×10^{-1}	2×10^{-3}	6×10^{-3}–2×10^{-1}
Zn	7×10^{-2}	2×10^{-1}–4×10^{-1}	4×10^{-2}	3×10^{-1}–2	3×10^{-2}–2	3×10^{-2}	2×10^{-1}–1
Sr	5×10^{-3}–3×10^{-2}	2×10^{-2}–1	1×10^{-2}		4×10^{-2}–3×10^{-1}	4×10^{-3}	2×10^{-2}–2×10^{-1}
Tc		6×10^{-3}–1×10^{-1}		1–4	9×10^{-2}–40		2×10^{-1}–29
Cs	4×10^{-2}–6×10^{-1}	3×10^{-4}–6×10^{-2}	2×10^{-3}–8×10^{-3}	8×10^{-3}–1×10^{-1}	1×10^{-3}–8×10^{-1}		5×10^{-3}–$\times10^{-1}$
Pb			8×10^{-4}–1×10^{-3}	7×10^{-4}–2×10^{-3}	8×10^{-4}–2×10^{-3}		5×10^{-4}–4×10^{-3}
Ra	4×10^{-1}–2	1×10^{-3}	1×10^{-3}	1×10^{-3}–4×10^{-3}	1×10^{-3}–3×10^{-2}	3×10^{-4}–4×10^{-3}	-2×10^{-4} –1×10^{-2}
U	4×10^{-5}–7×10^{-5}	5×10^{-4}	1×10^{-4}–9×10^{-4}	2×10^{-4}–9×10^{-3}	5×10^{-4}–1×10^{-2}	1×10^{-4}– 8×10^{-3}	2×10^{-4}
Np		8×10^{-3}–7×10^{-2}	2×10^{-3}	1×10^{-3}–4×10^{-3}	5×10^{-4}–6×10^{-2}	4×10^{-4}–1×10^{-3}	9×10^{-4}–1×10^{-2}
Pu	8×10^{-5}–1×10^{-2}	2×10^{-4}–3×10^{-2}	4×10^{-5}–1×10^{-3}	9×10^{-6}–4×10^{-4}	2×10^{-4}–3×10^{-4}	1×10^{-4}–1×10^{-3}	3×10^{-4}–2×10^{-1}
Am		2×10^{-5}–3×10^{-3}	4×10^{-5}	2×10^{-5}–2×10^{-4}	2×10^{-4}	3×10^{-6}–4×10^{-5}	2×10^{-5}–4×10^{-4}
Cm	1×10^{-3}–4×10^{-3}	2×10^{-5}–4×10^{-3}	4×10^{-5}	1×10^{-5}–3×10^{-4}	6×10^{-4}	3×10^{-6}–6×10^{-5}	3×10^{-5}–6×10^{-4}

[a]Uchida and Okabayashi, 1988b.

The concentration of ^{129}I in the root zone can be estimated by the following equation: where C_r is the concentration of the radionuclide in the cultivated soil layer (Bq kg^{-1}) as a function of time t, D_p is deposition rate per unit time and unit area (Bq m^{-2} y^{-1}), λ_R is

$$\frac{dC_{r(t)}}{dt} = \frac{D_p}{P_D} - \lambda_R \times C_{r(t)} - \frac{TF_j \times C_{r(t)} \times M_j \times n}{P_D} \tag{4.4}$$

the radioactive decay constant (y^{-1}), P_D is surface density for soil (assumes a 15-cm plow layer, expressed in dry weight) (kg m^{-2}), TF_j is the transfer factor for part j of rice plants, M_j is the amount of biomass per unit area (kg m^{-2}) at harvest, and n is the number of harvests per year (y^{-1}).

In this analysis, the fraction of the rice plant that is not removed by harvesting was assumed to be returned as litter and recycled in the root zone; i.e., rice plants could take up the radionuclide in the litter. The fractional loss rate λ_j (y^{-1}) of radioiodine from the soil layer, resulting from plant uptake and harvesting, was defined by

$$\lambda_j = \frac{TF_j \times M_j \times n}{P_D} \tag{4.5}$$

Using the boundary condition C_r (t=0) = 0 and assuming a constant deposition rate D_p, the solution of Equation 4.4 is

$$C_{r(t)} = \frac{D_p}{(\lambda_R + \lambda_j) \times P_D} \times \left\{1 - \exp\left[-(\lambda_R + \lambda_j)t\right]\right\} \tag{4.6}$$

To estimate the removal of ^{129}I by plant-root uptake from the paddy fields, four cases were considered:

Case 1: No consideration of removal by harvesting of the plants.
Case 2: Only grains (hulled) were removed by harvesting.
Case 3: Grains (hulled) and hulls were removed by harvesting.
Case 4: All parts of the plant above the ground were removed by harvesting.

The parameters used in this analysis are listed in Table 4.13. The *TF* values used for each organ of the rice plants were those for Andosol obtained in tracer experiments. The

Table 4.13. Transfer Parameters Used in the Model[a]

Case	*TF*[b]	*M* (kg m^{-2})[c]	P_D (kg m^{-2})[d]	*n* (y^{-1})
1	0	0	240	1
2	0.006	0.6	240	1
3	0.067	0.7	240	1
4	1.3	1.2	240	1

[a] Uchida and Muramatsu, 1995.
[b] *TF* values used for each organ of rice plants were those for Andosol obtained in the tracer experiments.
[c] The biomass yields (M) of the plants used were Japanese averages in 1989 (Sakumotsu-toukei kyokai, 1991).
[d] P_D is the surface density for soil (assumes a 15-cm plow, expressed in dry weight) (U.S. NRC, 1997).

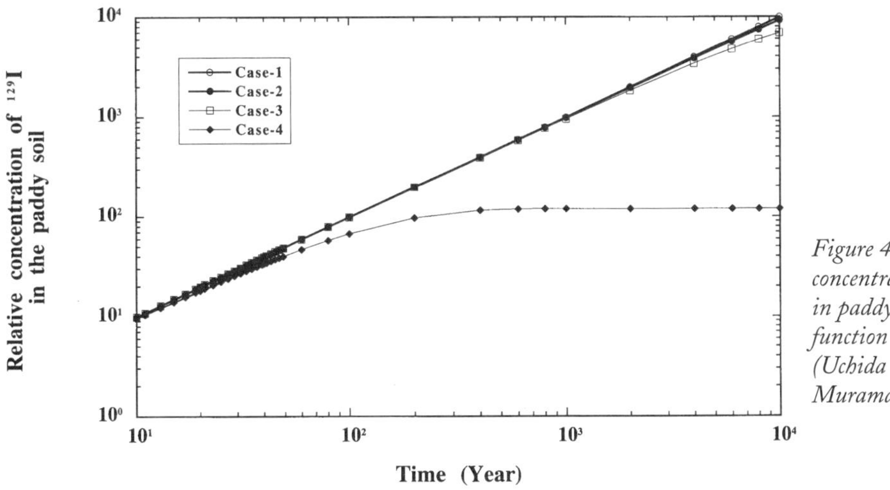

Figure 4.17. Relative concentration of ^{129}I in paddy soil as a function of time (Uchida and Muramatsu, 1995).

biomass yields (M) of the plants used were Japanese averages in 1989 (Sakumotsu-toukei kyokai, 1991). Figure 4.17 presents the relative concentration of ^{129}I in the soil of the root zone for cases 1 to 4 resulting from a continuous deposition of ^{129}I. Cases 1 and 2 showed the same trend, and case 3 was similar, during the period up to 1000 years. For these cases, the concentrations of ^{129}I in the soil increased gradually with time.

For case 4, depletion of the concentration in the root zone was clearly observed. These results were expected since more ^{129}I accumulated in stems and blades of rice plants than in hulls and brown rice. Subsequently, the concentration of the radionuclide in the root zone was constant for continuous deposition after 300 years. From these results, the harvesting of the upper parts of rice plants on the ground was judged effective in reducing ^{129}I in the root zone for the continuous release for more than 300 years; e.g., the release of the radionuclide from waste disposal sites into groundwater that is subsequently used to irrigate paddy fields.

This estimation can be adopted to evaluate the concentration of other long-lived radionuclides in agricultural land having high *TF*s like ^{99}Tc. In paddy fields, it is also important to estimate the leaching rate of radionuclides with percolation.

SUMMARY AND CONCLUSIONS

Japan is a very narrow but long island country located at about 24° to a little less than 46° of north latitude. The main island is separated geographically and meteorologically into two areas, the Japan Sea side and the Pacific Ocean coast, by mountains running from north to south throughout the archipelago. Thus, its meteorological characteristics, which are important for understanding the behavior of radionuclides in the terrestrial environment, vary greatly with the district.

Fallout due to U.S.S.R. nuclear weapon tests was higher along the north Japan Sea coast than on the Pacific Ocean side, partly because of the high annual rainfall in the former. The maximum annual depositions of radionuclides such as ^{90}Sr, ^{137}Cs, and ^{239,240}Pu were recorded in 1963, following the large-scale atmospheric thermonuclear tests conducted during 1961–1962. After cessation of atmospheric tests by the U.S.A. and U.S.S.R. in 1962, the fallout of radionuclides decreased steadily for four years. During the period from 1967 to 1981, however, the annual deposition showed no decrease because of the tests conducted by the People's Republic of China. A temporary high monthly deposition was observed in May 1986 due to the Chernobyl accident, but was followed by a rapid decrease. The radioactive contamination in Japan is mostly due to these atmospheric weapon testings.

The cumulative amounts of fallout radionuclides are comparatively low in Japan, so present radionuclide cleanup is not an urgent problem domestically. Because of the necessity of safety assessment carried out in advance of the construction of nuclear power plants and related nuclear fuel cycle facilities, intensive efforts have been directed to investigation of the migration of radionuclides in the environment, such as transfer factors (*TF*), distribution coefficients (K_d), etc.

After the Chernobyl incident, remediation of contaminated environments is one of the most important research subjects. Decontamination of radionuclides from soil by harvesting plants is one of the promising procedures. Disposal of high-level radioactive wastes remains an urgent and serious problem for the world.

Safety analysis is needed to estimate the potential individual and collective radiation doses to humans, based on hypothetical release and transport from a high radioactive waste disposal facility.

ACKNOWLEDGMENTS

We wish to express our thanks to Dr. Y. Nakamura and Dr. Y. Muramatsu, National Institute of Radiological Sciences, for their valuable comments.

REFERENCES

Aoyama, M., K. Hirose, Y. Suzuki, H. Inoue, and Y. Sugimura. High Level Radionuclides in Japan in May. *Nature*, 321, pp. 819–820, 1986.

Aoyama, M., K. Hirose, and Y. Sugimura. Deposition of Gamma-emitting Nuclides in Japan after the Reactor-IV Accident at Chernobyl. *J. Radioanal. Nucl. Chem.*, 115, pp. 291–306, 1987.

Aoyama, M., K. Hirose, and Y. Sugimura. The Temporal Variation of Stratospheric Fallout Derived from the Chernobyl Accident. *J. Environ. Radioactivity*, 13, pp. 103-115, 1991.

Brookins, D.G. *Eh-pH Diagrams for Geochemistry*. Springer-Verlag, Berlin, 1988, pp. 97–99.

Harley, J.H. Plutonium in the Environment: A Review. *J. Radiat. Res.*, 21, pp. 83–104, 1980.

Hirose, K., M. Aoyama, Y. Katsuragi, and Y. Sugimura. Annual Deposition of Sr-90, Cs-137 and Pu-239,240 from the 1961–1980 Nuclear Explosions: A Simple Model. *J. Meteor. Soc. Japan*, 65, pp. 259–277, 1987.

Hisamatsu, S., Y. Takizawa, and T. Abe. Fallout Pu in the Japanese Diet. *Health Phys.*, 51, pp. 479–487, 1986.

Hisamatsu, S., Y. Takizawa, and T. Abe. Radionuclide Contents of Leafy Vegetables; Their Reduction by Cooking. *J. Radiat. Res.*, 29, pp. 110–118, 1988.

Hisamatsu, S. and Y. Takizawa. Ingestion of Chernobyl Cs-137 in Akita City, Japan. *J. Environ. Radioactivity*, 11, pp. 267–278, 1990.

Hoffman, F. O. A Reassessment of the Deposition Velocity in the Prediction of the Environmental Transfer of Radioiodine from Air to Milk. *Health Phys.*, 32, pp. 437–441, 1976.

Ichikawa, R., M. Eto, and M. Abe. Strontium-90 and Cesium-137 Absorbed by Rice Plants in Japan, 1960. *Science*, 135, p. 1072, 1962.

International Atomic Energy Agency. Generic Models and Parameters for Assessing the Environmental Transfer of Radionuclides from Routine Releases. IAEA Safety Series No.57, IAEA, Vienna, 1982, p. 61.

Kamata, H. Soil Profiles of Radionuclides, in *Environmental Radioactivity*, Saiki, M., Ed., Soft Science Inc., Tokyo, 1984, pp. 181–182.

Katsuragi, Y. A Study of ^{90}Sr Fallout in Japan. *Pap. Meteor. Geophys.*, 33, pp. 277–291, 1983.

Kobayashi, H., M. Komamura, and A. Tsumura. The Behavior of Sr-90 from Environment to Cereal Crops by Analyzing the Radioactivity Survey Data in Japan., *Bulletin of the National Institute of Agro-Environmental Sciences*. Series B, No. 36 (in Japanese), 1984.

Komamura, M., K. Yuita, and Y. Koyama, Concentration of Fallout Sr-90 in Soil, Rice and Wheat, in *Proceedings of the 30th Meeting on Radioactivity Survey Data in the Environment*, Science and Technology Agency, Japan, pp. 21-23, 1988 (in Japanese).

Komamura, M., K. Yuita, and Y. Koyama, Concentration of Fallout Cs-137 in Soil, Rice and Wheat, in *Proceedings of the 31st Meeting on Radioactivity Survey Data in the Environment*, Science and Technology Agency, Japan, pp. 43–44, 1989 (in Japanese).

Lindstrom, F. T., R. Haque, V.H. Freed, and L. Boersma. Theory on the Movement of Some Herbicides in Soils—Linear Diffusion and Convection of Chemicals in Soils. *Environ. Sci. Technol.*, 1, pp. 561–565, 1967.

Mahara, Y. and S. Miyahara. Residual Plutonium Migration in Soil of Nagasaki. *J. Geophysical Res.*, 89, No.B9, pp. 7931–7936, 1984.

Mclendon, H.R. Soil Monitoring for Plutonium at the Savannah River Plant. *Health Phys.*, 28, pp. 347–354, 1975.

Ministry of Agriculture, Forestry and Fisheries; Division of Research, Secretariat of Minister, Shokuryo Jyukyu Hyo [Tables of Food Supply and Demand], Nohrintokei Kyokai Press, Tokyo, 1988 (in Japanese).

Miyake, Y., K. Saruhashi, Y. Katsuragi, and T. Kanazawa. Seasonal Variation of Radioactive Fallout. *J. Geophys. Res.*, 67, pp. 189–193, 1962.

Morisawa, S., Y. Inoue, H. Kamada, and S. Uchida. Prediction of Fallout Sr-90 Transport and Vertical Profile in Layered Aerated Zone. *J. Atomic Energy Soc.* Japan, 25, pp. 1020–1034, 1983 (in Japanese).

Morita, S., K. Tobita, and M. Kurabayashi. Determination of Tc-99 in Environmental Samples by Inductively Coupled Plasma Mass Spectrometry. *Radiochimica Acta*, 63, pp. 63–67, 1993.

Muramatsu, Y., S. Uchida, M. Sumiya, Y. Ohmomo, and H. Obata. Tracer Experiments on Transfer of Radioiodine in the Soil-Rice Plant System. *Water, Air, Soil Pollut.*, 45, pp. 157–171, 1989.

Muramatsu, Y., Y. Ohmomo, S. Uchida, and M. Sumiya. Transfer of Elemental Iodine and Methyliodide from the Atmosphere to Rice Plants and Leaf Vegetables, in *Proceedings of the International Symposium on Radioecology*, 12–16 October, Znojmo, Czechoslovakia, 1992.

Muramatsu, Y., S. Uchida, and Y. Ohmomo. Root-uptake of Radioiodine by Rice Plants. *J. Radiat. Res.*, 34, pp. 214–220, 1993.

Muramatsu, Y., S. Uchida, M. Sumiya, and Y. Ohmomo. Deposition Velocity of Gaseous Iodine from the Atmosphere to Rice Plants. *Health Phys.*, 71, pp. 757–762, 1996.

Nakamura, Y. and Y. Ohmomo. Factors Used for the Estimation of Gaseous Radioactive Iodine Intake through Vegetation: I. Uptake of Methyliodide by Spinach Leaves. *Health Phys.*, 38, pp. 307–314, 1980a.

Nakamura, Y. and Y. Ohmomo. Factors Used for the Estimation of Gaseous Radioactive Iodine Intake through Vegetation: II. Uptake of Elemental Iodine by Spinach Leaves. *Health Phys.*, 38, pp. 315–320, 1980b.

Namiki, A., H. Katagiri, and O. Narita. Transfer of Stable Iodine from Soil to Polished Rice and Leafy Vegetables, in *Proceedings of the 30th Meeting on Radioactivity Survey Data in the Environment*, Science and Technology Agency, Japan, pp. 56–59, 1987 (in Japanese).

National Astronomical Observatory. *Rika Nenpyo [Chronological Scientific Tables]*, 65th ed., Maruzen, Tokyo, pp. 198–199, pp. 208–209, 1992 (in Japanese).

National Institute of Radiological Sciences. Radioactivity survey data in Japan. National Institute of Radiological Sciences, Chiba, No. 40–83, 1975–1988.

Noguchi, H. and M. Murata. Physicochemical Speciation of Airborne ^{131}I in Japan from Chernobyl. *J. Environ. Radioactivity*, 7, pp. 65–74, 1988.

Okabayashi, H. and S. Uchida. Distribution Coefficient during 1980s, in Radionuclides Distribution Coefficient of Soil to Soil-Solution.; Environmental Parameters Series 2 (RWMC-90-

P-13), Saiki, M. and Y. Ohmomo, Ed., Radioactive Waste Management Center, Tokyo, 1990 (in Japanese).

Okajima, S., T. Shimasaki, and T. Kubo. Measurement of ^{239}Pu in Soil and Plants in the Nishiyama District of Nagasaki. *Health Phys.*, 58, pp. 591–596, 1990.

Sakumotsu-toukei kyokai. Sakumotsu-toukei, [The statistics data for crops in Japan], No. 32, Nourinsunsansyou, Tokyo, 1991 (in Japanese).

Schwartz, G. and F.O. Hoffman. An Examination of the Effect in Radiological Assessments of High Soil-plant Concentration Ratios for Harvested Vegetation. *Health Phys.*, 39, pp. 983–986, 1980.

Sehmel, G.A. Particle and Gas Dry Deposition: A Review. *Atmos. Environ.*, 14, pp. 983–1011, 1980.

Sumiya, M., S. Uchida, Y. Muramatsu, Y. Ohmomo, S. Yamaguchi, and H. Obata. Transfer of Gaseous Iodine from Atmosphere to Rough Rice, Brown Rice and Polished Rice. *Hoken Butsuri*, 22, pp. 265–268, 1987.

Tagami, K. and S. Uchida. Separation Procedure for the Determination of Tc-99 in Soil by ICP-MS, *Radiochimica Acta*, 63, pp. 69–72, 1993.

Tagami, K. and S. Uchida. Determination of Tc-99 in Rain and Dry Fallout by ICP-MS. *J. Radioanal. Nucl. Chem. Art.*, 197, pp. 409–416, 1995.

Tagami, K. and S. Uchida. Microbial Role in Immobilization of Technetium under Waterlogged Conditions. *Chemosphere*, 33, pp. 217–225, 1996.

Tagami, K. and S. Uchida. Concentration of Global Fallout ^{99}Tc in Rice Paddy Soils Collected in Japan. *Environ. Pollut.*, 95, pp. 151–154, 1997.

Takagi, H., T. Kimura, H. Kobayashi, K. Iwashima, and N. Yamagata. Transfer of Iodine from Paddy Soil to Rice Grain. *Hoken Butsuri*, 20, pp. 251–257, 1985 (in Japanese).

Tensho, K. and K. L. Yeh. Radioiodine Uptake by Plant from Soil with Special Reference to Lowland Rice, *Soil Sci. Plant Nutr.*, 16, pp. 30–37, 1970.

Uchida, S., M. Sumiya, Y. Muramatsu, Y. Ohmomo, S. Yamaguchi, H. Obata, and M. Umebayashi. Deposition Velocity of Gaseous I to Rice Grains, *Health Phys.*, 55, pp. 779–782, 1988a.

Uchida, S. and H. Okabayashi, Transfer Factors, in Transfer Factors of Radionuclides from Soils to Agricultural Products.; in *Environmental Parameters Series 1 (RWMC-88-P-11)*, Saiki, M. and Y. Ohmomo, Eds., Radioactive Waste Management Center, Tokyo, 1988b (in Japanese).

Uchida, S. Behaviour of Radio-strontium and -cesium in the Environment, in *Proceedings of Japan-USSR Seminar on Radiation Effects Research*, Tokyo, 1990, National Institute of Radiological Sciences, Chiba, pp. 79–93, 1990.

Uchida, S., Y. Muramatsu. M. Sumiya, and Y. Ohmomo. Biological Half-life of Gaseous Elemental Iodine Deposited onto Rice Grains. *Health Phys.*, 60, pp. 675–679, 1991.

Uchida, S. and Y. Muramatsu. Behaviour of Iodine-129 in Rice Paddy fields, in *Proceedings of Material Research Society Symposium on the Scientific Basis for Nuclear Waste Management*, Kyoto, 1994, Murakami, T. and R. C. Ewing, Eds., Materials Res. Soc., Pittsburgh, pp. 141–147, 1995.

Ueda, T., Y. Suzuki, and R. Nakamura. Transfer of Cs-137 and Sr-90 from the Environment to the Japanese Population via Marine Organisms, in *Proceedings of the Seminar on Radiological Safety Evaluation of Population Doses and Application of Radiological Safety Standards to Man and the Environment*, IAEA-SM-184/7, IAEA, Vienna, pp. 501–511, 1974.

U.S. Nuclear Regulatory Commission. Regulatory Guide 1.109, Calculation of Annual Doses to Man from Routine Releases of Reactor Effluents for Purpose of Evaluating Compliance with 10 CFR Part 50, Appendix 1, Revision 1, 1977.

Watanabe, T. Water Budget in Paddy Field Lots, in *Proceedings of the International Workshop on Soil and Water Engineering for Paddy Field Management*, Murty, V. V. N. and Koga K., Eds., Bangkok, Thailand, pp. 1–12, 1992.

Yamamoto, M., H. Kofuji, A. Tsumura, S. Yamasaki, K. Yuita, M. Komamura, K. Komura, and K. Ueno. Temporal Feature of Global Fallout ^{237}Np Deposition in Paddy Field through the

Measurement of Low-level ^{237}Np by High Resolution ICP-MS. *Radiochim. Acta, 64*, pp. 217–224, 1994.
Yanagisawa, K., Y. Muramatsu, and H. Kamata. Tracer Experiments on the Transfer of Technetium from Soil to Rice and Wheat Plants. *Radioisotopes*, 41, pp. 397–402, 1992.
Yasuda, H. and S. Uchida. Statistical Approach for the Estimation of Strontium Distribution Coefficient, *Environ. Sci. Technol.*, 27, pp. 2462–2465, 1993.
Yasuda, H., S. Uchida, Y. Muramatsu, and S. Yoshida. Sorption of Manganese, Cobalt, Zinc, Strontium and Cesium onto Agricultural Soils: Statistical Analysis on Effects of Soil Properties, *Water Air Soil Pollut.*, 83, pp. 85–96, 1995.
Yuita, K. Transfer of iodine from soil to rice grains, in *Proceedings on the 30th Annual Meeting of the Japan Radiation Research Society*, the Japan Radiation Research Society, Tokyo, p. 38, 1987.
Yuita, K., T. Yoshimura, K. Shoji, and H. Fukushima. Concentration of Stable Iodine in Rice Plants and Paddy Soils, in *Proceedings of the 31st Annual Meeting of the Japan Radiation Research Society*, the Japan Radiation Research Society, Hiroshima, p. 89, 1988.

CHAPTER 5

EFFECT OF ANTHROPOGENIC ORGANIC COMPOUNDS ON THE QUALITY OF SOIL AND GROUNDWATER AND REMEDIATION STRATEGIES IN JAPAN

Osami Nakasugi and Tatemasa Hirata

INTRODUCTION

Large-scale groundwater pollution due to hazardous chemicals was first discovered in a study by the Japan Environment Agency in 1982, which traced the pollution to the organochlorines like trichloroethylene and tetrachloroethylene. The survey was conducted to explore the groundwater pollution situation with regard to 18 substances in 1360 well waters sampled in 15 cities across the country. Following the results of the nationwide survey, a large number of local governments carried out their own investigations on the state of groundwater pollution with regard to three substances, the above two chemicals plus 1,1,1-trichloroethane. These surveys focused on areas with a high risk of pollution such as those surrounding industrial-commercial users of the chemicals.

Systematic groundwater pollution surveys have been conducted regularly since 1989 as part of the Inspection Program of Water Quality under the Water Quality Control Law. This program comprises three types of studies: (1) a general situation survey to investigate the state of pollution by dividing the survey area in a mesh grid, (2) a survey of the well waters in the surrounding locations to determine the spread of the pollution that has been identified, and (3) a groundwater pollution monitoring program to follow variations in the pollution conditions with time. However, hundreds of thousands of chemicals are being manufactured industrially and these as well as other compounds pose potential risks to human health through groundwater pollution. The General Inspection Survey of Chemical Substances on Environmental Safety, a program conducted since 1974 to establish the general situation with respect to chemical pollution of surface water and air, did not include groundwater in its scope. Since 1984, however, the Japan Environment Agency has carried out annual inspections to look into the cases of groundwater pollution due to a number of unregulated chemicals.

The discovery in 1963 of Cr(VI) pollution in the Greater Tokyo area prompted many local authorities to pass regulations and demand the implementation of soil pollution studies in their city area for construction projects such as the redevelopment of factory premises. Following the establishment of the Standard for Soil Environment in 1992, surveys have been carried out in connection with events that involved the changing of land use. These surveys are expected to reveal a clear picture of the pollution situation for the whole of Japan. On the basis of the three-year plans, the Environment Agency has conducted surface soil pollution studies since 1988 at industrial-commercial premises by focusing on sites in which groundwater pollution has been discovered.

NATIONWIDE SURVEY RESULTS

Volatile Organochlorines

The Environment Agency launched a survey in 1982 to assess the state of groundwater pollution at that time. The survey was triggered by the discovery of groundwater pollution due to volatile organochlorines in developed countries throughout the world and the detection of volatile organochlorines in Japan, when the general main supply water was analyzed for trihalomethanes. The survey covered 12 types of volatile organochlorines, and some or all of these were detected in all of the wells investigated. Chloroform, carbon tetrachloride, trichloroethylene, tetrachloroethylene, and 1,1,1-trichloroethane were detected in well waters accounting for more than 10% of the total sample (Table 5.1). Though there were only a small number of wells reporting chloroform and carbon tetrachloride concentration levels in excess of the guideline values laid down by the World Health Organization (WHO) for potable water, there was a high incidence of 3–4 % at which the guideline values for trichloroethylene and tetrachloroethylene were exceeded.

Considering the serious state of the survey results, local authorities began to conduct surveys of their own and discovered new evidence of groundwater pollution due to trichloroethylene and tetrachloroethylene in concentrations greater than the permissible levels in a few percent of the well waters sampled each year. It has thus been confirmed that the groundwater pollution exists on a nationwide scale (Fig. 5.1). Since 1989, the statistical incidence of levels in excess of permissible standards has been reduced. However, this is totally attributed to the fact that in accordance with the amendment of the Water Quality Law in 1989, areas with few industrial-commercial users or producers of these chemicals are now also being investigated.

An Inspection Survey of 1988 revealed the existence of widespread groundwater pollution by dichloroethylenes that had earlier been discovered in relatively high concentrations in a 1982 survey. The incidence at which cis-1,2-dichloroethylene exceeded the standard for water environment, in particular, was at the same level as that for trichloroethylene. In the 1982 survey, there had been no evidence of 1,1-dichloroethylene exceeding the permissible standards. The 1988 survey, however, showed that these chemicals were present in more than 1% of the wells examined. Similarly, it was found that in some wells 1,2-dichloroethane was also present in excess concentrations. In the 1991 inspection survey, methylene chloride was discovered in groundwater at levels above the standard values. Its presence is due to its increasing use as a substitute for trichloroethylene and 1,1,1-trichloroethane.

Dichloroethylenes are not used and produced industrially, with practically all of these chemicals being consumed as raw materials. It seems unlikely that any of these chemicals that are produced and used for specific purposes have entered the groundwater zone.

Table 5.1. Detection Rate of Volatile Organochlorines in Nationwide Groundwater Survey by Japan Environment Agency[a]

Organochlorines	1982(1360)		1988(359)		1991(135)		Maximum Concentration (μg L^{-1})	Standard or Guideline (μg L^{-1})
	Detection Rate	Excess Rate	Detection Rate	Excess Rate	Detection Rate	Excess Rate		
Methyl chloride	2	—	—	—	—	—	2	—
Methylene chloride	6	0	—	—	18	3	121	20*
Chloroform	305	0	83	0	—	—	31	60**
Carbon tetrachloride	131	4	22	1	—	—	2,200	2*
1,1-Dichloroethane	29	—	—	—	—	—	175	—
1,2-Dichloroethane	16	6	4	2	—	—	33	4*
1,1,1-Trichloroethane	186	1	87	1	—	—	5,900	1,000*
1,1-Dichloroethylene	13	0	57	7	—	—	56	20*
cis-1,2-Dichloroethylene	119	25	112	32	—	—	2,030	40*
trans-1,2-Dichloroethylene	20	0	45	0	—	—	15.8	40**
Trichloroethylene	379	40	106	28	—	—	4,800	30*
Tetrachloroethylene	372	53	131	56	—	—	23,000	10*

[a]The number in the parentheses after year means the sample size in each year. The superscripts of * and ** denote the standard and the guideline for water environment. The standard for water environment is always set up, based on the standard for drinking water; therefore, it is basically equal to that for drinking water.

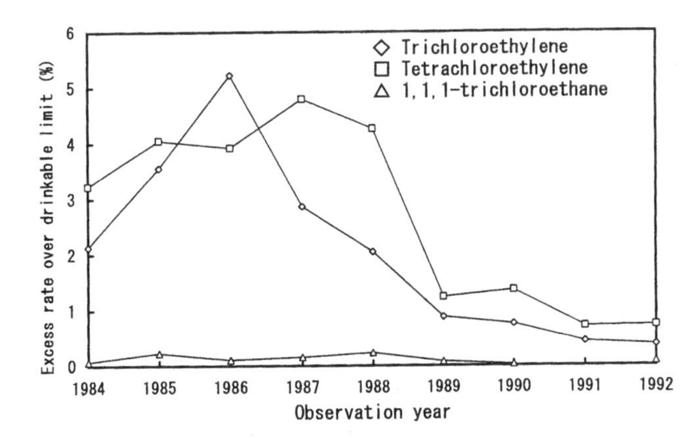

Figure 5.1. Temporal trend of excess rate of well water samples over the standard for water environment. The total sample size of the nationwide groundwater pollution survey until 1992 amounts to 47,000, approximately 2% of which cannot meet the standard for trichloroethylene and tetrachloroethylene.

Many of the wells in which dichloroethylenes have been discovered showed that trichloroethylene and tetrachloroethylene were both present simultaneously (Table 5.2).

Other Chemicals

The Japan Environment Agency's 1982 survey led to the detection of three types of aromatic hydrocarbons (benzene, toluene, and xylene) and two types of phthalates (dibutyl and di[2-ethylhexyl] phthalates) in groundwater. Four of these substances, except for dibutyl phthalate, were added to the standard for water environment and guideline, but only one well showed the presence of benzene in excess levels as compared with their standards and guidelines (Table 5.3).

Benzene, toluene, and xylene were tested for in the 1989 inspection survey, and only toluene was found to be present, and that was at levels below the standard values. In an incident in which gasoline had leaked from a tank into the ground, aromatic hydrocarbon pollution was observed to have spread, with both benzene and toluene clearly exceeding the standard levels (Nakaguma et al., 1994). Since benzene is lighter than water, it floats on the groundwater table and migrates in undiluted liquid form as suggested by Schwille (1984). As a result, the high-concentration range is likely progressively spread with groundwater migration.

In contrast, phthalates were also discovered in groundwater at a high incidence during the 1991 inspection survey (Table 5.3). Although there is no evidence of di(2-ethylhexyl) phthalate pollution in excess of the guideline values, this chemical has been widely detected at concentrations exceeding 1/10th of the guideline values. This survey also explored the presence in groundwater of di(2-ethylhexyl) adipate, a compound used as a plasticizer in the same manner as phthalates.

In the 1992 inspection survey, 1,4-dioxane, one of the Designated Chemicals under the Chemical Substance Control Law, was discovered at a high incidence. This substance has also been detected at a high rate of incidence in surface waters. The maximum concentration found in groundwater is higher than that in surface water. The maximum level exceeded 0.07 mg L^{-1}, a concentration the USEPA considers as being equivalent to a 10^{-5} cancer-causing risk. This chemical has a high solubility in water and is therefore difficult to separate from water.

Table 5.2. Simultaneous Detection Rate of Volatile Organochlorines in Well Waters, Where the Dichloroethylenes Are Detected

Organochlorines	1,1,1-Trichloroethane	Trichloroethylene	Tetrachloroethylene	Trichloroethylene+ Tetrachloroethylene[a]
1,1-Dichloroethylene	47/57	43/57	44/57	—
cis-1,2-Dichloroethylene	56/112	81/112	99/112	106/112
trans-1,2-Dichloroethylene	25/45	36/45	43/45	43/45

[a]The right end column of (trichloroethylene + tetrachloroethylene) means the detection rate of trichloroethylene or tetrachloroethylene being found.

Table 5.3. Detection Rate of Organic Compounds in Nationwide Groundwater Survey by Japan Environment Agency

	1982(1360)[a]		1989(242)		1991(185)			
Chemicals	Detection Rate	Excess Rate	Detection Rate	Excess Rate	Detection Rate	Excess Rate	Maximum Concentration (μg L^{-1})	Standard or Guideline (μg L^{-1})
Benzene	3	1	0	—	—	—	11	10*
Toluene	12	0	15	0	—	—	42	600**
Xylene	5	0	0	—	—	—	17	400**
Dibutyl phthalate	27	—	—	—	55	—	48	
Di(2-ethylhexyl) phthalate	40	0	—	—	60	0	42	60**

[a]The number in the parentheses denotes the sample size and the superscripts of * and ** are the same as in Table 5.1.

Areal Extent of Pollution

Groundwater pollution due to volatile organochlorines tends to extend in narrow bands along the groundwater flow. Figure 5.2 illustrates a pollution plume of tetrachloroethylene discovered in shallow groundwater. The pollution-causing site is a laundry firm located at the leading edge of the plume. The tetrachloroethylene plume, which exceeds the 0.01 mg L^{-1} standard for the water environment, can be recognized to spread over 1 km in the groundwater flow direction. Yet, though the plume has a 1-km stretch in the downgradient direction, it has not diffused more than 100 m in the lateral direction perpendicular to the main current. Even in the biggest groundwater pollution incident, in which the pollution plume reached a length of over 10 km, the spread in the lateral direction remained within 2 km. While the pollution plumes show certain seasonal changes, the pollution pattern remains virtually unchanged for many years. This is because pollution pools accumulate below the pollution sources in almost all of the incidents, and gradual dissolution of pollutant from sources results in making the groundwater quality worse.

Figure 5.3 displays streamwise plume length vs. the maximum concentration of trichloroethylene and tetrachloroethylene observed in polluted groundwater (Hirata and Nakasugi, 1992). The plume length is defined as the distance between two wells situated at the most upstream and downstream locations among polluted wells with concentra-

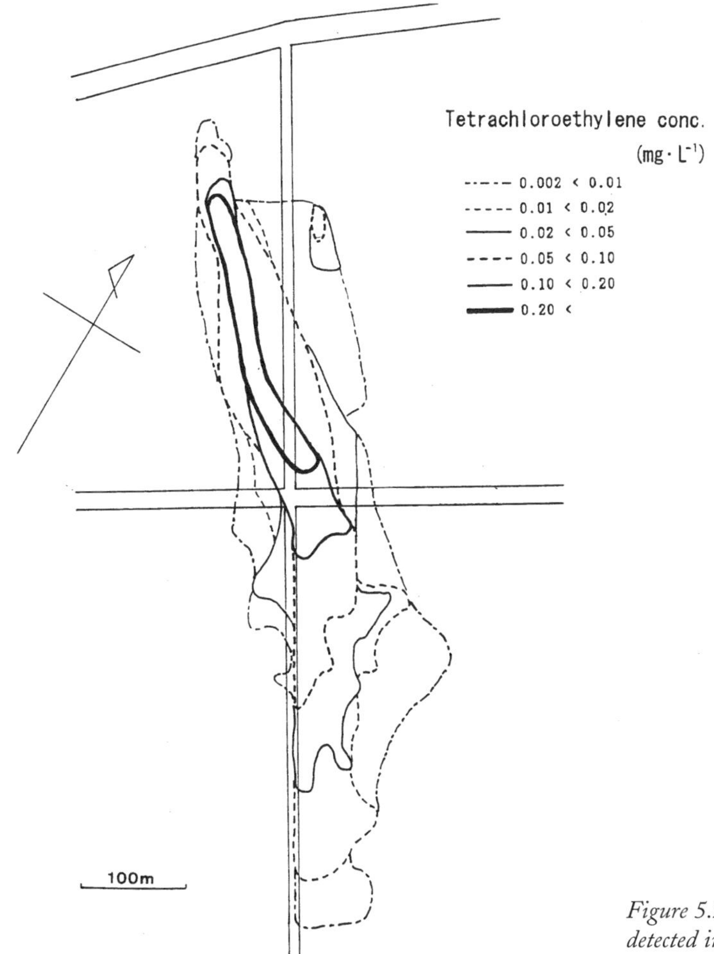

Figure 5.2. A tetrachloroethylene plume detected in a shallow groundwater.

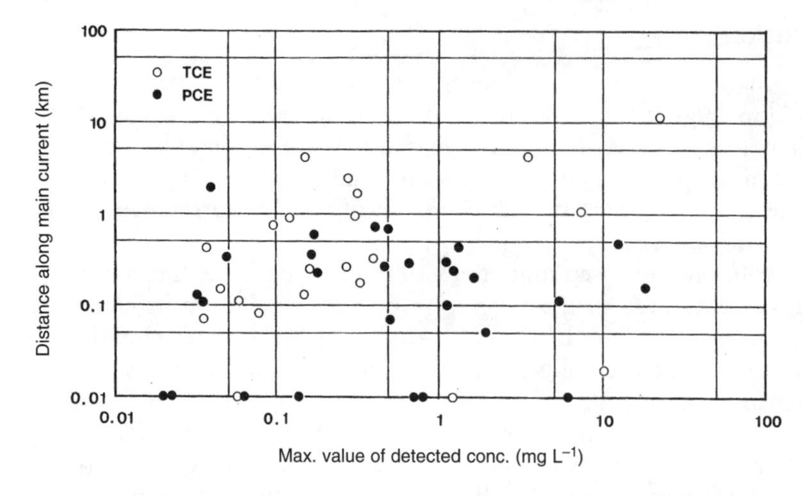

Figure 5.3. Streamwise plume length versus the maximum concentration of trichloroethylene (TEC) and tetrachloroethylene (PCE).

tions above the permissible standard, and the maximum concentration is employed as an alternative indicator for pollutant quantity staying in the subsurface environment. With respect to trichloroethylene, the areal extent of pollution becomes larger with increasing the maximum concentration, and with respect to tetrachloroethylene it is relatively small approximately within 1 km. This is probably due to the usage difference of organochlorines; i.e., trichloroethylene tends to be widely used for cleansing fine products in big high-tech industries where the annual consumption amount in a firm sometimes rises up to a hundred tons. On the other hand, the majority of tetrachloroethylene is likely to be used in relatively small companies including textile dry cleaning firms with monthly maximum consumptions as much as one ton.

SITE INVESTIGATION FOR VOLATILE ORGANOCHLORINES

Migration Path

Organochlorines such as trichloroethylene and tetrachloroethylene are highly volatile, so that most of the organochlorines consumed as solvents are released into the atmosphere with emission gas. The trichloroethylene concentration of the air, however, is not high, about 1 ppmv, and that of the rainwater is 0.001 mg L^{-1}. This is because the organochlorines are degradable in atmosphere with free OH radicals. In addition, organochlorines washed into soil with rainwater easily return to the atmosphere due to their high volatility.

On the basis of the groundwater surveys conducted by national and local authorities, many cases of large-scale pollution were revealed to have been caused by accidental leakage of nearly undiluted liquids from cracks in storage tanks, unacceptable waste, and casual treatment during the cleansing process. In such cases, the subsurface environment is subjected to the threat from percolation of undiluted liquids. Organochlorine liquids are heavier and lower in surface tension and viscosity than water; therefore, they are rather mobile in unsaturated soil compared to water.

From a column test using glass beads, trichloroethylene liquid is observed to readily migrate through the unsaturated zone to the groundwater table. When pore space is small with glass beads diameter of 1 mm, the test liquid tends to be stagnant on the groundwater table. On the other hand, with glass beads diameter being bigger than 3 or 5 mm, it

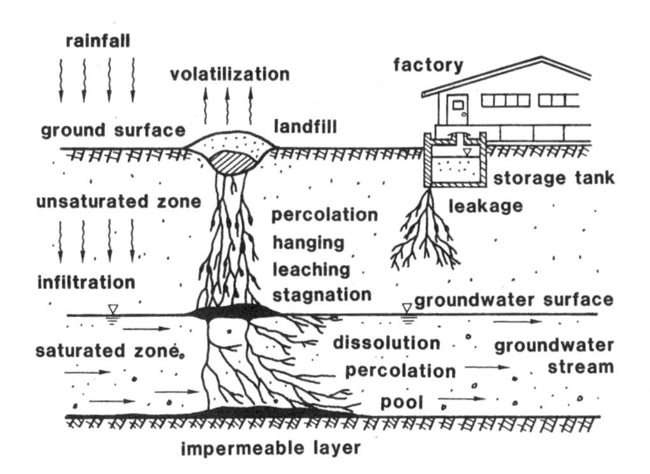

Figure 5.4. Schematical illustration of soil and groundwater pollution mechanism due to volatile organochlorine.

intrudes into the saturated zone, and reaches the bottom of the column, stagnating in the pore space of saturated zone (Hirata and Muraoka, 1988). Such migration features are illustrated in Figure 5.4. Basically organochlorines are immiscible with water; however, the solubilities of trichloroethylene and tetrachloroethylene in water are 1100 and 150 mg L^{-1}, respectively, which amounts to 37,000 and 15,000 times as high as the permissible standards for water environment (Table 5.4). Consequently, organochlorine liquids stagnant both in unsaturated and saturated zones are going to dissolve into water and distribute widely with groundwater flow.

Existing Form in Subsurface Environment

A pollutant source survey revealed evidence of soil concentrations in excess of 10,000 mg kg^{-1}. In most of the pollution incidents, contained organochlorines exceeded 100 mg kg^{-1} in soil. Figure 5.5 refers to a shallow soil pollution observed in the neighborhood of a dry cleaning firm. The soil concentration depicted for the different depth zones reaches levels of several tens of thousands of mg kg^{-1}. It is evident that tetrachloroethylene is present in the undiluted form just below the ground surface. As can also be recognized from Figure 5.5, such extremely high pollutant residue is limited to a very narrow range, and readings imply that the soil concentrations are two or three digits smaller at a distance of only 1–2 m away from this range.

Table 5.4. The Standard and Solubility of Chemicals for Water Environment of Japan

	Standard (mg L^{-1})	Solubility in Water (mg L^{-1})	Solubility/Standard
Methylene chloride	0.02	20,000	1×10^6
Carbon tetrachloride	0.002	800	4×5^5
1,2-Dichloroethane	0.004	9,000	2.3×10^6
1,1,1-Trichloroethane	1	900	9×10^2
1,1,2-Trichloroethane	0.006	4,500	7.5×10^5
cis-1,2-Dichloroethylene	0.04	800	2×10^4
Trichloroethylene	0.03	1,100	3.7×10^4
Tetrachloroethylene	0.01	150	1.5×10^4
Benzene	0.01	1,780	1.8×10^5

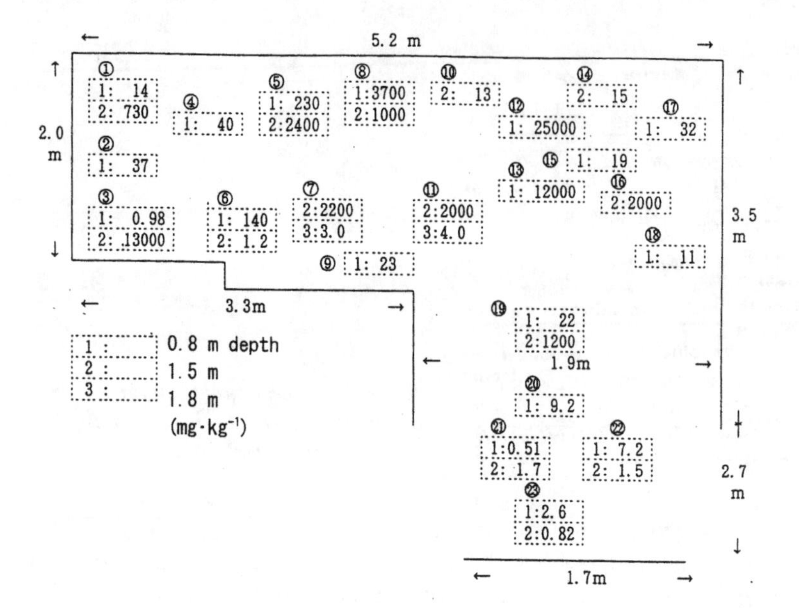

Figure 5.5. Tetrachloroethylene soil concentration observed in dry-cleaning premises. In this case the groundwater table appears at about 2-m depth.

Figure 5.6 illustrates evidence of trichloroethylene pollution reaching down to the deep soil layer (Hirata and Nakasugi, 1993). As there are volcanic ash deposits of up to 60 m with a uniform geological profile, the existing form of measurement provides a very clear picture of the migration behavior of organochlorines in soil. The maximum soil concentration reaches 138 mg kg^{-1} at a depth of 46 m below the ground surface. Similarly, the maximum for groundwater concentration has been measured as 295 mg L^{-1} at the same location. Taking an example on a contour line of 10 mg kg^{-1}, the range of spread will not reach more than 45 m even when trichloroethylene percolates over more than 40

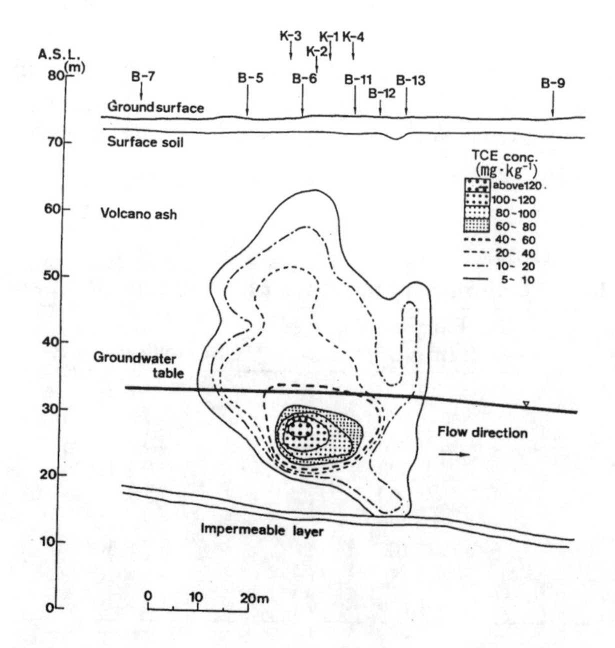

Figure 5.6. Soil concentration and existing form of trichloroethylene in case of relatively deep soil and groundwater pollution incident.

Table 5. 5. The Maximum Concentrations of Volatile Organochlorines in Soil, Groundwater, and Soil Gas Detected Around Pollutant Source with Their Soil Features

Incident	Organochlorines	Soil (mg kg^{-1})	Depth (m)	Soil Feature	Groundwater (mg L^{-1})	Soil Gas (ppm)
1	Trichloroethylene	865	2.5	Clay	137	1,500
2	Trichloroethylene	6,600	2	Silty sand	140	—
3	Trichloroethylene	40	25-27	Silt	360,000	—
4	Trichloroethylene	10	0.7	Surface soil	410	—
5	Trichloroethylene	138	46	Silty sand	455	9,400
6	Trichloroethylene	232	3	Gravel	1,390	—
7	Trichloroethylene	210,000	7-8	Silty sand	40	—
8	Trichloroethylene	4,300	4.5	Clay	28	—
	Tetrachloroethylene	14,000			75	—
9	Tetrachloroethylene	360	5.5	Silt	80	—
10	Tetrachloroethylene	8,100	2.1	Gravel	33	14,000
11	Tetrachloroethylene	25,000	0.8	Clay	22	3,000
12	Tetrachloroethylene	62,000	2.3	Silty sand	160	14,000

m. In the case of a 100 mg kg^{-1} contour, this range of spread becomes narrower, extending only about 10 m. It appears that trichloroethylene permeating into the soil migrates almost straight down without any significant transverse spread.

Table 5.5 gives the maximum values for the soil, groundwater, and soil gas concentrations collected to date. In some cases, with concentrations in the range of a few tens to a few hundreds of thousands of mg kg^{-1}, the pollutant is present in the soil in the form of undiluted liquid. The vertical soil concentration profile indicates that while maximum pollutant concentrations occur at different depths, in many cases the concentration is highest at or near the gravel/clay layer boundary at the bottom of the aquifer. The column experiments by Hirata and Muraoka (1988) demonstrated that volatile organochlorines stagnate just above the aquifer and descend to the bottom of the aquifer. It has been verified that such migration does actually take place in real soil.

The concentrations in groundwater for trichloroethylene and tetrachloroethylene were found to have exceeded their solubility in water of 1100 and 150 mg L^{-1}, respectively. The organochlorines form in pools of undiluted liquid mainly in the vicinity of the pollutant source and are present in groundwater at high concentration nearly at their solubility in water. In the presence of undiluted liquid, the concentrations measured for the soil gases should be equal to their vapor pressure, i.e., trichloroethylene: 76,300 ppm and tetrachloroethylene:18,400 ppm. Actual measurements for soil gas concentrations have shown evidence of high concentrations in excess of 10,000 ppm.

Table 5.6 summarizes the pollutant spread range for soil concentration of 10 mg kg^{-1}. Though the sample size is limited, the results emphasize that with every meter the organochlorine migrates down, the pollution spread widens approximately by 1 m. This measurement reconfirms the vertical percolation of organochlorines without spreading transversely in soil.

Transformations of Organochlorines in Soil and Groundwater

Field measurements have shown the presence in groundwater not only of trichloroethylene, tetrachloroethylene, and 1,1,1-trichloroethane but also of the lower chlorine-content substances like dichloroethylene and dichloroethane. The dichloroethylenes discovered in

Table 5.6. Pollutant Stretch in Subsurface Environment Over Soil Concentration of 10 mg kg^{-1}

Incident	Organochlorines	Maximum Conc. (mg kg^{-1})	Spatial Stretch (m × m)	Depth (m)
1	Trichloroethylene	6,600	7.5 × 9	7
2	Trichloroethylene	210,000	3 × 3	4
3	Trichloroethylene	138	45 × 40	46
4	Tetrachloroethylene	14,000	4 × 3	—
5	Tetrachloroethylene	2,600	<3 × <3	Surface soil

groundwater are not produced industrially in Japan, with the exception of 1,1-dichloroethylene. The latter is utilized almost entirely as a polymer feedstock and as an intermediate substance. It is therefore unacceptable to see its presence as a groundwater pollutant caused by the deliberate use as solvent, and it should be regarded as an intermediate product in the subsurface environment.

The general view is that trichloroethylene and tetrachloroethylene are strongly resistant to biodegradation in soil and the groundwater zone. Recent research, however, has shown clear evidence that volatile organochlorines are degradable both biologically and chemically in the subsurface environment. Vogel and McCarty (1985) have found in their experiments that tetrachloroethylene is transformed by methanogenic microorganisms under anaerobic conditions with the formation of trichloroethylene, dichloroethylene, and vinyl chloride. In view of these observations, they proposed a reductive dechlorination pathway. Though the type of dichloroethylene isomer as an intermediate differs according to the types of microorganisms, the most frequently detected isomer is cis-1,2-dichloroethylene. In fact, cis-1,2-dichloroethylene is the most frequently detected in polluted groundwater of Japan. Vogel and McCarty (1987) also found that 1,1,1-trichloroethane is decomposed by the reductive dechlorination which, in turn, is broken down further to chloroethane and ultimately carbon dioxide.

The reaction between the gaseous 1,1,1-trichloroethane and soil is known to take place in such a manner that the 1,1,1-trichloroethane adsorbed in the soil is decomposed with the catalytic action of soil Fe and Al to 1,1-dichloroethylene.

In a large portion of cases in which dichloroethylenes were detected in groundwater, it was possible to see the simultaneous presence of trichloroethylene, tetrachloroethylene, or 1,1,1-trichloroethane and to discover a correlation between the concentrations of these chemicals and dichloroethylenes. Table 5.2 lists the incidence at which trichloroethylene, tetrachloroethylene, and 1,1,1-trichloroethane were discovered as occurring simultaneously. In many of the wells in which cis-1,2- and trans-1,2-dichloroethylenes were detected, trichloroethylene and tetrachloroethylene were existing simultaneously. The wells in which 1,1-dichloroethylene was detected showed a high incidence of the simultaneous presence of 1,1,1-dichloroethane. This suggests that trichloroethylene, tetrachloroethylene, and 1,1,1-dichloroethane can be the original precursors of the dichloroethylene detected in the subsurface environment.

Uchiyama et al. (1989) discovered microorganisms capable of biodegrading trichloroethylene at a very high efficiency. The microorganisms feed and grow on methane as their nutrient source. Figure 5.7 demonstrates the biodegradation capability in a mixed methane and oxygen culture. At a trichloroethylene concentration of 0.035 mg L^{-1}, a value virtually equivalent to the permissible standard for water environment, more than 90% of the trichloroethylene disappeared in three days. It has been established that

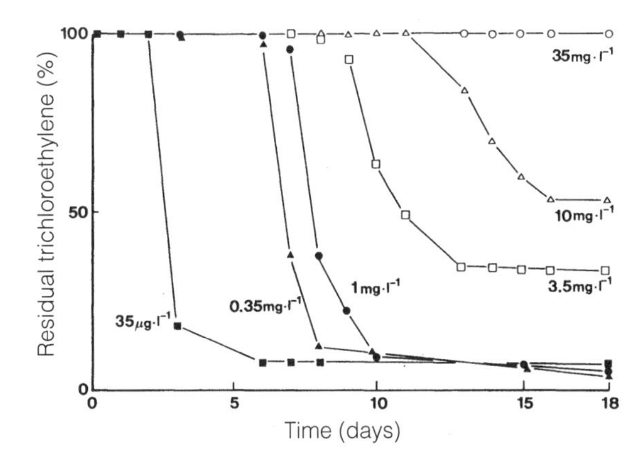

Figure 5.7. Trichloroethylene degradation at different initial concentrations. The average percentage of trichloroethylene reduction is determined to compare the respective control not included with the microorganisms discovered (from Uchiyama et al., 1989).

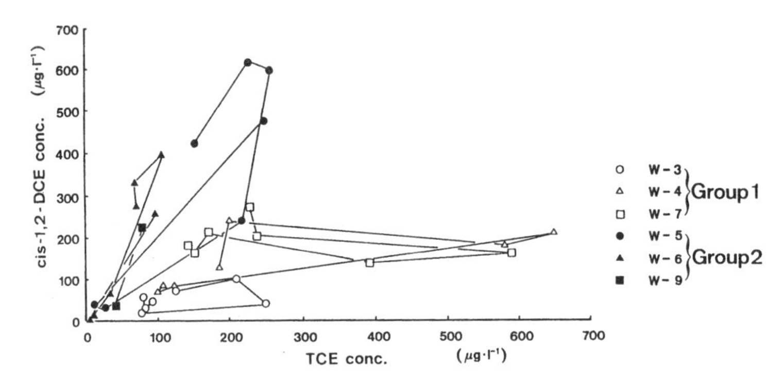

Figure 5.8. Seasonal behavior of cis-1,2-dichloroethylene against trichloroethylene (TCE) in individual well water through a year. In this case a pollutant source is an electric firm producing semiconductors, etc., and groundwaters were sampled from the shallow groundwater aquifer comprising sand and gravel between 7- and 10-m depth.

these microorganisms are capable of degrading trichloroethylene at concentration levels of up to 10 mg L^{-1}, which is 300 times the standard.

Laboratory experiments have thus demonstrated that dichloroethylene can occur as a result of microbial decomposition. There is mounting evidence to suggest that biodegradation reactions also take place in the real subsurface environment. An evidence refers to the seasonal changes of trichloroethylene and cis-1,2-dichloroethylene concentrations observed in a same area, as shown in Figure 5.8. The observations indicate that the cis-1,2-dichloroethylene concentration in one of two groups is not sensitive to the change of trichloroethylene concentration, while another one raises the cis-1,2-dichloroethylene concentration in a wide range proportional to that of trichloroethylene. In addition, the cis-1,2-dichloroethylene concentration in the latter is greater than that of trichloroethylene as the mother substance, which implies that microbial decomposition takes place during the migration of trichloroethylene through the groundwater.

With increasing distance away from the pollutant source, the dichloroethylene/trichloroethylene ratio in the groundwater increases, when the microbial activity for biotransformation is present. Figure 5.9 displays the variations in cis-1,2-dichloroethylene

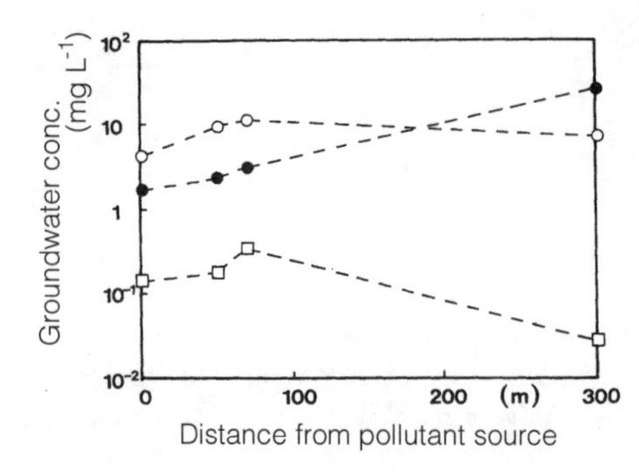

Figure 5.9. Changes of volatile organochlorine concentrations in groundwater in the downgradient direction from a pollutant source. The open circle is for trichloroethylene, a closed one for cis-1,2-dichloroethylene, and an open square for dichloroethane.

and trichloroethylene concentrations in the same aquifer at the pollutant source and at three locations within an approximately 300-m distance from the pollutant source. While the trichloroethylene concentrations virtually level off in the downgradient direction, the cis-1,2-dichloroethylene concentration rises to a level above the trichloroethylene concentration at 300 m from the pollutant source. For the same location, it can be seen that the ratio of cis-1,2-dichloroethylene and trichloroethylene concentration becomes higher with increasing the groundwater depth. This is further evidence to suggest that trichloroethylene can be transformed by microbial activity as it migrates with the flow of the groundwater.

Surface Soil Gas Survey as Remote Geochemical Monitoring

The success of a remedial operation depends on how much information concerning the contaminant, its location in soil and groundwater, and the areal extent of pollution can be obtained. This is because, from the cost-benefit point of view, the remedial operation must take place close to the pollutant source to reduce the volume of contaminated soil, water, and gas to be treated. However, making boreholes and covering the whole basin with investigation wells is very costly and not practical.

Organochlorines are basically volatile; therefore, organochlorines stagnant in the subsurface environment are thought to vaporize into pore air in the unsaturated soil. In this context, by taking a soil-gas sample from surface soil and analyzing volatile chemicals, it is possible to delineate the pollution boundary in groundwater and to identify the pollutant source, using comparison of the distribution pattern of soil-gas concentration. Figure 5.10 displays the relationship between tetrachloroethylene concentrations of soil and soil gas (Hirata and Nakasugi, 1993). There is a linear relationship on a full log-log scale, which means both concentrations are partitioned well. In addition, Figure 5.11 illustrates the concentration contours of tetrachloroethylene in surface soil-gas observed in a shallow aquifer, where the groundwater table was at 4-m depth and the aquifer depth was estimated to be 1 m. The contour lines are extending like concentric circles around the suspicious pollutant source denoted by a closed triangle. This result confirms that soil gas reconnaissance is a quicker and less expensive technique for identifying pollutant source and delineating the plume boundary (Marrin and Thompson, 1987).

Another result for the surface soil-gas monitoring is illustrated in Figure 5.12. The site investigation confirmed that the sandy soil extends to 3-m depth with a groundwater table at 1.5-m depth, overlying the clay layer. The tetrachloroethylene concentration in the surface soil gas was determined with a gas chromatography-ECD. In the n-

gas tube. Unless the concentration of the soil gas is on the order of a few ppm or more, the Gas Tube Method will not give a correct concentration reading. Since the detection chemical in the tube reacts with chlorine, it is not possible to identify the source substance, whether trichloroethylene, tetrachloroethylene, or some other chlorine-containing compounds. Among the various soil gas survey techniques, the Gas Tube Method is the most convenient and simplest to use.

Although the various survey methods differ in terms of the gas sampling and analysis procedures, they show good agreement with each other over their detection limit. While there is a high degree of correlation between the soil gas concentrations determined by the different methods, their detection sensitivities differ significantly. For the same area, investigations have been conducted to examine the relationship between the soil gas concentration and the distance from the pollutant source. The results have demonstrated that there is a linear correlation between these two values on a log-log scale, as shown in Table 5.7. Studies have also been carried out to determine the distance at which a pollutant can no longer be detected on the basis of the distance dissipation correlation from the pollution source and the corresponding detection sensitivities associated with the various soil gas monitoring techniques. The results show that the limit distance is approximately 25 m for the n-Hexane Method, 5 m for the Gas Tube Method, and 100 m for the Mobile Labo Method and the Fingerprint Method. In addition, the double-values of these distances make the interval in a mesh grid required for the detection of pollutants in the target pollution zone. It is clearly concluded that while high-sensitivity methods call for a wide spacing of the survey mesh graticule, the low-sensitivity methods need a large number of test points to obtain equivalent results to those achieved with the high-sensitivity methods.

REMEDIATION TECHNOLOGIES FOR SUBSURFACE POLLUTION

Classification and Availability of Technologies

The pollutant entering into the soil and groundwater results in the following conditions, depending on its chemical feature: (1) adsorption on the soil particles, (2) dissolution in the water, and (3) vaporization in the air of soil pores if the pollutant is a volatile substance. When a pollutant enters the soil in large quantity as undiluted liquid, another

Table 5.7. Log-Linear Regression Analysis for the Relation Between Surface Soil Gas Concentration and Distance from the Pollutant Source[a]

	$Y=aX^b$ (Y:concentration, X:distance)						
Method	**Site**	**a**	**b**	**Correlation Coefficient**	**Sample Size**	**Detection Limit**	**Distance (m)**
n-Hexane	A	13.55	−2.135	−0.8849	12	0.01	29.3
	B	33.88	−2.617	−0.8827	12		22.3
Gas Tube	A	113.7	−2.789	−0.8528	8	1	5.46
	B	106.7	−2.976	−0.8631	7		4.8
Mobile Labo	A	23.48	−2.184	−0.8894	12	0.001	100
	B	42.06	−2.725	−0.8786	12		49.8
Finger Print	A	786300	−2.386	−0.9261	12	10	113
	B	825200	−1.467	−0.8421	8	100	467

[a]The right end column means the distance, where the surface soil gas concentration is equal to the detection limit of each method. The site A is contaminated with trichloroethylene and the site B with tetrachloroethylene. The detection limit of Fingerprint is in ion count and the others in ppmv.

possibility exists; namely, (4) that the pollutant occupies the pore spaces in undiluted liquid form. In any of the existing forms, the determination of concentrations is totally attributed to partitioning characteristics among the gas, liquid, and solid phases. When the distribution is biased toward any particular phase, the most effective way to repair the pollution situation is to eliminate the high-concentration medium.

Remediation technology for subsurface pollution consists of removal of pollutants from the soil or groundwater and their detoxification. Generally it is classified into two categories: (1) diffusion control of the pollutants, and (2) removal and decomposition of pollutants as collected in Figure 5.13. In the pumping of groundwater, diffusion control comes about with the use of a barrier well. When it takes place close to the pollutant source, the element of pollutant removal technology predominates. These measures are thus closely interrelated and should be used in a proper combination to achieve an improved remediation effect.

Solidification technology as one of these techniques is not applicable to such liquid pollutants as trichloroethylene. Chemical reaction and vitrification techniques, however, are primarily reserved for heavy metals. Microbial decomposition is in widespread use as a decontamination and detoxification technique for hydrocarbons such as petroleum and petroleum products which are relatively readily degradable. Recently, however, it has become possible to discover microorganisms capable of decomposing at high efficiency volatile organochlorines. Their use is now being developed for commercial application, with the technology having reached the stage of site verification tests.

A large variety of different remediation technologies have been proposed, and the methods so far applied to pollution sites in Japan include such physical pollutant removal techniques as polluted soil excavation, groundwater extraction, and soil vapor extraction. The remediation technologies that are used in practice depend on such factors as the existing form of the pollutant in the soil and groundwater, the depth of pollution, the costs involved in carrying out the remediation measures, and the conditions of groundwater use. In practice, remediation at pollution sites should apply a combination of techniques, because none of remediation technologies works enough to repair the wide range situation of subsurface pollution.

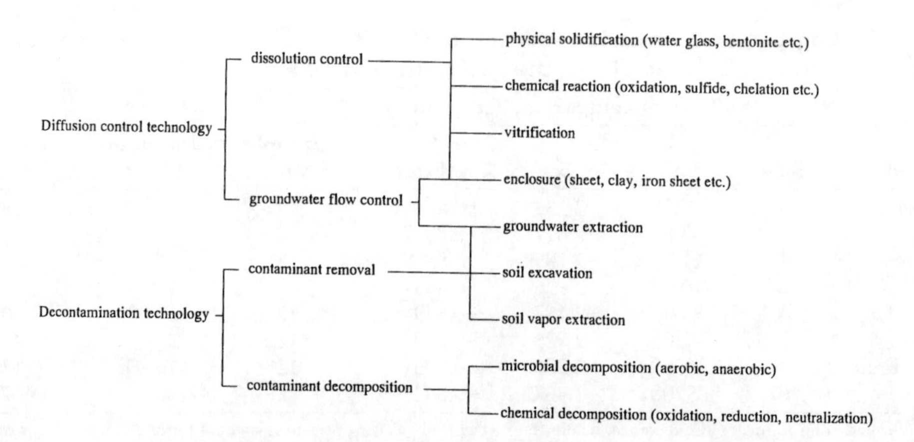

Figure 5.13. Classification of remediation technologies for subsurface pollution.

Remedial Operation and Evaluation

Polluted Soil Excavation and Groundwater Extraction

Experience with polluted soil excavation work carried out so far indicates that the quality of the groundwater is restored immediately after the excavation has been completed. Figure 5.14 compares the tetrachloroethylene concentrations in shallow groundwater before and after polluted soil excavation. As a result of the polluted soil excavation program, the tetrachloroethylene concentration has been reduced by one order of magnitude. This implies that 90% of the tetrachloroethylene has been recovered from the soil. In another polluted site with trichloroethylene, excavation of the polluted soil to 7-m depth has been found to lead to a reduction of nearly two orders of magnitude in pollutant concentration in the shallow groundwater as illustrated in Figure 5.15. Experience shows, however, that the restoration of the groundwater quality after operation still presents a problem; the groundwater does not easily return to a former clean condition although the pollutant concentrations in the shallow groundwater may tend to diminish. Due to the difficulty of removing polluted soil in the deep subsurface strata, the pollution removal effect of soil excavation is totally absent in the deep groundwater aquifer. If the pollution situation in this site is considered, groundwater extraction was continued even after the polluted soil had been excavated. It was not until the order of ten tons of trichloroethylene had been recovered that the groundwater began to show signs of improvement near the standard for water environment.

The extracted groundwater is subjected to an air-stripping treatment and the pollutants contained in the air injected into the stripping tower are adsorbed on activated carbon before releasing the spent air into the atmosphere. Similarly, the excavated polluted soil requires appropriate treatment, which may include drying by exposure to the sunlight, vaporization of the pollutants by applying heat, or extraction of soil vapor from the soil piled up. For this type of disposal operation, recovering the pollutants is also necessary by adsorption on activated carbon so as to avoid air pollution.

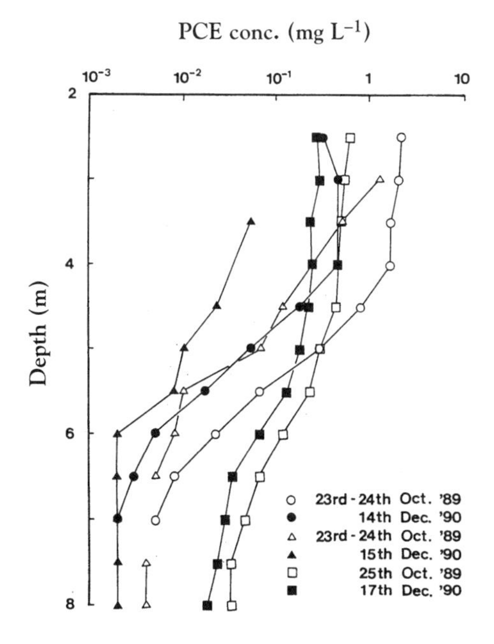

Figure 5.14. Comparison of vertical profile of tetrachloroethylene (PCE) concentrations in groundwater before and after the polluted soil excavation, which took place in May 1990. In this operation the soil excavated amounts to 500 m³.

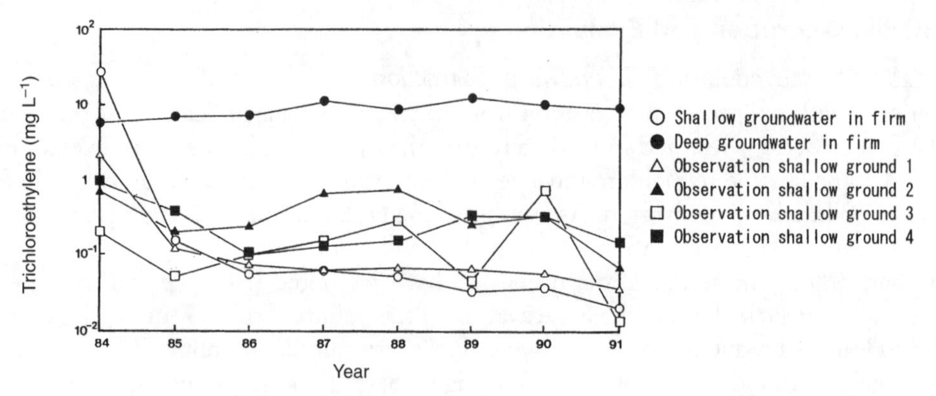

Figure 5.15. Temporal variation of trichloroethylene concentrations in shallow groundwater after the polluted soil excavation. The amount of the soil excavation over the concentration of being 1 mg kg^{-1} counts for 1007 m^3, and the continuous groundwater extraction after the operation took place in May 1984 results in removing 17 tons of trichloroethylene during a period of 1984 through 1989.

Dual Extraction

Soil vapor extraction technique is employed to remove pollutants that have vaporized in the soil vapors in the vadose zone. Such a remediation process is reported to have led to the elimination of from a few kilograms to one ton of volatile organochlorines in operations lasting for a period of a few months to a year. As the technique involving the removal of the pollutants by suction is basically available only for the vadose zone, it is not able to achieve remediation for the groundwater pollution. Soil vapor extraction technology therefore also creates the need to pump up groundwater in large quantities to repair the groundwater, plus the need to offset the rise of groundwater level in an extraction well associated with the suction operation.

Experience with the use of soil vapor extraction from deep soil layers of 30– to 40-m depth has demonstrated that in the initial stage the procedures have been successful in removing by suction 1 kg hr^{-1} of trichloroethylene as seen in Figure 5.16, which is one order of magnitude higher than the removal rate achieved with groundwater extraction. Yet the removal rate due to soil vapor extraction began to decline much earlier than in the case of groundwater extraction, so that the removal rates of these two methods develop inversely with the progress of remediation: as the remediation procedure advances, so the pollutant concentration in the soil and the existing form will change. This means that a flexible approach is needed that allows for a change to more efficient and more low-cost technologies in accordance with these changes in the remediation operation. Figure 5.16 gives the results obtained at a groundwater extraction rate of 2 m^3 hr^{-1} at the onset of remediation. After 7000 hours from the onset of remediation, however, the treatment groundwater capability is intensified to a maximum of 30 m^3 hr^{-1}, with a continuous regime of 25 m^3 hr^{-1} extraction being sustainable. As a result, it is possible to increase the pollutant removal rate with groundwater extraction to a level of 0.15 kg hr^{-1} and the trichloroethylene concentration, which was initially in excess of 300 mg L^{-1} in the observation well, has been brought down to nearly 10 mg L^{-1}.

In shallow soil pollution, simple methods such as the iron pipe striking and the well point methods may also be capable of removing pollutant substances. The well point method, in particular, eliminates both soil vapors and groundwater using a high extrac-

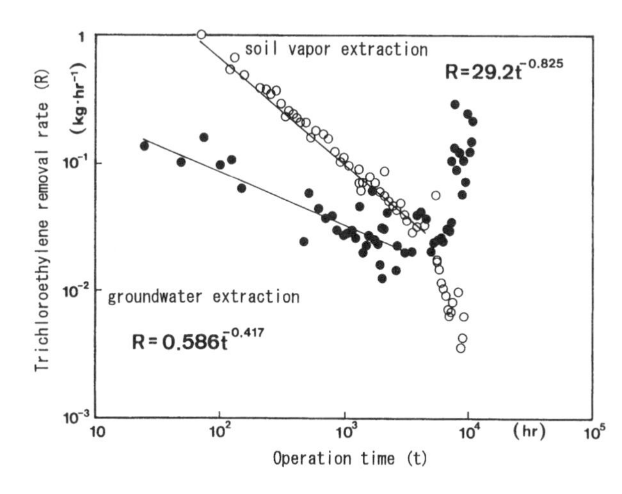

Figure 5.16. Comparison of trichloroethylene removal rates between soil vapor extraction and groundwater extraction. The trichloroethylene existing form in this site is depicted in Figure 5.6.

tion pressure. Though its remediation depth is limited to 6–7 m from the surface, it has proved capable of removing 90% of the organochlorines in soil.

As has been presented here, the polluted soil excavation and the removal by suction of soil vapors do not extend to repairing the groundwater, although they are capable of removing pollutants in the vadose zone. Substances such as trichloroethylene are little soluble in water so that it takes a long time to remedy this type of pollution by groundwater extraction. As a result of long-term operation, however, it is considered possible to recover pollutants in larger amounts by the groundwater extraction than those by polluted soil excavation and soil vapor extraction, so that the pumping up of groundwater may be fairly described as a basic remediation technology of essential importance.

Bioremediation

Physical remediation technologies such as soil vapor extraction and groundwater extraction involve the removal of pollutant substances from the subsurface zone and their ultimate vaporization and recovery by adsorption on activated carbon. Though these techniques are ideal for the treatment of low-boiling point chemicals, the issue still remains that the solutions desorbed from the activated carbon and the activated carbon itself will need to be eventually disposed of by incineration treatment. This implies that, in some sense, the pollution burden is shifted from the subsurface to the atmospheric environment. In contrast to this, bioremediation technology holds out significant hope for the possibility and practical availability of a process for the complete decomposition of the pollutant substances to carbon dioxide.

Biodegradation is already in practical use in the industrialized West (Europe and America) as a remediation technology for relatively easy-to-decompose hydrocarbons such as gasoline. Similarly, for substances such as trichloroethylene, it has been possible to discover microorganisms that are capable of decomposing pollutants at high efficiency under anaerobic conditions. In-situ bioremediation gives rise to certain problems that need to be resolved, such as the issues of the toxicity of the intermediate products and the public acceptance of the technique. Bioremediation is a promising measure in the final stage after the implementation of physical remediation procedures, such as soil vapor extraction and groundwater extraction technologies. In Japan, site verification of bioremediation technology for volatile organochlorines has now commenced with a view to developing it for practical application.

Suitable Procedure from Discovering Pollution through Applying Remediation Technology

To deal with pollution once it has been discovered, remediation measures are employed after the clear determination of the pollution situation. The sequence of events is this: (1) when the need for remediation measures has been established and the measures are implemented, (2) after the forms in which the pollutant substances exist in the subsurface zone have been clearly identified, and (3) the appropriate remediation techniques are selected.

The need and urgency for remediation measures depend on the conditions of groundwater use and on the trends for future pollution. These future pollution trends affecting the groundwater quality are assessed on criteria such as the determination of the polluted aquifer, the magnitude and scale of pollution, and the existing forms of the pollutant substance in the groundwater. To identify the polluted aquifer, the practice is to carry out surveys of the groundwater quality in the surrounding areas and geological surveys. The problem with wells created only for the purpose of groundwater extraction is that often screens are established in a number of aquifers, and though pollutants have been discovered, it is not clear which aquifer is polluted. Further, even though the polluted aquifer may be identified, it is not clear whether the pollution is transient or is continuously observed for many years. It will also be necessary to determine the stage in which the groundwater pollution is present (in other words, whether it is in the incipient stage or in the final stage) and to make a critical judgment of this it will be necessary to examine the changes in groundwater quality and the spatial distribution of the pollution with time. To explore the existing forms of the pollutant in the subsurface zone and the three-dimensional structure of the pollution, it is necessary to carry out borehole surveys to identify the pollution source. These detailed survey results will permit a judgment of the future groundwater pollution trends, but this is not an easy task. For this reason the practice in Japan is often to commence surveys on the premise that some sort of remediation measure or other will be implemented.

For the effective removal of pollutants from the soil or groundwater, remediation measures should take place as close to the pollution source as possible. The success of a remediation operation is attributed to determining the detailed existing form of the pollutant in soil and groundwater. In other words, it is required to delineate the areas with a pollution risk on the basis of groundwater quality surveys and soil vapor monitoring and to enforce detailed surveys in order to identify the pollutant intrusion locations to ground in the areas where the pollutant has been discovered. In locations with a high concentration of soil vapors, the pollutant intrusion is presumed to have occurred so that a borehole survey will be carried out to explore the vertical profile of pollution and to complete the three-dimensional structure of pollution.

After the existing forms have been revealed by a series of pollution surveys, the next step is to pick the remediation technology to be applied in response. The technique adopted depends on the scale and depth of pollution. Similarly, the aquifer depth also varies significantly from region to region and the existing forms differ widely. There is no effective remediation technology for every form of pollution. It is also difficult to apply a single remediation technique throughout the operation from the initial stage to the final stage of remediation. In accordance with the progress of remediation, it is essential to look for a flexible approach to change over to a more cost-efficient or low-cost technology as the measure proceeds. During the implementation of remediation measures, it is also prerequisite to monitor their effectiveness. If necessary, the remediation measure should be reimplemented after a review of the technology.

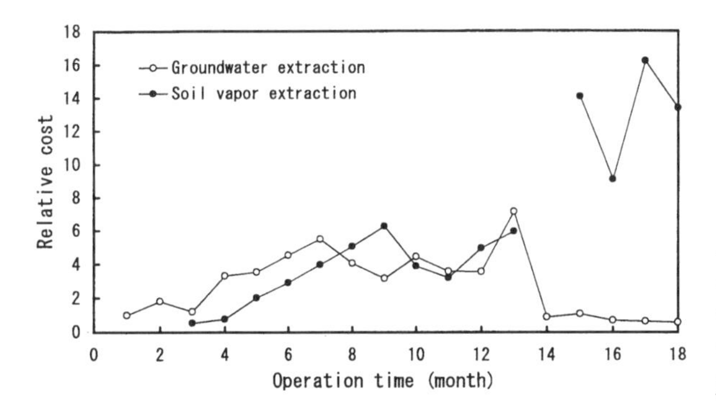

Figure 5.17. Monthly changes of relative cost efficiency between the operations of soil vapor extraction and groundwater extraction, based on the observed results presented in Figure 5.16.

As an example on the data presented in Figure 5.16, the running costs required for the groundwater extraction operation in the first month of the remediation program have been considered as unity and the monthly changes of the normalized relative costs for groundwater and soil vapor extractions have been plotted in Figure 5.17 (Hirata, et al., 1998). With the passage of time, the removal rate for both of these operations is declining and the relative costs for both extraction operations are increasing until the end of the eighth month. Conversely, the groundwater extraction volume increased substantially for 13 months and after this, the relative costs for soil vapor extraction have still continued to rise until about the 15th month, although the relative costs for groundwater extraction have fallen below unity. As a result, the remediation measures at the pollution site have led to the discontinuation of the soil vapor extraction operation because of the declining pollutant removal rate and the escalating relative costs, while the groundwater extraction operations are currently being continued.

SUMMARY AND CONCLUSIONS

The Japan Environment Agency and local governments have conducted nationwide surveys for subsurface pollution since 1982. From these data sets, the pollution mechanism and the general situation with regard to soil and groundwater pollution by volatile organochlorines have been generally delineated. For the remediation of soil and groundwater pollution by volatile organochlorines, innovative technologies are being developed and introduced, and their effectiveness in pollution removal has been demonstrated. Some technologies are naturally still in the in-situ or indoor verification stages as is the case with bioremediation using microbial activity. In general, however, it is emphasized that remediation technology has been approximately established. It is definitely possible to achieve polluted soil and groundwater remediation and restoration with the expenditure of large amounts of time and money. Remediation measures have thus been enforced in practice on large-scale operating sites. Still, in many pollution cases the operating sites are small in scale, and cost is the major problem in the implementation of remediation.

To facilitate positive advances in remediation, development of suitable survey methods and efficient pollutant removal technologies is necessary. With the benefit of soil and groundwater pollution surveys, it should definitely be possible to carry out effective measures on the basis of preliminary pollution survey. However, there are limits as to the funds that can be expended in remediation. A proper balance in the whole package of remediation operations should be sought. The need is to establish a completely integrated package of remediation measures starting with survey programs, including measures such as the planning of feasible remediation operations consistent with the uses of the groundwater and the regional characteristics. This should also include a clear definition of the objectives of the remediation program.

Apart from the types of soil and groundwater pollution presented above, there are many other forms of groundwater pollution such as chemical pollution due to agricultural practices, nitrogen pollution originating from inorganic fertilizer, and arsenic pollution believed to be of natural origin. This groundwater pollution is of the diffusive type with a spatial extension in pollutant source. There is also a concern that groundwater pollution due to antimony and boron may also be present. With the large amounts of oils used in our living sites, pollution due to petroleum such as benzene is also in widespread evidence. Since petroleum products are lighter than water, they float on the surface of the groundwater and follow the flow of the groundwater, so that the pollution source itself will move and the range of pollution expand. For these types of pollution, development of a whole series of measures beginning with the solution of the pollution mechanism is required.

REFERENCES

Hirata, T. and K. Muraoka. Vertical Migration of Chlorinated Organic Compounds in Porous Media. *Wat. Res.*, 22, pp. 481–484, 1988.

Hirata, T., N. Egusa, O. Nakasugi, S. Ishizaka and M. Murakami. Cost Efficiency of Subsurface Remediation Using Soil Vapor Extraction and Groundwater Extraction. *Wat. Sci. Tech.*, 37(8), pp. 161-168, 1998.

Hirata, T. and O. Nakasugi. Groundwater Pollution by Volatile Organochlorines in Japan and Related Phenomena in the Subsurface Environment. *Wat. Sci. Tech.*, 25(11), pp. 9–16, 1992.

Hirata, T. and O. Nakasugi. Surface Soil Gas Survey for Identifying Pollutant Source and Existing Form of Organochlorines in Subsurface Environment. *IAH Selected Papers*, 4, pp. 39–49, 1993.

Japan Environment Agency. Nationwide Survey Results for Groundwater Contamination in 1982, Japan Environment Agency, Tokyo (in Japanese), 1983.

Klusman, R.W., K.J. Voorhees, J.C. Hickey, and M.J. Malley. Application of the K-V Fingerprint Technique for Petroleum Exploration, in *Unconventional Methods in Exploration for Petroleum and Natural Gas I, V*, Davidson, M.J., Ed. Southern Methodist University Press, Dallas, 1986, pp. 219–243.

Marrin, D.L. and G.M. Thompson. Gaseous Behavior of TCE Overlying a Contaminated Aquifer. *Ground Water*, 25, pp. 21–27, 1987.

Nakaguma, H., I. Sato, S. Watanabe, S. Kawakami, H. Maruyama, and K. Tajima. Investigation of Groundwater Polluted with Gasoline. *Jpn. Soc. Wat. Environ.* 17, pp. 315–323 (in Japanese), 1994.

Nakasugi, O. and T. Hirata. Studies on Control of Groundwater Contaminated with Volatile Organohalogen Compounds, Special Res. Report, Natl. Inst. Environ. Stud. Jpn., Tsukuba, 1994.

Schwille, F. Migration of Organic Fluids Immiscible with Water in the Unsaturated Zone, in *Pollutants in Porous Media*, B. Yaron, G. Dagan and J. Goldshmid, Eds. Springer-Verlag, Berlin, 1984, pp. 27–48.

Uchiyama, H., T. Nakajima, O. Yagi, and T. Tabuchi. Aerobic Degradation of Trichloroethylene by a New Type II Methane-Utilizing Bacterium, Strain M. *Agric. Biol. Chem.*, 53, pp. 2903–2907, 1989.

Vogel, T.M. and P.L. McCarty. Biotransformation of Tetrachloroethylene to Trichloroethylene, Dichloroethylene, Vinyl Chloride, and Carbon Dioxide Under Methanogenic Conditions. *Appl. Environ. Microbiol.*, 49, pp. 1080–1083, 1985.

Vogel, T.M. and P.L. McCarty. Abiotic and Biotic Transformations of 1,1,1-Trichloroethane Under Methanogenic Conditions. *Environ. Sci. Technol.*, 21, pp. 1208–1213, 1987.

CHAPTER 6

HEAVY METAL POLLUTION IN SOILS AND ITS REMEDIAL MEASURES AND RESTORATION IN MAINLAND CHINA

H. K. Wang

INTRODUCTION

The rapid growth of industrialization and urbanization in the People's Republic of China has resulted in large quantities of waste gas, wastewater, and waste solid being discharged into the environment, thus giving rise to serious environmental pollution problems and deterioration of the agroecosystems.

The recently discharged amount of the municipal and national industrial waste gas, wastewater, and waste solid in China, not including the township enterprises, were 11,363 billion m^3, 36.5 billion tonnes, and 0.62 billion tonnes, respectively, in 1994. The treatment percentages of the above three wastes were 36.7%, 42.4%, and 12.4%, respectively (Environmental Protection Agency of China, EPAC, 1995). The amount of discharge of the three wastes of the township enterprises in China, mainly the rural industries, was approximately 10% of that of the whole nation. Their waste treatment percentages are much more lower than that of the nation.

The Chinese government, especially the officers of EPAC, clearly understand that waste treatment is of great importance before it is discharged into the environment, but at present China does not have a sufficient financial investment for the treatment of most of the wastes.

Soil pollution can be caused not only by municipal and industrial pollution sources, but also by agricultural measures themselves. Such agricultural measures include pesticides, chemical fertilizers, animal and poultry excrement, and sludge application on agricultural land, sewage irrigation, discarded residues of agricultural plastic membrane mulching, etc.

Of all these discharged pollutants, heavy metals are perhaps the most harmful. In this chapter, discussion is focused on the environmental impact of heavy metals in soils of China. Certain toxic heavy metals are taken up by crops, which can thus become a hazard to human and animal health, such as Cd, Hg, Pb, As, etc., and other heavy metals, such as Cu, Ni, Zn, Cr, etc., can first lead to reduced crop yields. Finally, the agricultural control measures are described for the remediation and restoration in heavy-metal contaminated agroecosystems.

Table 6.1. Background Levels of Trace Elements in Soils of Some Locations of China (mg/kg), 1986[a]

Location	Hg	As	Cr	Cu	Zn	Pb	Cd	Ni
Beijing cinnamon soil	0.042	8.41	58.7	12.2	43.1	13.3	0.13	21.8
Shanghai eutric cambisol	0.216	8.95	84.4	23.5	79.8	21.3	0.134	–
Zhejiang eutric cambisol	0.085	8.22	47.5	17.5	75.0	75.1	–	24.0
Guiyang acrisol	0.19	13.5	87.7	33.4	66.7	25.7	0.11	27.8

[a]Wang, 1994.

BACKGROUND LEVELS OF TRACE ELEMENTS IN SOILS OF CHINA

Background levels of trace elements are fundamental to the understanding of the natural environment. They provide clues for tracing soil pollution history, stipulating environmental protection criteria, and their control in contaminated soils of agroecosystems. They also help researchers to understand the significance of trace-element cycling in agriculture, and in the identification of the causes of many endemic diseases, as well as expediting mineral exploration.

In any attempt to determine the degree of pollution by heavy metals in soils, it is of primary importance to establish the natural background levels of these metals, and then subtract them from existing values for metal concentration to derive the total enrichment caused by anthropogenic influences. Such analysis can also tell us where treatment measures should be carried out in these contaminated soils. Background levels of heavy metals in soils of some regions in China are shown in Table 6.1 (Wang, 1994). More data concerning Mainland China are presented in the references (Wang, 1994; Chinese Environmental Monitoring Center, 1991).

SITE RESEARCH OF SOIL POLLUTION IN CHINA

In China, there are three main causes of soil pollution: (1) gas emission, (2) mine and smelter discharges, and (3) sewage irrigation and sludge application on agricultural land.

Gas Emission

Beijing has been a capital for more than 1000 years. The anthropogenic pollution in soil during the last 50 years, municipal and industrial, is significant. Pollutants are discharged via wastewater, waste gas, and gaseous emissions from coal combustion, especially for space heating in winter. Figure 6.1 shows Beijing's mercury pollution in soils, primarily caused by gas emission (Wang, 1994). It is actually a record of pollution history. A field survey of the natural background of chromium may be considered as an another example to evaluate soil pollution in Beijing (Wang, 1994). Probability graph paper is used to plot the date (20 soil samples) and determine a cumulative distribution of the analytical chromium results (Fig. 6.2). It is clear that background (11 samples), threshold (3 samples), and anomalous values (polluted, 6 samples) can be distinguished from this figure.

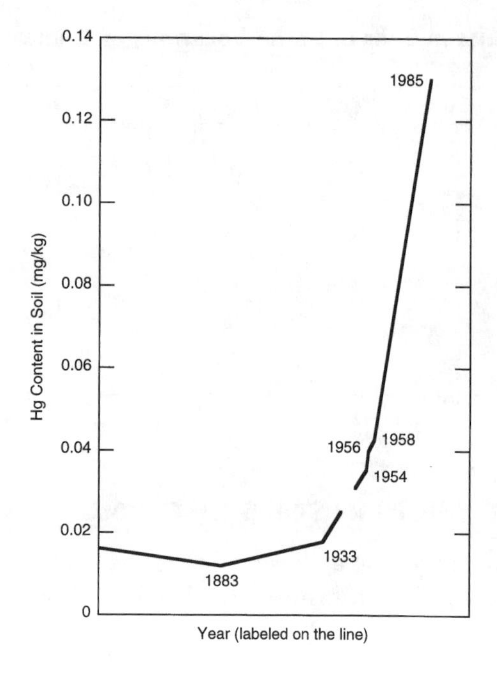

Figure 6.1. Mercury contents in soils of Beijing from different historical periods, including soil profile in the Great Wall (Li and Xu, 1987).

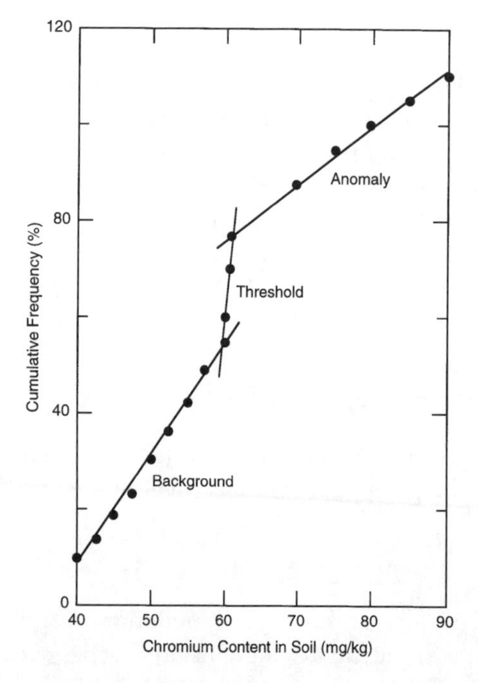

Figure 6.2. Cumulative frequency line for chromium contents in cinnamon soil, a zone soil in Northern China, 0–20 cm (Wang, 1994).

We do not know exactly what these pollution sources were for those anomalous values (especially high in Cr contents), over such a long historical period in Beijing.

The contaminated soil that resulted from gas emission was greater than 5 million ha in China. Included are acidic rain, photochemical smog, F, As, and heavy metal pollution. Among heavy metal pollution, for example, Hg pollution was reported in Beijing, Sichuan (accompanied with wet and dry deposition), nearby Hg and Au mines and

Table 6.2. Nickel Contaminated Soils and Crops Near Two Mines and Smelters in China (kg/kg), 1985[a]

	Soils		Crops	
Mines and Smelters	**Background**	**Contaminated**	**Background**	**Contaminated**
			Rice Grain	
Panshi, Jilin	21.4	29.3–65.3	0.155	13.3
			Wheat Grain	
Jinchuan, Gansu	33.7	53.6–1050	0.42	2.67

[a]Wang, 1994.

smelters in Guizhou, Hebei, etc., with many factories using Hg as a raw material (e.g., thermometers, etc.). Fluorine pollution was especially serious in Baotou, Inner Mongolia, near a steel smelter, and in Jiangsu and Zhejing, near a big chimney for coal combustion of a brick manufacturer. Fluorine pollution is very harmful to bone and teeth metabolism, and silkworm growth is impeded by their eating of F-enriched mulberry leaves (Wu, 1992).

Polluted by Mines and Smelters

China has a big nickel production base which is located in Jinchuan, Gansu. Another smaller mine is at Panshi, Jilin. The nickel content of soils and crops near these two mines and smelters is presented in Table 6.2. It is apparent that the values of all polluted soils and crops are much higher than that of the background (Wang, 1994). We show here only a typical example as a site research. Actually, soil pollution that came from mine and smelter discharges is especially serious and ubiquitous all over China. A summary of this will be presented very briefly in the last paragraph of this section.

Sewage Irrigation and Sludge Application on Land

It has been estimated that more than 30 industrial districts, amounting to an area > 500,000 ha have been polluted due to sewage irrigation.

An investigation in 1980–1983 (Wang, 1984) showed that in the 37 sewage irrigation regions of China, the pollution index P values of 32 regions had been found to be greater than 1, thus indicating that most water did not meet the requirements of water quality from the environmental point of view. As a whole, according to a survey of the irrigation water in 15 regions of China in 1985–1988, the following percentages were reached: Cr 20%, As 14.4%, Hg 33.3%, Pb 6.7%, Zn 14.7%, and Cu 12.5% (Chinese Agr. Environ. Monitoring Center, 1991).

In addition to valuable plant nutrients, sewage sludge contains considerable concentrations of heavy metals. In China, the contents of many elements in the sludge applied on agricultural land often exceeded the sludge criteria in the past decade, thus giving rise to serious soil pollution (Wang, 1991; Table 6.3).

Until now, we haven't had a whole survey of soil pollution nationwide. Some information mainly came from background research, surveys of sewage irrigation, and some local reports concerning contaminated soils in China.

In the last 15 years, heavy metal pollution has been reported by many workers (Wang,

Table 6.3. Elementary Contents of Sludges (mg/kg) in Some Regions of China, 1978–1988[a]

Location	Cd	Pb	Cu	Zn	Hg	Cr	As
Shanghai (Chuansha)	25 to 48	135 to 339	220 to 792	2985 to 8480	6 to 19.8	–	–
Guangzhou	65 to 624	–	–	456 to 5360	0 to 4.7	30 to 1000	–
Beijing (Gaobeidian)	1.9 to 54	136 to 260	350 to 508	605 to 1250	17 to 600	250 to 447	7.8 to 11
Xian	450	2912	2912	–	–	660	–
Tianjin	0.095 to 45	44 to 1680	78 to 2960	312 to 3120	2.13 to 9.10	190 to 950	–
Sludge criteria (EPA,China)	5.0 to 20	330 to 1000	250 to 500	500 to 1000	5.0 to 15	600 to 1000	75

[a]Wang, 1991.

1984; Chinese Agr. Environ. Monitoring Center, 1991; He et al., 1991; Gao, 1992; Wang, 1994). It is especially serious from the pollution sources of the mines and smelters nearby, such as V, Ti pollution in Panzhihua, Sichuan; Pb and Zn pollution in a Shuikoushan mine, Hunan; Pb, Cd, and As pollution in Shenyang smelters, Liaoning; Hg pollution in mercury mines, Hubei and Guizhou; Cu pollution in copper mines, Jiangxi; and Mo and Cd pollution in tungsten mines, Jiangxi, etc.

HEAVY METAL POLLUTANTS AND THEIR INFLUENCING FACTORS

Chemical Forms of the Heavy Metals

The last 15 years or so have seen an increasing interest in the chemical forms of trace metals in environmental media. Many analytical procedures involving sequential extractions have been developed for the partitioning of the trace elements (Tessier et al., 1979; Sposito et al., 1982; Hickey, 1983; Salmons and Forstner, 1980). Though there are many different classified methods, heavy metals generally can be divided into five fractions: exchangeable, bound to carbonate, bound to Fe-Mn oxides, bound to organic matter, and residue. In China, many researchers have reported their results in the field of the speciation of heavy metals in soils (Tang, 1982; Wang, 1993; Kong, 1993, Han et al., 1990). Fractionation and availability of added Cd with two loaded levels in three soils environment (acidic, neutral, and calcareous) were studied by Han et al. (1990). Results showed that exchangeable Cd (Exc-Cd) is about 40%–50%, 60–80%, and 10–20% in acidic, neutral, and calcareous soils, respectively. Cd weakly bound to organic matter (WBO), Cd residue in primary and secondary minerals (Res-Cd), and Cd bound to carbonate (Carb-Cd) were accounted to be 5–20%, 10–70%, and 10–20%, respectively.

Wang and Yan (1990) reported that the distribution of Cu speciation in the unpolluted calcareous cinnamon soil of Beijing was as follows: copper residue (70.9%)>organic bound (14.9%)>carbonic bound (12.8%)>exchangeable (1.28%). In the same soil now polluted by Cu (polluted for 0.5 year), the speciation order was changed: carbonate bound (49.1%)>organic bound (34.2%)>residue (16.1%)>exchangeable (0.52%). It concluded that copper was the metal significantly associated with the organic fractions (Hickey,

Table 6.4. Some Heavy Metal Criteria Recommended in Calcareous Soil (mg/kg) of Northern China[a]

	Hg	Cd	Cu	Ni	Zn	Pb
Total amount	1.5	2.0	120	165	270	260
Available (DTPA)	–	1.5	42	25	61	120

[a]Wang, 1993.

1983), with the exception of the residue in unpolluted soils and carbonate bound in new polluted calcareous soils (Wang and Yan, 1990).

Determination of the total metal content of soils has a valuable place in characterizing the properties of those materials in an absolute sense, but it has long been understood that only the metal soluble fractions have any biological significance. In general, dilute strong acid in acidic soil and DTPA in calcareous soil were recommended as the reagents to extract available trace elements. These include water and weak soluble acid, and exchangeable and weakly bound cations, which are adsorbed at the specific adsorption point of those soil particles. Wang (1993) has recommended both on total and available amount of the heavy metals for the stipulation of maximum permissible levels in calcareous soils of northern China (Table 6.4). The use of total and available amounts of the heavy metals is more simple and practicable for the evaluation of environmental quality than their detailed speciation.

In China, many researchers have studied different reagents to extract heavy metals in soils (Wang, P.X., 1993; Xia et al., 1993; Yang and Wang, 1992). Mercury is not an essential element, and it is toxic, especially to animals and humans. EDTA or DTPA cannot extract available mercury from soil samples (Wang, 1984). Yuan et al. (1982) selected thiomalic acid (TMA) and Wang et al. (1983) recommended thioglycolic acid (TGA) as the extracting reagent for the determination of available mercury in soils. Both of them discovered that there is a much higher correlation to ion mercury concentration between R-SH extracted in soils and crop grains accumulated than to the total amount in soils and crop grains accumulated.

Soil Properties

Many soil properties are important in considering the soil pollution. Among these are pH, oxidation potential (Eh), organic matter, cation exchange capacity, etc. Of these, pH is perhaps the most important, since most heavy metals are more available under acidic conditions than in neutral and alkaline condition. The cation exchange capacity of a soil is a reflection of its organic matter content, as well as the nature and content of clay and sesquioxides. In general, the higher the cation exchange capacity, the more heavy metal a soil can accept without potential hazard.

Generally, the anthropogenic heavy metals remain mainly in the upper layer of the soil profile or migrate on a limited basis. The mechanical composition of soil may influence metal concentration in soils. Wang and Li (1987) reported that heavy metal (Cu, Zn, Ni, and Pb) contents in the soil profiles increased in accordance with its clay fractions. In addition, it appears that metal concentrations in crops are determined more by the amount of the metal applied to soil in the initial application than upon the cumulative total quantity of metal applied over a period of years (Page and Chang, 1978).

MANAGEMENT PRACTICES FOR THE MITIGATION OF HEAVY METAL POLLUTION IN SOILS OF CHINA

The remediation and restoration of soils contaminated by pollutants have recently become a great practical problem. Remedial measures of the soil require, insofar as possible, a full understanding of pollutants, soils, and crop properties. Soil pollution is not like air and water pollution which are, comparatively speaking, easy to dilute by many meteorological factors. However, pollution in soils contaminated with heavy metals is usually long lasting. Therefore, it is necessary to emphasize that a heavily contaminated soil is likely to be the sink of pollutants, resulting in decreased crop yields, or the risk of a contaminated food chain, as well as their relative uptake by crops for a long time.

Many agricultural measures have been recommended for the remediation and restoration of the heavy metals contaminated soils in China. Most of them are described in the following sections.

Soil Dressing

Soil dressing is one of the good measures for restoration of contaminated soils. In Japan, there are many good experiences from soil dressing (Kakuzo and Ichiro, 1981). For example, the dressing of soil free from As contamination on top of polluted soils was adopted as the general method to improve As polluted soil. For reclamation and improvement of Cd contaminated soil, Kakuzo and Ichiro (1981) concluded that (1) reduction of the Cd levels in topsoils by means of mixing with unpolluted soil is not so effective; (2) covering over the polluted topsoils with unpolluted soil is effective, as well as replacement of polluted top soils; and (3) it is desirable that the depth of the unpolluted soil to be replaced or covered should exceed 30 cm to secure a lasting reclamation.

There are some difficulties and problems in the reclamation or improvement of the polluted arable land by soil dressing. In China, though many scientists have recommended soil dressing for improvement of soil pollution, no information has been reported. Most probably, the reasons are due to China's great need for a large area of arable topsoils, and its need for considerable funds for capital investment for soil dressing.

Control the Oxidation Potential in Soils

Redox reactions are highly significant in soil environments. These reactions take place not only chemically, but also by the participation of many microorganisms. It can also be readily seen that the types, rates, and equilibria of redox reactions largely determine the nature of the solute species in soil solutions.

Many heavy metals are transition elements. One of the characteristics of most transition elements is the ability to exhibit more than one oxidation state. They participate in redox reactions very actively. For instance, when rice paddies are drained before harvest, the behavior of C, N, S and especially most of the heavy metals (e.g., Fe, Cr, Mn, As, etc.) generally follow redox potential rise, and the reducing state of those heavy metals decrease, and C, N, S oxidize. When the soils are flooded again, the reactions reverse.

In contaminated soils, coupled surfur oxidation (SO_4^{2-}/S^{2-}) plays an important role for the mitigation of heavy metal pollution in soils. Though the sulfur (generally, 100–500 mg/kg in soils) and heavy metal concentration in soils are rather dilute, the product of the S^{2-} ion and the cations of heavy metals concentration can easily exceed their solubility product (SP), because their solubility products are quite small (e.g., $SP_{HgS} = 10^{-47}$), and

Table 6.5. E_h Changes in Paddy Fields and Its Effect on Heavy Metals Content of Rice (Location, Shengyang, Meadow Brown Soil, China) [a]

Elements	Cd Contents in Soil (mg/kg)	Treatment	Soil E_h (mV) Tilling	Jointing	Milky Maturing	Cd Contents in Rice (mg/kg)
Cd	2.0	Drained	303	293	233	0.278
		Flooded	234	271	224	0.145
Pb	500	Drained	289	305	261	0.400
		Flooded	255	270	255	0.225
As	30	Drained	265	268	271	0.435
		Flooded	255	250	258	0.402
Cr	100	Drained	264	266	274	0.230
		Flooded	258	261	266	0.210

[a]Zhang et al., 1988.

metal sulfide precipitates. Therefore, it decreases both heavy metals and sulfide toxicity for crops and, at the same time, the crops' uptake.

Many researchers have reported their results to maintain low oxidation potential for the alleviation of heavy metal pollution in soil by keeping paddy fields waterlogged for as long a period as possible during the growing season (Cao et al., 1993a; Liu, 1990; Zhang et al., 1988; Chen et al., 1981). Zhang et al. (1988) reported that the oxidation potentials are generally about 100–200 mV in paddy fields. It will increase to 314 mV after water is drained and decrease to 54 mV after it is waterlogged for a very long time. The effects on heavy metal contents of rice through waterlogging or draining in control of the oxidation potential in paddy fields are shown in Table 6.5 (Zhang et al., 1988). It appears that low oxidation potentials have a significant effect on the decrease of most heavy metals uptake by crops.

Arsenic is very toxic to both crops and animals. As a phytotoxicity pollutant in soils, it often occurred in China (Xia, 1988; Wu et al., 1994; Li et al., 1986; Zou, 1986). On the contrary, for example, Zou (1996) reported that the critical value of the maximum permissible limit of arsenite is 8–12 mg/kg in paddy soils (AsO_3^{a-} predominates) for rice growing, and can change to as high as 160 mg/kg (AsO_4^{a-} predominates) in soils for wheat. Therefore, it is much better to maintain the agricultural land drained all the time (high Eh), and at the same time to select rather drought-resistant crops, such as wheat, corn, etc., in arsenic–contaminated soils rather than rice.

By the Application of Soil Amendments

Among all of the measures for remediation and restoration in heavy metal contaminated soils in agroecosystems, the application of soil amendments is perhaps the most practicable. This is not only very effective for the remediation of soil pollution, but it is also abundant in natural resources, with the advantages of lower price, and easier for application on land.

Lime

Lime has been applicated as a soil amendment, especially in acidic soils for many years. In former times, lime application was only to neutralize the acid in soils. Recently, it has been discovered that lime application is of great significance to mitigate the heavy metals toxicity in contaminated soils.

Most of the cations precipitate out as hydroxides, thus decreasing their solubility, and

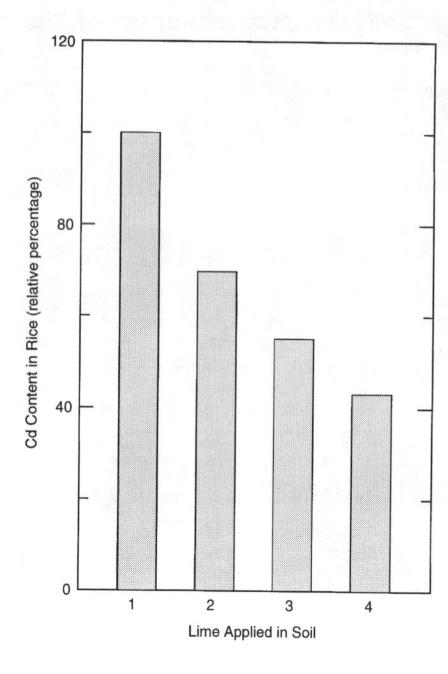

Figure 6.3. Effect on Cd contents in rice after the application of different amounts of lime in soil (Cao et al., 1993a):
1) Control
2) 1.87 ton / ha
3) 3.75 ton / ha
4) 7.5 ton /ha

at the same time, the free ion concentration, and detoxification results. Zinc is perhaps rather complex, because it is amphoteric. The other exceptions are anion formation elements, such as molybdenum (MoO_4^{2-}), arsenic (AsO_3^{3-}, AsO_4^{3-}), selenium (SeO_3^{2-}, SeO_4^{2-}), and chromiun (Cr^{VI}). For example, contrary to metal cations, anionic molybdenate uptake by plants is increased as soil pH is raised. Plants tolerate quite high levels of Mo without causing symptoms or substantial yield reduction. This causes a severe Mo-induced Cu-deficiency by the formation of insoluble $CuMoS_4$ precipitate (Logan and Chaney, 1983). This Cu deficiency in cattle due to Mo pollution in soil has been reported in Mo-contaminated soils nearby the tungsten mine areas of China (Dai et al., 1993).

Lime application to mitigate the toxicity in heavy metals contaminated soils has been reported by many workers (Cao et al., 1993a; Deng et al., 1991; Li, 1990; Liu, 1990; Chen et al., 1981). Toxicity is easily corrected by liming. Chen et al. (1981) have extended their research results on 400 ha of agricultural land by the application of lime 3.0 to 7.5 kg/ha per year in meadow brown soil (pH = 6.9–7.3) in the suburb of Shengyang, China. The lime application decreased the Cd content about 50% in the rice grains. The effect of the application of lime on the decrease of Cd content in rice grain is shown in Figure 6.3. Cao et al. (1993a) suggested that it is much better not to apply too much lime on land (not greater than 3.75 ton/ha); otherwise yield reduction will take place.

Organic Matter

Organic fertilizer has been very important in traditional agriculture for a long time. In the last two decades, we have understood that organic fertilizer can also mitigate the heavy metals toxicity in contaminated soils. Their reactions with heavy metals are very complex. It depends on the organic matter's cations, their stability constants, and environmental pH value, and whatever other competitive anions and cations exist.

Because of its highly complex nature, advances in the knowledge of metal-organic matter reactions have been slow. In general, the molecular weight of humic acid is greater than fulvic acid. Only those metal-chelates of low molecular weight are soluble (Wang, 1993). Metal-fulvic acid chelates usually are more soluble. It can reduce sorption of metals from the soils and improve essential micronutrients availability, and correct trace elements deficiency. The solubility of fulvic metal chelates is strongly controlled by the ratio [FA]/[M]; when this ratio is lower than 2, the formation of water insoluble chelates is favored (Kabata-Pendias and Pendias, 1989). The insoluble chelates play an important role in detoxifying the heavy metals in contaminated soils, because they combine with toxic cations, thus reducing their availabiltiy.

In China, many researchers have applied organic matter, mainly manure, green manure, humic acid, sludges, etc., for remediation in heavy metals contaminated soils (Li, 1982; Liu, 1990; Tian and Liu, 1990; Deng et al., 1991). Deng et al. (1991) have reported the effect of organic matter fertilizer on the decrease of Cr toxicity. It is clear that the main effect came from reducing the property of the organic matter by the reduction of more toxic Cr^{VI} to the less toxic Cr^{3+}, and then the chelate effect takes place in which the humic substances combine with Cr^{3+} for further detoxicity. At the same time, Cr^{3+} will quite easily form hydroxide precipitate in soils. Tian and Liu (1990) have reported that there is a significant mitigatory effect on Al toxicity produced by the application of lime and organic fertilizer in acidic soil. Zhang et al. (1988) expressed their research results by the multiregression equations: the dependent variables, metal contents in rice (Y_1, Y_2); independent variables, soil pH (X_1), and organic matter content in soil (X_2) as follows:

$$Y_1(\mathrm{Cd}) = 0.43 - 0.07X_1 - 0.02X_2 \quad 0.99 \tag{6.1}$$

$$Y_1(\mathrm{Pb}) = 3.08 - 0.32X_1 - 0.43X_2 \quad 0.99 \tag{6.2}$$

Phosphate Fertilizers

Utilization of phosphate fertilizers, especially calcium magnesium phosphate (Ca-Mg-P, P_2O_5 14–20%, CaO 25-30%, MgO 15–18%), for the alleviation of heavy metals toxicity in soils has been applied very often in China. Xiong (1993) has maintained that the precipitation of metal phosphate does not occur in soils. Actually, phosphate is easily fixed as $Ca_3(PO_4)_2$ precipitate in calcareous soils, and as $FePO_4$ and $AlPO_4$ precipitates in acidic soils. The solubility product of many heavy metals and phosphate are even much smaller than $Ca_3PO_4)_2$ (S.P. = 2.010^{-29}). There are many possibilities to form $M_3(PO_4)_2$ precipitates in the heavy metal-contaminated soils by the application of more phosphate fertilizer. In addition, Ca-Mg-P fertilizer is alkaline, and it is helpful, just as a lime, for the formation of metal precipitates. Many researchers have reported their practice of the application of phosphate fertilizer for the alleviation of heavy metal toxicity in soils of China (He, 1994; Cao et al., 1993a; Liu, 1990; Tian and Liu, 1990). He (1994) concluded that the Ca-Mg-P fertilizer can decrease Sb toxicity in soils for rice, and also its uptake by rice tissues. It can decrease the Cd contents in rice too (Cao et al., 1993a, 1993b,). It is revealed that the decrease of rice Cd is in accordance with the increase of soil pH, and at the same time, the decrease of available Cd in soils (Table 6.6). Table 6.7 shows its effect on tissue accumulation of Cd by the application of Ca-Mg-P fertilizer (Cao et al., 1993b).

Silicon Fertilizer

It has recently been recognized that silicon fertilizer has a significant effect on the increase of rice yield in meadow and red soils and resistance to crop lodging (Deng, 1989). A Japanese scientist has recommended silicon fertilizer to decrease the Cd content in rice (Tadaaki, 1985). Cao et al. (1993a) have studied the control of Cd pollution in paddy red

Table 6.6. The Effect on pH and Available Cd in Soils After the Application of Ca-Mg-P Fertilizer (Location, Red Soil, Hubei, China) [a]

Treatment	pH in Soils	Available Cd (mg/kg) [b]
1. 0 (CK)	5.27	3.58
2. 1.5 ton/ha	5.52	2.87
3. 7.5	6.40	1.72
4. 15	7.29	0.38
5. 30	7.68	0.32

[a] Cao et al., 1993b.
[b] ($CaCl_2$ extracted).

Table 6.7. Effect of Cd Uptake on Rice After the Application of Ca-Mg-P Fertilizer (Location, Acrisol, Hubei, China)[a]

	Cd Content (mg/kg)		
Treatment	Roots	Stems and Leaves	Rice Grains
1. 0 (CK)	11.8	2.77	0.39
2. 1.5 ton/ha	11.3	2.48	0.30
3. 7.5	11.8	2.38	0.24
4. 15	7.18	1.26	0.12
5. 30	3.91	0.45	0.06

[a] Cao et al., 1993b.

soil by the application of silicon fertilizer. They concluded that it can increase pH, and at the same time, decrease the available Cd in soils. The results are shown in Figure 6.4.

Other Soil Absorbent Amendments

Other soil absorbent amendments have also been recommended for the mitigation of heavy metals contamination in soils of China. Among all of these amendments, it is clear

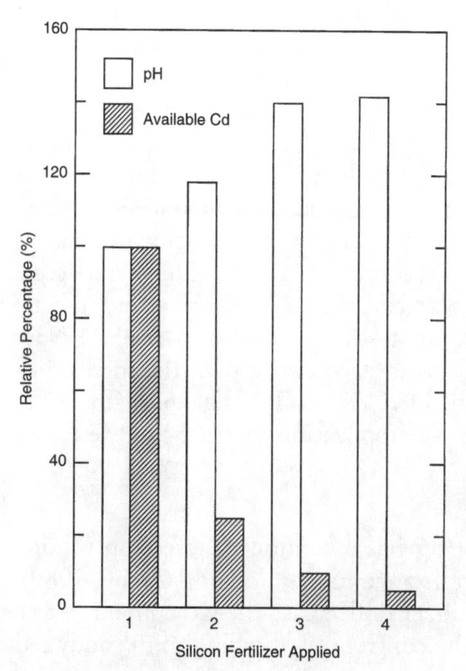

Figure 6.4. The pH and relative available Cd changes in soils after the treatment of different amounts of silicon fertilizer, $CaSiO_3$ (SiO_2 45%, CaO 41%) (Cao et al., 1993a).
1) Control
2) 1.5 ton / ha
3) 15 ton / ha
4) 30 ton / ha)

that the detoxification mechanism mainly comes from their adsorption property, though at the same time, cation exchanges and chemical reactions will take place in soils. Wu et al. (1991) have reported that many absorbents have different detoxicity toward Cu, Cd, and Pb. The detoxifying order is as follows:

$$\text{Furnace Dust} > \text{Active Carbon} > \text{Peat} > \text{Dry Activated Sludge} \qquad (6.3)$$

Liu and Zhang (1993) reported that mercury accumulation in wheat tissues is decreased by the application of sludge and fly ash together. Zang et al. (1989) developed a study on the inhibition of Cd absorption by crops in Cd-contaminated soils with the application of furnace slag.

Among the most important framework silicates are the zeolites. Their chief characteristic is the openness of the $[(Al, Si)O_2]_n$, which is useful as ion-exchangers and selective absorbents. In China, Chen and Lu (1995) reported their research results by the application of zeolite for the decrease of heavy metals uptake by crops in calcareous soil. For instance, they concluded that at different Pb-contaminated levels (1500 and 3000 mg/kg Pb in soils, respectively), the Pb content in spinach is highly negatively correlated with the zeolite amount applied in soil (15 and 30 ton zeolite per ha, respectively). The regression equation with the dependent variable y, Pb content (mg/kg) in spinach, and the independent variable x, zeolite applied (ton/ha) are as follows (Chen and Lu, 1995):

$$y = 1.5433 - 0.0002x \qquad (r = -0.9305, P = 0.01) \qquad (6.4)$$

Liu et al. (1989) reported their results that the application of zeolite at a rate of 15,000 kg/ha can inhibit the accumulation of Cd in the turnip.

Zeolites are an abundant resource in China and easy to exploit. The advantages of zeolite application are not only its high effectiveness for retention of heavy metals in soils, but also its low price and convenience.

Setting Up a New Crop Distribution System

Crops differ remarkably in metal uptake and in tolerance of soil metal. Many pollutants may accumulate in sufficient concentrations in crops to pose a hazard to the health of animals and humans, but without any visible effect on the growth of crops. The tolerance of plants to the levels of these pollutants in soils varies widely with plant species, as well as with soil properties. The biochemical basis for resistance to toxicity is complicated by a great variety of reactions and interreactions at the molecular and cellular levels, even in closely related organisms and tissues (Cumming and Tomsett, 1992; Wood and Wang, 1983).

Uptake is strongly a function of plant species and of plant organs. For example, the order of absorption and accumulation of mercury by the edible parts of crops is as follows (Wang, 1984):

$$\text{Rice} > \text{Chinese Cabbage} > \text{Turnip} > \text{Maize} > \text{Sorghum} > \text{Wheat} \qquad (6.5)$$

Therfore, it is much better to select wheat rather than rice in mercury-contaminated soils. On the contrary, the order for absorption and accumulation of cadmium is (Dong and Chen, 1981):

$$\text{Turnip} > \text{Chinese Cabbage} > \text{Wheat} > \text{Rice} \qquad (6.6)$$

Obviously, it is much better to select rice, rather than other crops mentioned in the discussion of cadmium-contaminated soils.

The potential for breeding crop cultivars with reduced metal concentration or greater metal tolerance was suggested in 1973 (Logan and Chaney, 1983). Some very promising results were reported by Hinesly et al. (1978) for differences in Cd and Zn uptake among maize inbreds and maize hybrids. It is unfortunate that not much research and practice with the selection or breeding of crop cultivars has been reported in China.

Growing nonedible plants in the heavy metals contaminated soils has been recommended by many workers (Chen et al., 1986; Huang et al., 1986; Gao, 1992; Long et al., 1994).

There were approximately over 300 ha of Cd-contaminated lands in the western suburb of Shenyang, which was caused by wastewater irrigation. An investigation in 1975 showed that the Cd contents in soils was 5.08 mg/kg and in rice grain, 1.09 mg/kg in that area. Since 1978, agricultural ecology engineering has been carried out by:

(1) controlling the total amount of Cd discharge in industrial wastewater, from originally 10 tons decreased to 0.38 ton a year; the Cd content in sewage, from 230 μg/L, decreased to 10–15 μg/L;
(2) growing nonedible crops in the slightly polluted lands, by breeding crop variety such as sorghum, castor, etc. Sorghum was utilized for the manufacture of alcohol, and castor, for soap. At the same time, irrigation at that region was carried out by sewage and clean water mixed together;
(3) growing ornamental plants, such as flowers, grasses, and woody plants in heavily polluted lands.

All of these have obtained very good ecological and financial effects. As a biological control, growing trees in heavy metals contaminated soils is very promising. *Populus pekinensis, populus canadensis,* and *populus xenrama robusta* were all planted for experiments. All the poplar trees assayed enriched Cd, and at the same time, removed Cd from soils (Table 6.8; Huang et al., 1986). The Cd content in the leaves of the control poplar tree are 0.50 to 0.76 mg/kg, but those in Cd-contaminated soils can accumulate as high as 11.9 to 29 mg/kg (Huang et al., 1986).

Other Remediative Measures

One of the other techniques for the mitigation of heavy metal pollution in soils is the bioremediation, i.e., biological treatment, of pollutants in situ. Biotechnology for decreasing pollutants in soils, surface, and groundwater is growing rapidly (Shen, 1993). For ex-

Table 6.8. Variation of Cd Content (mg/kg) in the Soil by Cultivation of Beijing Poplar Young Trees (Line Distance, 1.5 x 2.0 m; 50 Trees/0.6 ha) (Location, Meadow Brown Soil, Shenyang, China) [a]

Soil Depth	1983			1984		
(cm)	Initial	Final	Decreased	Initial	Final	Decreased
0-20	4.70	3.50	1.20	4.69	3.71	0.98
20-50	1.30	1.30	–	1.67	1.62	0.05
50-100	0.23	0.23	–	0.87	0.86	–

[a] Huang et al., 1986.

ample, microbial volatilization of selenium may have application for detoxifying Se-contaminated soils, if high rates of volatilization can be maintained (Karlson and Frankenberger, 1988). This method is full of promise for the removal of heavy metals, such as Se, Hg, As, etc., because their alkylated compounds are highly volatile. It is regretted that there are no reports of the utilization of this technique in China.

CONCLUSIONS

We attempted to follow a path of scientific inquiry into how natural systems function, explored the past influences, and presented problems and future prospects for people and the environment. Our ultimate goal is to propose a systematic approach to the management of the environment with a view to improving the quality of life.

Now we understand that environmental pollution consists not only of scientific and technical problems, but also ethical, moral, and social problems. We only have one world. Further economical development is still of great importance, but it must be sustainable. We should take into account that our surroundings only have a limited capacity to degrade wastes; when this limit is exceeded, the ecological equilibrium is deteriorated.

Environmental protection is a comprehensive new systematic engineering, because it covers many topics and makes many interdisciplinary connections that are not generally brought out in traditional works. At first, continual environmental education for all people is necessary. Control of the total amounts of pollutant discharged and treatment of all the waste legislation and its strict implementation, are of primary importance. In sustainable, or ecological agriculture, integrated pest management (IPM), rather than only chemical pesticides is advocated, and fertilization is recommended with an emphasis on organic matter cycling.

Research conducted over the last 15 years in China has created a reservoir of professional expertise and information upon which to base state and local guidelines for identification and rational resolution of concerns. Good management practices are a prerequisite to give assurance that growing crops in contaminated soils can be accomplished safely with minimal risk and attendant benefits.

We hope that this research may prove as useful to others as it has to us and that it may play some role in helping to solve our environmental problems.

REFERENCES

Cao, R.L., W.R. Huo, Z.L. He, A.T. Hu, and B. Deng. A Study on Controlling Cd Pollution in Paddy Soils in Luo Qiao Township, Dayie County. *Chinese Environ. Sci.* (Beijing, China), 13, 6, pp. 433–439, 1993a.

Cao, R.L., W.R. Huo, Z.L. He, Y. Zhou, J.J. Wang, and Y.L. Li, Effect of Calcium Magnesium Phosphate Fertilizer on Chemical Forms and Translocation and Uptake of Cd by Rice in Soil. *Chonging Environ. Sci.* (Chongqing, China). 15, 6, pp. 6–9, 1993b.

Chen, J.M. and J. Lu. Zeolite as Soil Amendment for the Alleviation of Cd and Pb Uptake in Spinach. *Beijing Agr. Sci.* (Beijing, China) 3, pp. 23–25, 1995.

Chen, T., Y.Y. Wu and Q.X. Kong. Residual Effect of Liming Amelioration on Cd Soil at Changshi Irrigation Area. *China J. Environ. Sci.* (Beijing, China). 2, 3, pp. 40–42, 1981.

Chen, T., Y.Y. Wu, Q.X. Kong, F. Tan, and X. Wang. Reclamation of Cd-Contaminated Farmland. *Acta Scientiae Circumstantiae* (Beijing, China). 6, 4, pp. 384–399 1986.

Chinese Agriculture Environmental Monitoring Center. A Survey of Chinese Agr. Environ. Quality and Devel. Trends, A Report to Chinese Agr. Ministry, 1991.

Chinese Environmental Monitoring Center. Chinese Elementary Background in Soils. Chinese Sci. Publ., Beijing, China, 1990.

Cumming, J.R. and A.B. Tomsett. Metal Tolerance in Plants: Signal Transduction and Acclimation, in *Biogeochemistry of Trace Elements*, Adriano, D.C., Ed. Lewis Publishers, Chelsea, MI, 1992, pp. 329–364.

Dai, Q.W., Z.M. Zeng, J.M. Wang, Z.L. Wu, and P. Fan. The Effect of Metal Mine on Animal Husbandry in Jiangxi Province. *Agro-Environ. Protection*. (Tianjin, China). 12, 3, pp. 124–126, 1993.

Deng, B.E., T.C. Liu, and H.Y. Li. A Study on the Effect of Liming and Manuring on the Improvement of Soil Pollution by Cr. *Soil and Fertilizer* (Beijing, China) 6, pp. 19–22, 1991.

Deng X.Y. The Preliminary Report of the Application of Silicon Fertilizer in Hubei. in *Study on Chinese Chemical Fertilizer Application*. Lin, B., Ed. Beijing Sci. and Tech. Publ. (Beijing, China), 1989, p.436.

Dong K.Y. and J.M. Chen. Regularity of Absorbing and Accumulating Cd in the Crops. *J. Environ Sci*. (Beijing, China). 3, pp. 6–11, 1981.

Environ. Protection Agency of China (EPAC). Chinese Environmental Anouncement. Environmental Work Newsletter. (Beijing, China). 17, pp. 16–17, 1995.

Gao, L. Soil Pollution and Its Control. *Agro-environ. Protection*. 11, 6, pp. 272–273, 1992.

Han, F.X., A.T. Hu, and H.Y. Qin. Fractionation and Availability of Added Soluble Cd in Soil Environment. *Environ. Chem*. (Beijing. China). 9, 1, pp. 49–53, 1990.

He, M.C. Study on the Effect of Soil Sb on the Growth and Residue in Rice. *Agro-environ. Protection* (Tianjin, China) 13, 1, pp. 18–22, 1994.

He, Z.Y., Z.J. Ye, and F.Z. Wu. The Introduction of Agricultural Environment Science. *Shanghai Sci. Tech. Publ*. pp. 242–246, 1991.

Hickey, M.G. Partitioning of Heavy Metals in Contaminated Soils. *Diss. Abst. Int. P.T.–Sci. Eng.* 43, 9, Mar. p. 79, 1983.

Hinesly, T.D., D.E. Alexander, E.L. Ziegler, and G.L. Barrett. Zinc and Cadimiun Accumulation by Corn Inbreds Grown on Sludge Amended Soils. *Agro. J.* 70, pp. 425–428, 1978.

Huang, H.Y., D.M. Jiang, Z.X. Zhang, and Y.B. Zhang. Study on Biological Control in Cd Polluted Soils in Shenyang, in *Study on Pollution Ecology of Soil-Plant Systems*, Gao, Z.M., Ed. China Sci. and Tech. Publ. Beijing, China. 1986, pp. 85–87.

Kabata-Pendias, A. and H. Pendias. Trace Elements in Soils and Plants. CRC Press, Boca Raton, FL, 1989, p. 7.

Kakuzo Kitagishi and Ichiro Yamane. *Heavy Metal Pollution in Soils of Japan*. Japan Sci. Soc. Press. Tokyo, Japan, 1981, pp. 121, 187.

Karlson, U. and W.T. Frankenberger. Effect of Carbon and Trace Elements Addition on Alkylselinite Production by Soil. *Soil Sci. Soc. Am. J.* 52., pp. 1640–1644, 1988.

Kong, Q.X. Distribution of Some Forms of Heavy Metals in Soils. *Agr.-environ. Protection* (Tianjin, China). 12, 6, pp. 267–273, 1993.

Li, C.L. and Q. Xu. The Variation Trend and Cause of the Background Values of Mercury in Soils of Beijing Area at Various Times, *China Environ. Sci.* (Beijing, China), 7(1), pp. 35–37, 1987.

Li, G.F. A Study on the Mechanism of the Translocation and Enrichment of the Heavy Metals in Sewage Irrigation Soils of Baiyin Region, Gansu. *Environ, Research* (Lanzhou, China). 1, pp. 28–36. 1982.

Li, H.Y. Soil Contaminated by Cr and its Improvement by Measures. *Environ. Introduction News* (Nanjing, China). 2, pp. 5–7, 1990.

Li, J.R., H.Y. Sun, T.H. Yu, and F.H. Lian. An Investigation of Soils of Yinshanban. Shaoxing. *Agro-environ. Protection*. (Tianjin, China). 5, pp. 34–39, 1986.

Liu, G.X. and G.Y. Zhang. Study of Hg Accumulation in Wheat Plant Applied with Sewage Sludge and Fly Ash. *Agr-environ. Protection*. 12, 2, pp. 201–203, 1993.

Liu, G.Y., K.Q. Bo, and W.M. Zhang. Effect of Cd Uptake on Turnip by the Application of Zeolite in Soils. *Beijing Agr. Sci.* (Beijing), 6, pp. 20–23, 1989.

Liu, L.J. The Control of Pathway of Cd Contaminated Soils in Southern Jiangxi. *The Resources Development and Protection*. (Sichuan, China). 6, 2, pp. 100–102, 1990.

Logan, T.J. and R.L. Chaney. Metals, in *Utilization of Municipal Waste Water and Sludge on Land,* Page, A.L., (Ed.). Univ. of California, Riverside, 1983.

Long, Y.T., S.F. Liu, J.P. Xiang, J.F. Yang, and K.Q. Li. Study on the Purification Effect on Paddy Soil Mercury by Ramie. *Agro-environ. Protection,* 13, 1, pp. 30–33, 1994.

Page, A.L. and A.C. Chang. Trace Elements Impact on Plants During Cropland Disposal of Sewage Sludge. *Proceedings of 7th National Conference on Municipal Sludge Management.* Washington D.C. Information Transfer Inc., 1978, p. 84.

Salmons, W. and U. Forstner. Trace Metal Analysis on Polluted Sediments, *Environ. Technol. Lett.* 1, pp. 506–512, 1980.

Shen, D.Z. Bioremediation: Biological Treatment of Contaminants In Situ. *Advances in Environ. Sci.* (Beijing, China). 5, pp. 56–60, 1993.

Sposito, G., L.J. Lund, and A.C. Chang. Trace Metal Chemistry in Arid Zone Soils Amended with Sewage Sludge. I. Fractionation of Ni, Cu,Zn, Cd and Pb in Solid Phase. *Soil Sci.Soc.Am. J.* 46, pp. 260–264, 1982.

Tadaaki, Y. Countermeasures of Heavy Metal Polluted Soils (1). *J. Agr. Sci.* (Tokyo, Japan). 40, 10, pp. 443–447, 1985.

Tang, H.X. New Development in the Field of Water Pollution Chemistry, *Environ. Chem.*, (Beijing, China). 1, 2, pp. 93–101, 1982.

Tessier, A., P.G.C. Campbell, and M. Besson. Sequential Extraction Procedures for the Speciation of Particulate Trace Metals. *Anal. Chem.* 51, 7, pp. 844–851, 1979.

Tian, R.S. and H.T. Liu. Aluminum in Acidic Soils and its Toxicity. *China J.Environ. Sci.* (Bejing China). 11, 6, pp. 41–46, 1990.

Wang, H.K. Sewage Irrigation in China. *International J. Devel Technol.* 2, pp. 291–301, 1984.

Wang, H.K. and T. Li. The Evaluation of Agricultural Environmental Quality in Chang Ping Experiment Station of Beijing, *Agr. Unv. Acta. Agriculturae Universitatis Pekinensis* (Beijing, China). 13, 3, pp. 309–318, 1987.

Wang, H.K. and S.C. Yan. Study on Copper Pollution from Fertilization of the Soil with Sludge. *Chinese J. Environ. Sci.* (Beijing, China). 11, 3, pp. 6–10, 1990.

Wang. H.K. *Water Pollution and Its Control.* Beijing Agr. Univ. Publ. Beijing, China, 1991.

Wang, H.K. Environmental Quality Criteria of Some Toxic Elements in Soils. *Agro-environ. Protection.* (Tianjin, China). 12, 4, pp. 162–165, 1993.

Wang, H.K. Research Methodology for Determining Background Levels of Trace Elements, in Soils of China, in *Biogeochem. of Tree Element,* Adriano, D.C. et al., Eds., Sci. and Tech. Letters, Northwood, England, 1994, pp. 397–413.

Wang, H.K., Z. Wang, and P. Li. Determination of Effective Mercury, in *Soil, Environ. Sci.* (Beijing, China), 4, 4, pp. 61–63, 1983.

Wang, P.X. The Chemical Forms and Translocation of Cu and Ni in Loessial Soils Amended with Sludge. *Agro-environ. Protection.* (Tianjin, China). 12, 1, pp. 14–16, 1993.

Wood, J.M. and H.K. Wang. Microbial Resistance to Heavy Metals. *Environ, Sci. Technol.* 17, 12, p. 582 A-590A, 1983.

Wu, F.Z. The Introduction of Air Pollution, *Agr. Publ.*, Beijing, China, pp. 5–6, 1992.

Wu, L.S., Z.L.Gu, S.Q. Xie, and D.Z. Zhou. Factors Affecting Biological Toxicity of Heavy Metals in Soils and Their Regulations. *Chinese J. Environ. Sci.* (Beiijing, China). 12, 3, pp. 12–18, 1991.

Wu, Y.Y., X. Wang, Y.Q. Ma, and T.Z. Wu. The Combined Pollution of As with Other Metals and Their Prevention. *Agro-environ. Protection* (Tianjin, China). 13,3, pp. 109–114, 1994.

Xia, Z.L. *Soil Environmental Capacity and Its Application.* Meteorologic Publ. (Beijing, China), 1988.

Xia, Z.L., W.Q. Meng, and S.Z. Li. The Choice of Extractants of Available Cd in Purple Soils. *China Environ. Sci.* (Beijing, China), 13, 1, pp. 49–53, 1993.

Xiong, L.M. Fertilization and Plant Heavy Metal Uptake. *Agro-environ. Protection.* (Tianjin, China). 12, 5, pp. 217–222, 1993.

Yang, Z.Y. and H.K. Wang. A Study on the Available Pb Extracted by NH_4Ac and DTPA in

Soils, in *The Characteristics of Soil Resources and Its Application*. Li, X.L., Ed. Beijing Agr. Univ. Publ., China, 1992, pp. 105–109.

Yuan, E.L., S.F. Song, X.F. Zhao, and Q.T. Jiang. Thiomalic Acid as an Extracting Reagent for the Determination of Soluble Hg. *Chinese J. Soil Sci.* (Shenyang, China), 6, pp. 26–29. 1982.

Zang, H.L., C.R. Zheng, and H.M. Chen. Study on the Inhibition of Cd Absorption by Crops in Cd Contaminated Soils. *Agro-environ. Protection*. S, 1, pp. 33–34, 1989.

Zhang, X.X., X.Z. Xong, Y.S. Wang, L.P. Wang, S.H. Zong, and H. Ren. Environmental Capacity of Meadow Brown Soil for Heavy Metals. *Act Scientiae Circumtantiae*. 8, 3, pp. 296–306, 1988.

Zou, B.J. Arsenic in Soils. *Progress in Soil Sci.* (Nanjing, China). 2, pp. 8–13, 1986.

Table 7.2. Comparison of Background Values (Average Values) in Soils Between China and Other Countries[a]

Element	China	USA	Japan	UK
	mg/kg			
As	11.2	7.2	9.02	11.3
Cd	0.097	–	0.413	0.62
Co	12.7	9.1	10	12
Cr	61.0	54	41.3	84
Cu	22.6	25	36.97	25.8
Hg	0.065	0.0891	0.28	0.098
Mn	583	550	583	61
Ni	26.9	19	28.5	33.7
Pb	26.0	19	20.4	29.2
Se	0.29	0.39	–	0.40
V	82.4	80	–	108
Zn	74.2	60	63.8	59.8

[a]Modified from Wei et al., 1991.

Table 7.3. Background Values of Rare Earth Elements in Soils of China[a]

	Soils of China				
Element	Range	Midvalue	AM[b]	GM[c]	Range of 95% Degree of Confidence
Sc	0.03–61.7	10.8	11.1	10.6	5.5–20.2
Y	0.50–130	22.1	22.9	21.8	11.4–41.6
La	0.26–242	36.8	39.7	37.4	18.5–75.3
Ce	0.02–265	65.2	68.4	64.7	33.0–127
Pr	0.10–40.5	6.17	7.17	6.67	3.1–14.3
Nd	0.05–100	25.2	26.4	25.1	13.0–48.4
Sm	0.004–20.1	4.90	5.22	4.94	2.53–9.65
Eu	0.01–5.15	1.00	1.03	0.98	0.52–1.86
Gd	0.19–16.8	4.44	4.60	4.38	2.31–8.30
Tb	0.005–3.10	0.59	0.63	0.58	0.25–1.33
Dy	0.07–14.4	4.03	4.13	3.93	2.08–7.43
Ho	0.04–3.04	0.84	0.87	0.83	0.44–1.56
Er	0.13–9.37	2.47	2.54	2.42	1.29–4.55
Tm	0.04–1.40	0.36	0.37	0.35	0.19–0.65
Yb	0.02–7.68	2.35	2.44	2.32	1.25–4.32
Lu	0.002–1.90	0.35	0.36	0.35	0.19–0.62
Ce	15.4–492	142.8	143.2	136.9	74.0–253.3
Y	2.6–185	37.9	37.2	35.0	19.8–65.8
TR	180–582	181.1	187.6	179.1	97.1–330.2

aModified from Chen, 1990.
[b]AM: Arithmetic mean.
[c]GM: Geometric mean.

Background Values of Rare Earth Elements in Soils of China

In 1990, the China National Environment Protection Agency (NEPA) formally published a *List of Background Values of Rare Earth Elements in Soils of China* (see Table 7.3). In addition to the Lanthanides, the list includes Sc and Yb, which are similar in chemical property to the rare earth elements of the lanthanide series. The NEPA also separately established the background values both for all rare earth elements and for those in the Ce and Gd groups in soils. Like the background values of heavy metal elements in soils, the background values of rare earth elements in soils can serve as the basis for establishing the criteria for soil environmental quality, exploring mineral deposits, and determining a possible correlation between the abundance or deficiency of rare earth elements in soil and quality of farm products, and their effect on animal health.

Application of Background Values of Elements in the Soil

Determining the Soil Environmental Quality

In general, to work out the criteria of soil environmental quality, both the background values of the soil and the ecological effect should be taken into account. Based on this principle, Wei et al. (1991) put forward the suggested values for Hg, Cd, and Pb in soils of China as the criteria for soil environmental quality (Table 7.4).

Using Background Values of the Soil to Determine the Threshold Value of Endemic Diseases Caused by Deficiency of Nutrients

In the Loessial Plateau of China, the prevailing Keshan disease and Kaschin-Beck disease are closely associated with a low content of selenium in soil (Table 7.5). It can be seen from the table that the amount of total Se, water-soluble Se in the soil, and Se in the hair

Table 7.4. Suggested Values of Environmental Quality for Hg, Cd, Pb, and As[a]

Grade	Name	Level	Based on[b]	Recommended Value
I	Background value	Ideal level	GM_D	Hg 0.10, Cd 0.15, Pg 30, As 10-15
II	Standard value	Adaptable	GM_D	Hg 0.20, Cd 0.30, Pg 60, As 20(25)
III	Warning value	Tolerable	Normal ecologic effect	Hg 0.50, Cd 0.50, Pb 100, As 27
IV	Critical value	Beyond	Severe ecologic effect (or an average of high background values)	Hg 1.0, Cd 1.0, Pb 300, As 30

[a]Wei et al., 1991.
[b]GM: Geometric mean; D: Geometric standard deviation.

Table 7.5. Background Values of Soil Se for the Diseased and Nondiseased Areas of the Loessial Plateau[a]

Item	Seriously Diseased Area	Moderately Diseased Area	Slightly Diseased Area	Average	Nondiseased Area	T Value Test
	(mg/kg)					
Total Se in soil	75.6	88.0	101.2	88.26	145.5	t=6.45 P < 0.001
Water-soluble Se in soil	1.51	1.98	2.27	1.92	3.82	t=7.23 P < 0.001
Se in man's hair	57.0	74.0	98.0	76.33	229.0	t=8.14 P < 0.001
Se in wheat	6.13	9.53	11.75	9.12	33.7	t=66.2 P < 0.001

[a]Wei et al., 1991.

of the local people and food (wheat) they eat are all lower than those in the nonaffected area.

The different forms of Se in the soil also tally with the level of the Se content in the human body and hair, as well as with the severity of the diseases. According to Wei et al. (1991), 90.6% of the samples from the disease-stricken areas with background values of soil Se are below 110 μg/kg, whereas 87.7% of the examples from nonaffected areas are above 110 μg/kg. For this reason, the background threshold value for the soil Se of the diseased-stricken areas (referring to the Keshan and Kaschin-Beck diseases) and nonaffected areas of the Loessial Plateau was set at 110 μg/kg.

SOIL LOADING CAPACITY FOR HEAVY METALS

Definition and Research Methods

Research results obtained from environmental studies show that ensuring the long-term stability of the ecological environment is impossible by only restricting the concentrations of pollutants that are drained off from their sources. Rather, the total amount of the pollutants should be controlled. Thus, the study of the environmental capacity began in the early 1970s and soil loading capacity became one of the important topics.

Soil loading capacity for heavy metals refers to the maximum amount of heavy metals that the soil is capable of holding within a given environmental unit and a given duration of time by following the environmental quality criteria that ensure the yields and biological quality, e.g., yields, nutrients, and the content of toxic elements of agricultural products, without causing environmental pollution.

Soil loading capacity for heavy metals is studied by:

1. Investigating the regional basic conditions such as natural conditions, social production status, and soil distribution types and characteristics;
2. Evaluating the situation and regionalization of soil heavy metal pollution in the region so as to ascertain the existing capacity;
3. Investigating the ecological effect of heavy metal pollution through examining the effects of heavy metals of different pollution concentrations on the physiology, ecology, and biomass of various indicator organisms in the ecological system of soils and the residual and accumulation of pollutants in the organisms;
4. Investigating the effect of soil heavy metal pollution on the environment, especially the secondary pollution of heavy metals to surface and underground water after their entering into the soils;
5. Ascertaining the critical contents of heavy metal pollutants in the soils, as they are key factors affecting the determination of soil environmental capacity and are themselves affected by many factors such as soil properties, indicator difference, pollution course, environmental factors, compound types, and organic matters;
6. Probing into the pathways of heavy metals in soils. This is because heavy metals are less variable in soils, as far as the total amount is concerned, even though their fixation and release vary with the environmental condition and time; i.e., their "available forms" are influenced by many factors which sometimes have no certain relationships with the total amount. It is necessary to understand output of heavy metal through surface runoff, underground percolation, and crop uptake.
7. Modeling the soil loading capacity of heavy metals. A mathematical model of soil loading capacity is a quantitative relationship composed of many parameters in the soil ecological system and its boundary environment. Soil loading capacity as depicted by the model may be divided into static capacity and dynamic capacity (Xia, 1988).

Soil static capacity is the maximum loading capacity of contaminative heavy metals contained in soils within a fixed environment unit and a fixed duration of time, when assuming that heavy metals in soils would not take part in the cycling in the environment, and it can be written as (Xia, 1988):

$$Q_{si} = W(C_{ci} - C_{oi}) \quad (7.1)$$

where Q_{si} is the static loading capacity of heavy metal i (mg/ha), W is the weight of soil in the plow layer (kg/ha), C_{ci} is the critical content of heavy metal i (mg/kg), and C_{oi} is the original content of heavy metal i. When C_{oi} is equal to the background value of soil, it is the static capacity of regional background soil, and the static capacity may be obtained by dividing Q_{si} by the fixed number of years t. Although the soil's static capacity is not the real capacity, it is of a certain value in application since its parameters are simple.

Soil dynamic capacity refers to the maximum loading capacity of contaminative heavy metals contained in soils within a fixed environment unit and a fixed duration of time when heavy metals are assumed to participate in the material cycling in pedosphere, and it may be expressed as:

$$Q_d = W\{C_{ci} - [C_{pi} + f(I_1, I_2, \ldots I_n) - f'(O_1, O_2, \ldots O_n)]\} \quad (7.2)$$

where Q_d is the quantity of some heavy metals possibly contained in the soil (mg/ha), C_{ci} is the critical content of heavy metal i (mg/kg), C_{pi} is the concentration of contaminative heavy metal i in the soil mg/kg, I is the input item, and O is the output item. The soil permissible quantity within various fixed numbers of years, i.e., soil dynamic capacity, may be calculated by using a computer.

Factors Affecting Soil Loading Capacity of Heavy Metals

Soil environmental capacity is affected by many factors such as soil properties, indicators, pollution course, and compound types.

Effect of Soil Properties

Soil is a very complex, uneven system. The effect of the type of a soil on environmental capacity is quite obvious. Even for the same type of soils derived from the same parent materials located in different regions, their effects on the chemical and biological behavior of heavy metals in soils differ significantly though their properties have no great difference. Research on the chemical behavior of heavy metals in three kinds of yellow-brown soils, derived from Xiashu yellow soil in Xuyi and Nanjing of Jiangsu Province and Xiaogan of Hubei Province, showed that soil properties had significant effects on heavy metal forms, microbes, and plant yields.

Effect on the Forms of Heavy Metals

In the case of the same contaminative Cd concentration, the exchangeable, organically bound, oxides coated, and residual forms of Cd were quite different in the three soils. In the soils of Xuyi, Nanjing, and Xiaogan they were 57.5, 65.3, and 51.5%, respectively, suggesting that the yellow-brown soil in Nanjing had a weaker retention of Cd than that in Xuyi or Xiaogan.

Effect on Microbes

Differences in soil properties resulted in the different effects of heavy metals on the digestion activity of pot-culture paddy soil, the biomass of soil microbes, and the activity of soil enzymes. For example, the effect of Cu (added at the rate of 100 mg/kg) on the digestion activity of pot soil was 109, 42, and 57% in the three soils of Nanjing, Xuyi, and Xiaogan, respectively, as compared with that in the check, indicating that under the experimental conditions, Cu could stimulate the digestion activity of Nanjing soil to a certain degree, but obviously it inhibited that of Xuyi and Xiaogan soils. On the other hand, Pb (500 mg/kg) obviously stimulated the digestion activity of Xuyi and Xiaogan soils (158% and 110%, respectively) against the control, but markedly inhibited that of Nanjing soil (63%). Cd (1 mg/kg) evidently inhibited the digestion activity of all three kinds of soils, and its inhibition was in the order of Nanjing soil (55%) > Xiaogan soil (46%) > Xuyi (35%). Arsenic (40 mg/kg) had a remarkable ability to inhibit the digestion activity of Xiaogan soil (62% compared with the check), but did not show any effect on that of Nanjing soil and Xuyi soil.

Effect on Plant Yield

Among the three soils, there existed obvious differences in the concentration of contaminative Cd, As, and Cu when rice yield was decreased by 10%. For example, in the case of rice yield decreasing by 10%, the amount of Cd added to Nanjing soil was almost half of that added to Xuyi soil or Xiaogan soil.

Thus it may be concluded that yellow-brown soils derived from the same parent material but located in different regions would have different critical values of heavy metals contents. Therefore, attention must be paid to both the typical as well as the representative properties of soils in the studies on soil loading capacity for heavy metals.

Effect of Indicators

In order to find the loading capacity of a soil for heavy metals, a given reference matter should be selected as an indicator for attaining a special aim. The loading capacity obtained may vary greatly with the indicator.

Different Effects on Rice and Wheat

The amount of heavy metals found in brown rice and in wheat seed was apparently different when the same concentration of heavy metals was added into the soil. The following are examples of the results obtained: the amounts of Cd and Pb are higher in wheat seeds than in brown rice.

Different Effects on Microbes

Heavy metals have quite different effects on various types of microbes. For example, the addition of Cd at 0.5 to 100 mg/kg into a soil could very significantly inhibit the fungi but had no effect on actinomyces and bacteria.

Effect of Pollution Course

Heavy metals, like any other soil element, may be dissolved in soil solutions, adsorbed on colloid surfaces, occluded in soil minerals, or they may precipitate with other compounds in the soils. All these activities are related to the pollution course.

Time and Concentration

The amount of heavy metals in water drained from a field plot varied greatly with time. In an experiment conducted by the Institute of Soil Science, Academia Sinica, from 4 July to 15 August, 1990, it was found that the content in the field drainage water decreased from 4.49 to 0.18 μg/L for Cd (Cd added into the soil: 3 mg/kg), from 175 to 1.6 μg/L for Pb (Pb added: 240 mg/kg), from 2.8 to 0.9 μg/L for As (As added: 30 mg/kg), and from 1.7 to nondetectable for Cu (Cu added: 150 mg/kg). With time through soil retention, the heavy metal concentrations in the drainage water become lower and lower, and their harmful effect on the organisms also decreased accordingly.

Changes of Forms

The effect of pollution course is also reflected in the changes of heavy metal forms in soils. For example, the amount of adsorbed As (extracted with 1 mol/L NH_4Cl), Fe-type As (extracted with 0.1 mol/L NaOH), and Al-type As (extracted with 0.5 mol/L NH_4F) tended to decrease with time, whereas the amount of occluded As apparently increased, rising from 6.4% to 33% during a 30-day waterlogging balance process. Changes of forms inevitably influence plant uptake, thus bringing about evident effects on the soil critical values.

Pollutant Accumulation

Uptake of heavy metals by plants tends to increase with increasing concentration, as long as it is within a certain range. When the concentration goes beyond the range, the uptake will decrease because plant roots are injured, thus loading to a lower absorbing ability. Therefore, it is easy to make errors if the soil pollution status of an area is determined simply from the contents of pollutants in the seeds.

Effect of Environmental Factors

Temperature. Research on the mechanism of plants uptaking heavy metals has shown that in some cases it is affected by temperature. Temperature changes will certainly affect water evaporation, thus in turn affecting heavy metal uptake by plants. For example, a single cropping of midseason rice takes up obviously more Cd than the late rice. When Cd in the soil was 10 mg/kg, Cd content in brown rice was about 0.5 mg/kg for late rice, but could reach 2.3 mg/kg for midseason rice. The cultivation season has an obvious effect on As content in brown rice. When As in the soil was 40 mg/kg, As content in brown rice was 0.67, 0.43, and 0.33 mg/kg for early rice (the mean monthly temperature at ripening stage: 27.8 to 28°C), midseason rice (the mean monthly temperature at ripening stage: 16.9 to 22.7°C) and late rice (the mean monthly temperature at ripening stage: 10.5 to 16.9°C) respectively, suggesting that As uptake decreases apparently with the dropping of the temperature.

pH and Eh. Generally speaking, fixation of heavy metal cations by the soil increases with the rise of pH. For example, an experiment on Pb adsorption by yellow-brown soil from Nanjing showed that soil retention of Pb increased markedly with the rise of pH (Chen et al., 1994). Arsenic is a variable valence element. Increasing waterlogging time, raising pH, and decreasing Eh make water-soluble As increase within a certain time.

Effect of Compound Types

Compound types have a remarkable effect on soil environmental capacity. For example, within a certain concentration range, $CdCl_2$ and $CdSO_4$ resulted in decrease in mean rice yield by 33% and 17.8%, respectively. Different Pb compounds exert different effects on rice yield and plant uptake, due obviously to the role of the anions (Chen et al., 1994).

Combined pollution factors had remarkable effects on soil loading capacity for heavy metals (Zheng and Chen, 1989; 1990; 1995; Chen and Zheng, 1992; Chen et al., 1994).

Application of Soil Loading Capacity of Heavy Metals

Soil heavy metal loading capacity may be used for exerting total quantity control of regional heavy metal pollutants. In an area where sewage water is used for irrigation and sludge is used as manure, soil loading capacity for heavy metals is an important parameter for controlling farmland pollution and for predicting the progress and result of the pollution. In some sewage and sludge utilizing areas, heavy metal pollution has already oc-

Table 7.6. Soil Pollution Index Grades

Grade	Index Range	Pollution Degree
0	0	Background area
1	0–0.7	Safe area
2	0.7–1.0	Warning area
3	1.0–1.5	Lightly polluted area
4	1.5–2.0	Moderately polluted area
5	2.0–2.5	Heavily polluted area
6	> 2.5	Severely polluted area

curred and the continued input of the pollutants must be strictly controlled. In the yet uncontaminated areas, it is also necessary to develop plans to control the total amount of pollutants so as to prevent the development of any pollution problems.

Soil loading capacity for heavy metals may be applied in the following aspects.

Evaluation of Regional Soil Quality

The index method in the following equation may be used in the evaluation:

$$Pi = Ci - Bi/C_{ci} - Bi \tag{7.3}$$

where Pi is the soil pollution index of heavy metal i (mg/kg), Ci the measured value of heavy metal i in soil (mg/kg), and Bi the soil background value of heavy metal i(mg/kg). Soil pollution status may be graded according to the pollution indexes shown in Table 7.6.

For example, based on annual Pb input of 3.9 mg/kg and 19.1 mg/kg, respectively, and assuming that the soil critical content of Pb is 213 mg/kg, changes in Pb content in the soils of two Pb-polluted areas are given in Table 7.7. It can be seen from the table that in order to control regional environmental Pb pollution, the input of Pb must strictly be controlled. For some Pb-polluted areas, if all the pollutant sources are cut off, the restoration of the polluted areas may be predicted as shown in Table 7.8. Once the soil is polluted, restoration is very difficult. A total of 2770 years will be required to restore the soil Pb content to the critical value for the soil given in the example.

Establishing the Water Quality Criteria for Regional Sewage Irrigation

The following rules should be followed in establishing the water quality criteria for regional sewage irrigation:

(i) Soil is a quite precious resource, so its lifetime should be lengthy;

(ii) When many pollutants enter the soil simultaneously, the most important and the most influential or restrictive element should be selected as the basis of ascertainment, and their interactions should be taken into consideration as well;

(iii) In case there are two or more pollutant sources, the total amount of restrictive elements emanating from them should be controlled and the criteria may be calculated using the following equation:

$$C_W = (Q\ d\ /\ t\text{-q})/M_W \tag{7.4}$$

where C_W is the content of a certain heavy metal element in the irrigation water (mg/L), $Q\ d\ /\ t$ the annual dynamic capacity of the heavy metal element in the soil (mg/ha), q is the amount of the heavy metal pollutants entering into the soil (mg/ha), and M_W is the amount of irrigation water (L/ha). Calculated permissible concentrations of heavy metal

Table 7.7. Prediction of Pb Contents in Soils (mg/kg) [a]

Area No. (Annual Input)	Present		Prediction (years)							
			15		30		50		100	
	Content (mg/kg)	Grade	Content	Grade	Content	Grade	Content	Grade	Content	Grade
1 (3.9 mg/kg)	27	0	85.1	1	143	1	218	2	404	4
2 (19.1 mg/kg)	27	0	311	3	593	6	964	6	1872	6

[a]Zheng and Chen, 1995.

Table 7.8. Prediction of Restoration of Pb-Polluted Soil (Soil Pb Concentration: mg/kg) [a]

Present Concentration	Year						
	10	20	30	50	100	500	2700
	(mg/kg)						
1460	1450	1441	1431	1412	1366	1045	213

[a]Zheng and Chen, 1995.

Table 7.9. Calculated Admissible Concentrations of Heavy Metals in the Sewage Irrigation Water of Yellow-Brown Soil Areas (calculated by 100 Years) [a]

Element	Calculated Value (mg/L)	State Criteria (mg/kg)	
		1st Type	2nd Type
Cd	0.002	≤0.002	0.003 (farmlands) 0.005 (afforestation lands)
Pb	1.71	≤0.5	1.0
As	0.16	≤0.5	0.1 (paddy fields) 0.5 (uplands)
Cu	0.86	≤1.0	1.0 (soil pH<6.5) 3.0 (soil pH>6.5)

[a]Zheng and Chen, 1995.

elements for regional sewage irrigation water in yellow-brown soil areas are listed in Table 7.9. It shows that the permissible concentration of Cd is approximate to the national criteria. However, the permissible concentrations of Pb, As, and Cu are higher than the national criteria, suggesting that the soils have a considerably larger heavy metal carrying capability, and that the heavy metal pollution wouldn't exceed the soil loading capacity for such elements within 100 years.

Establishing Regional Criteria for Sludge Application

In establishing a criteria for regional sludge application, it is necessary to consider the suitable range and rate of sludge application based on the water quality criteria for regional sewage irrigation. The annual rate of sludge application is generally no more than 30 t/ha (dry weight). When sludge contains heavy metal elements, its continuous application in the same field should not last more than 20 years. The regional sludge application criteria may be calculated by

$$C_s = (Q\, d\, /\, t\text{-}q)/M_s \tag{7.5}$$

where C_s is the concentration of a certain element in the sludge, $Q\, d\, /\, t$ the annual dynamic capacity of the heavy metal element in the soil (mg/ha), q the amount of the heavy metal pollutant entering into the soil (mg/ha), and M_s the applied amount of sludge (kg/ha). Table 7.10 presents the calculated permissible concentrations of heavy metal elements in sludge applied in yellow-brown soil areas. At soil pH > 6.5 and after 20 years' continuous application of sludge at 30 t/ha · year, the calculated Cd content is lower than the national criteria in acid, neutral, and alkaline soils. The amount of calculated Cu, how-

Table 7.10. Calculated Admissible Concentrations of Heavy Metals in Sludge Applied in Yellow-Brown Soil Areas (Application of 20 years at 30 t/ha yr)[a]

Element	Calculated Value (mg/kg)	State Criteria (mg/kg)	
		Acid Soil (pH < 6.5)	Neutral and Alkaline Soil (pH < 6.5)
Cd	1.18	≤5	≤20
Pb	>1500	≤300	≤1000
As	152	≤75	≤75
Cu	289	≤250	≤500

[a]Zheng and Chen, 1995.

ever, is approximate to the national criteria in the acid soil but lower in both neutral and alkaline soils, which shows that Cd in the sludge applied should be strictly controlled.

Besides, soil loading capacity for heavy metals can also be used as the basis for establishing the criteria for the agricultural use of fertilizers, coal ash, and other solid wastes as well as controlling the total amount of regional pollutants.

Problems Existing in the Studies on Soil Heavy Metal Loading Capacity

At present, studies on soil loading capacity for heavy metals are still based on the black box theory, in which attention is only paid to the input and output. The chemical processes occurring among the heavy metals, however, are not taken into account, although they are important factors affecting the soil loading capacity for these elements.

Soil loading capacity of heavy metals depends on or is controlled by the many chemical processes in the "black box" (Chen et al., 1988). It is a pity that the present research models of soil loading capacity of heavy metals still lack some of the parameters concerning these processes. Working under the guidance of the above theories, of course, one may relatively soon obtain the parameters for establishing the soil loading capacity ("capacity values") for the heavy metals but would not reveal the theoretical basis and suitable soil conditions of this kind of parameters due mainly to the presence of the following problems:

1. Lack of long-term experimental results. The "capacity values" obtained at present by means of special reference are the results obtained under special conditions and they would vary greatly with the changes in conditions.
2. Certain limitations in the selection of compound types. When different types of compounds are added into soils, their mobility in and ecological effect on soils might be different due to the differences in their own properties and in their interactions with soils (Chen et al., 1991a, b). An experiment on different types of As compounds (Li et al., 1986) showed that the toxicity of arsenides was apparently higher than that of arsenates, and various arsenates also differed significantly in the degree of toxicity because of their combination with different metal cations.
3. Lack of combined pollution experiments. The capacity values obtained so far all come from the experiments with a single pollutant. However, in nature, pollution by a single heavy metal element is rarely seen. Rather, the damage is mostly the combined pollution by many heavy metal elements. The capacity values obtained from single element experiments may be named "apparent capacities" and it is not suitable as the basis for establishing criteria. The "apparent capacity" should be calibrated into "practical capacity," otherwise this value could not serve as a criterion for benign circulation of the soil's ecological

environment because interactions among elements are rather obvious in both crop yields and element contents (Zheng and Chen, 1989, 1990; Chen and Ponnamperuma, 1982).

4. Calibration of "apparent capacity." When used as a reference criterion for sewage irrigation, land treatment, and agricultural utilization of sludge and solid wastes, etc., "apparent capacity" should be calibrated; otherwise, it could result in poor environmental consequences. The calibrated "apparent capacity" is called "practical capacity." Further studies are required to find a better way to calibrate it so that it tallies with the actual situation. According to the research results of combined pollution studies (Chen and Zheng, 1992), the following calibration method is put forward:

$$\text{Practical capacity} = \text{apparent capacity} \times C_s^{1/n} / \sum C_i^{1/n} \qquad (7.6)$$

where C_s is the "apparent critical value" of some element calculated by the method of single element experiment, n the oxidation number of the elements, and C_i the concentration of heavy metals, such as Cu, Pb, Zn, Cd, Ni, and Co, in the soil. The basis for selecting the ion concentration in the calculation is open to question. In this chapter, the following rules are adopted:

(1) The background values of Cu, Pb, Zn, Cd, Ni, and Co are selected to carry out the calibration when no concrete pollutant is dealt with. For example, the "apparent critical content" of Pb in Nanjing yellow-brown soil was 586 mg/kg, but the "practical critical value" of Pb obtained from calculation was 206 mg/kg. The capacity value calculated by using the practical critical value is the "practical capacity."

(2) When there are concrete pollutants, the element concentrations in the concrete pollutants plus the background value of soil, i.e., the total concentration in the soil, are taken as the basis of calculation. For example, if the contents of elements in a waterlogged compost are 315 mg/kg for Zn, 1190 mg/kg for Pb, and 116 mg/kg for Cd, the annual additions to the soil are 4.2 mg of Zn/kg, 15.9 mg of Pb/kg, and 1.5 mg of Cd/kg according to the annual application of 30 t waterlogged compost/ha and 0–20 cm soil weight of 2,250,000 kg/ha. So, the calibrated practical critical content of Pb is 200 mg/kg after a one-year application of the compost in Nanjing yellow-brown soil and 176 mg/kg after a 10-year application.

To sum up, soil loading capacity for heavy metals is affected by many factors including soil properties, indicators, pollution course, environmental factors, compound types, combined pollution by many heavy metals, etc. The capacity values vary greatly with the changes of conditions. In fact, soil loading capacity for heavy metals is not a definite value. Rather, it is a value of the acceptable limits and its lower limit should be the restricting value of capacity. As a systems engineering study, studies on soil loading capacity for heavy metals are relatively new endeavors. By strengthening research work on the influencing factors, recognizing some problems concerning the "black box," introducing the factors into the models and solving the relationship between "available forms" and "total amount," the study on soil loading capacity would certainly be enhanced.

FORMS OF HEAVY METALS IN SOILS

So far there has been no approach that can strictly distinguish between the different forms of chemical elements in soils. In recent years some methods of sequential extraction have developed for determining the chemical form of heavy metal microelements in soils (Mckeague and Day et al., 1966; Chao, 1972; Gupta and Chen, 1975; Tessier et al., 1979; Shuman, 1985). Although the microelement forms distinguished by this method are only for the definition of operation, one can get a better idea of the chemical behavior and the fate of all the heavy metal microelements, including those originally present in the soil and those just entering the soil. So this approach is of great practical significance, no matter whether from the standpoint of environment protection or from that of the trace nutrients of plants.

With respect to the environment and plant nutrition, much research work has been done concerning the forms of some heavy metal microelements (Cu, Zn, Co, Ni, Cr, V, and Pb) in soils of China (Shao et al., 1993; Shao and Xing, 1994a, 1994b).

Form and Distribution of Heavy Metal Elements in Soils

Form and Distribution of Cu and Zn in Soils

The content of total Cu in China's major types of soil ranges from 7.2 to 45.2 mg/kg, averaging 22.5 mg/kg. On the average, the exchangeable Cu accounts for 0.4% of the total Cu. For calcareous soils, the proportion of carbonate bound Cu with reference to total Cu is not very significant either; i.e., 0.9–4.2% only. However, the organic-bound Cu constitutes as high as 26.8% of the total Cu on the average, showing that the organic matter has a stronger binding force with Cu. The proportions of Mn oxides-bound Cu and of amorphous Fe oxide-bound Cu are not high, averaging 1.2% and 7.8%, respectively. Crystalline Fe oxides bound Cu makes up 20.5% of the total Cu, on the average. And 41.9% of the copper remains in aluminosilicate minerals (Shao and Xing, 1994a).

The content of total Zn in the soils studied is between 30.6 and 98.2 mg/kg, averaging 62.8 mg/kg. The proportion of exchangeable Zn is higher compared with that of Cu. Except for a few of the soils, exchangeable Zn accounts for 5-10% of the total Zn. Like Cu in calcareous soils, the content of carbonate bound Zn is not high, as it accounts for only 2.4 to 6.1% of the total Zn. Organic bound Zn is much less than the organic bound Cu. The content of Mn oxides is 3.4% of the total Mn. Similar to Cu in behavior, amorphous ferric oxide bound Zn, which averages only 6.6% of the total, has a considerably lower proportion than the crystalline Fe oxides bound Zn, which has an average of 26.4%. Most of the Zn, on the average 53.0% of the total Zn, is bound with the residues (Shao and Xing, 1994a). This is in agreement with the results of Singh et al. (1988) and Shuman (1985).

Form and Distribution of Co and Ni in Soils

The content of total Co in the soils tested ranges from 4.4 to 25.1 mg/kg, with an average of 12.1 mg/kg. On average, exchangeable Co accounts for 9.4% of the total Co, which is higher than the corresponding figures for Cu and Zn. In calcareous soils, carbonate-bound Co accounts for 9.4% of the total Co, on the average. The proportion of Mn oxide-bound Co with an average of 11.1% is higher compared with that of Cu and Zn. This is consistent with the conclusion that Mn oxides have a greater affinity for Co (Taylor and Mckenzie, 1966; Mckenize, 1970). Organic-bound Co forms 8.9% of the total Co. The average proportion of amorphous Fe oxide-bound Co is only 3.6%, whereas crystalline Fe oxides-bound Co has an average of 14.8%. Most of the Co is bound in a form of residue, which constitutes 48.2% of the total Co (Shao et al., 1993)

The content of total Ni in the soils studied is 7.2 to 45 mg/kg, with an average of 22.5 mg/kg. Exchangeable Ni, carbonate bound Ni in calcareous soils and organic bound Ni make up 11.1%, 10.6%, and 9.9%, respectively, of the total Ni. Like Co, these three forms of Ni have fairly high proportions with reference to the total Ni in soils. But Mn oxide-bound Ni with an average of 4.1% is lower in content than Mn oxides bound Co. The content of amorphous Fe oxide-bound Ni, only 3.0% on the average, is also low. Crystalline Fe oxide-bound Ni, similar to Co, has an average content of 14.0%. Fifty-five percent of the nickel is in the form of a residue.

Except for Mn oxide-bound Ni, nickel is similar to cobalt in its form and distribution in the soils. This because these two elements are quite similar in their crystal chemical and geochemical properties (Shao et al., 1993).

Form and Distribution of Cr and V in Soils

Cr and V in the soils differ greatly with Cu, Zn, Co, and Ni in form and distribution. The content of total Cr in the soils studied is between 20.0 and 107 mg/kg, averaging 59.8 mg/kg. On average, exchangeable Cr, carbonate-bound Cr, Mn oxide-bound Cr, organic-bound Cr and amorphous Fe oxide-bound Cr constitute 0.29%, 2.5%, 0.31%, 5.2%, and 2.5% m of the total Cr, respectively. The content of crystalline Fe oxide-bound Cr with 9.8% on the average is not high either, with reference to total Cr in soils. In the soils, Cr is bound chiefly with clay minerals in residue, its relative percentage being as high as 81.5% (Shao et al., 1993).

The content of vanadium in the soils studied is 51.7 to 236 mg/kg, averaging 100 mg/kg. In soils, V tends to be similar to Cr in form and distribution. Exchangeable V, carbonate-bound V in calcareous soils, Mn oxide-bound V and organic-bound V forms, on average, 0.34%, 3.8%, 1.7%, and 2.3%, respectively, totaling 8.1%. The content of amorphous Fe oxide-bound V and crystalline Fe oxide-bound V, in soils are present mainly in undisturbed silicate minerals (Shao et al., 1993).

Effect of Different Solid Components on Heavy Metal Enrichment in Soil

By using the solid components naturally existing in soils such as pure clay minerals, oxides, organic substances, carbonate, etc., many researchers have made extensive studies on the retention, desorption, complexing, and fixation of heavy metal ions in soils. However, the soil as a whole is a very complicated system consisting of varied solid components that are present in the form of organic and inorganic complexes. Hence, the various solid components in soils retain or fix heavy metal ions in a much more complicated way than do pure minerals, oxides, or organic materials. Shao and Xing (1994b) investigated, by using the method of sequential extraction, the effect of different solid components on the enrichment of heavy metal elements in different representative soils of China. They worked out the enrichment coefficient and enrichment capacity for different solid components and heavy metal elements in soils, thus providing a theoretical basis for controlling soil pollution by heavy metals.

Enrichment Coefficient and Enrichment Capacity

"Enrichment coefficient," referring to the enrichment of a unit volume of solid component in soil, is not affected by the content of the soil component, but rather depends only on the property and soil conditions of the component itself. Enrichment coefficient can be calculated through the following formula:

$$\begin{array}{r}\text{Enrichment coefficient}\\ \text{of a component}\end{array} = \frac{\begin{array}{c}\text{Amount of a given heavy metal}\\ \text{bound with the component (mg/kg)}\end{array}}{\begin{array}{c}\text{Amount of the component}\\ \text{in soil (mg/kg)}\end{array}} \times 10^4 \qquad (7.7)$$

"Enrichment capacity" denotes the magnitude of enrichment a certain component really acts on under specific soil conditions, influenced also by the content and enrichment capacity of the component. It can be calculated by the following formula:

$$\text{Enrichment coefficient of a component} = \frac{\text{Amount of a given heavy metal bound with the component (mg/kg)}}{\text{Total amount of the component in soil (mg/kg)}} \times 100 \quad (7.8)$$

Relations Between Different Soil Solid Components and Enrichment Coefficient and Enrichment Capacity of Heavy Metals in Different Soils

The relations between different soil solid components and the enrichment coefficient and enrichment capacity of heavy metal elements (Cu, Zn, Co, Ni, Cr, and V) in different soils are shown in Table 7.11.

As can be seen from Table 7.11, Cu has both the highest enrichment coefficient and the highest enrichment capacity for organic matter as a solid component, indicating that organic matter is most important for the enrichment of Cu. Co has the lowest enrichment coefficient, while V has the lowest enrichment capacity for organic matter. For clay minerals, the enrichment coefficient and the enrichment capacity of V and Cr are higher than those of the other elements, indicating that the enrichment of V and Cr in soils is controlled mainly by clay minerals, and the enrichment coefficient and enrichment capacity of Co and Cr are rather low. For amorphous Mn oxides and amorphous Fe oxides in acidic and neutral soils, Zn and V have a high enrichment coefficient. For amorphous Mn oxides in all the soils, Co has the highest enrichment capacity, followed by Ni, suggesting that amorphous Mn oxides contribute greatly to the enrichment of Co and Ni. In neutral and calcareous soils, crystalline Fe oxides are quite important for the enrichment of Zn.

Concentration of Heavy Metal Ions in Soil Solution

In soil solution, the concentrations of heavy metal ions are in equilibrium with the concentrations of other forms of heavy metal elements, especially the exchangeable heavy metal ions. The concentration of heavy metal ions in soil solution may be influenced by many factors such as ionic solubility product, pH, Eh, soil moisture, temperature, etc. The heavy metal ions in soil solution form the active fraction, which is very important both for plant trace nutrition and for environmental protection. With reference to Minnich and McBride (1987) and Elkhatib, et al. (1987), Xu and Xing (1995) studied the concentrations of such ions as Mn, Zn, Cu, Co, Ni, Cr, and Pb in solutions of different Chinese soils that were under conditions of saturated soil moisture retention.

Concentrations of Mn, Zn, Cu, Co, Ni, Cr, and Pb in Unpolluted Soil Solutions

The concentrations of Mn, Zn, Cu, Co, Ni, Cr, and Pb in solutions of different types of unpolluted soils of China are given in Table 7.12. Of the seven elements determined, only Mn is of the mg/L grade and the remainder are of the μg/L grade, and Ni, Cr, and Pb are the lowest in concentration. As far as the relative concentration is concerned, there may be the following two cases: (1) the relative concentration of Zn, Cu, Co, Ni, Pb, and Mn is in the range of a part per thousand; (2) the relative concentration of Cr is in the range of a part per ten thousand.

Concentration of Heavy Metal Ions in Polluted Soil Solutions

In China, little has been done on the concentration of heavy metal ions in polluted soil solutions. Moreover, the concentration of heavy metal ions may vary greatly with the difference in the degree of pollution and in the kind of heavy metals. The following study il-

Table 7.11. Enrichment Order of Heavy Metals for Soil Solid Components[a]

Soils		Amorph.Mn	OM	Amorph.Fe	Crystalline Fe	Clay Mineral	$CaCO_3$
Acidic	Enrichment coefficient	Zn>V>Ni> Co>Ca>Cr	Cu>Cr>Zn= V>Ni=Co	Zn>V>Cr>Cu >Co>Ni	V>Zn>Cr>Cu >Ni>Co	V>Cr>Zn Cu=Ni>Co	–
	Enrichment capacity	Co>Ni>Zn >Cu>V>Cr	Cu>Co>Ni> Cr>Zn>V	V>Zn>Cu> Co=Cr>Ni	V=Zn>Cu >Cr>Co>Ni	Cr>V>Ni =Zn>Co>Cu	
Neutral	Enrichment coefficient	Zn>V=Co >Ni>Cu>Cr	Cu>Ni>Zn >Cr>V>Co	V>Zn>Cu >Ni>Cr>Co	Zn>V>Cr >Cu>Ni>Co	V>Cr>Zn >Ni=Cu>Co	–
	Enrichment capacity	Co>Ni>Zn >Cu>V>Cr	Cu>Ni>Co >Cr>Zn>V	Cu>V>Ni >Zn>Co>V	Zn>Cu>Co >Ni>V=Cr	Cr>V>Zn >Co>Ni>Cu	
Calcareous	Enrichment coefficient	V>Co>Zn >Ni>Cu>Cr	Cu>V>Cr >Ni>Zn>Co	V>Cu>Ni >Zn>Cr>Co	Zn>V>Cu >Cr>Ni>Co	V>Cr>Zn >Ni=Cu>Co	V>Ni>Zn> Cu>Co>Cu
	Enrichment capacity	Co>Ni=V >Zn>Cr>Cu	Cu>Ni>Co >Cr>Zn>V	Cu>V>Ni >Cr>Co>Zn	>Zn>Co=Ni >V>Ca>Cr	Cr>V>Zn >Ni>Co>Cr	Ni>Co>Zn =V>Cr>Cu

[a]Shao and Xing, 1994b.

Table 7.12. Concentrations of Heavy Metal Ions in Solutions of Different Unpolluted Soils of China[a]

Element	Range (mg/L)	Arithmetic Mean (mg/L)	% of Total Amount of Elements in Soil (n = 32) (mg/L)
Zn	0–1.16	0.29	5.12 ± 6.18
Cu	0.006–0.69	0.14	6.55 ± 8.95
Co	0–0.92	0.14	7.13 ± 9.11
Ni	0.009–0.18	0.03	1.34 ± 1.36
Mn	0–3.2	1.04	2.29 ± 3.03
Cr	0–0.07	0.01	0.26 ± 0.29
Pb	0–0.56	0.06	4.48 ± 5.71

[a]Xu and Xing, 1995.

Table 7.13. Concentrations of Heavy Metal Ions in the Soil Solution with Pollutants from a Copper-Smelting Plant (mg/L) [a]

Element	Polluted Soil	Unpolluted Soil (averages)
Zn	108.0	0.29
Cu	1175.0	0.14
Co	119.9	0.14
Ni	17.3	0.03
Mn	76.0	1.04
Cr	0.24	0.01
Pb	0.71	0.06

[a]Xu and Xing, 1995.

lustrates the difference in heavy concentrations between a severely polluted soil and a normal soil. The sample of the polluted soil was taken from a paddy field near a copper-smelting plant in a county of Jiangxi Province. Due to serious pollution, the land now lies in waste.

The concentrations of Zn, Cu, Co, Ni, Mn, Cr, and Pb in the polluted soil solution is 372, 8392, 850, 576, 73, 24, and 12 times higher, respectively, than that with the unpolluted soil solution (Table 7.13). Heavy metal elements that are carried in sewage or sludge to the farmland are mostly soluble. In normal soil solution, the concentration of heavy metal ions commonly are at a level of μg/kg. In soil solution, the concentration of heavy metal ions relative to their total content is commonly a part per thousand, a part per ten thousand, or a part per hundred for a few elements. The concentration of a heavy metal ion in soil solution and its concentration relative to its content in soil can serve as an important index for determining the pollution level of the soil.

Form and Distribution of Rare Earth Elements in Soils

By using the improved method of sequential extraction (Zhu and Xing, 1992b), studies were done on the form and distribution of rare earth elements in the major soils of China (Zhu and Xing, 1992a).

Concentration Distribution of Individual REE

The percentages (extraction rate) of the concentrations of various forms of REE (rare earth elements) in the total REE varied greatly (Table 7.14). Generally speaking, the ele-

Table 7.14. Average Percentages of REE in Forms[a]

	Form					
Element	**Water Soluble**	**Exchangeable**	**Carbonate and Specifically Adsorbed**	**Organically Bound**	**Fe-Mn-Oxides Bound**	**Residual**
La	0.10	1.25	8.90	6.75	10.17	72.83
Ce	0.12	1.05	8.17	3.75	18.07	68.84
Pr	0.45	1.19	8.70	5.76	13.04	70.86
Nd	0.17	0.44	9.11	5.33	14.09	70.86
Sm	0.52	0.64	10.26	5.18	15.92	67.48
Eu	0.17	0.45	10.74	4.88	16.00	67.76
d	0.35	0.67	11.07	4.96	15.08	67.87
Tb	1.21	1.72	10.65	4.73	14.82	66.87
Dy	0.13	0.34	10.14	3.65	13.86	71.88
Ho	1.09	2.64	10.12	3.90	8.50	73.75
Er	0.29	0.47	8.79	2.61	10.84	76.99
Tm	2.98	3.72	8.91	6.51	10.86	67.02
Yb	0.30	0.71	7.27	2.28	9.93	79.51
Lu	0.38	0.67	6.90	1.90	8.54	81.61
Sc	0.01	0.02	0.22	0.16	3.92	95.67
Y	0.06	0.44	11.06	3.16	10.84	74.45

[a]Zhu and Xing, 1992a.

ments with an odd atomic number had higher extraction rates of water-soluble and exchangeable forms than those with an even atomic number. The same result was also found in the extraction of organically bound REE in some soils, such as the alkaline soils which were lower in OM content but higher in $CaCO_3$ content. As for other forms, the extraction rates of various elements had a relatively small difference. It may be seen from Table 7.14 that the residual Sc form accounted for 95.7% of the total Sc in soils. In fact, Sc is much different from other REE in properties, and it is usually not included in REE in geochemistry. Fractionation results showed Sc was less active than other REE in soils. Besides Sc, the proportions of Yb and Lu in residual form were also higher than those of other REE. Tb (terbium) and Tm (thulium) were found to have the lowest proportion in the residues. Tm, however, had the highest proportion in both water-soluble form and exchangeable form, and its proportion in organically bound form is only next to that of La, suggesting that Tm is a relatively active element, though its content is low in soils.

Distribution of Total REE in Various Forms

The content of water-soluble REE was very low, ranging from 0.02 to 1.08 mg/kg, with an average extraction rate of 0.17%. Correlation analysis shows that the content and extraction rate of water-soluble REE were negatively correlated with pH of the soil at a significant level. Meanwhile, the content of water-soluble REE was found to have a positive correlation with the contents of Fe and Mn in the soil. There also existed a certain relationship between the content of water-soluble REE and the soil parent materials, with the water-soluble REE content tending to be lower in the base rocks than in the acid rocks.

The exchangeable REE content varied within 0.06 to 26.04 mg/kg, with an average value of 2.04 mg/kg and an average extraction rate of 0.82%. Correlation analysis results show that the content of REE in this form was positively correlated with the contents of clay, exchangeable cations, and Fe and Mn in the soil. For soils derived from the same parent material in the same region, different utilization patterns could cause an obvious difference in the content of exchangeable REE. For example, the average exchangeable REE content of six upland soil samples was 3.67 mg/kg, while the average value of six corresponding paddy soil samples was 1.28 mg/kg.

The carbonate-bound and specifically adsorbed REE content was 0.29 to 75.19 mg/kg, with an average value of 16.55 mg/kg, and an average extraction rate of 8.4%. The analysis results of 11 $CaCO_3$-containing soil samples indicated a positive correlation between the REE content in this form and the $CaCO_3$ content. It was found from the analytical results of all 34 soil samples that the content of REE in this form also had a positive correlation with the soil CEC (cation exchange capacity) and an extremely significant correlation with the iron content.

The organically bound REE content ranged from 0.24 to 25.12 mg/kg, with an average value of 7.52 mg/kg, and its extraction rate varied from 0.18 to 15.8%, with an average of 4.32%. These values of extraction rate were coincident with the soil OM content, which ranged from 0.41 to 14.34%, with 3.8% as the average for the 34 soil samples.

The amount of REE bound to reducible Fe and Mn oxides, which was 2.69 to 86.30 mg/kg (25.65 mg/kg on average) and with an extraction rate of 2.59 to 27.45% (13.66% on average), was found to be the highest among other forms of REE extracted. Hu and Xing (1992) also had an extremely significant positive correlation with the soil Fe content, showing a tendency of higher extraction rate in soils developed on basic rocks than in soils on acid rocks. This is attributable to the fact that the Fe content of basic rocks was higher than that of acid rocks. It may be seen from Table 7.14 that the REE content in every form except the organically bound form was correlated with the soil Fe content. Therefore, the Fe content in the soil is the most important factor affecting the distribution of REE in various forms.

The content of REE remaining in the solid residues after extraction of the five forms discussed above accounted for 45.4% to 92.9% (70% on average) of the total REE in the soil. This indicates that the REE in soils were relatively fixed in the primary minerals and that little translocation occurred during the whole soil-forming processes. REE have been classified as weathering resistant elements in a few research papers. Anthropogenic activities could affect the REE content in soils.

Through studying the fractionation of REE forms in soils, many useful results were obtained. The proportion of REE in the active state was small in soils, but the total amount of soil REE was high. Judged by the concentration range of microelements affecting biological activities, the content of REE in the active state is rather considerable in some soils. As far as the water-soluble and exchangeable REE are concerned, their content has been as high as dozens of mg/kg in some soils, such as the latosol collected from Guangdong Province. It has also been reported that the effective concentration range of REE ions could be 0.1 to 0.5 mg/kg in the experiment of water culture (Wu, 1988), indicating that the soils probably have not been deficient in REE.

HEAVY METAL POLLUTION OF SOILS IN CHINA

Generally speaking, soil pollution is primarily caused by sewage irrigation, sludge application, mining, and smelting.

Heavy Metal Soil Pollution Caused by Irrigation with Sewage Water

In China, irrigation of farmlands and vegetable lands with sewage water has been practiced for thousands of years, even though large-scale sewage irrigation did not develop until the 1950s. Statistics (Wu, 1992) show that the sewage irrigation area was 42,000 ha in 1962; 93,000 ha in 1972; 180,000 ha in 1976; 333,000 ha in 1979; and more than 670,000 ha in 1988. To date, it has reached over 1,400,000 ha. The upland crop regions of Northern China have more than 90% of the total sewage irrigation area of the country, including Beijing, Tianjin, Shenyang, Jinan, Xian, Shijiazhuang, Zhengzhou, Qiqihar, Luoyang, Baoding, and Harbin. In the South, Wuhan, Chengdu, and Changsha account for 6% of the general irrigated areas. The remaining 4% are scattered in regions, such as the Northwest and the Qinghai-Xizang Plateau.

As a result of relatively weak economic conditions and poor industrial techniques in China, a great quantity of untreated industrial and urban sewage water was drained into farmlands, causing about 45% of the total sewage irrigation area to become polluted (Wu, 1992). A more serious issue is the pollution by heavy metals, especially by Hg and Cd. In China, the Cd-contaminated cultivated lands amount to about 13,000 ha, involving 25 regions of 11 provinces and municipalities. There are about 32,000 ha of cultivated lands that are subject to Hg pollution. Rice produced at 21 sites in 15 provinces and municipalities was detected to be above the acceptable limit of Hg content (> 0.02 mg/kg). Eleven areas were found to produce grains containing > 1 mg Pb/kg; and grains produced in six regions contained > 0.7 mg As/kg.

Heavy metals enter into human bodies via the food chain. In the Daye region of Hubei, each person takes in 2.2 to 5.95 mg Cd/week, which is higher than the permissible criterion of 0.4 to 0.5 mg/week stipulated by the World Health Organization. In south Tianjin, where sewage is drained into the river from which water is used for irrigation, about 50% of the grain-producing soils are subject to contamination by heavy metals; and 100% of the vegetable lands in south Tianjin that were irrigated with a large quantity of sewage water and applied with sludge are contaminated by heavy metals. The Shanghai Mayibang Sewage-Irrigated Area is one of the more seriously heavy-metal-contaminated areas in China (Zhou et al., 1993), where Cd, Zn, and Cu content is 3.0 to 5.0, 194.5 to 240.7, and 57.6 to 101.2 mg/kg, respectively, in the heavily polluted areas; 1.38 to 2.87, 140.0 to 198.9, and 34.1 to 67.1 mg/kg, respectively, in the moderately polluted areas; and 0.67 to 1.16, 125.3 to 171.4, and 24.7 to 45.9 mg/kg, respectively, in the lightly polluted areas. Wang et al. (1993) measured the amount of Cr, Cd, Pb, As, and Hg contained in wheat, rice, maize, Indian millet, and spiked millet grains in the Taiyuan Sewage–Irrigated Area and found that, except for Hg, all other elements exceeded the acceptable limits. The amount of Cd was especially high. It was found that 83.7% of wheat, 99.3% of rice, and 100% of maize, Indian millet, and spiked millet, exceeded the acceptable Cd limit, and 70–100% of all the five kinds of food grains also contained Pb amounts that exceeded the tolerable limit.

Heavy Metal Pollution of Soils in Industrial and Mining Areas

Heavy metal pollution of soils in industrial and mining areas is primarily caused by mining and smelting, especially in the south of China. Dai et al. (1993) investigated the effect of heavy metal pollution around eight main mining areas in Jiangxi Province. The Cu, Zn, Mo, and Cd concentrations from 139 farmland soil (0–30 cm) and fodder samples collected are shown in Table 7.10. The background values (mg/kg) are 20.3 ± 10.3 for Cu, 69.4 ± 28.5 for Zn, 0.5 ± 0.39 for Mo, and 0.108 ± 0.136 for Ca (CSBEP, 1990). It can be seen from Table 7.15 (Dai et al., 1993) that Mo content in the nearby regions of various tungsten mines far exceed the soil background values, suggesting severe Mo pollution. Similarly, Mo content in fodders produced in the same regions also greatly exceeds the safety limit. It is generally considered that when Mo content in forage grass is lower than 1 mg/kg, there is no Mo poisoning (Liu, 1991). However, if the Mo content is

Table 7.15. Microelement Contents of Soils and Fodders in Primary Metal Smelting Plants and Mines of Jiangxi [a]

Element	Xihuashan Tungsten Mine	Xialongtang Tungsten Mine	Dangpine Tungsten Mine	Dajishan Tungsten Mine	Pangushan Tungsten Mine	Yangping Copper Mine	Dexing Copper Mine	Guixi Smelting Plant
	Dry weight of soil (mg/kg)							
Cu	20.50 (±10.83)	32.20 (±14.01)	21.77 (±7.81)	9.30 (±1.56)	22.01 (±9.73)	383.2 (±378.1)	2081 (±1472)	700.2 (±208.8)
Zn	82.75 (±38.77)	81.60 (±18.02)	79.17 (±34.95)	205.5 (±83.80)	218 (±76.26)	262.8 (±102.2)	1286 (±760)	116.9 (±23.35)
Mo	10.98 (±4.88)	27.56 (±0.48)	32.60 (±10.59)	103.0 (±18.38)	10.56 (±0.97)	0.58 (±0.41)	12.98 (±2.52)	1.48 (±0.25)
Cd	1.77 (±0.55)	1.70 (±0.44)	1.62 (±0.93)	3.43 (±0.21)	1.85 (±0.66)	1.11 (±0.56)	1.47 (±0.77)	21.94 (±11.08)
	Dry weight of fodder (mg/kg)							
Cu	30.29 (±22.15)	57.50 (±16.73)	33.00 (±15.26)	27.97 (±9.98)	34.91 (±18.74)	194.2 (±106.5	862.9 (±413.4)	444 (±112.6)
Zn	156.5 (±88.90)	41.22 (±13.99)	32.17 (±12.45)	1.53 (±0.97)	188.6 (±57.13)	180.6 (±21.40)	368.6 (±0.70)	113.5 (±171.8)
Mo	61.08 (±24.30)	68.50 (±18.43)	91.94 (±41.03)	46.58 (±43.35)	10.37 (±2.85)	0.77 (±0.42)	1.58 (±0.70)	1.64 (±0.43)
Cd	2.28 (±0.60)	1.12 (±0.80)	0.96 (±0.85)	5.05 (±2.27)	0.87 (±0.63)	0.73 (±0.05)	1.23 (±0.08)	4.09 (±5.40)

[a]Modified from Dai et al., 1993.

higher than the background value and the Cu:Mo ratio is smaller than 2:1, Mo poisoning likely occurs. The very high Mo content and severe imbalance of Cu/Mo ratio in the fodders shown in Table 7.15 are consistent with the Mo poisoning of ruminants widely occurring in the areas. As Mo and Cd synergize in animal bodies, cattle that are affected in these areas develop white skin and wool, suffer from diarrhea, and lose weight (Dai et al., 1993). Table 7.15 also shows that the soils and fodders near the Cu mines and smelting plants have relatively high Cu, Zn, Mo, and Cd contents, with the maximal soil Cu content of 3755 mg/kg and maximal fodder Cu content of 1572 mg/kg in some typical areas. The Cd content is as high as 29.89 mg/kg in soil and 16.1 mg/kg in forage grass, but up to now there appears no Cu or Cd poisoning of cattle or pigs due to relative high Zn content in the area. However, the nutrition status and reproductive capacity of domestic animals are relatively poor and the incidence of cancer, especially human liver cancer, is very high in these areas.

The 1993 data statistically indicate that village and town enterprises made up 40% of the gross value of the industrial output in China. The relatively backward techniques and low efficiency in the use of raw materials of these enterprises result in severe environmental pollution, thus making a major impact on the originally very weak rural ecology and environment of China. Guizhou Province, for example (Peng, 1988), has about 37,900 t of Hg in the province, but the recovery of Hg is only 30-40% because local methods and technology adopted in Hg smelting are backward and the equipment used is simple and crude. Consequently, more than 50% of Hg in the smelting process is lost into the environment. In 1986, Guizhou Province reported that the total Hg vapor emitted was as high as 146.8 t from the village and town enterprises; the Hg content of some soils affected by Hg smelting through local methods was up to 55.64 mg/kg, which was 184.5 times the value of uncontaminated soils. The Hg content reached 0.48 mg/kg in the rice grains and 0.23 mg/kg in the pepper produced, which were both far higher than the state food health limits. The recovery rates of Pb and Zn ores by village and town enterprises were only about 38%, and about 60% of the ores were abandoned into the environment. Over 520 Zn-smelting ovens were installed at Heizhangmagu Town and about 2.23 t of Pb and 61 kg of Cd were discharged with wastewater into the Magu River, resulting in severe pollution of the water and farmlands, which in turn strongly affected the health of the residents and animals living in the area. According to a survey done in 1986 (Peng, 1988), among 86 tested Hg-smelting workers, 31 persons had gingival bleeding, 43 had odontoseisis or odontoptosis, 41 had hand tremors, 40 had ophthalmospasm, and the urine of 79 persons was found to contain > 0.02 mg Hg/L.

Ecological Cycling of Heavy Metals in Farmland Systems

Heavy metals in the environment cannot be decomposed by organisms, but they may accumulate and be transformed in the bodies of organisms, thus bringing about potential harm to human beings. Heavy metals have obvious effects on rice microbes, enzymatic activities, and the animal population (Chen, 1992; Chen and Wang, 1992). The Cd pollution in the Zhangshi Sewage-Irrigated Area of Shenyang is taken as an example to show the ecological cycling of Cd in farmland systems.

The Zhangshi Sewage-Irrigated Area lies in western Shenyang, where sewage water has been used to irrigate 2800 ha of rice fields since 1962. Cd in the sewage comes mainly from the Shenyang Smelting Plant (Wu et al., 1986) which drains off 2.6 to 2.7×10^4 t sewage water per day. Before measures were taken to control the sources of pollution, the sewage water contained 1.16 to 6.0 mg Cd/L. Thus, the amount of Cd discharged from the plant was as high as 10 t/year, of which 25.26% occurred in the summer, the peak period of irrigation of paddy soils, thus resulting in considerable Cd accumulation in the soils (Table 7.16) and a relatively high Cd content in the rice grains (Table 7.17). Research has indicated that because of the higher Cd content in the fodders such as rice brans and husks in the sewage-irrigated areas, the internal organs of pigs also contained significantly higher amounts of Cd than these in the control area (Table 7.18). An investigation on the diet (including staple foods, nonstaple foods, and drinking water) of residents in the irrigated areas showed that Cd taken in by each resident was 558 μg/day, which was 32 times more than that in the control area. The Cd taken up by the residents came primarily from staple foods in the sewage irrigated areas, where the rice produced in 1982 contained 1.04 mg Cd/kg of rice. But in the control area, Cd accumulation came from nonstaple food. Long-term eating of Cd-polluted farm and other products produces significant results. Test results in 1980 indicated that the amount of Cd in the blood and urine of residents in the sewage-irrigated area were 1.06 and 13.26 μg/L, respectively, while those in the control area were 0.42 and 2.13 μg/L, respectively. Cd pollution may

Table 7.16. Cd Contents of Soils in Zhangshi Irrigated Area (extracted with 0.1 mol/L HCl) [a]

Site	Year				
	1978	1980	1983	1984	1985
			(mg/kg)		
Yizha	7.19	4.52	7.75	1.60	2.60
Erzha	5.06	3.90	7.06	4.81	5.63
Sanzha	5.5	2.26	4.95	6.09	4.60
Desheng	0.96	0.49	2.11	3.33	1.33
Shaling	0.85	2.20	1.23	1.25	1.75

[a]Modified from Wu et al., 1986.

Table 7.17. Cd Contents of Brown Rice in Zhangshi Irrigated Area[a]

Site	Year				
	1978	1980	1983	1984	1985
Yizha	0.53	0.88	2.06	1.60	1.07
Erzha	0.20	0.42	0.26	0.52	0.48
Sanzha	0.24	0.36	0.17	0.22	0.41
Desheng	0.11	0.04	0.24	0.19	0.42
Shaling	0.06	0.06	0.07	0.12	0.04

[a]Modified from Wu et al., 1986.

Table 7.18. Comparison of Pig Viscera Cd Between Sewage-Irrigated Area and Check Area[a]

Organ	No. of Samples		Average Value (mg/kg)	
	Control Area	Polluted Area	Control Area	Polluted Area
Meat	36	35	0.04 ± 0.02	0.32 ± 0.17
Liver	32	35	0.17 ± 0.07	1.62 ± 0.92
Kidney medulla	35	35	0.85 ± 0.25	6.18 ± 4.73
Cortex renis	35	35	2.34 ± 0.82	17.9 ± 9.85

[a]Modified from Wu et al., 1986.

affect the second generation via the mother's body. Analytical results show that the Cd content in the placenta of those in the sewage-irrigated area is 2.4 times higher than that in the control area. Similarly, the amount of Cd in the liver and kidney in the former area is 2.2 times higher than in the latter area.

Countermeasures to Prevent and Control Heavy Metal Pollution of Soils

The most effective methods for preventing and controlling soil pollution should be to cut off the pollutant sources or to reduce the pollutants to the acceptable criteria before they are discharged from their sources, and at the same time, to control the total amount of pollutants, so as to avoid new pollution. Since the 1980s, China has made many efforts in controlling the sources of pollutants. Such endeavors focus on prevention and control of heavy metal pollution, as well as attempts to improve soils that have been affected. The following are some of the measures that have been adopted.

Conversion from Paddy to Upland Cropping and Vice Versa

Some pollutants that may be harmful to the rice plants or cause the rice product to exceed the food health criteria might not affect the upland crops. The Department of Environmental Protection of the Institute of Soil Science, Academia Sinica (1977), found that on an As-polluted soil, planting wheat would not result in As pollution of the wheat grains, whereas growing rice on the same land would cause a very high As content in the brown rice. Generally speaking, As exists mostly in the forms of arsenates in upland soils or dry soils, but arsenides increase with the dropping of the redox potential under the flooded condition. Arsenides have greater toxicity and As injury is significant at Eh < 50 mV. Thus, the formation of arsenates is generally considered to possibly begin at Eh of about 100 mV (Zou, 1986). Therefore, As-polluted soils are suitable for growing upland crops. In areas polluted by pesticides containing calcium arsenate, growing upland crops in accordance with the local conditions has been very effective.

A study carried out in Hangu, Tianjin, showed that there was obvious Hg pollution for rice grown on a Hg-polluted soil. However, Hg was not detected in most seeds of wheat, maize, and sorghum grown on the same soil. Furthermore, the amount of Hg detected in small parts of the seeds was also below the tolerable food health limits and also not significantly different from that in the check soil (Yang and Rong, 1982).

Contrary to As and Hg, Cr-polluted soils are good for growing rice because Cr under an oxidation condition exists in a hexavalent form (Chen, 1992), which has a higher toxicity than Cr (III). Hence, Cr-polluted soils may be converted from upland fields to paddy fields in order to decrease the harmful effects of Cr. Because rice has a higher Cr uptake from irrigation water than from the soil, Cr content in the Cr-bearing sewage water should be controlled when it is used for irrigation (Chen et al., 1978).

Addition of Modifiers

Modifiers are added into soils with the primary purpose of fixing heavy metals or changing their forms so as to eliminate their harmful effects on organisms. For example, salts of iron may play a certain role in reducing As harm to rice, because soil adsorption capacity for As^{5+} increases with the decrease of As^{3+} concentration as a consequence of oxidation-reduction of Fe^{3+} in flooded soils. Application of iron sulfate to the soil in combination with other agricultural measures, such as water control, may increase the rice yield of As-polluted fields by 63% compared with that of the control field.

The application of some modifiers can obviously decrease exchangeable Cd in the soils. Chen et al. (1981) and Wu et al. (1986) found that application of appropriate amounts of lime and fused calcium-magnesium phosphate could reduce 77.5% of the metal in the soil. Combined application of lime and humic acid could decrease 60% of the exchangeable Cd, and a single application of lime could decrease 30% of the same metal in the soil, thus decreasing the Cd uptake by rice. As a result of these three different applications, the Cd content in brown rice decreased 42.4%, 33.3%, and 18.2%, respectively. A large-area liming experiment conducted in the severely and moderately polluted areas of the Zhangshi Irrigated Area showed that liming at a rate of 1500–1875 kg/ha decreased Cd in the seeds by 50%. Application of blast furnace slag into red soils could inhibit Cd uptake by rice, wheat, and vegetables (Zang et al., 1987; 1989), with a Cd decrease of more than 90% in the brown rice, about 50% in the wheat berries, and about 85% in the cabbage. In addition to Cd reduction, the yield of these crops also increased. The effect of blast furnace slag may be attributed to its component silicon (SiO_2, 32%), because active Si may raise soil capability to fix Cd (Chen, 1988). Some other modifiers, like bentonites, humic acids, and limestone, could also improve heavy metal polluted soils to a certain degree (Huang, 1982). For calcareous soils, chrome-bearing calcium-magnesium phosphate applied as a modifier was most effective in inhibiting Cd uptake by rice (Huo et al., 1989).

Application of Organic Manures

Organic manures affect the fixation of many metals in soils. Application of milk vetch into red soils results in a remarkable decrease of Cd in the soil solution. However, the effect of organic manures on Cd varies with the soils. Thus, application of organic manures should be adopted to suit different soils (Chen et al., 1988). Organic manures play an important role in decreasing Cr (VI) damage. Cr (IV) disappeared in a soil 50 days after application of organic manure (Wang et al., 1986).

Soil humus also affects the accumulation of Hg in rice (Huang, 1982). Soil Hg accumulation increased with increase of the humus content. However, Hg content in brown rice decreased with the increase of humus, which might be attributed to the decrease of Hg migration due to chelation of Hg with humus, suggesting that appropriate application of organic manures could prevent Hg pollution of crops. For example, in soils with a Hg content of 10 mg/kg, application of organic manure significantly decreases the Hg content in brown rice.

Agroecological Engineering Measures

Agroecological engineering measures are good methods for proper utilization and improvement of severely heavy-metal-polluted farmlands. As an example, such measures brought economic, social, and environmental benefits to the Cd-polluted farmlands in Zhangshi Irrigated Area (Chen et al., 1986). The main measures include the following:

(1) Agricultural improvement; for example, breeding good quality strains of rice and maize and growing sorghum. In the severely polluted areas, improved different varieties of regular rice, hybrid rice, and hybrid maize are grown. Rice is irrigated with both sewage and nonsewage water, and the seeds obtained are sown next year in the nonpolluted areas. Upon harvesting, sorghum stalks are used for extracting alcohol, and the residues for producing fiberboard, extracting furfural, or producing biogas as a kind of energy source. The sorghum seeds may be used as seed source or for extraction of alcohol.

(2) Soil purification by growing trees, shrubs, grass, and ornamental seedlings as well as sod. Poplar tree seedlings of 1–2 years old are transplanted from nonpolluted nursery gardens to severely Cd-polluted areas and irrigated once or twice every year. Transplantation and reproduction of the grown sod may not only beautify the environment but also purify soils. Castor oil plants can be grown as the raw materials in soap-making and other chemical industries. Such practices in the Zhangshi Irrigated Area proved that these measures can reduce Cd pollution of the food chain and at the same time benefit the economy.

Industrial and Agricultural Application of Rare Earth Elements in China and Consequent Potential Pollution Problems

Rare earth elements are not only used widely in metallurgy, glass and ceramics making, petrochemical industry, and in production of luminous and magnetic materials, they also contribute to plant growth as well as yield and quality of the farm products (Guo, et al., 1988).

China is the richest in rare earth elements in the world. Its reserves of rare earth elements constitute 70–80% of the world's total. In China, they have been extensively applied to agriculture and industry. Chinese scientists have done much research on the background values and forms of rare earth elements in soils of China and also on the effect of their application on agriculture. However, they failed to pay due attention to the environmental problems that may arise as a result of extensive use of rare earth elements in industrial and agricultural production of China.

Like other countries, China has applied its rare earth elements to industrial production such as iron smelting, metallurgy, petroleum splitting, glass and ceramics making, and production of luminous and magnetic materials. In the early 1980s, China began to use rare earth elements as fertilizers on a relatively large scale and, through a technological process, produced "Nongle" as a commodity specifically for agricultural use. According to the statistical data for 1981–1987, in China the acreage of farmlands receiving rare earth element fertilizers was expanding year by year and came up to about 930,000 ha in 1987. By the same year, the amount of rare earth elements applied to the farmlands in the whole country had reached 250 tons (calculated in terms of R_2O_3). By 1990, the area of farmlands supplied with rare earth element-bearing fertilizers had reached 2 million hectares, and the rare earth elements added to farmlands had amounted to 500 tons (Guo et al., 1988).

In considering the use of rare earth elements and the kind and distribution of their deposits in China, much attention must be paid to preventing possible pollution from the following three aspects:

1. As rare earth elements are used in metallurgical and chemical industries, glass and ceramics making, and in production of luminous and magnetic materials, they may enter into the environment in the form of wastewater, waste residues, or dust in such a manner as do heavy metal elements.
2. In agriculture, rare earth elements go directly into the environment and even enter into the food chain through the soil.
3. Seventy to eighty percent of the proven rare earth element deposits in China are in the Baiyun'ebo mine in Baotou in the Inner Mongolia Autonomous region, and the remainder are distributed in some provinces or areas in southern China. The rare earth elements are deposited into the soil mostly through ion absorption. When they are mined, they may enter the water environment and cause pollution.

In short, environmental problems should not be overlooked when rare earths elements are employed in industrial and agricultural production and when their deposits are mined.

EFFECT OF THE AMOUNT OF DISSOLVED HEAVY METALS AND ENRICHMENT OF OTHER INORGANICS IN GROUNDWATER ON ECOLOGICAL ENVIRONMENT

In general, heavy metal elements move very little in soil. Even under conditions of submerged reduction and strong leaching, there is no evidence that heavy metal elements in soil, whether of an endogenic or exogenic source, move downward (Yang, 1994). In China, only in the northern parts where some light-textured soils are irrigated with wastewater do a few heavy metal elements such as Hg and Cd move downward and enter the groundwater, so that content of these two elements in the groundwater rises slightly. Such a situation has not been found in sewage-irrigated areas in southern China for the reason that the soils are commonly heavier in texture (AMC, MAAHF, 1985).

Effect of the Amount of Dissolved Heavy Metal Elements from Soil on Water Environment

Metal elements dissolving in soil solutions is an important way of element migration into the water environment. To assess the impact of the Three Gorges Reservoir on the Yangtze River after it is completed, Chinese researchers studied the amount of the heavy metals of soils that would be dissolved and their effect on the water environment of the reservoir (Huang et al., 1994). Results show that the concentrations of all the heavy metal elements studied will be higher compared with those of the corresponding current values of the Yangtze River water (Table 7.19). Under simulated conditions, the concentrations of soluble Mn, Co, Cr, Ni, Cu, Pb, Zn and Cd are 9.6, 24, 5.5, 10.4, 26, 59.7, 38, and 281 times, respectively, the corresponding values of the trunk stream water of the Yangtze River. The concentration of total soluble P is 0.4 to 11 times as high as that in the river water along the Three Gorges. Therefore, regardless of the influence of river water dilution and suspended solids, in a bend of the reservoir where there is a gentler flow of water, the concentrations of soluble elements in the interstitial water, especially those of Cu, Cd, Zn, and Pb, will increase so significantly that they will exceed the water quality criteria for fishery and thus detrimentally influence the water environment of the areas. The increase in phosphorus and other elements in the water will stimulate the growth of algae. At the same time, the concentration of mercury in the interstitial water stored at the bottleneck of the Three Gorges Reservoir bend will increase, thus enriching the existence of this element in the aquatic organisms.

Effect of NO_3 in Groundwater on Quality of Drinking Water

NO_3 in groundwater is derived mainly from chemical nitrogen fertilizer, organic manure, and domestic wastes. During the transformation of nitrogen or nitrogen-containing organic substances, NO_3 is formed. China is one of the countries where enormous quantities of N fertilizer are used each year. According to the 1991 statistical data, China's consumption of chemical N fertilizer covers one-fifth of the world's total, and it tends to

Table 7.19. Comparison Between the Soluble Concentrations of Soil Elements and Their Corresponding Current Values of the Yangtze River Water[a]

	Element								
Source	**Mn**	**Co**	**Cr**	**Ni**	**Cu**	**Pb**	**Zn**	**Cd**	**P**
Soils	69.0	3.77	3.78	6.15	29.1	31.6	116	6.22	93.3
River water along the Three Gorges	9.64		1.23	1.80	2.32	0.54	5.50	0.06	7.64
Yangtze River water	6.53	0.15	0.58	0.54	1.06	0.52	2.97	0.022	

[a]Modified from Huang et al., 1994.

further increase. China is also a country where the largest amount of organic manure is used in the world. In city, suburbs, and farmlands of China, there have been quite a few areas where the concentration of NO_3 in groundwater (e.g., in wells for drinking water) exceeds the acceptable criteria for drinking water. The suburbs of Beijing and the rural area of Suzhou can be used as examples to illustrate the NO_3 growth in the groundwater of city suburbs and rural areas in China.

According to the report of the investigation in 1981 (Wan et al., 1989), in the southeastern part of Beijing City Proper and in its southwestern and southeastern suburbs, the NO_3^--N content in the in-use layer of groundwater exceeded 10 mg/L, and that in the southeastern part of the city ranged from 30 mg/L to 90.3 mg/L.

Between 1976 and 1980, the groundwater NO_3^--N content in the city proper and in its extensive suburbs increased by more than 1 mg/L, whereas in the eastern and southern parts of the city and in the nearby suburbs it rose by over 5 mg/L. In 1980, areas which had a NO_3^--N content exceeding 10 mg/L almost doubled in Beijing as against the figures in 1976.

A random sampling survey in 1993 in the rural area of Suzhou, where the village and town industries were well developed and high rates of N fertilizer were used in the farmlands, found that of the 22 wells tested, 14 had a NO_3^--N content above 5 mg/L, and 5 had a NO_3^--N content of up to 32–78 mg/L. This was 3–8 times higher than the allowable concentration of NO_3^--N specified in the nation's criteria for drinking water (Xing, 1993) .

Effect of the Groundwater Which is Rich in Fluorine but Low in Iodine on Endemic Diseases

Fluorine and iodine are two chemical elements that are closely associated with the people's health. Generally, when the F content in the drinking water exceeds 1 mg/L, fluorosis develops as a result of excessive fluorine being absorbed. When the F content is over 3 mg/L, some will suffer from a F-induced endemic bone disease; and if it is above 7 mg/L, a great number of persons will suffer from a very serious bone disease. The F-caused disease is one of the major endemic diseases in China. Statistics show that fluorine diseases caused by excessive fluorine are widespread in China, covering 28 provinces (regions) and cities with a total population of about 120 million people. The number of persons suffering from stained teeth is up to 21 million, and about 1 million persons are suffering from the F-induced bone disease (Chen and Yu, 1990). Areas where such diseases are prevalent are closely connected with the fluorine in the natural environment, such as groundwater that is rich in fluorine. Beijing and Tianjin are the typical areas where groundwater is high in fluorine and where the population of the affected areas accounts for 6.8% of the agricultural population of the suburbs. In the Tianjin area, the Tanggu district is the most striking example of fluorine poisoning. It was found through a random sampling survey that among the 6–15 year olds, the incidence of a fluorine-stained tooth disease was as high as 93.3%; and for persons over the age of 20, the average incidence of a fluorine-induced bone disease was 1.99%, or even as high as 14.42%. The fluorine content of groundwater used as drinking water in the area was high, and was generally above 4 mg/L (Wan et al., 1989).

Goiter is also one of the endemic diseases found in the outer suburbs of Beijing. The cause of goiter is linked with the loss of iodine through leaching and with iodine-deficiency in groundwater. In the mountainous area in the outer suburbs of Beijing where there are rock strata of granite, gneiss, volcanic conglomerate, and limestone, the iodine content in the groundwater is as low as 3 μg/L, particularly at the site of Nankou, where the content is only 0.7 to 0.9 μg/L. In this area the incidence of goiter is above 10% (Wan et al., 1989). In seriously affected areas, cretinism is also found. The sufferers are mostly the descendants of the goiter patient.

SUMMARY AND CONCLUSIONS

The environmental background values of certain environment-pollution-and-human-health-related heavy metal and nonmetal elements in the soil of China stay at the same order of magnitude as those of the U.S., Japan, and the U.K. Except for the values of Hg and Cr, which are lower than those in Japan, the U.K., and the U.S., the background values of all the other elements, discussed in the chapter, in the soils of China are rather approximate to those of the three countries.

The background values of some elements in soil can be taken as indicators of regions of certain endemic diseases. In the western part of China, for instance, the distribution of areas with high incidence of Kersan disease and Kaschin-Beck's diseases is closely associated with the low content of total Se and water-soluble Se in the soil. When total Se falls below 110 μg/kg and water soluble Se below 2.4 μg/kg in the soil, the incidence of Kersan and Kaschin-Beck diseases will go up significantly.

Soil loading capacity for heavy metals refers to the maximum load of heavy metals the soil is capable of holding within a given environmental unit and a given duration of time without the risk of exceeding the criteria for environmental quality, affecting the yield and biological quality of agricultural products, polluting the environment. There are many factors affecting soil loading capacity; for instance, soil properties, indicators, pollution course, environmental factors, chemical compounds, etc.. Though soil loading capacity has found application in regional soil quality evaluation, establishment of regional water quality criteria for sewage irrigation, and establishment of regional criteria for sludge application, in the present studies on soil loading capacity, there remain a number of problems urgently calling for solution; e.g., lack of long-term experimental results, lack of compound pollution tests, and limited types of compound pollutants for test.

To distinguish forms of heavy metal (including rare earth) elements existing in the soil in micro amounts, the geochemical sequential extraction method is adopted. Although it is only distinction in terms of operation, it helps in acquisition of certain useful knowledge concerning the chemical behaviors of the heavy metal elements originally existing in the soil or coming externally from other sources, and the effects of different soil solid components on adsorption and enrichment of heavy metal elements in the soil, which can be used as a basis for controlling heavy metal pollution of the soil.

Human activities, such as mining, can significantly change the concentrations of water-soluble heavy metal elements in the soil. A comparison study revealed that the proportions of the concentrations of the water-soluble Zn, Cu, Co, Mn, Cr, and Pb to their totals in the soil polluted from copper mining are 372, 8392, 850, 576, 73, 24, and 12 times higher, respectively, than those in the unpolluted soil in the neighboring farmland.

Some of the industrial and mining areas of China are rather seriously polluted by heavy metals. For instance, in Jiangxi Province, the soils in the neighborhood of its major metal smelting plants and mines produced fodder that was significantly contaminated, particularly by Cu, Mo, and Cd. In the neighborhood of tungsten minings, Mo and Cd are the major pollutants, and the serious disproportion between Cu and Mo in the fodder produced leads to extensive incidence of Mo and Cd poisoning syndrome on farm cattle. In the farmland in the vicinity of Cu minings and smelteries, the soils may contain as much as 3766 mg Cu/kg and the fodder produced thereupon as much as 1572 mg Cu/kg. The farmland soil there contains 29.5 mg Cd/kg and the fodder produced thereupon,

16.1 mg Cd/kg. In some industrial and mining areas of Guizhou, the soils are severely polluted. And the soil in Zhangshi Irrigation Area of Liaoning is also rather seriously polluted by Cd-containing sewage.

Management of heavy metal pollution of the soil has been a major concern of the people. Although it is very hard to remedy the pollution when heavy metals enter into the soil, some positive results have been achieved in China by turning paddy field into upland field, applying organic manure, using modifiers, and adopting a comprehensive series of agroecological engineering measures.

China is the country that is the richest in rare earth resources, possessing 70% or so of the world's total rare earth reserves. Rare earths have been made use of in various aspects of the industry of China. Though the researchers in China do not now fully understand the physiological functions of the rare earths, they have been used as fertilizer in agriculture for the past dozen of years because of the proved fact that rare earths benefit certain crops in yield and quality. By 1990, the acreage of farmland using rare earths as fertilizer had reached 2 million hectares. In that year, the amount of rare earths in the R_2O_3 form applied to the farmland totaled 500 tons. Because of the mining, refining, and industrial or agricultural use of the rare earths, China has sensed the risk of rare earth pollution of her water bodies and soils. Nevertheless, little has been understood concerning the biogeochemical behaviors of the rare earths after they enter into the soil or water body

China is also a country using the highest amount of chemical N fertilizers in the world. According to the Agricultural Yearbook of China, 1995 witnessed the application of 22.2 million tons of chemical N fertilizers, which accounted for one-quarter of the world's total consumption of the fertilizer. In many high-yielding regions, overuse of chemical N fertilizers has significantly changed the N concentration in the surface and groundwater. Of the latter, the concentration of NO_3^- has well surpassed the concentration criterion for drinking water.

Fluorine and iodine are two chemical elements that are closely associated with human health. Since groundwater has been used as one of the drinking water sources in some areas, too much F in the groundwater causes fluorosis to become one of the major endemic diseases in China, and causes it to spread over 28 provinces and regions. The number of sufferers of stained teeth has reached well over 21 million and of F-induced bone disease, about one million.

In most areas of China, the drinking water lacks iodine. Especially, in areas of granite, gneiss, and limestone mountains, the iodine content in the groundwater is 0.7 to 0.9 μg/L only. Goiter is a typical iodine-deficiency-induced endemic. In areas with severe I deficiency, cretinism is also found. Recently, the Chinese government has been effectively promoting use of I-added salt to control iodine-induced diseases.

REFERENCES

Agroenvironment Monitoring Center, Ministry of Agriculture, Animal Husbandry and Fishery (AMC, MAAHF), *Reports on China's Agroenvironmental* Quality (in Chinese), 1985.

Chao, T.T. Selective Dissolution of Manganese Oxides from Soils and Sediments with Acidified Hydroxylamine Hydrochloride. *SSSAJ*, 36, pp. 764–768, 1972.

Chen, C.Q., Y.L. Xu, Z.J. Ye, X.X. Yu, H.Y. Su, and J.T. Wu. Study on Regularities of Cr Changes in Agricultural Crops and Wastewater Using Neutron Activation Technique. *Envi-romental Science* (in Chinese). 1, pp. 13–16, 1978.

Chen, T., Y.Y. Wu, and Q.X. Kong. Pot-Culture After Effect of Liming for Amendment of Cd-Bearing Soil in Zhangshi Irrigated Area. *Environmental Science* (in Chinese). 2, pp. 40–42, 1981.

Chen, H.M. and F.N. Ponnamperuma. Yield and Cadmium Concentration of Wetland Rice

Grain as Affected by Addition of Cadmium, Phosphorus and Zinc Compounds. *Philip. J. Crip. Sci.*, 7, pp. 109–113, 1982.

Chen, T., Y.Y. Wu, Q.X. Kong, F. Tan, and X. Wang. Preliminary Study on Effect of Cd-Bearing Industrial Wastewater on Prevention and Control of Farmland Contamination and Land Use—Agro-Ecological Engineering, in *Studies on Pollution Ecology of Soil-Plant System* (in Chinese), Z. M. Gao, Chief Ed., 1986. China Science and Technology Press, Beijing, pp. 86–91.

Chen, J.F., R.W. Xu, and H.M. Chen. Prospect of Studies on Soil Chemical Behavior and Environmental Contamination of Chemical Substances. *Environ. Sci.* (in Chinese). 7, pp. 8–12, 1988.

Chen, H.M. Several Factors Affecting Cd Adsorption on Soils. *Soils* (in Chinese). 20, pp. 131–136, 1988.

Chen, H.M. and C.R. Zheng. Effect of Plant Materials on Physico-Chemical Properties of Flooded Soils. *Soils* (in Chinese). 21, pp. 234–238, 1989.

Chen, H.M. Present Situation, Development Tendency and Countermeasures of Soil Pollution in China. *Prog. in Soil Sci.* 18, pp. 53–56, 1990.

Chen, G.J. and D.F. Yu. *Fluorine in Environment* (in Chinese), Science Press, Beijing, 1990, pp. 153–163.

Chen, H.M., C.R. Zheng, and X.H. Sun. Effect of Anions on Absorbability and Extractability of Lead Added in Soil. *Pedosphere.* 1, pp. 51–62, 1991a.

Chen, H.M., C.R. Zheng, and X.H. Sun. Effect of Different Lead Compounds on Growth and Heavy Metal Uptake of Wetland Rice. *Pedosphere.* 1, pp. 253–264, 1991b.

Chen, H.M. and C.R. Zheng. Discussion About Soil Environmental Capacity Research. *Acta Pedological Sinica* (in Chinese), 29, pp. 219–225, 1992.

Chen, H.M. and H.K. Wang. Research Progress in Biological Effect of Heavy Metal Pollution in Agricultural Environment, in S*elected Papers on Pollutants and Their Biological Effect in Environment* (in Chinese). Science Press, Beijing, 1992, pp. 24–31.

Chen, H.M. Paddy Soil Pollution and its Prevention and Control, in *Paddy Soils of China*, Li Q.K., Chief Ed., (in Chinese) Science Press, Beijing, 1992, pp. 529–545.

Chen, Y.X. Soil Chemistry of Chromium. *Progress in Soil Science.* 5, pp. 8–13, 1992.

Chen, H.M., C.R. Zheng, and X.H. Sun. Factors Affecting Soil Environmental Capacity of Pb. *Soils* (in Chinese). 26, pp. 189–195, 1994.

China State Bureau of Environment Protection (CSBEP). The Background Values of Elements in Soils of China (in Chinese), China Environmental Science Press, Beijing, 1990, pp. 2, 87–91.

Dai, Q.W., Z.M. Zeng, J.Y. Wang, Z.L. Wu, and P. Fan. Preliminary Investigation on Effects Of Main Metal Smelting Plants and Mines on Animal Husbandry in Jiangxi Province. *Agro-Environ. Prot.* (in Chinese). 12, pp. 124–126, 1993.

Department of Environmental Protection, Institute of Soil Science Academia Sinica. Studies on Soil Pollution and Control, *Environ. Sci.* (in Chinese), (3), pp. 41–46, 1977.

Elkhatib, E.A., J.L. Hern, and T.E. Staley. A Rapid Centrifugation Method for Obtaining Soil Solution. *SSSAJ*, 51, pp. 57–583, 1987.

Guo, B.S. *Rare Earths in Agriculture* (in Chinese). China Agricultural Science and Technology Press, Beijing, 1988, pp. 1–23.

Gupta, S.K. and K.Y. Chen. Partitioning of Trace Metals in Selective Chemical Fractions of Nearshore Sediments, *Environ. Lett.* 10, pp. 129–158, 1975.

He, Y., Z.J. Ye, and F.Z. Wu. *Outline of Agroenvironmental Science* (in Chinese). Shanghai Science and Technology Publishing House, Shanghai, 1991, pp. 1–13.

Huang, Y.S. Study on Effect of Humus, Sulfur and Sodium Chloride on Rice Hg Accumulation. *Acta Ecologica Sinica* (in Chinese). 2, pp. 1–10, 1982.

Huang, S.D., X.Q. Xu, and S.Y. Lu. *The Project of the Three Gorge Reservoir Versus Environmental Pollution and People's Health* (in Chinese). Science Press, Beijing, 1994, pp. 80–101.

Huo, W. R., R.L. Chao, Z. L. He, Y. Zhou, and J.J. Wang. Role of Different Modifiers in Inhibiting Rice Cd Uptake. *Agro-Environ. Prot.* (in Chinese). 8, pp. 38–40, 1989.

Li, X.G., X.P. Li, F.Q. Chen, Z.Q. Xia, and Z. Tan. Effects of Different Arsenic Compounds on

Rice Growth and As Uptake. *Studies on Soil, Environmental Capacity* (in Chinese). Meteorological Press, Beijing, 1986, pp. 75–83.
Liu, Z. *Agrochemistry of Microelements* (in Chinese), Agricultural Press, Beijing, 1991.
Mckeague, J.K. and J.H. Day. Dithionite and Oxalate-Extractable Fe and Al as Aids in Differentiating Various Classes of Soils. Canada, *J. Soil Sci.* 46, pp. 13–22, 1966.
Mckenzie, R.M. The Reaction of Cobalt with Manganese Dioxide Minerals. *Aust. J. Soil Res.* 8, pp. 97–106, 1970.
Minnich, M.M. and M.B. McBride. Copper Activity in Soil Solution: I. Measurement by Ion-elective Electrode and Donnan Dialysis, *SSSAJ*, 51, pp. 568–572, 1987.
Peng, G.S. Development of Town and Township Enteritises in Guizhou Province. *Environ. Sci.* (in Chinese). 9, pp. 62–67, 1988.
Shao, X.H., G.X. Xing, and W.X. Yang. Distribution of Chemical Forms for Co, Cr, Ni and V in Typical Soils of China. *Pedosphere*, 3, pp. 289–298, 1993.
Shao, X.H. and G.X. Xing. Study on Chemical Behaviours of Cu and Zn in Soils, in Xi'an, '94: *Proceedings of the Fifth National Workshop on Soil Science for Young Soil Scientists of China*, (in Chinese). Agricultural Science and Technology Press, Beijing. 1994a, pp. 207–216.
Shao, X.H. and G.X. Xing. in Nanjing '93: Effect of Soil Solid Components on Enrichment of Heavy Metals, in *Pedosphere–Proceedings of the Second Symposium on Material Cycling in Pedosphere*, Zhao, Q.G., Chief Ed., Nanjing University Press, Nanjing, 1994b, pp. 527–535.
Shuman, L.M. Fractionation Method for Soil Microelements. *Soil Sci.* 140, pp. 11–12, 1985.
Singh, J.P., S.P.S. Karwasra, and Singh Machendra. Distribution and Forms of Copper, Iron, Manganese and Zinc in Careous Soil of India. *Soil Sci.*, 146, pp. 359–366, 1988.
Taylor, R.M. and R.M. Mckenzie. The Association of Trace Element with Manganese Minerals in Australian Soils. *Aust. J. Soil Res.*, 4, pp. 29–39, 1966.
Tessier, A.,P. G.C. Campbell, and M. Bisson. Sequential Extraction Procedure for Speciation of Particulate Trace Metals. *Anal. Chem.*, 51, pp. 844–851, 1979.
Wan, G.J. Environmental Evolution of the Jing-Jin-Bo Region and the Route of Its Development and Protection (in Chinese), Science Press, Beijing, 1989, pp. 80–94
Wang, C.B., A.C. Ning, X.B. Chang, and J.K. Wu. Monitoring and Evaluation of Five Food Crops in Fenhe River Basin of Taiyuan. *Agro-Environ. Prot.* 5, pp. 208–212, 1993.
Wang, X.P., X.Y. Zhao, W.X. Jin, F.D. Zhang, M.X. Zeng, and Q. Guo. Effectiveness of Organic Manures in Cr-Polluted Soil Amendment. *Environ. Sci.* (in Chinese). 7, pp. 18–21, 1986.
Wei, F.S., J.S. Chen, Y.Y. Wu, and G.J. Zheng. Study on the Background Values of Soil Elements of China. *Environ. Sci.* (in Chinese), 12, pp. 12–19, 1991.
Wu, Y.Y., T. Chen, and X.X. Zhang. Study on Cd-Pollution Ecology in Zhangshi Irrigated Area of Shenyang, in *Studies on Pollution Ecology of Soil-Plant System,* Z.M. Gao, Ed., China Science and Technology Press, Beijing, 1986, pp. 295–301.
Wu, Z.M. The Effect of Rare Earth Elements on Rooting of Plants Cutting. *J. Chinese Rare Earth Soc.* (in Chinese), 6, pp. 62–70, 1988.
Wu, Y.S. *Urban Sewage Disposal-Investment and Policy Making* (in Chinese). China Environmental Science Press, Beijing, 1992.
Xia, Z.L. *Soil Environmental Capacity and Its Application* (in Chinese), Meteorological Press, Beijing, 1988.
Xing, G.X., unpublished data, 1993.
Xu, L.Y and G.X. Xing. Concentration of Heavy Metal Ions in Soil Solution. *Soils* (in Chinese), 27, pp. 245–246, 280, 1995.
Yang, G.Z. and J. Rong. Evaluation Problems in Soil Hg Pollution of Hangu Area. *Soils* (in Chinese), 14, pp. 17–19, 1982.
Yang, W.X. "Form transformation and migration of heavy metal elements in submerged soils, and effect of heavy metal elements on greenhouse gases." Thesis. Presented to the Institute of Soil Science, Academia Sinica, Nanjing, in partial fulfillment of the requirement for the degree of Doctor of Philosophy, 1994.
Zang, H.L., C.R. Zheng, and H.M. Chen. Study on Controlling Cadmium Absorption by Crops on Cadmium Contaminated Soil, *Agro-Environ. Prot.* (in Chinese), 6, pp. 28–29, 1987.

Zang, H.L., C.R. Zheng, and H.M. Chen. Study on the Inhibition of Cadmium Absorption by Crops in Cadmium Contaminated Soil, *Agro-Environ. Prot.* (in Chinese), 8, pp. 33–34, 1989.

Zheng, C.R. and H.M. Chen. Effect of Combined Pollution on Rice Growth. *Soils* (in Chinese), 21, pp. 10–14, 1989.

Zheng, C.R. and H.M. Chen. Transportation of Pollutant Heavy Metals in Soil-Rice System and Its Effect on Rice. *Acta Scientiae Circumstantiae* (in Chinese), 10, pp. 145–152, 1990.

Zheng, C.R. and H.M. Chen. Studies on Loading Capacity of Soil for Heavy Metals. *Prog. Soil Sci.* (in Chinese), 23(5), pp. 21–28, 1995.

Zhou, G.T., P.L. Sheng, and Q.Y. He. Effect of Agricultural Production on Soil Heavy Metal Contents. *Agro-Environ. Prot.* (in Chinese), 12, pp. 274–275, 1993.

Zhu, J.G. and G.X. Xing. Forms of Rare Earth Elements in Soil: I. Distribution. *Pedosphere*. 2, pp. 125–134, 1992a.

Zhu, J.G. and G.X. Xing. Study on the Sequential Extraction of Rare Earths in Soils. *Soils* (in Chinese), 24, pp. 215–218, 1992b.

Zou, B.J. As in Soils. *Progress in Soil Science* (in Chinese), 2, pp. 8–13, 1986.

CHAPTER 8

ENVIRONMENTAL PROBLEMS IN SOIL AND GROUNDWATER INDUCED BY ACID RAIN AND MANAGEMENT STRATEGIES IN CHINA

Guoliang Ji, Jinghua Wang, and Xiaonian Zhang

INTRODUCTION

Environmental problems induced by acid rain first attracted the concern of scientists in China in the late 1970s. It was observed that in southwestern China, notably in the vicinity of Chongqing and Guiyang, some forests were damaged without any apparent reason. Because the pH of the rainwater in these areas was low due to the emission of large amounts of sulfur oxide coming from the burning of high sulfur-containing coal by industrial plants and also because the soils were already strongly acid, it was inferred that such damage was caused by acid rain. Since then, a multidisciplinary study has been initiated by the Academia Sinica, in cooperation with regional institutions. Since 1986, a national research program involving a great number of institutions was developed to study the relationship between acid rain and the environment. In 1987, an extensive survey on acid rain-related problems was organized under the auspices of the Association of Science and Technology. A total of 19 societies, including the Soil Science Society of China, participated in the survey. From these research efforts, it was realized that the extent of the impact of acid rain in China was much larger than previously thought, and that acid rain-affected areas are increasing in extent.

The acid rain-induced problems in soil and groundwater in China show certain characteristic features. The acid rain is of the sulfate type. Most of the soils that are likely to suffer from the effect of acid rain have small buffering capacities against acid input. These soils can undergo a series of chemical reactions with both hydrogen ions and sulfate ions. These features shall be dealt with in the present chapter.

ACID RAIN

Formation

Coal combustion is responsible for most of the atmospheric pollution in China, especially in the big cities. In 1982, coal accounted for about 74% of the total energy production in

China, with about 18%, 3%, and 5% contributed by oil, gas, and hydropower, respectively (Zhao and Xiong, 1988). An estimated 40% of the coal consumed was for industrial, commercial, and domestic use in cities, with an area totaling less than 0.5% of the whole country. Most of the coal used in cities is burned in medium-sized and small furnaces with low stacks without devices for removing SO_2 from flue gas. Thus, under unfavorable meteorological conditions, heavy pollution occurs in these areas. Furthermore, in regions where the ecosystem, including soil, is more sensitive to acid rain, the coal consumed is mostly high in sulfur content. In the south and the southwest, the S content of the coal is generally 2–3%, and in some areas it may be as high as 5% (Li, 1989; Zhang, 1985). The emitted SO_2, after undergoing chemical reactions in the air, transforms into H_2SO_4 when dissolved in rainwater. This is the main reason for the appearance of sulfate-type acid rain. Besides, statistical data show that the percentage of nitrate ions in total anions in acid rain is slightly higher than that in nonacid rain, although the percentage is less than 15% of the total anions (Wang et al., 1989). This implies that NO_x is also a contributing factor to the formation of acid rain in China.

Distribution

Table 8.1 shows the average pH of rainwater in some cities. It can be seen that the rainwater in the north is neutral in reaction while that in central China is neutral to slightly acid. On the other hand, in southern and southwestern China, nearly all the rainwaters have a pH much lower than the critical value of 5.6 for acid rain. In some large cities in the north, such as Beijing and Tianjin, there is also a very high emission of SO_2, especially during the winter. In fact, the occurrence of acidic precipitation is closely related to the distribution of suspended soil particles in air. As shall be seen later, soils in the north are mostly calcareous and in the south mostly acid, with those in central China neutral or slightly acid in reaction. It is the neutralizing action of alkaline substances contained in suspended soil particles that makes the rainwater nonacid in the north.

Table 8.1. Average pH of Rainwater in Some Cities of China

Region	Locality	Year	pH	Reference
North	Beijing	1981	6.80	Zhao & Xiong, 1988
	Tangshan	1981	7.12	Wang et al., 1983
	Shijiazhuang	1981	6.81	Wang et al., 1983
	Lanzhou	1981–1982	6.85	Zhao & Xiong, 1988
Central	Wuhan	1980–1981	6.95	Zhang & Kuo, 1983
		1983	6.44	Zhao & Xiong, 1988
	Nanjing	1980	5.60	Mo & Xie, 1982
		1981	6.38	Wang & Li, 1984
	Shanghai	1980	6.78	Zhang & Kuo, 1983
	Hangzhou	1981	5.10	Zhao & Xiong, 1988
Southwest	Chongqing	1979–1980	4.65	Wang, 1983
		1982	4.14	Zhao & Xiong, 1988
	Guiyang	1982	4.02	Zhao & Xiong, 1988
	Ermei	1988–1990	4.53	Dou & Yan, 1992
South	Fuzhou	1982	4.49	Zhao & Xiong, 1988
	Nanning	1981	5.74	Wang & Li, 1984
		1988	4.76	Cao et al., 1993
	Liuzhou	1988	4.21	Cao et al., 1993
	Shaoguan	1988	3.94	Cao et al., 1993
	Guangzhou	1988	4.39	Cao et al., 1993

Within a given region, the distribution pattern of acid rain is determined by local geographic and meteorological conditions. Two main types of distribution patterns may be distinguished. If the region is surrounded by mountains and the air transport is weak, the occurrence of acid rain is confined to the local area and the acidity of the rainwater decreases from the center of the emission source outward. This is the case near Chongqing and Guiyang. Within the Chongqing region, the average pH of rainwater during 1982–1984 was 4.14 in urban areas and 4.44 in rural areas, and within the Guiyang region, it was 4.07, 4.42, and 4.58 in urban, suburban, and rural areas, respectively (Zhao and Xiong, 1988).

On the other hand, if meteorological conditions permit the transport of air to a long distance, the acidity of rainwater varies along the path of the wind direction. In some regions of Guangdong and Guangxi provinces, such cases can be observed. There, the distribution of acid rain is along the prevailing wind direction and air-flow transportation path from north to south; namely, the acid rain is distributed along Shaoguan-Yingde-Guangzhou-Foshan-Jiangmen in Guangdong and along Guilin-Liuzhou-Nanning in Guangxi (Qi et al., 1993). In the Ermei Mountain region of Sichuan Province, the acid rain has also been transported from a long distance (Dou and Yan, 1992).

Because of the rapid development in the industries, particularly in the wide establishment of small-scale factories in many regions, both the acidity of rainwater and the extent of acid rain-affected areas have increased in recent years (Ding and Wang, 1997). This can be clearly seen in Figure 8.1. In 1986, the pH = 5.6 contour was in a position approximately along the Yangtze River. In 1993–1995, it moved to the Yellow River or even to the northeast of the mainland. In 1986 there were only small local areas where the pH of the rain was lower than 4.5, whereas in 1993–1995 large areas within a pH = 4.5 con-

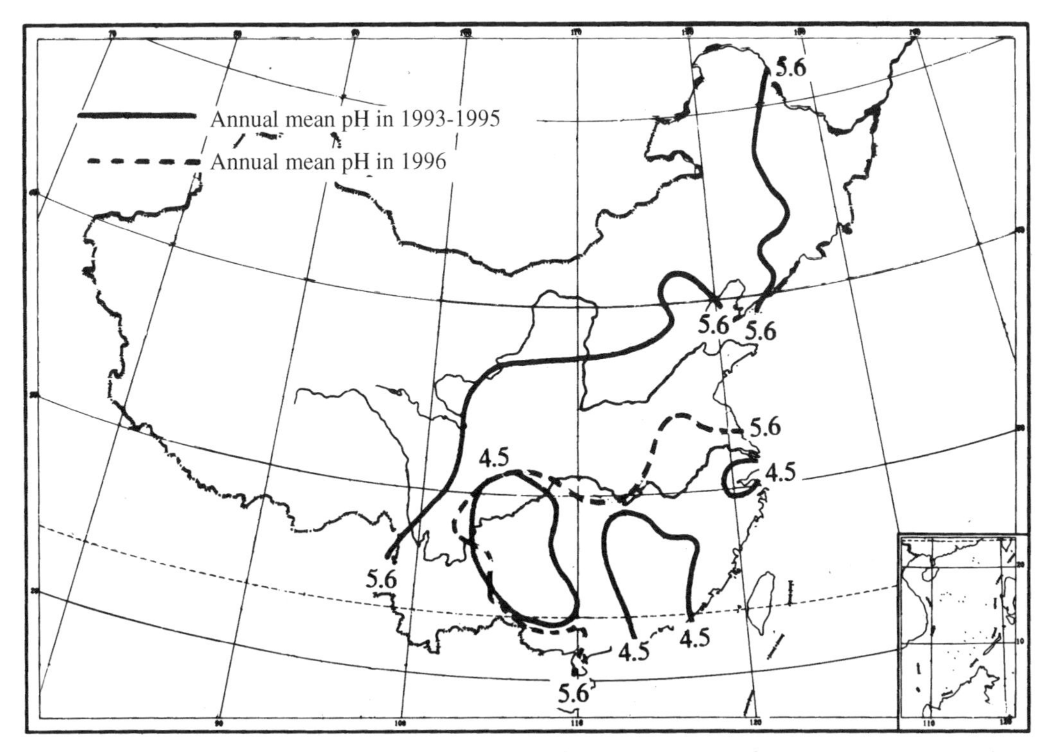

Figure 8.1. Change in the distribution of pH = 4.5 and pH = 5.6 contours from 1986 to 1993–1995 in China (Ding and Wang, 1997, with permission).

tour occurred (Fig. 8.1). It was reported that in many cities of central and southern China, the frequency of acid rain during the year had increased markedly. In the Pearl River delta where acid-forming substances came chiefly from the northern part of Guangdong Province, the acidity of rainwater has increased, and the pH = 4.8 and pH = 5.0 contours moved southward rapidly (Qi et al., 1993). In Guiyang region, the average pH of rainwater in 1983, 1985, and 1989 was 4.53, 4.20, and 3.98, respectively, and areas with strongly-acid rain are expanding (Xiong et al., 1993). This is also the case when viewed throughout the whole country.

Composition and Properties

The composition of acid rain varies greatly among the regions. Even within the same region the composition changes markedly during the year. Nevertheless, some regularities have been observed. The concentrations of major ions in rainwater of some regions are given in Table 8.2. For anions, the most notable feature is the high percentage of sulfate, except in the rainwater of the north in which chloride ions are balanced by alkali and alkaline earth metal cations. It can be calculated that, for the acid rain, the SO_4/NO_3 ratio is as high as 5.4 to 19.6, with an average of 10.9 on an equivalent basis, much larger than the ratio in most acid rains in northern Europe and North America where the ratio ranges from 0.8 to 4.9, with an average of 2.8 (n = 9) (Wang et al., 1989). For cations, the concentrations of NH_4^+, Ca^{2+}, and Mg^{2+} are higher than those in the USA by 5–10 times (Wang et al., 1989; Wei et al., 1989). The concentrations of these cations in nonacid rain are higher than those in acid rain by 2.9 times. The total concentration of five major ion species ($SO_4^{2-} + NO_3^- + NH_4^+ + Ca^{2+} + Mg^{2+}$) in the north is 824 $\mu mol_c\ L^{-1}$, while it is 378 $\mu mol_c\ L^{-1}$ in the south. Among the three major cation species, NH_4^+ accounts for 30–55% of the total in acid rain, whereas it is 10–30% in nonacid rain (Wang et al., 1989). It was assumed that this difference in percentage was caused by more intensive biological activities in the south and the effect of calcareous soil particles in the north.

In summary, the level of hydrogen ions in many rainwaters in China has reached those found in developed countries. The concentration of other cations is comparatively high, and sulfuric acid accounts for the major part of the acidity of acid rain. These features would have important relevance to the effect of acid rain on properties of the soil, as shall be discussed in the following sections.

Table 8.2. Concentration of Major Ions in Rainwater[a]

		Cations ($\mu mol_c\ L^{-1}$)					Anions ($\mu mol_c\ L^{-1}$)		
Locality	Year	H	NH_4	Na	K	Ca	Cl	NO_3	SO_4
Beijing	1981	0.2	141	141	40.2	184	157	50.2	273
Tianjin	1981	0.5	126	175	59.2	287	183	29.2	318
Guiyang urban	1982–1984	84.5	78.9	10.1	26.4	231	8.2	21	411
rural	1982–1984	26.3	50.6	5.9	7.0	87.7	21.1	15.9	167
Chongqing urban	1982–1984	72.4	106	51.4	7.4	110	15.0	31.6	307
rural	1982–1984	36.3	64.1	45.4	23.4	42	23.9	18.0	165
Fanyu	1986	65.6	274	26.1	17.3	72.5	36.9	36.6	350
Guangzhou	1989	45.7	149	—	—	206	—	24.0	230
Nanning	1989	17.4	131	—	—	150	—	17.0	244
Guilin	1986	14.8	50.0	37.5	23.0	67.2	20.2	19.7	107
Wuzhou	1988	37.2	46.1	—	—	32.5	—	7.3	74

[a]Qi et al., 1993; Tang et al., 1993; Wang et al., 1989, with permission.

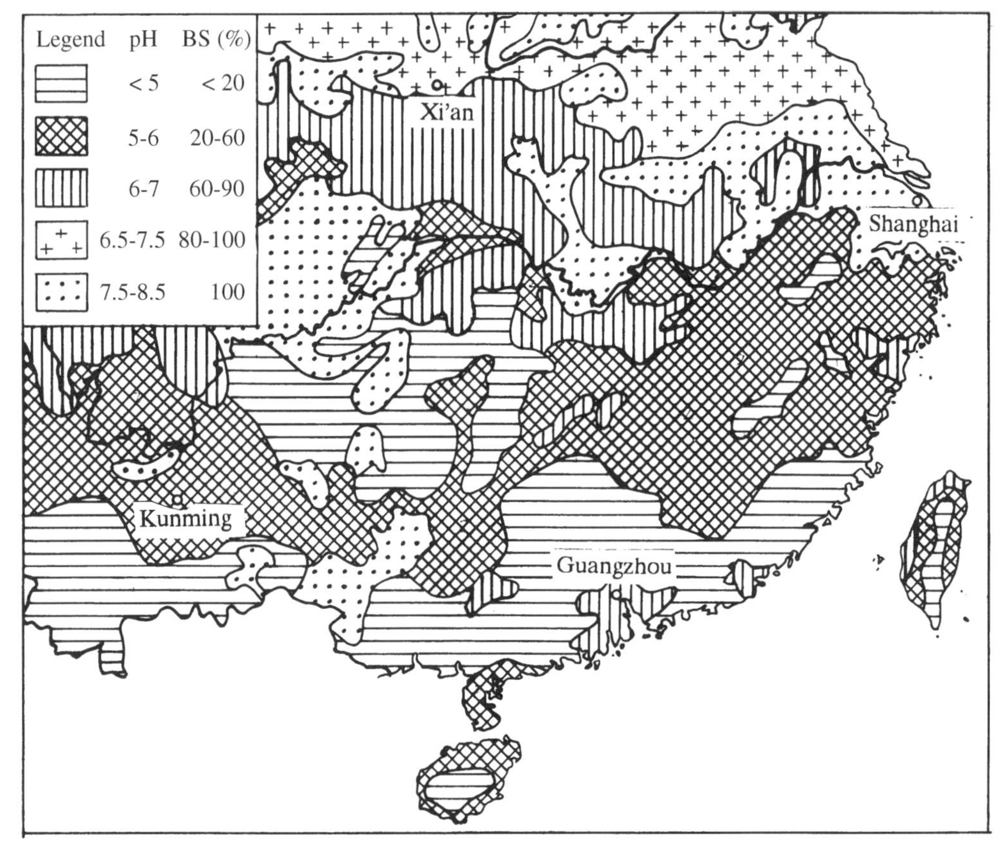

Figure 8.2. Generalized soil pH map of central and southern China (BS is base saturation in percentage) (Yu and Zhang, 1990, with permission).

ACIDITY AND ACIDIFICATION OF SOILS

Distribution and Properties of Acid Soils

Distribution

The distribution of acid soils in China shows certain geographical regularities. Except in mountain regions of the northeast where brown forest soils and podzolized soils predominate, nearly all acid soils are found south of the Yangtze River (Changjiang). Figure 8.2 is a generalized soil pH map of central and southern China. Four main types of acid soils may be distinguished. Latosols, with a pH of 5–6, are mainly distributed on Hainan Island and Leizhou Peninsula. Lateritic Red Earths, with a pH of less than 5, occupy large areas to the south of the Nanling Mountains. North of this region the soils are predominantly Red Earths with pH 4.5 to 5.5. In mountain regions of southern China where the humidity is relatively high, especially in the southwest, Yellow Earths occur widely. This particular type of soils, except those derived from sandstone, although strongly acid in reaction (pH 4 to 4.5), generally possesses a relatively high buffering capacity against acid input. Therefore, as far as the effect of acid rain on soil acidification in China is concerned, the most important soil types are Latosols, Lateritic Red Earths, and Red Earths. This is in sharp contrast to northern Europe and North America where the most important acid soils suffering from the effect of acid rain are podzolized soils. For this reason, in this chapter we shall limit our discussions chiefly to the properties of these three types of soils. Chemically, there is one important characteristic feature in common for these soils; i.e., because of the high contents of iron and aluminum oxides, they not only carry a large proportion of variable negative surface charge, but also carry a large amount of positive surface charge, which is variable depending on environmental conditions, especially pH. Hence comes the name variable charge soils, in contrast to constant charge soils of temperate regions with a nearly constant CEC. The chemical properties of these variable charge soils have been dealt with in detail by T.R. Yu and his colleagues (Yu, 1997). As shall be seen later, this characteristic feature is of significance in properties relating to soil acidity and acidification.

Criteria of Soil Acidity

Two parameters: the intensity factor, pH; and the capacity factor, the quantity of exchangeable acidity; have to be considered in discussing soil acidity. For a given soil, both of these parameters are determined primarily by the exchangeable cation status. In addition, the mineralogy of the soil is also important.

For pH, three critical values can be distinguished. These can be seen from Figure 8.3 in which the change in pH of the electrodialyzed clay fraction of two variable charge soils and a constant charge soil (Yellow-brown Earth) with the increase in added NaOH is presented.

The pH at which the soil is devoid of adsorbed base ions and where all the cation-exchange sites are occupied by hydrogen and aluminum ions may be called the "ultimate pH." This is the lowest pH a soil can have, except when there is the presence of free organic or inorganic acids. As seen in the figure, this pH is 5.0 for the Latosol, 4.5 for the Red Earth, and 3.9 for the Yellow-brown Earth. This difference is related to differences in clay minerals and other chemical composition of the soil. The high ultimate pH of the Latosol is caused by the predominance of kaolinite and the high content of iron oxides, while the low ultimate pH of the constant charge soil Yellow-brown Earth is related to the predominance of hydrous mica.

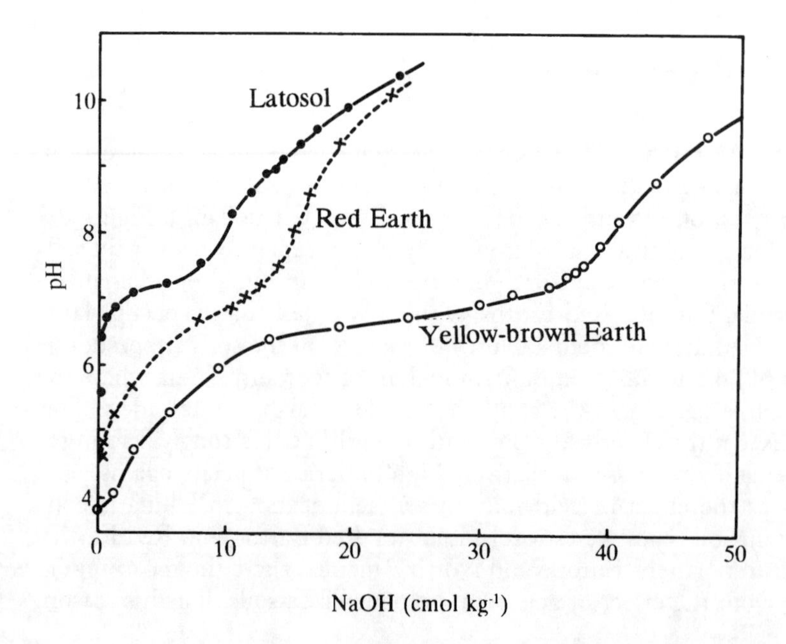

Figure 8.3. Change in pH of the electrodialyzed clay fraction of two variable charge soils and a constant charge soil with the increase in added NaOH (Kong et al., 1997, with permission).

When all the cation-exchange sites are balanced by base cations, the pH is called the "neutralization pH." This is equivalent to the complete neutralization of all the acid groups of a weak acid. In Figure 8.3, this pH is 8.2 ± 0.1 where lies the inflection point of the titration curve. In principle, the quantity of NaOH required to raise the pH to this point corresponds to the quantity of exchangeable acidity of the given soil, which is 9.5, 13.5, and 41.0 cmol kg^{-1} for the Latosol, Red Earth, and Yellow-brown Earth, respectively. For field soils where calcium and magnesium ions generally account for more than 90% of the exchangeable base cations, the neutralization pH is around 7.

The pH at which half of the cation-exchange sites is occupied by base cations may be called the "half-neutralization pH," because this is equivalent to a weak acid half-neutralized by a base. This pH is equivalent to the pK_a value of the weak acid. For the three soil colloids examined, the values are 7.6, 6.7, and 6.5, respectively, reflecting the difference in their acid strength. For a given soil, this pH is affected by the kind of the base cations. If the cation species is calcium instead of sodium, this pH would be lower than that shown in Figure 8.3. At this pH, the buffering strength of the soil is the strongest, as can also be seen in the figure.

The quantity of acid, hydrogen and aluminum ions, replaced by the metal cations of a neutral salt, is the exchangeable acidity. This capacity factor of soil acidity is of significance when it is required to decrease the acidity to a desired pH by the application of an alkaline material, such as lime.

Effect of Submergence

In the southern and central part of China, many fields are cultivated for paddy rice. Because of submergence during the rice-growing season, reduction processes will occur. In all the reduction processes there is invariably the participation of protons:

$$MnO_2 + 4H^+ + 2e \rightarrow Mn^{2+} + 2H_2O \tag{8.1}$$

$$Fe_2O_3 + 6H^+ + 2e \rightarrow 2Fe^{2+} + 3H_2O \tag{8.2}$$

$$NO_3^- + 2H^+ + 2e \rightarrow NO_2^- + H_2O \tag{8.3}$$

This would consume hydrogen ions, resulting in the rise in pH.

However, the effect of submergence on soil pH is complicated. The direction, speed, and extent of change in pH during submergence is determined by such factors as the initial pH, the content of easily decomposable organic matter, and temperature, among others (Yu, 1985). This can be illustrated by the examples shown in Figure 8.4. For strongly acid soils containing little organic matter, the pH rises rapidly after submergence, and then attains a steady value after about two weeks (curve 1). If the soil contains a high amount of organic matter, the pH may decrease slightly following a rapid rise within the first several days, due to the production of carbon dioxide and organic acids. Then it rises again gradually from the minimum at about one week, and finally attains a steady value (curve 2). If the soil is neutral originally, the pH may remain nearly constant (curve 3). For alkaline soils, the pH may decrease at first and eventually attain a steady value (curve 4). Thus, for acid paddy soils with a pH of 4.5 to 5.5 when dry, the pH may be maintained at 6–7 during the rice-growing period due to submergence. This would have great practical significance as far as the effect of soil acidity on plant growth is concerned (Yu, 1985). This may be one of the reasons for the wide cultivation of rice in southern China, even on strongly acid soils, although experiments have shown that the growth of the majority of rice species generally decreased when the pH of the medium was lower than

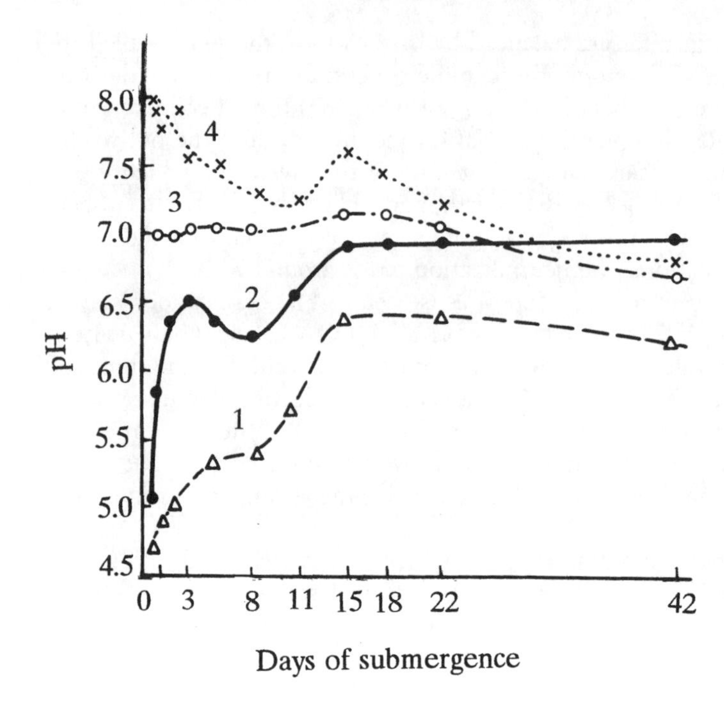

Figure 8.4. Change in pH of soils during submergence (curve 1, strongly acid soil containing little organic matter; curve 2, acid soil with a high content of organic matter; curve 3, neutral soil; curve 4, alkaline soil) (Yu and Li, 1990, with permission).

about pH 5 (Yu, 1985). To date, no evidence of the damage to rice plants caused by the effect of acid rain on soil acidification has been reported.

On the other hand, the long-term effect of alternate submergence and drainage on soil reaction is acidification. This can be explained as follows. Under reduced conditions, ferrous and manganous ions produced in the soil can compete with originally adsorbed base cations such as calcium and magnesium, thus enhancing their leaching loss. When oxidation conditions resume, these ferrous and manganous ions precipitate in their oxidized state, and disappear from the solution and the exchange sites. Acidity is produced. Then, these sites are occupied by hydrogen ions, which subsequently attack the clay minerals and increase the exchangeable aluminum. Therefore, repeated submergence and drainage would lead to a gradual acidification of the soil. Such a phenomenon can be commonly observed under field conditions.

Buffering Against Acid Input

Having been intensively weathered, variable charge soils generally do not contain primary minerals, except quartz. Needless to say, calcareous materials are also absent. Therefore, for these soils there are only two kinds of substances that can have a buffering ability against hydrogen ions entering the soil from external sources: exchangeable base cations and secondary minerals, including free oxides. This is in contrast to soils of the north where carbonates and primary minerals may be responsible for the buffering ability against acid input. For most of these variable charge soils, exchangeable base cations account for less than 25% of the total adsorbed cations (Yu and Li, 1990). Besides, the quantity of negative surface charge carried by the soil, the site for exchangeable cations, is generally low. These two factors together make the buffering effect of exchangeable base cations against hydrogen ions much weaker than that in constant charge soils of the north. Thus, the principal fate of input hydrogen ions would be to react with the solid phase of the soil, as shall be explained in the next section.

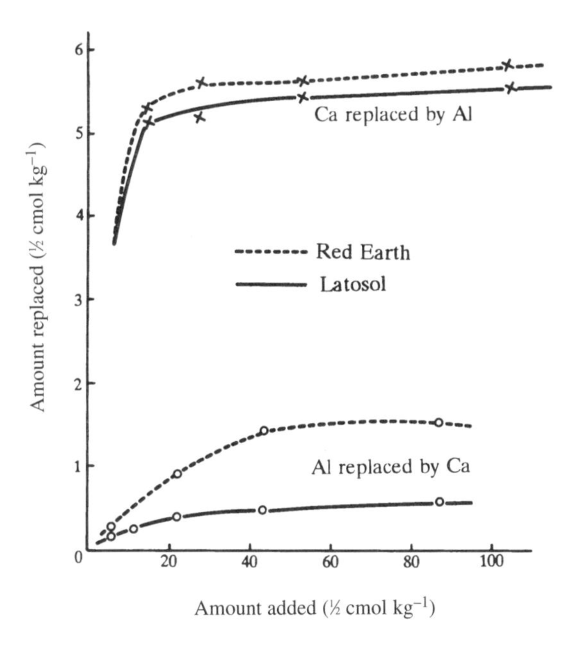

Figure 8.5. Comparison of replacing power between aluminum ions and calcium ions (Kong et al., 1997, with permission).

Acidification Processes

The first step of acidification would be the replacement of adsorbed base cations by hydrogen ions and the subsequent leaching loss of these cations. As a result, the base-saturation percentage decreases. Later, because hydrogen-saturated soils would transform into aluminum-saturated soils spontaneously, as was discussed in detail by Kong et al. (1997), these aluminum ions would enhance the leaching of base cations, because aluminum ions possess a very strong replacing power when compared with base cations, such as calcium, as is evidenced in Figure 8.5.

For variable charge soils in particular, there may be other reactions occurring when hydrogen ions are introduced. The most remarkable one among these reactions is the direct protonation at the surface of soil particles, leading to the formation of positive surface charge. This is because, on the surface of hydrated oxides of iron and aluminum and on the edges of clay minerals such as kaolinite, there are OH-containing groups, which are capable of accepting protons.

If the soil has already been highly base-unsaturated, especially when it is practically devoid of exchangeable bases, the input of hydrogen ions may lead to the release of water-soluble aluminum, due to hydrogen-aluminum transformation. This is different from the case for constant charge soils of the north in which primary minerals are first attacked, resulting in the release of basic metals.

For those soils with kaolinite as the predominant mineral, the presence of a large amount of interlayer hydroxyl aluminum would be unlikely. Nevertheless, experiments have shown that the input of hydrogen ions can result in the increase of exchangeable acidity even when the soil has already been completely H^-, Al-saturated (Zhang and Yu, 1997). It is supposed that this is caused by the demasking of exchange sites originally covered by iron oxides and the activation of nonexchangeable hydroxyl aluminum and its polymers.

The contributions of the three fates to the consumption of hydrogen ions, the increase in positive surface charge, the increase in soluble aluminum, and the increase in ex-

Table 8.3. Contribution of Three Fates to Consumption of Hydrogen Ions Added to Variable Charge Soils[a]

	H^+ (cmol kg^{-1})		Equil.	In H^+ consumption (%)			
Soil	Added	Cons.	pH	A[b]	B	C	Total
Red Earth	0.575	0.49	4.46	16.2	34.4	20.5	71.1
	1.150	1.00	4.22	17.3	56.2	43.0	116.5
	1.725	1.53	4.11	13.4	64.1	28.1	105.6
	2.300	2.06	4.02	12.3	73.8	4.4	90.5
	2.875	2.59	3.95	10.3	78.3	9.7	98.3
Lateritic Red Earth	0.575	0.51	4.55	43.6	7.5	45.6	96.7
	1.150	0.99	4.20	39.2	35.1	54.4	128.7
	1.725	1.55	4.16	30.9	31.5	32.2	94.6
	2.300	2.02	3.94	27.3	55.3	22.3	104.9
	2.875	2.50	3.83	25.9	60.2	27.2	113.3
Lateritic Red Earth+6.5%Fe_2O_3	1.150	1.05	4.40	43.8	3.5	22.9	70.2
	1.725	1.53	4.11	48.8	16.6	28.8	94.2
	2.300	2.07	4.03	46.0	17.7	30.0	93.7
	2.875	2.50	3.82	42.2	36.5	22.8	101.5
Latosol	0.575	0.53	4.77	24.8	2.4	63.9	91.1
	1.150	1.07	4.47	38.9	12.1	76.1	127.1
	1.725	1.58	4.23	35.8	26.6	67.8	130.2
	2.300	2.09	4.07	32.4	35.4	55.5	123.3
	2.875	2.62	3.99	29.2	42.9	47.4	119.5

[a]Zhang and Yu, 1997, with permission.
[b]A – increase in positive surface charge; B – increase in soluble aluminum; C – increase in exchangeable acidity.

changeable acidity, differ with the composition of the soil and the amount of hydrogen ions added. Table 8.3 shows some relevant data in this respect. For the Red Earth, the percentage in consumption by the release of soluble aluminum is the highest while that by the increase in positive surface charge is the lowest. This should be related to the relatively higher content of amorphous aluminum oxides, which can be easily attacked by acid, and the low content of iron oxides, which is responsible for most of the acceptance of protons. For the Latosol, the increased consumption of hydrogen ions due to increased exchangeable acidity plays an important role, whereas the increase in soluble aluminum is of less significance. This must be caused by the high content of iron oxides and the predominance of kaolinite and gibbsite, because these Al-containing minerals are relatively difficult to be attacked by acid. The properties of the Lateritic Red Earth are intermediate between the two above-mentioned soils. For this soil, as expected, the addition of iron oxides results in an increase of positive surface charge, as is shown in Table 8.3.

Another complicated factor is that sulfate ions from the acid rain can be adsorbed by these soils through a mechanism known as ligand exchange with coordinately adsorbed hydroxyl groups, resulting in the release of OH^- ions. These OH^- ions would neutralize some H^+ ions of the soil solution, thus lessening the effect on soil acidification.

To what extent this effect may be exerted under field conditions cannot be exactly ascertained. This should be determined by the kind of soil and the composition of the acid rain. Here two factors are involved. Not all the sulfate ions adsorbed will lead to the release of equivalent amounts of hydroxyl ions, because they can replace hydroxyl ions as well as coordinately adsorbed water molecules. In the former case, their adsorption can also result in the increase of negative surface charge and the decrease of positive surface charge of the soil (Zhang and Yu, 1997). Besides, in acid rain, sulfate ions are balanced not only by hydrogen ions, but also by base cations, as has been shown in Table 8.2. How-

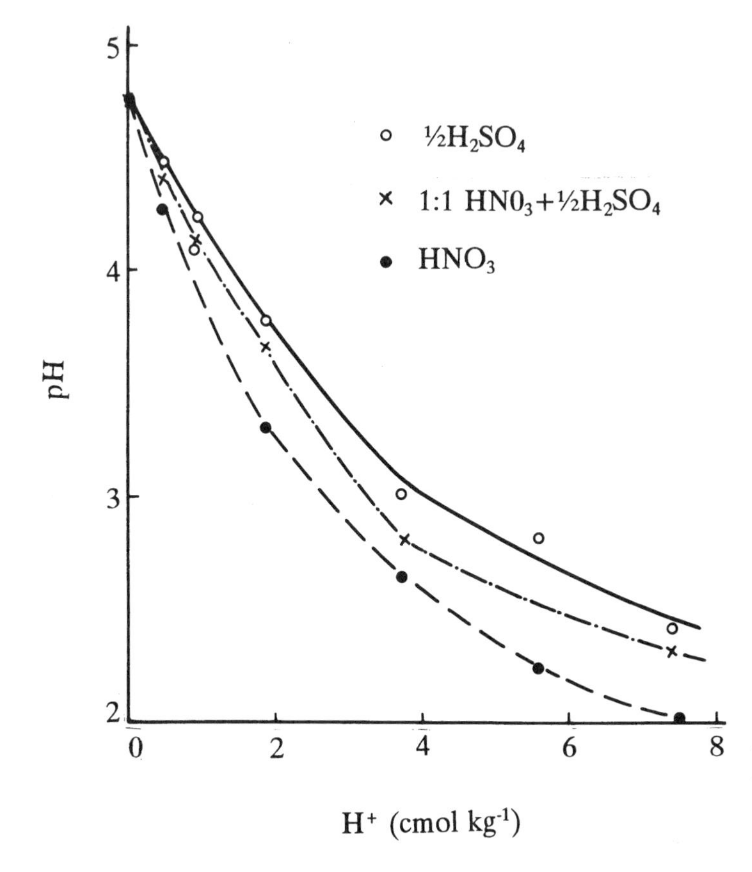

Figure 8.6. Decrease in pH of a Red Earth caused by H_2SO_4 and by HNO_3 (data of J.H. Wang).

ever, this concordant presence of sulfate-type acid rain and sulfate-adsorbing soils in southern China should be of practical significance in problems relating to soil acidification. A laboratory experiment showed that, for a Red Earth, about 1.6 $cmol_c$ kg^{-1} of HNO_3 was required to acidify the soil from its original pH 4.8 to pH 3.5, while the required amount of H_2SO_4 was about 2.6 $cmol_c$ kg^{-1}, i.e., with a factor of 1.6 (Fig. 8.6). It was estimated (Wang and Yu, 1996) that when acid rain chiefly containing H_2SO_4 was deposited on variable charge soils the acidification rate might be slower by 20%–40% than that when the acid rain chiefly contained HNO_3 for soils with a high organic matter content, and that the rate might be half of that caused by HNO_3 for soils with a low organic matter content, especially for Latosols.

Thus, acidification processes in soils of southern China present certain features different from those in the northeast of the country and northern Europe and North America, where both the consumption of hydrogen ions due to the increase in positive surface charge and the ligand exchange of sulfate ions with adsorbed hydroxyl groups are not of significance.

Consequences of Soil Acidification

Change in Soil Surface Properties

It has been mentioned that surface protonation may lead to the creation of positive surface charge of the soil. Actually, both the quantities of positive surface charge and nega-

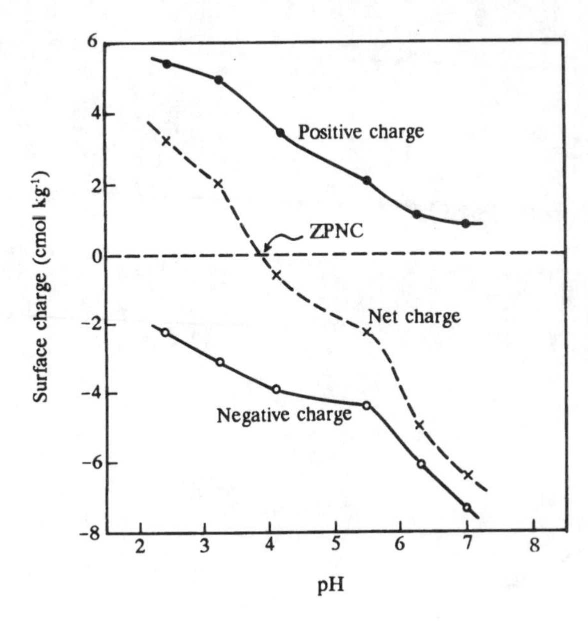

Figure 8.7. Effect of pH on surface charge of a Latosol (Zhang and Zhao, 1997, with permission).

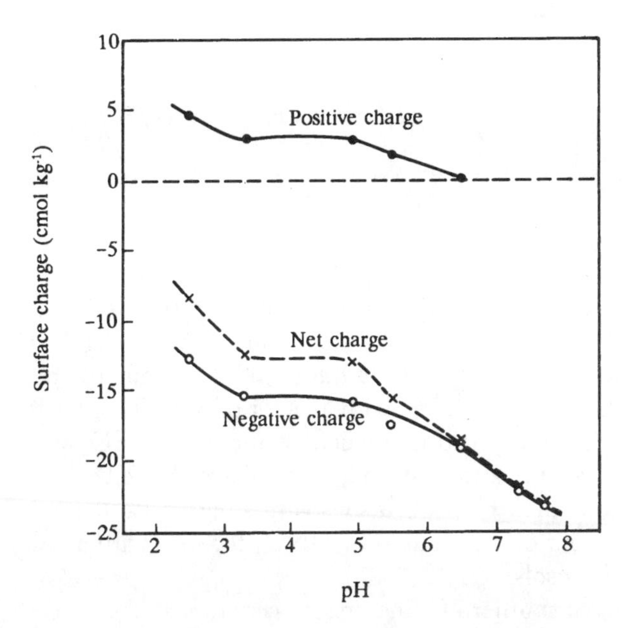

Figure 8.8. Effect of pH on surface charge of a Red Earth (Zhang and Zhao, 1997, with permission).

tive surface charge can be affected by hydrogen ions. Therefore, hydrogen ion concentration is the determining factor in affecting the surface charge of variable charge soils. Figures 8.7 and 8.8 show two examples. For the Latosol, the quantity of positive surface charge may exceed that of negative surface charge when the pH is sufficiently low. Hence, there appears a zero point of net charge (ZPNC) at a critical pH.

Because both quantities of negative surface charge and positive surface charge change with the change in hydrogen ion concentration, the adsorption of cations and anions through coulombic forces varies with pH. For cations (K^+) shown in Figure 8.9, the ad-

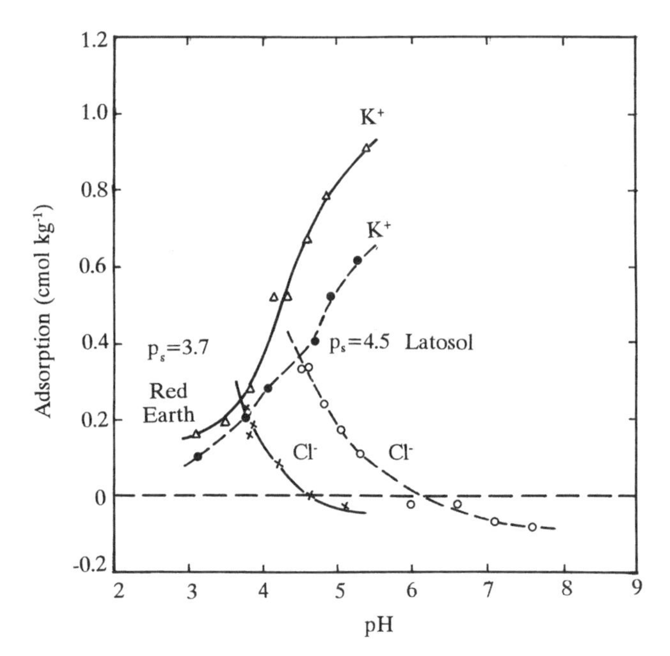

Figure 8.9. Potassium adsorption and chloride adsorption by two variable charge soils as affected by pH (0.001 mol L^{-1} KCl) (p_s is iso-ionic point) (Yu and Zhang, 1990, with permission).

sorbed amount decreases with the decrease in pH, while the opposite is true for anions (Cl^-).

Furthermore, other properties of the soil, such as specific adsorption of cations and anions, ion diffusion, electrokinetical properties, and oxidation-reduction reactions, etc., are all affected by pH (Yu, 1997). These complicated phenomena are, to a large part, related to the variability of surface charge of these soils.

Thus, soil acidification not only induces an increase in soil acidity per se, but also affects a series of other properties of variable charge soils through affecting surface charge. By chance, these soils are distributed in regions where the occurrence of acid rain is most common.

Enhanced Leaching Loss or Immobilization of Nutrients

Because of the replacement of exchangeable base cations by hydrogen and aluminum ions, these metal ions will be leached out from the soil solum through percolating water. Of course, the lower the pH of the rainwater, the more pronounced is this effect. If the rainwater is sufficiently acid, the minerals of the soil may also be attacked. Many experiments have shown such an effect. The extent of this effect is related to the type of the soil. In Table 8.4, some examples in this respect are given. Under field conditions, the intensified leaching loss of nutrient cations K, NH_4, Ca, and Mg may be one important reason for the gradual decline of soil fertility when affected by acid rain.

On the other hand, soil acidification may also cause the immobilization of phosphate. Figure 8.10 shows the fixation of added phosphate by a Red Earth as a function of soil pH. As can be seen, when the pH is lower than about 6, the percentage of fixation invariably increases with the decrease in pH, irrespective of the kind of accompanying cation species. Here, two mechanisms may be involved. The lower the pH, the stronger is the coulombic attracting force due to the increase in positive surface charge density of the soil. Besides, variable charge soils with a high content of free iron oxides possess a strong

Table 8.4. Effect of Simulated Acid Rain on Concentration of Cations in Percolating Water from Soils (average value of 10 collections)[a]

Soil	pH of Input Water	Concentration ($\mu mol\ L^{-1}$)			
		Al	Mn	Ca	K
Latosol	4.1	0	3	—	9.4
	2.6	0	4	—	9.8
	1.9	0	4	—	10.3
Lateritic Red Earth	4.1	4.7	21	213	22.9
	2.6	5.9	39	227	23.6
	1.9	14.6	213	679	46.3
Red Earth	4.1	4.3	308	373	12.1
	2.6	7.3	444	436	19.5
	1.9	23.4	886	724	19.1

[a]Unpublished data of J.H. Wang.

specific adsorption ability for phosphate, which is also pH-dependent. However, when the pH is higher than about 6, chemical precipitation due to the formation of calcium phosphate may occur, if the soil is saturated by exchangeable calcium. The enhanced immobilization of native and applied phosphate following soil acidification is of great practical significance, because these soils are generally highly deficient in phosphate. Analytical data based on 281 samples show that the amount of available phosphate is very low in soils with a pH of lower than 5.5 (Yu and Li, 1990).

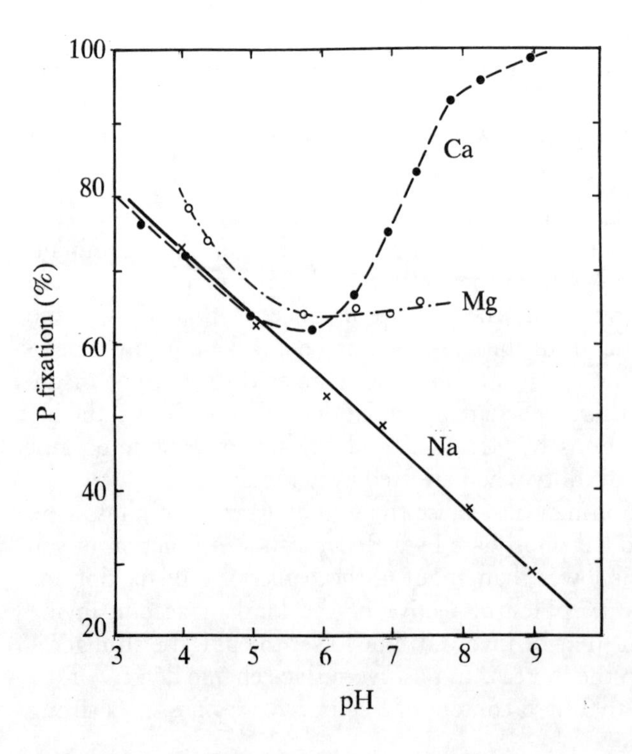

Figure 8.10. Fixation of phosphate by a Red Earth as a function of pH in the presence of different exchangeable cations [the soil samples were incubated with different amounts of NaOH, $MgCO_3$ or $Ca(OH)_2$, and then the P-fixation experiment was conducted] (Yu and Li, 1990, with permission).

Liberation of Toxic Elements

Table 8.4 also shows that in the percolating water there are Al and Mn ions, and that their concentrations increase when the pH of the simulated rainwater is decreased. Many studies conducted in China revealed that the quantity of exchangeable aluminum of soils is strongly affected by pH. As can be seen from Figure 8.11, for a large number of soils of southern China, most of which are variable charge soils, exchangeable aluminum increases markedly in quantity with the decrease in pH when the pH is lower than about 5.5.

For soils polluted by heavy metals, the mobility of these metals is also strongly affected by pH. Generally, within a pH range of as narrow as 1 pH unit, the critical value of which varies with the kind of the metal and to a less extent with the type of the soil, the adsorption of metal ions by soil increases sharply from a rather low percentage at low pH to nearly 100% at high pH. Figure 8.12 shows such a tendency for Pb, Zn, and Cd when added to two Latosols, one of China and another one of Brazil. Thus, the mobility of these metals is very sensitive to pH. It is for this reason that in many countries, such as in China, when evaluating the environmental quality with respect to heavy metals such as Hg, Cr, Cu, Pb, Zn, and Ni, soil pH is considered as one important criterion in addition to CEC. This is of particular significance in soils of southern China where the pH is already low and the effect of acid rain on further soil acidification is most likely.

Effect on Plant Growth

In soil science, it is generally considered that the toxicity of aluminum, the toxicity of manganese, and the insufficiency of calcium are three principal causes for the poor growth of plants on acid soils. Among these factors, the toxicity of aluminum may perhaps be the most important one. Many observations have definitely correlated the poor growth of plants with the content of exchangeable or water-soluble aluminum of the soil and have also confirmed the injury of plant roots by aluminum ions in solution. Evidence has also shown that among different forms of soluble aluminum only monomeric aluminum ions are responsible for the toxicity.

With regard to the effect of manganese on plant growth in acid soils, the data shown in Figure 8.13 provide some hints. Within the pH range of about 4.5 to 6, both the

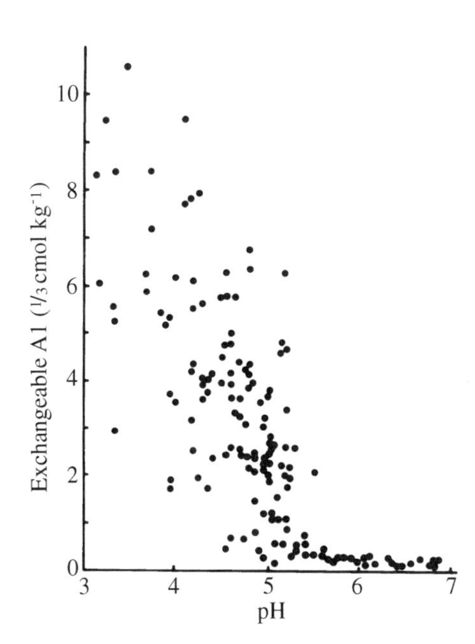

Figure 8.11. Relationship between exchangeable aluminum and pH in soils of southern China (Kong et al., 1997, with permission).

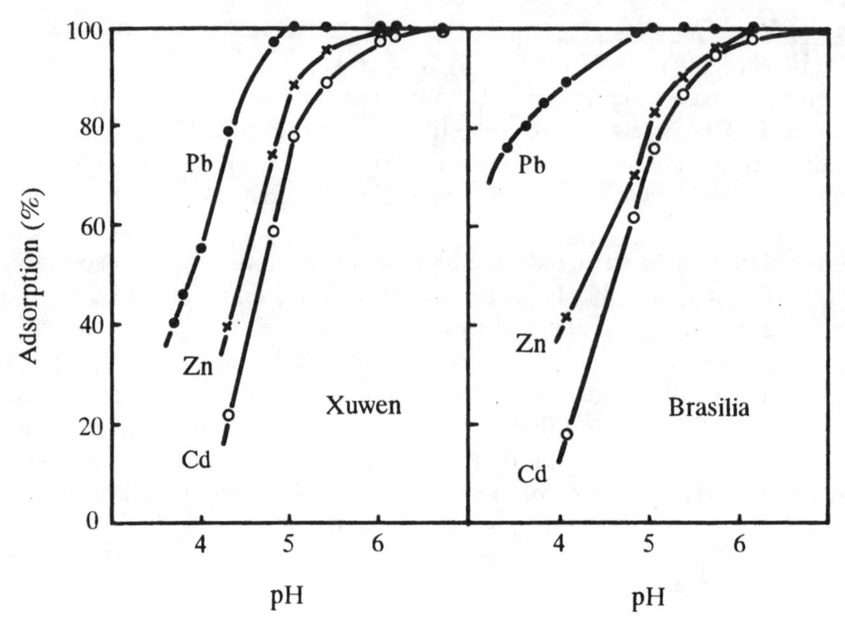

Figure 8.12. Adsorption of Pb, Zn, and Cd by two Latosols at different pH (data of G.Y. Zhang).

amount of exchangeable manganese of the soil and the manganese content of pea plant are linearly correlated with hydrogen ion concentration. Interestingly, within this pH range, the dry weight of pea also decreases with the decrease in soil pH. The coincidence of the three curves within this pH range is not accidental.

Under field conditions, relating the poor growth or damage to plants to a single factor of soil acidity is generally difficult, because a variety of soil factors are usually intermingled. This is also the case for the effect of acid rain on plant growth through soil acidification. Nevertheless, observations in some regions of southern and southwestern China do imply that the growth of some trees, notably masson pines, has been adversely affected by soil acidification in recent years. For example, in the Liuzhou region of Guangxi Province where acid rain pollution is quite severe, soil pH in areas where pine trees grow abnormally is lower by 0.3 to 0.5 unit than that in soils with healthy pines (Wang, 1988). At the southern suburb of Chongqing where masson pines have been severely damaged in the past years, the pH of the soil (Yellow Earth) is as low as 3.6 to 3.9. Meteorological data show that, in this region, the atmosphere has been heavily polluted by SO_2 emission.

Prediction of Soil Acidification

Sensitivity of Soil to Acid Input

Just as soil acidity can be expressed in terms of both intensity (pH) and capacity (the quantity of exchangeable acidity), the sensitivity of soils to acid input can also be evaluated with two parameters, namely, the change in pH after the input of a certain amount of acid, and the quantity of acid required to decrease the pH to a given value.

For the latter parameter, van Breemen et al. (1983), by applying the principles of water

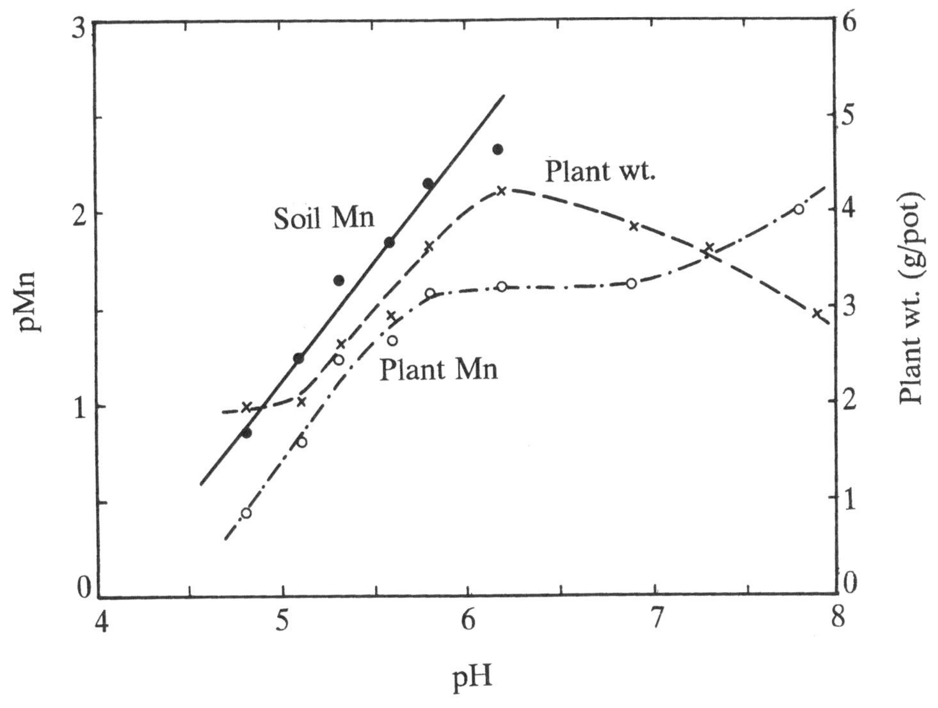

Figure 8.13. Effect of soil pH on soil exchangeable Mn, plant Mn, and plant (pea) weight grown on a Red Earth (Yu and Li, 1990, with permission).

chemistry to soils, introduced the concept "acid-neutralizing capacity." For the calculation of this quantity in a soil, it is necessary to make a total analysis of all of the elements. Then, based on a specified "reference pH," the sum of the equivalent of all the acid-forming elements is subtracted from that of all the alkali-forming elements. If pH 5 is chosen as the reference pH, it is necessary to consider calcium, magnesium, potassium, sodium, manganese, sulfur, and phosphorus. If a reference pH of 3 is selected, aluminum must also be considered. Apparently, in doing so the operation would be very tedious.

For practical purposes, two parameters may be introduced.

The change in pH after the introduction of 20 $\mu mol_c\ kg^{-1}$ of sulfuric acid may be termed the sensitive value ΔpH_{sv}. This quantity of acid is equivalent to the annual input of 1000 mm of acid rain with a pH of 4.0 for 45 years. This value is not only determined by the kind of the soil, but is also affected by the base-saturation status for a given soil. As can be seen from Figure 8.3, three base-saturation ranges can be distinguished. When the base-saturation (BS) percentage is less than about 30% or within the range of about 70–100%, a small change in BS may induce a large change in pH. Theoretically and in practice, the change in pH at BS 50% is the smallest. What is of interest here is that for most soils likely suffering from the effect of acid rain in southern and southwestern China, the base-saturation percentage generally lies lower than 30%. Hence, these soils are in their sensitive state as far as their sensitivity to acid rain is concerned.

The quantity of acid required to change soil pH to 3.5 may be termed the acid-tolerant limit of the soil. In principle, this is the maximum of acid a soil can tolerate beyond which soil acidity is toxic to nearly all plant species. The selection of pH 3.5 is based on the consideration that this pH is the lowest known value for natural soils of China, except for some acid-sulfate soils along the seashores. Actually, for most plants, the growth will be adversely affected if grown on soils with a pH much higher than 3.5.

For different kinds of soils with the same base-saturation percentage, the numerical value of acid-tolerant limit is chiefly determined by the quantity of negative surface charge. Therefore, CEC is commonly used as one of the important indices for evaluating the sensitivity of soils to the effect of acidification.

Gradation of Soil Sensitivity

Based on a large number of analytical data obtained for soils of southern China, the sensitivities are classified into four grades according to both ΔpH_{sv} and acid-tolerant limit (Table 8.5). These two parameters, one in terms of intensity and the other in terms of capacity, although of different meaning and different significance, are interrelated. Generally, soils with a high acid-tolerant limit are less sensitive to the effect of acid on ΔpH_{sv}. When the four grades of sensitivity are compared, it would be logical to conclude that soils of grades I and II, especially of the very sensitive grade, should be seriously considered when adopting practical measures for avoiding or diminishing environmental problems related to acid rain. As can be seen from the pH–H^+ input curves for the four soil examples shown in Figure 8.14, each representing one sensitivity grade, the acid-tolerant limits are about 5, 12, 33, and 83 mmol kg^{-1} of hydrogen ions, respectively, differing by more than one order of magnitude between grade I and grade IV.

Regionalization of Soil Sensitivity

A generalized map of sensitivity to acid rain for soils of China, based on the acid-tolerant limit, is shown in Figure 8.15.

Grade I includes Lateritic Red Earths and Red Earths derived from sandstone, granite, and Quaternary red clay. They generally have a relatively light texture. The clay mineralogy is predominated by kaolinite and is accompanied by a small amount of hydrous mica and vermiculite. The CEC range is 5–12 $cmol_c$ kg^{-1}. The soil contains 2–6 $cmol_c$ kg^{-1} of exchangeable aluminum, and has a pH of 4–5 and a lime potential (pH–0.5 pCa) of lower than 2.5.

Grade II includes Lateritic Red Earths, Red Earths with granite, sandstone, and schist as the parent materials and some paddy soils. Besides the predominance of kaolinite, the contents of hydrous mica and vermiculite are higher than those for soils of grade I. The CEC range is 7–15 $cmol_c$ kg^{-1} and the content of exchangeable aluminum is 2–5 $cmol_c$ kg^{-1}.

Table 8.5. Gradation of Sensitivity of Soils to Acid Rain According to Sensitive Value ΔpH_{sv} and Acid-Tolerant Limit for Soils of Southern China[a]

Grade	Sensitivity	ΔpH_{sv}	Acid-Tolerant Limit (mmol kg^{-1})
I	Very sensitive	>1.2	<5
II	Sensitive	0.8–1.2	5–20
III	Slightly sensitive	0.5–0.8	20–50
IV	Insensitive	<0.5	>50

[a]Wang et al., 1994, with permission.

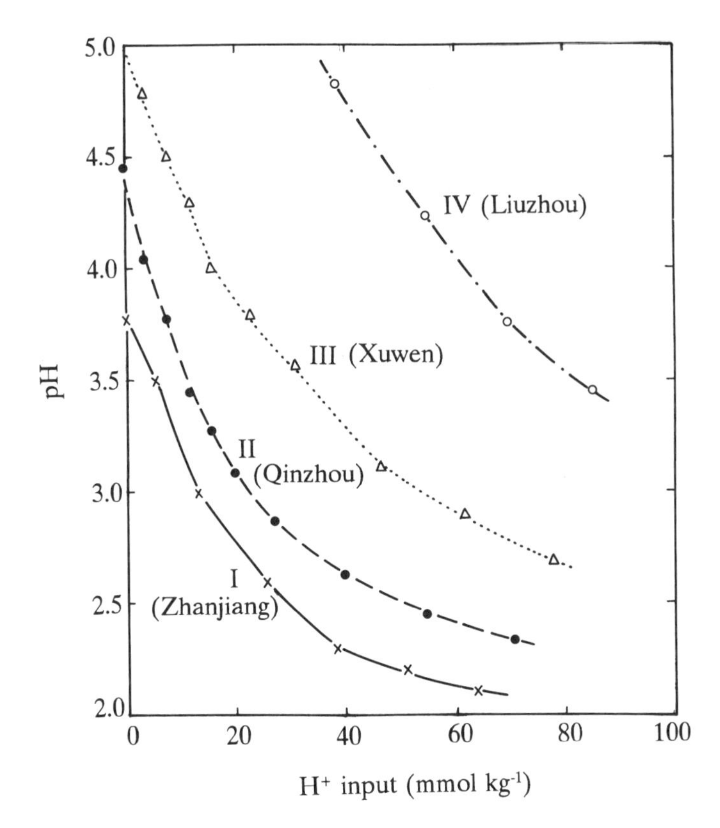

Figure 8.14. pH–acid input curve of soils with different acid-tolerant limits (Wang et al., 1994, with permission).

Grade III includes Latosols, Lateritic Red Earths, and some alluvial soils. The clay content of these soils may be as high as 70%. The CEC is generally within the range of 9–18 $cmol_c\ kg^{-1}$. Soil pH and lime potential lie at 5–6 and about 3.5, respectively. Among these soils, two types are representative. Latosols derived from basalt and distributed on Leizhou Peninsula contain kaolinite, gibbsite, and iron oxides as the dominant clay minerals. Paddy soils in the Pearl River delta generally have a slightly acid reaction.

Soils in grade IV are closely related to their parent material in properties. Most of them are developed from limestone or nonacidic sandstone-shale. The pH and lime potential are 6.5 to 7 and 5.5 to 7, respectively. In the clay mineralogical composition, both vermiculite and kaolinite occupy a large part. Therefore, the CEC is generally 16–25 $cmol_c\ kg^{-1}$.

An inspection of Figure 8.15 reveals that soils of grade I and grade II occupy most of the areas in southern China. This is the basic reason why in this region soil acidification is truly a problem.

ACIDIFICATION OF GROUNDWATER

Effect of Soil Properties on Composition of Groundwater

Except in the plain areas with a high water table, groundwater comes chiefly from the percolation of rain through soils (Li et al., 1990). This is particularly true for most of the acid soils likely suffering from the effect of acid rain in China where the annual precipi-

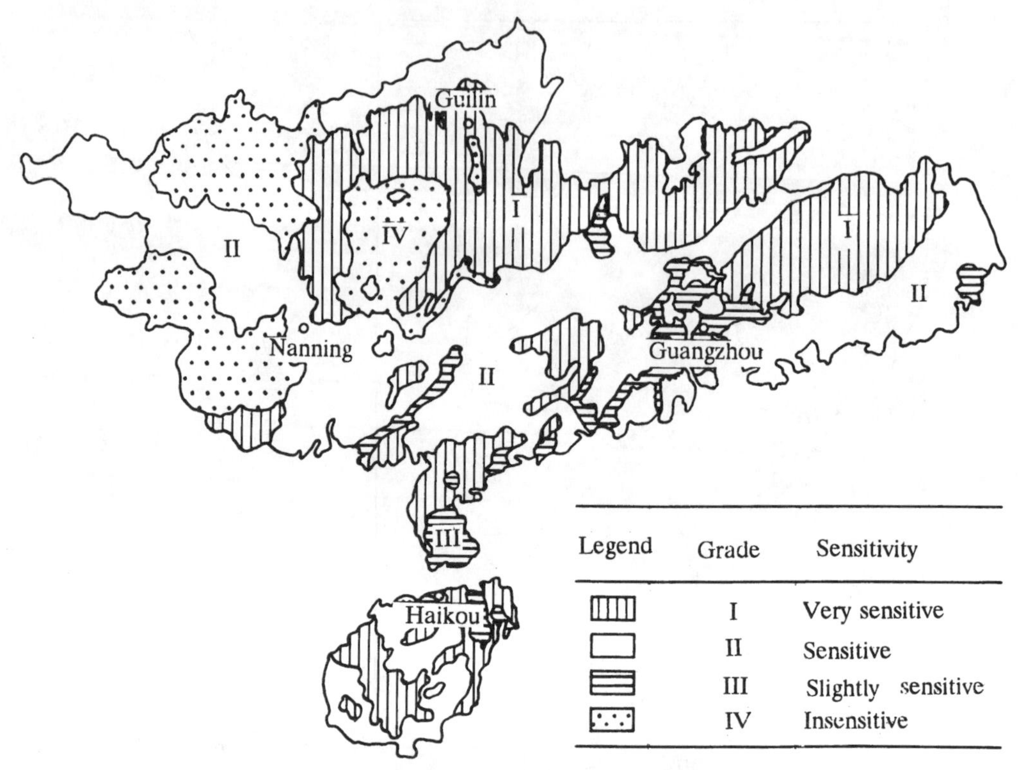

Figure 8.15. Generalized map of sensitivity to acid rain for soils of southern China based on acid-tolerant limit (Wang et al., 1994, with permission).

tation is generally over 1000 mm. Therefore, the composition of groundwater should be closely related to that of the percolating water of soils.

The concentrations of some elements in percolating water from representative soils is given in Table 8.6. Two features may be noticed. The concentrations of the major nutrient elements calcium, potassium, and phosphorus in percolate from the paddy soil are especially high. This should be caused by fertilization in the cultivated layer. In the percolate there are certain amounts of iron and aluminum. This can occur only in acid soils. This point is of significance with regard to the quality of groundwater, because water-soluble aluminum is a toxic element to organisms, and soil acidification should intensify the leaching of aluminum to groundwater, which in turn may influence river water and lake water.

Influence of Acid Rain

Just like the effect of acid rain on soil acidification, generally it is difficult to find direct evidence of the influence of acid rain on groundwater within a short time, because it is a long-term effect that needs years or even decades before any significant damage is evident. This is true in the majority of areas of southern China where industrialization is still in an initial stage. However, acid input into soils can induce the increase in soil acidity, thereby enhancing the leaching loss of nutrients, and particularly the liberation of toxic elements. Such changes in soil properties would influence the composition of percolating water, which in turn would affect the quality of groundwater. A simulation for a watershed in Guangxi Province (Xiong et al., 1993) showed that under the present

PROBLEMS
Kong, X.L., X.N. Zh
Yu, T.R., Ed., C
Li, H.Z. Regionalizat
Environmental
Li, Y.S., J.Q. Xia, and
Press, Beijing, 1
Mo, T.L. and K.L. Xi
viron. Sci., pp. 5
Qi, L.W., Y.P. Pang,
Trend of Acid R
lution, Nationa
648–653.
Tang, X.Y., M.R. Wa
in Spring in Gu
ences of China,
van Breemen, N., J. M
Soil, 75, pp. 283
Wang, J.H. Soil Acidi
In: *Acid Rain a*
91–96.
Wang, J.H. and T.R. Y
ification of Vari
Wang, J.H., X.N. Zh
Pedol. Sinica, 31,
Wang, M.L. and H.Z
and Hofei. *Env*
Wang, P., L.J. Jiang, Z
in Rainwater in
Wang, R.B., F.S. Wei
position and Ch
on Acid Rain, S
Beijing,. 1989, p
Wang, W.P. Prelimina
lut. Protec., 2, pp
Wei, F.S., M.X. Wang
istics in Distrib
lections of Papers
Science Press, B
Xiong, F., X.P. Zhou,
tion Study at La
tion, National E
Xiong, J.L., Z.H. Fan
mation in Guizh
Protection Bure
Yu, T.R. *Physical Chem*
Yu, T.R. *Chemistry of*
Yu, T.R. and C.K. Li.
jing, 1990, pp. 5
Yu, T.R. and X.N. Zh
ence, Science Pr
Zhang, F.S. and T.R.
ford University
Zhang, G.H. Approa
3, pp. 25–29, 19

Table 8.6. Concentration of Some Elements in Percolating Water from Soils of Southern China

	Concentration (μmol L^{-1})							
Soil	Ca	Mg	K	Na	Fe	Al	Si	P
Latosol (Xuwen)	180	226	90	390	3.2	7.0	246	1.0
Latosol (Zhanjiang)	290	794	41	300	4.5	7.4	379	0.7
Lateritic Red Earth	85	70	51	178	4.5	7.8	239	0.3
Red Earth	113	49	15	222	tr.	tr.	270	0.8
Paddy soil	1360	786	79	165	5.0	11.0	207	14.8
Irrigation water	285	54	38	26	—	—	25	0.1

[a]Gong and Xu, 1990; Zhao et al., 1990, with permission.

acid load, soils are in the process of being significant acidified, and that the acidification rate of surface water will be significantly accelerated, if the acid deposition load increases continuously.

STRATEGIES FOR CONTROLLING THE ACIDIFICATION OF THE SOIL ENVIRONMENT

An increasing awareness of the realities and potential effects of acid rain on the environment lead scientists and the government to adopt the following strategies:

Limitation on Emission of Acid-Forming Elements to the Atmosphere

This is the fundamental measure for controlling acid rain. Regulations have been adopted that stipulate the necessity of limiting the emission of SO_2 to the atmosphere from factories. A variety of methods for the removal of sulfur from coal have been developed. Besides, attempts are being made to change the energy source nationwide. A very large hydropower station in the middle of the Yangtze River is currently under construction. The government also plans to build more nuclear energy stations.

Remediation Through Soil Management Measures

For agricultural lands, a variety of management measures can control or remedy soil acidification. For a long time, farmers have applied ashes from straw, grasses, and wood to the land. The application of lime or ground limestone is quite common in local areas where liming materials are available. A phosphate fertilizer named fused calcium-magnesium phosphate, manufactured by fusing rock phosphate with serpentine, is quite effective in neutralizing soil acidity. And, as mentioned before, the cultivation of paddy rice can decrease soil acidity temporarily.

Selection of Acid-Tolerant Plants

The sensitivity of crops to soil acidity differs greatly. For example, sweet potato and peanut are much more tolerant to soil acidity than cotton and barley. Among economically important trees grown in southern China, tea and rubber are rather acid-tolerant. The former is even acidophilic. Because animal husbandry is not well-developed in southern China, forage grasses are not common. On the other hand, the cultivation of

Zhang, G.Y. and T.R. Yu. Coordination Adsorption of Anions, in *Chemistry of Variable Charge Soils,* Yu, T.R., Ed., Oxford University Press, New York, 1997, pp. 171–215.

Zhang, X.N. and A.Z. Zhao. Surface Charge, in *Chemistry of Variable Charge Soils,* Yu, T.R., Ed., Oxford University Press, New York, 1997, pp. 14–60.

Zhang, Y.Y. and D.H. Kuo. Preliminary Study on Acidity of Acid Rain in Wuhan Region. *Environ. Sci.*, 4, pp. 57–59, 1983.

Zhao, D.W. and J.L. Xiong. Acidification in Southwestern China, in *Acidification in Tropical Countries*, Rodhe, H. and Harrema, R., John Wiley, New York, 1988, pp. 317–346.

Zhao, Q.G., Z.T. Gong, C.Q. Hou, and G.C. Zhou. Tropical Soils, in *Soils of China*, Institute of Soil Science, Science Press, Beijing, 1990, pp. 43–73.

green manures is
such as lespedeza
applied to farmla

SUMMARY AN

The potential en
China are discuss
from the combus
uted mostly in th
to undergo a seri
results are an inc
trients, and the li
ification of soils
acid in reaction.

A combinatio
rence of variable
rain-induced env
ate regions in no
in acid rain and s

Most of the s
small buffering
droxyl ions durin
by factors other t
tion effect by aci

Because indus
veloping stage, e
not serious on a
does exist, if con

ACKNOWLED

We are grateful t
search was suppo
Science and Tec

REFERENCES

Cao, H.F., J.M. S
Forest Produ
Protection o
Press, Beijin
Chen, Z.H. Trend
in *Studies on*
Science Pres
Ding, G.A. and
Chinese Scie
Dou, J.Y. and Y.J.
12(3), pp. 7
Gong, Z.T. and Q
jing, 1990, p

CHAPTER 9

DYNAMICS, FATE, AND TOXICITY OF PESTICIDES IN SOIL AND GROUNDWATER AND REMEDIATION STRATEGIES IN MAINLAND CHINA

Dao Ji Cai and Zhong Lin Zhu

INTRODUCTION

In modern agriculture, pesticide plays a significant role in preventing diseases of insects, pests, and weeds, in enabling stable and high production yields of crops, and in raising the quality of agricultural products. Although alternative methods for integrated management practice (IMP) are increasing, and the pesticide use rate will be decreased with the development of science and technology in the future, pesticide at present is still an important component of IMP. The rate of pesticide production and usage per year is over 2,000,000 tonnes a.i. (active ingredients) in the world, and is increasing at a rate of 5% per year. It is estimated that the rate of pesticide production and usage will reach 3,000,000 tonnes by the year 2000.

China is a large agricultural country, with about 100 million hectares of cultivated land, a high cropping index, and a very large and extensive pesticide use rate. The total number of the registrated pesticide formulations was up to 1,786 in 1993. The pesticide use rate is about 200,000 tonnes a year in China. The large and extensive use of pesticides in one way has brought enormous economic benefits. It is shown that pesticide use can prevent about 7% of grain and 18% of cotton loss, which, are equivalent to about 30 million tons and 500,000 tonnes, respectively, according to a typical survey conducted in China. On the other hand, it has caused environmental pollution and ecological damage. It is estimated that the economic loss was over $1 billion US in 1983 due to the residue of organochlorine pesticides in grains exceeding the national pesticide residue standards.

With progress in modernization, Chinese people have increased greatly their awareness of the environment. The viewpoint of integrated planning, simultaneously carrying out and coordinately developing the economy with emphasis on environmental protection, has been accepted by more and more people. Environmental protection has become one of the fundamental national policies in China. Although pesticides are necessary in agricultural production, they can also be toxic chemicals, and more and more people have become highly concerned by their effect on the ecosystem. Much money has been put into research work related to this area in China, and thousands of pages of literature related to the effects of pesticides on the ecosystem were published in chemistry, pesticide, environmental science, toxicology, biology, ecology, plant protection, weed science, and many other special journals in the past decade. This chapter is neither a collection nor a compendium of research reports, but a summarization on the dynamics, fate, and toxicity of pesticides in soil and groundwater and remediation strategies in China.

HISTORY OF PESTICIDE PRODUCTION AND USAGE IN CHINA

History of Pesticide Production

Although the pesticide industry in China started late, it developed very quickly. The annual production of pesticides was only 500 tonnes in 1950, but reached 537,200 tonnes in 1980, the highest yearly record of pesticide production in Chinese history. High residual pesticides such as the organochlorine pesticides BHC and DDT were originally the main kinds used and amounted to more than 50% of the total amount of pesticide production. Since BHC and DDT were banned in 1983, the actual annual amount of pesticide production in China had decreased sharply and was reduced to 210,400 tonnes in 1990. However, the areas of cultivated land where problems are solved by pesticide use have increased. This is attributed to the large numbers of substitute kinds of pesticides with high efficiency and low residue. The annual amount of pesticide production in China is shown in Table 9.1.

The law that pesticides should be registered before they are allowed to be widely used came into effect in China in 1982. Since then, the total registration number of pesticides was up to 1786 in 1993, and the total amount of pesticide production was up to 223,000 tonnes in the same year, which was ranked the third in the world.

The development of the Chinese pesticide industry is shown not only in the increase in the amount of pesticide production, but also in the increase of the types of pesticides. There were a few kinds of inorganic insecticides such as lead arsenate, sodium fluoride, etc., which could be produced in China before 1950. However, with the rapid development of synthesized organic pesticides during the 1950s to the 1980s, this type of pesticide developed quickly so that there were 70 kinds in 1980. Nevertheless, the main kind

Table 9.1. Annual Pesticide Production in China (unit: 10,000 tonne, a.i.)[a,b]

Year	Amount	Year	Amount	Year	Amount	Year	Amount
1950	0.05	1961	9.20	1972	40.20	1983	33.10
1951	0.07	1962	8.80	1973	45.60	1984	29.85
1952	0.19	1963	10.80	1974	37.10	1985	20.43
1953	0.50	1964	12.90	1975	42.20	1986	20.30
1954	1.03	1965	19.30	1976	39.10	1987	19.10
1955	2.60	1966	26.20	1977	45.77	1988	21.30
1956	5.50	1967	22.40	1978	53.30	1989	24.50
1957	6.50	1968	17.10	1979	53.67	1990	21.04
1958	8.60	1969	26.20	1980	53.72	1991	21.46
1959	13.70	1970	32.10	1981	48.43	1992	21.89
1960	16.20	1971	38.70	1982	45.69	1993	22.30

[a]a.i. - active ingredient.
[b]Reprinted from Cai and Tang, 1992.

remained as organochlorine pesticides, such as BHC and DDT, which amounted to approximately 4,900,000 and 400,000 tonnes, respectively, before they were banned. Today China can produce more than 140 kinds of pesticides.

In China, the percentage of insecticide production has always been high when compared to entire pesticide production (Table 9.2). Before the 1950s, China could produce only some inorganic insecticides and fungicides, each about 50% of the total produced. During the 1960s to the 1980s, the production of organochlorine pesticides (e.g., BHC and DDT) and organophosphorus pesticides (e.g., trichlorphon, dichlorovos, dimethoate, parathion) sharply increased, and the proportion of insecticides produced reached 96.7% in 1970. After the 1980s, with the banning of BHC and DDT, the amount of insecticides produced decreased greatly, and the organophosphorus insecticides had substituted for the organochlorine insecticides to become the main kinds used (Table 9.3). Since then, a large number of highly effective insecticides, such as carbamate, aldicarb, carbofuran, carbaryl, pyrethriod, tetramethrin, permethrin, and deltamethrin, and biological pesticides such as BT have been developed and put into production. Up to 1993, the total amount of insecticides production was about 150,000 tonnes, which was 68% of the total pesticides produced in China.

Fungicides, such as copper sulfate, sulfur, ethylmercury chloride, and asozine, were widely used in China during the 1950s and 1960s. Because they have long residual periods in the environment, the use of mercurial and arsenical pesticides were banned or restricted in the 1970s in China. Since then, the synthetic organic fungicides, such as carbendazim, methyl-thiophanate and kitazine were quickly developed. Up to 1993, there were about 20 such kinds, and the amount of fungicide production was about 40,000 tonnes, over 18% of the total pesticides produced. The history of production and application of herbicides in China is very short. There were only about 10 kinds of herbicides before 1978, and the application region was concentrated in parts of northeastern China.

Table 9.2. Percent of Insecticides, Fungicides and Herbicides of the Total Amount of Pesticide Production in China (%)[a]

Year	Insecticides	Fungicides	Herbicides
1950	50.0	50.0	—[b]
1960	89.0	11.0	—
1970	96.7	2.8	0.5
1980	93.2	5.7	0.8
1990	78.5	14.0	7.1
1993	68.0	18.7	12.5

[a]Reprinted from Cai and Tang, 1992.
[b]Very few.

Table 9.3. Amount of Different Kinds of Insecticides in China (unit: tonne, a.i.)[a]

Year	Inorganic	Organo-chlorine	Organo-phosphous	Carbamate	Pyrethroid	Others	Total
1950	213	—	—	—	—	—	213
1960	1568	76410	2389	—	—	—	80367
1970	—	190664	26780	—	—	—	217444
1980	—	258041	73993	1978	—	3073	337085
1990	—	—	114687	4303	1400	25000	145090

[a]Reprinted from Cai and Tang, 1992.

Table 9.4. Pesticide Use Level in Different Provinces of China (1989, unit: a.i. g/ha)[a]

Province	Use Level	Province	Use Level	Province	Use Level
Anhui	910	Hunan	2100	Shandong	2070
Fujian	3550	Jilin	825	Shanghai	4500
Gansu	540	Jiangsu	2670	Shanxi	480
Guangdong	4190	Jiangxi	780	Sichuan	690
Guangxi	2400	Liaoning	2145	Tianjin	1950
Guizhou	780	Neimeng	300	Tibet	1200
Hebei	530	Ningxia	730	Xinjiang	600
Heilongjiang	660	Beijing	1860	Yunnan	810
Henan	960	Qinghai	670	Zhejiang	2450
Hubei	1500	Shanxi	465		

[a]Reprinted from Hua and Shan, 1996.

Since 1980, the kinds and the production amount of herbicides have developed very quickly. The development rate of herbicides was the quickest in the Chinese pesticide industry. Today, there are more than 40 kinds, and the production amount was up to 28,000 tonnes in 1993, about 12% of the total pesticides produced. The application districts have extended to all areas of China.

Pesticide Use Levels and Regional Differences

China is a country with a vast territory and consequent large differences in climatic conditions, agricultural cropping systems and economic development levels among regions. The use of pesticides in different regions therefore is dissimilar. In southeast regions, where the population is dense and the agricultural cropping index is also high, the pesticide use level in these regions is naturally larger than those in other regions of China. Among these districts, Shanghai is the largest in pesticide use level (Table 9.4), followed by Guangdong, Fujian, Jiangsu, Zhejiang, and Guangxi. Hunan, Shandong, Beijing, Tianjin, Hubei, Liaoning, and some regions that produce cotton, fruits, and other economic crops also have a high rate of pesticide application. The pesticide use level of these regions is between 1.5 to 2.2 kg/ha, slightly higher than the average value of the whole country, which is 1.5 kg/ha. The region where pesticide use level is the lowest is Neimeng, only 0.3 kg/ha, which is one-fifth of the average value of the whole country.

Environmental problems caused by the application of pesticides are closely related to the kind, application method, application rate of pesticides, environmental conditions, etc. The main pesticides-induced environmental problems in China are insecticides that have high toxicity to rats, silkworms, honeybees, or aquatic organisms, and herbicides that have high selectivity or super-high efficiency. These problems mainly occur in southeastern China and in regions where the pesticide application rates are high.

DYNAMICS, FATE, AND TOXICITY OF PESTICIDES IN SOIL

Factors Affecting Pesticide Residues in Soil

The period during which a pesticide remains persistent in the soil has great practical importance in determining how long it has residual biological activities and how well it performs its intended tasks. In general, the longer the pesticide persists in soil, the longer the pesticide effectively controls insect pests and weeds, the greater are the potential hazards on the ecosystem. For example, some organochlorine pesticides, such as BHC and

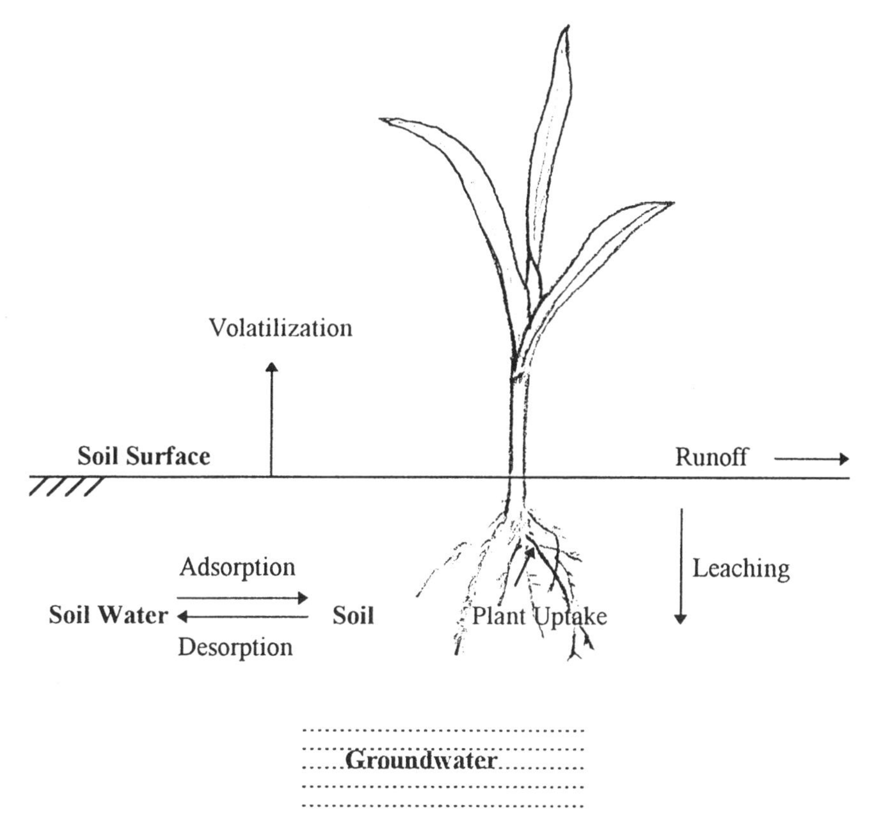

Figure 9.1. Schematic of the behavior of pesticides in soil.

DDT, may persist in the soil for a long time, from several years to more than 10 years. They could still be detected in the soil and in many other environmental media although they had already been banned in China in 1983 (Cai and Tang, 1992). Persistence of a pesticide beyond the critical period for control may lead to residue problems. These are readily apparent in the injury of sensitive crops by carryover of herbicides from preceding crop rotations (Cai, 1989). Jiang et al. (1992) showed that chlortoluron, a herbicide, can effectively control weeds and other broad-leaved weeds of cereal crops, especially *Alopecurus myosuriodes* in wheat fields, as it still persisted in the soil after the crop was harvested. Consequently, its residual bioactivity always posed hazards to the successive crops. Ideally, the pesticide should retain biological activity during the growing season of a crop, then decompose to innocuous products before the successive crops are planted, so that the pesticide may not only effectively play its role in pest and weed control, but also have no undesirable impacts on crops and the environment. The behavior of a pesticide in soil may be illustrated as Figure 9.1. Pesticide residue in soil is influenced by many factors, such as physiochemical properties of itself, soil factors, and agricultural and environmental factors. Table 9.5 gives a detailed description of the factors that theoretically influence the residue of pesticide in soil.

Degradation of Pesticides in Soil

Soil is not only a storehouse, but also a collecting and distributing center for pesticide in the environment. A major fraction of any pesticide, no matter how applied, eventually

Table 9.5. Factors That Theoretically Influence the Pesticide Residues in the Soil[a]

Properties of pesticide:	Water solubility, vapor pressure, pKa, pKb, stability, polarity, ionizability
Soil factors:	Soil texture and structure, content of organic matter, salinity, moisture content, porosity, temperature, pH, cation exchange capacity (CEC), permeability, kind and content of heavy metal ion, kind and population of microorganism, hydraulic conductivity
Agricultural factors:	Cropping pattern, cropping practices, crop type, pesticide formulation, application method, time and rate, frequence and times, irrigation time and volume
Environmental factors:	Rainfall, air temperature, evapotranspiration, illumination intensity and time, wind

[a]Reprinted from Zhu, 1994.

finds its way to the soil. Many pesticides are even applied directly to or incorporated into the soil at time of application. Many complex interactions will take place between soil and the entered pesticide. Among these, degradation of pesticide in soil is the most important factor, which not only influences its effectiveness and persistence in soil, but is also one decisive factor in evaluating its potential hazards on the ecosystem. The degradation of a pesticide in natural soil includes two parts, biological degradation and nonbiological degradation. And the latter includes chemical degradation and photodecomposition.

The biological degradation of a pesticide in soil mainly refers to microbial degradation. The dissipation rate of most pesticides in soil would be obviously lower in sterilized soil than that in nonsterilized soil and microbial degradation critically influences the fate, residue, and behavior of pesticide in soil (Zheng, 1990). Since the study of microbial degradation of pesticides began in 1940, the important role of microbial in the dissipation of pesticides in the environment has been confirmed. A large number of environmental problems caused by extensive application of pesticides during the middle 1960s to 1970s, and the practical perspectives of microbial degradation of pesticides make research on this area active. It had been discovered that many kinds of microorganisms have the ability to degrade pesticides in varying degrees (Zheng, 1990; Han and Li, 1991; South China Agricultural University, 1990; Zhang and Cai, 1990). Many kinds of microorganisms that have the ability to degrade pesticides had been obtained by separation. Table 9.6 lists parts of these microorganisms and pesticides which can degrade.

Chemical degradation includes hydrolysis, oxidation-reduction reaction, etc. Hydrolysis reactions of pesticides, in some cases, such as the chloro-s-triazine herbicides and the organophosphate insecticides, occur more rapidly in the soil than in aqueous solutions due to catalysis of reactions by the soil sorption. In other cases, they occur more slowly, such as in the case of carbofuran (Huang, 1987). Oxidation or reduction reactions may be important for particular pesticides. For example, oxidation is important to aldicarb, disulfoton, and phorate, whereas DDT, DNOC, guintozene, etc., depend on reduction reactions (Han and Li, 1991).

Chemical degradation of pesticides in soil is influenced by many factors. Han and Li (1991) showed that the degradation rate of pesticide may be 2–8 times higher at 35°C than at 25°C. Yang et al. (1987) showed that the degradation rates of carbofuran, parathion-methyl, and lindane were obviously higher in a paddy field than in a dry field. The influence of soil pH is very different on different kinds of pesticides. For example, its influence on aldicarb is very high, whereas its influence on PCP is very low.

In addition to microbial and chemical degradation, photodecomposition plays a role in the dissipation of pesticides in soil (Han and Li, 1991). There are many reports on

Table 9.6. Microorganisms and Degradability of Pesticides[a]

Microorganism	Pesticides That Can Be Degraded by the Microorganism
Achromobacter	DDT, Carbaryl, 2,4-D, MCPA
Agrobacterium	DDT, Dalapon
Arthrobacter	DDT, Malathion, Diazinon, 2,4-D, MCPA, Simazine, Propanil
Bacillus	DDT, EPN, Parathion, Methyl- Parathion, Fenitrothion, Toxaphene, Dalapon, Linuron, Monuron, Lindane
Corynebacterium	DDT, 2,4-D, MCPA, Dalapon, Dinoseb, Paraquat, Diquat
Flavobacterium	Parathion, Methyl- Parathion, Malathion, Diazinon, 2,4-D, MCPA, Dalapon, Chlorpyrifos
Pseudomonas	DDT, Toxaphene, Malathion, Parathion, Dichlorovos, PCP, Diazinon, Phorate, Carbaryl, 2,4-D, MCPA, Dalapon, Dinoseb, Monuron, Simazine, Paraquat, Lindane
Xanthomonus	DDT, Parathion, Fenitrothion, Monuron
Aerobacter	DDT, Methoxychlor, Lindane, Toxaphene
Esherichia	DDT, Lindane, Prometryne, Amitrole
Streptococcus	DDT, Diazinon, Simazine, Dalapon
Nocardia	DDT, 2,4-D, 2,4-DB, Dalapon, Maleic hydrazide
Streptomyces	Diazinon, Dalapon, Simazine
Aspergillus	DDT, Trichlorphon, Linuron, Carbaryl, MCPA, Atrazine, 2,4-D, Dalapon, Monuron, Simazine, Simetryne, Prometryne, Trifluralin
Cephalosporium	Atrazine, Prometryne, Simetryne
Cladosporium	Atrazine, Prometryne, Simetryne
Fusarium	DDT, Trichlorphon, Fenitrothion, Carbaryl, Simazine, Atrazine, Chlordimeforn, Lindane
Peniciolium	DDT, Carbaryl, Trichlorphon, Parathion, Atrazine, Prometryne, Simazine, Propanil
Rhizopus	DDT, Fonofos, Carbaryl, Atrazine, Trichlorphon,
Trichoderma	DDT, Lindane, Dalapon, Atrazine, Simazine, Dichlorovos, Parathion, Malathion, PCP
Chlamydomonas	Metobromuron, Atrazine
Chlorella	Phorate, Parathion

[a]Fan and Huang, 1982; Li et al., 1985a; South China Agricultural University, 1990; Zheng, 1990; Han and Li, 1991; Lang, 1992.

photodecomposition of pesticides in different environmental media, especially in aqueous solution (Chen, 1985; Jiang and Cai, 1987a; Kong et al., 1991; Wan, 1991; Yue and Hua, 1992), but there are few reports of what takes place in soil (Jiang and Cai, 1987a; Yue, 1989). Jiang and Cai (1987a) studied the photodecomposition of lindane, carbofuran, and parathion-methyl in soil and in aqueous solution, and found that pesticides degrade much more slowly in soil than in aqueous solution (Table 9.7). They concluded that the photochemical processes mainly take place in the soil surface. In soil solution, soil adsorption decreased the amount of pesticides that photodegradation takes place in the soil. Yue (1989) had studied the effects of soil moisture content, temperature, and thickness on the photodegradation of ethylenethiourea (ETU). The results showed that temperature had little effect, and that moisture content and thickness were very significant factors, thus concurring with the conclusions of Jiang and Cai (1987).

Photochemical reactions of pesticides may be accelaterated or decelerated by substances separately named as photosensitizer and photoquencher. Yue and Hua (1992) studied the effects of pyridaphenthion, carbendazim, and carbofuran on photodegradation rates of cypermethrin, deltamethrin, and fenvalerate. They found that the photolytic rates of the three pyrithroid insecticides were accelerated significantly by pyridaphenthion and carbendazim; carbofuran delayed the photolytic rates of cypermethrin and deltamethrin, but had little effect on fenvalerate (Table 9.8). They observed a significant positive correlation between the dosage of photosensitizers and their photosensitive efficiency (Fig. 9.2).

The dissipation rate of pesticides in field soil may be influenced by biological, chemi-

Table 9.7. Photodegradation Rate of Some Pesticides (half-life time, hour)

Pesticide	Half-Life Time	Media	Reference
Carbofuran	83.1	Soil	
	10.2	Water	Jiang and Cai, 1987
Lindane	1746.0	Soil	
	59.2	Water	Jiang and Cai, 1987
Methyl-Parathion	628.5	Soil	
	27.7	Water	Jiang and Cai, 1987
Azinophos-methyl	1.2	Hexane	Wan, 1991
Dimethoate	9.4	Hexane	Wan, 1991
Fenitrothion	24.8	Hexane	Wan, 1991
Malathion	49.0	Hexane	Wan, 1991
Quinalphos	0.43	Hexane	Wan, 1991
Cypermethrin	41.1	Glass film	Yue and Hua, 1992
Deltamethrin	34.9	Glass film	Yue and Hua, 1992
Fenvalerate	107.3	Glass film	Yue and Hua, 1992
Ethylenethiourea	1.5	Soil	Yue, 1989
Aldrin	113.49	Glass film	Chen, 1985
Atrazine	58.67	Glass film	Chen, 1985
α-BHC	91.09	Glass film	Chen, 1985
β-BHC	151.80	Glass film	Chen, 1985
δ-BHC	153.84	Glass film	Chen, 1985
Carbaryl	51.66	Glass film	Chen, 1985
Carbofuran	70.64	Glass film	Chen, 1985
Chlorpyrifos	52.45	Glass film	Yue, 1989
α-Cypermethrin	179.82	Glass film	Chen, 1985
p,p′-DDE	206.66	Glass film	Chen, 1985
p,p′-DDT	192.31	Glass film	Chen, 1985
Deltamethrin	150.14	Glass film	Chen, 1985
Dichlorovos	7.25	Glass film	Chen, 1985
Dicofol	143.63	Glass film	Chen, 1985
Dieldrin	153.84	Glass film	Chen, 1985
Dimethoate	64.10	Glass film	Yue, 1989
Ethion	85.47	Glass film	Chen, 1985
Fenitrothion	57.69	Glass film	Chen, 1985
Fenthion	55.83	Glass film	Chen, 1985
Lindane	103.95	Glass film	Chen, 1985
Marathion	51.28	Glass film	Chen, 1985
Methomyl	48.41	Glass film	Chen, 1985
Mevinphos	23.79	Glass film	Chen, 1985
Methyl-Parathion	41.20	Glass film	Yue, 1989
Oxamyl	55.38	Glass film	Chen, 1985
Parathion	88.19	Glass film	Chen, 1985
cis-Permethrin	177.51	Glass film	Chen, 1985
trans- Permethrin	121.03	Glass film	Chen, 1985
Phosalone	71.37	Glass film	Chen, 1985
Phosmet	53.25	Glass film	Chen, 1985
Propazine	108.17	Glass film	Chen, 1985
Quinalphos	44.10	Glass film	Chen, 1985
S-5439	168.85	Glass film	Chen, 1985
Simazine	53.25	Glass film	Chen, 1985
Trietazine	161.00	Glass film	Chen, 1985

Table 9.8. Rate Constants and Half-Life of Sensitive Photodegradation for Thin Films of Three Pyrithroids ($10^{-2}h^{-1}$, h)[a]

Treatment	Dose µg/cm²	Cypermethrin		Deltamethrin		Fenvalerate	
		Rate	Half-Life	Rate	Half-Life	Rate	Half-Life
Single	0.27:0	1.685	41.1	1.983	34.9	0.646	107.3
+Pyridaphenthion[b]	0.27:0.27	3.166	21.9	3.701	18.7	0.994	69.7
+Carbendazim	0.27:1.33	2.895	23.9	2.805	24.7	0.965	71.8
+Carbofuran	0.27:0.27	0.735	94.3	1.158	59.8	0.595	116.5

[a]Reprinted from Yue and Hua, 1992.
[b]Means plus.

cal, and photochemical degradation, volatilization, plant uptake, temperature, soil texture and organic matter content, and agricultural and environmental factors, etc. Large differences in the dissipation rate may exist for the same pesticide. Table 9.9 lists the range of dissipation half-lives of parts of pesticides in different districts of China. In general, the dissipation rates of pesticides in soil are usually expressed in the following formula:

$$Ct = C_o \times e^{-kt} \tag{9.1}$$

where C_o is the initial concentration of pesticide in soil, t is time, Ct is pesticide concentration at time t, and k refers to the dissipation rate constant.

The formula is fitted well to the results obtained in laboratory degradation experiments of pesticides. However, it does not reflect the field results. Wang et al. (1992a, 1992b) showed that great errors may exist between the actual field results and that calculated from the above formula, and the errors would be narrowed if the formula was replaced by the following equation:

$$f(t) = a_0 + a_1t + a_2t^2 + \ldots + a_nt^n \tag{9.2}$$

where $f(t)$ is the pesticide concentration in soil at time t, and $a_0, a_1, a_2, \ldots, a_n$ refer to the constants.

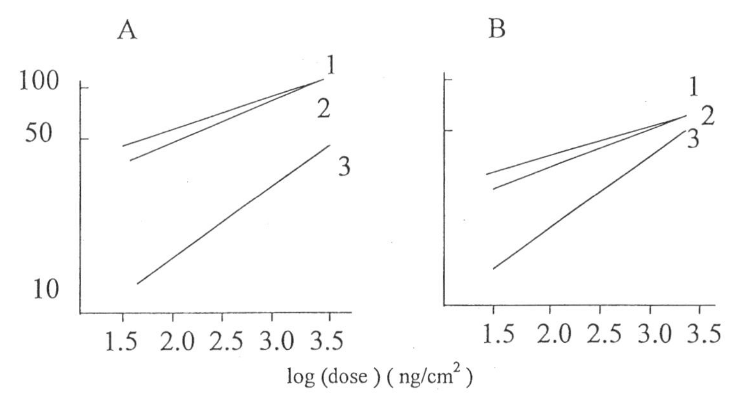

Figure 9.2. Relationship between photosensitive efficiency and dose of sensitizer. A; phridaphenthion as photosensitizer. B: carbendazim as photosensitizer. 1—Cypermethrin 2—deltamethrin. 3—Fenvalerate (Reprint from Yue and Hua, 1992)

Table 9.9. Degradation of Pesticides in Field Soil (half-life, d)

Pesticide	Degradation Half-Life			Reference
Azocyclotin	4.1–13.8			[a]
Benzoximate	4.7– 7			[a]
BHC	30–300	flooded soil	10–20	Han and Li, 1991
Bifenthrin	50			[a]
Buprofezin	20–54			[a]
Carbaryl	27	flooded soil	13	Qian, 1993
Carbofuran	20–56	flooded soil	20	Qian, 1993
Carbosulfan	2.3–8.4	flooded soil	5.9–6.9	[a]
Cartap	0.7–4.2			[a]
Chlorfluszuron	15.3–25.5			[a]
Chlorothalonil	2.6			[a]
Chlortoluron	43.6			[a]
Clofentezine	16.9–20.5			[a]
Clomazone	10.4–17.8			[a]
Cyfluthrin	2.5			[a]
DDT	45–480	flooded soil	7–45	Han and Li, 1991
Deltamethrin	11			[a]
Diclofop-methyl	2.4–3.2			[a]
Dieldrin	360	flooded soil	>90	Han and Li, 1991
Diflubenzuron	15–30			[a]
Dimethoate	10–23	flooded soil	10–20	Han and Li, 1991
Esfenvalerate	11.9–17.6			[a]
Ethofenprox		flooded soil	4.1	[a]
Fenitrothion	8–30	flooded soil	4–6	Han and Li, 1991
Fenothiocarb	15.5			[a]
Fenpyroximate	5.7–36.7			[a]
Fenthion	16–21	flooded soil	2–4	Han and Li, 1991
Fluazifop-butyl	1.1–1.6			[a]
Flufenoxuron	5.6–28.8			[a]
Fluvalinate	14–16.5			[a]
Haloxyfop	9.4–26			[a]
Hexythiazox	18.4–23.4			[a]
Isazophos	27.2–32.8	flooded soil	2–11	[a]
Leptophos	78–128	flooded soil	30–37	Han and Li, 1991
Methomyl	1–5.4			[a]
Metolachlor	18.5–22.6			[a]
Parathion	49	flooded soil	2–9	Qian, 1993
Pendimethalin	5.8–11			[a]
Phorate	7.3–48.5			[a]
Phoxim	14			[a]
Pursuit	3–9			[a]
Pyrazosulfuron-ethyl		flooded soil	9	[a]
Quinalphos	5–6			[a]
Quinclorac		flooded soil	2–11.1	[a]
Quizalofop-ethyl	0.6–12.9			[a]
Sebufos	39–58			[a]
Tebufenpyrad	8.1–18.1			[a]
Triallate	60–70			[a]
Tribenuron-methyl	7.6–24.8			[a]
Vinolozolin	32.3			[a]

[a]Abstracted from the Chinese pesticides registration data in 1990–1994.

Adsorption of Pesticides on Soil

The behavior of pesticides in the soil is controlled to a great extent by soil adsorption and desorption. The action of adsorption influences not only movement and volatilization, but also biological and chemical degradation of pesticides in soil (Jiang and Cai, 1987b; Huang, 1987).

The adsorption mechanisms of pesticide on soil have four forms, which are identified as physical, electrostatic, hydrogen bonding, and coordinating interaction. In physical interaction, since adsorption is actually due to the van der Waals force, it is usually named van der Waals attraction. The magnitude of the van der Waals force decreases sharply with the increase of intermolecular bonding distance, similarly affecting the adsorption capacity. For example, adsorption of urea herbicides is weakened with the lengthening of molecular bond distance, and ranks in the order of neburon >linuron > dinuron >monuron >fenuron. Van der Waals attraction contributes to a limited extent to the adsorption of all pesticides on soil among the four interactions. The low isosteric heats and low magnitude of adsorption are characterization of such interaction.

The prerequisite of electrostatic interaction is that the pesticide exists in the form of ion, either cation or anion in soil. Examples are paraquat, diquat, and dinoseb, which are adsorbed by their interactions with soil ion.

Hydrogen bonding is a special kind of electrostatic interaction in which the hydrogen atom serves as a bridge between two electronegative atoms. Pesticides which have the atomic groups COOH or NH, etc., may interact with oxygen atom in soil to form the hydrogen bond. For example, carbamate or urea pesticides separately interact with carboxyl or phenoxyl to form hydrogen bond in soil. The heavy metal ions in soil solution, such as Cu^{++}, Ag^{+}, etc., usually interact with amidogen of pesticides, as chloro-s-triazine herbicides, to form a coordinated amine (Cai, 1993b).

Jiang and Cai (1987b), Wang et al. (1989), and Li et al. (1985a) showed that the adsorption of pesticides on soil is mainly influenced by the content of soil organic matter for nonionic pesticides, and the adsorption constant K_d of the same nonionic pesticide may vary greatly in soils with different content of organic matter. Jiang and Cai (1987b) studied the adsorption of carbofuran in six kinds of soils and found that the adsorption constant K_d was well correlated with the organic carbon content of soil. The regression formula was as follows:

$$K_d = 0.0205 + 44.26 \times \%OC \; (n = 6, r = 0.35) \qquad (9.3)$$

where $\%OC$ is the percentage of soil organic carbon, n is the sample number, and r is the regression coefficient. Dividing K_d by the percentage of soil organic carbon, we obtain another value of K_{oc}. Since K_{oc} is largely independent on the properties of the soil, the adsorption property of a pesticide is thoroughly represented by the value of K_{oc}. Table 9.10 lists the K_{oc} values of some pesticides.

Movement of Pesticides in the Soil

Movement of pesticides in the soil is very complex. It may disappear into the atmosphere by volatilization, into surface water body by soil runoff, and into groundwater by percolation (Zhou, 1990; Yang et al., 1992). As the movement of pesticides may have great effects on the behavior of pesticides in the soil and on the scope of pesticide pollution, it is a matter of great concern (Yu and Chen, 1986; Zhou, 1990).

The amount of pesticide that goes into the surface water body is closely related to the

Table 9.10. Adsorption Constants of Some Pesticides on Soil

Pesticide	Adsorption Constant (K_{oc})	Reference
Dichloropropane	26	Han and Li, 1991
Dibromoethane	32	Han and Li, 1991
Aldicarb	36	Han and Li, 1991
Monuron	83	Han and Li, 1991
Dibromochloropropane	130	Han and Li, 1991
Atrazine	170	Han and Li, 1991
EPTC	280	Han and Li, 1991
Lindane	1300	Han and Li, 1991
Trifluralin	3900	Han and Li, 1991
Methyl-parathion	10000	Han and Li, 1991
DDT	24000	Han and Li, 1991
Carbofuran	72	Jiang and Cai, 1987

amount of pesticide residue in the soil and the amount of runoff, generally not more than 10% of the applied dose. It was reported that a river in Zhejiang Province was polluted by BHC because of runoff from an agricultural field. It was found that the BHC of residue concentration in the river was 0.5 to 2.0 μg/L (Huang, 1991). Since the application of BHC, DDT, and other high residual organochlorine pesticides were banned in 1983, the kinds of pesticides used now mostly have lower residues that can easily disappear from surface water by degradation, thus causing no cumulative pollution problems.

Pesticide residues in the soil entering into groundwater have been reported previously (Cai et al., 1983; Lou and Li, 1984). The mobility of pesticides in the soil can be measured by methods of soil thin-layer chromatography (TLC) and column drip-washing. Zhang et al. (1994) measured the mobility of jiajiyiliulin, midinyanglin, kecaoan, and danjiami using the TLC method, and found that their R_f values were 0.23, 0.28, 0.44, and 0.68, respectively. Shan et al. (1994a) studied the movement of aldicarb, carbofuran, and alachlor (Table 9.11) in different soils using a column with a length of 35 cm and a radius of 3.2 cm, and found that the mobility of a pesticide was correlated with its solubility in water. The higher the solubility, the greater the mobility. Pesticides therefore are more mobile in soil with a high sandy content than in soil with a low sandy content. The sequence of pesticides mobility in the soil is thus sand > sandy loam > loam > clay.

Volatilization of Pesticides from Soil

The volatilization process may be one of the important ways whereby pesticides disappear from the soil. The magnitude of pesticide loss by volatilization during or after application can range from a few percent to more than half of its applied amount (Cai, 1993a). The volatilization process not only can influence the partition and fate of pesticides in the soil system, it can also pollute the adjacent fields and be hazardous to nontarget organisms.

The volatilization rate of pesticides from the soil is closely related to the properties of the pesticide, the formulation and application method used, and the soil type and climate conditions, etc. (Cai, 1993a). Generally, the volatilization rate of a pesticide in the soil is lower than that in water solution or in glass film. As well, it is greater in warm, moist, and sandy soil than that in cold, dry, and clayed soil (Cai, 1993a).

Jiang and Cai (1990) studied the volatilization of trifluralin, lindane, parathion-methyl, and carbofuran from water solution, glass film, and soil (Table 9.12), and con-

Table 9.11. Mobility of Pesticides in Column Soil (Percent of the Added Pesticide)[a]

Pesticide	Profile(cm)	Sand	Sandy Loam	Loam	Clay
Alachlor	0–5	1.38	4.26	7.64	19.30
	5–10	2.07	16.37	29.51	66.82
	10–15	10.11	31.61	39.17	13.18
	15–20	33.79	32.94	19.43	0.49
	20–25	40.23	12.33	4.02	–
	25–30	10.57	2.01	–	–
	Leached water	1.84	0.40	–	–
	Total	99.99	99.92	99.77	99.79
Carbofuran	0–5	–	–	–	–
	5–10	–	–	0.49	0.41
	10–15	–	–	0.54	0.50
	15–20	–	0.40	1.54	1.20
	20–25	–	1.20	3.09	5.96
	25–30	–	4.46	8.02	13.23
	Leached water	94.60	86.40	78.60	71.10
	Total	94.60	92.46	91.25	91.49
Aldicarb	0–5	–	–	–	–
	5–10	–	–	–	–
	10–15	–	–	–	–
	15–20	–	–	–	–
	20–25	–	–	–	2.44
	25–30	–	–	1.58	4.78
	Leached water	97.40	93.40	88.20	88.00
	Total	97.40	93.40	89.78	95.22

Soil Type	Clay (%)	Sand (%)	pH	O.M. (g/kg soil)	CEC (meq/100g soil)
Sand	3.3	94.4	9.0	1.9	2.34
Sandy loam	7.0	81.0	8.7	4.7	5.06
Loam	19.9	49.6	8.6	8.6	9.01
Clay	8.2	22.2	8.5	15.8	16.29

[a]Reprinted from Shan et al., 1994a.

cluded that the volatilization of pesticides from the glass film mainly depends on their molecular weight and vapor pressure. Pesticides with high vapor pressure would volatilize more readily. Volatilization of pesticides from aqueous solutions was not only affected by their vapor pressure, but also by their water solubility. Although the vapor pressure of carbofuran is comparatively higher among the four tested pesticides, it evaporated lower than others, as it has high water solubility. Besides the molecular weight, vapor pressure, and water solubility, volatilization of pesticides from soil may be influenced by the sorption of pesticides on the soil. The volatilization of trifluralin was the highest from glass film or from aqueous solution, but was the lowest from the soil.

Residues and Bound Residues of Pesticides in Soil

Among the pesticide residues in the soil, some can be extracted conventionally by organic solutions, while some cannot. The latter type, known as bound residues of pesticides, generally can be measured by the technique of liquid scintillation counting with the radiola-

Table 9.12. Volatilization Rate of Pesticides from Glass Film, Water, and Soil (%)[a]

Pesticide	Molecular Weight	Vapor Pressure (mm Hg)	Water Solubility (mg/L)	Glass Film	Water	Soil
Triflualin	335	65.0	0.3	99.5	92.6	6.2
Lindane	290	9.4	10.0	88.0	89.4	8.6
Methyl-parathion	263	9.7	60.0	23.8	15.8	14.4
Carbofuran	221	20.0	500.0	97.9	3.6	15.6

[a]Reprinted from Jiang and Cai, 1990.

beled pesticides. Lang (1992) measured the residue of radiolabeled ^{14}C parathion in soil, and found that at 28 days after application, the residual amount extracted by organic solution was only 32% of the applied, while bound residual amount was 48% and the total residual amount was 80% of the applied. This indicated that the residue measured by the conventional method is only part of the actual residues. Bound residues of pesticides in the soil are influenced by many factors such as the properties of the pesticide itself, the contents and components of soil humus, and clay content and moisture, etc. The amounts of bound residues of different pesticides in the soil are different. Shan (1994) showed that bound residues for insecticides and herbicides range from 3% to 80% and 10% to 90%, respectively. Table 9.13 shows the bound residues of some pesticides in the soil.

The bound residues of pesticides mainly exist in the soil humus. There is very little that exists in the soil clay particles. Zhang (1986) studied the distribution of deltamethrin

Table 9.13. Bound Residues of Pesticides in Soil[a]

Pesticide	Time of Incubation (d)	Soil Type	Bound Residues (% of Applied Dose)
Parathion-methyl	28	sandy loam	26–33
	28	clay loam	35–45
Parathion	13	loam	29.1
	13	sand	6.0
	28	sand	48.5
Phorate	13	loam	13.1
	13	sand	4.5
p,p′-DDT	13	loam	2.8
	13	sand	0.7
Lindane	13	loam	4.6
	13	sand	1.3
Chlorfenvinphos	32	sandy loam	15.6
	32	clay loam	12.8
Fonofos	13	loam	15.6
	28	loam	30.8
Flufenoxuron	32	sandy clay	57
Deltamethrin	180	organic soil	19.2
	14	sand	13.8
Chlortoluron	100	paddy soil	5.1
	100	black soil	15.7
Trifluralin	100	paddy soil (flooded)	16.5
	100	black soil	16.6
	100	paddy soil (dry)	1.7

[a]Reprinted from Shan, 1994.

Table 9.14. Distribution of Bound Residues of ^{14}C-Chlorotoluron among the Components of Soil Humus in Soils[a]

Soil Type	Condition of Incubation	Time of Incubation (d)	Bound Residues (% of Applied)	Ratio of Distribution among the Components of Humus (% of the Bound Residues)		
				Fulvic Acid	Humin Acid	Humin
	arid	50	4.59	16.89	20.95	62.84
Paddy	land	100	5.11	27.72	29.35	40.76
	paddy	50	5.06	8.20	28.57	62.64
soil	field	100	6.89	14.11	14.11	71.77
	arid	50	8.78	9.49	27.85	65.92
Black	land	100	15.67	4.96	21.63	71.99
	paddy	50	14.86	3.18	11.96	83.74
soil	field	100	16.03	4.16	12.30	80.94

Soil Type	pH	CEC Cmol(+) /kg	Clay (%)	O.M. (g/kg)	Constitution of Humins (g/kg dry soil)			
					Total Carbon	HAC	FAC	HAC/FAC
Paddy soil	5.24	7.61	15.3	2.63	0.445	0.187	0.258	0.72
Black soil	7.47	24.70	24.4	3.20	0.419	0.223	0.196	1.14

[a]Reprinted from Chen et al., 1992.

in different components of soil and found that 97% of the total residues was in soil organic humus, and only 3% was in the inorganic part of the soil. Chen et al. (1992) studied the residues of chlorotoluron in paddy and dry soil, and concluded that the ratio of bound residues to total residues increased with time. Although the amount of bound residues of pesticides in different soils may vary greatly, their general distribution tendency among the components of soil humus are the same, i.e., most is distributed in humin, followed by humic acid and the least in fulvic acid (Table 9.14). The distribution of bound residues in soil humus is closed related to the content and proportion of each component of soil humus, and the moisture content of the soil. The larger the ratio of humic acid content to fulvic acid content in black soil, the larger is the ratio of bound residues of chlorotoluron in humic acid to that in fulvic acid.

Plant Uptake of Pesticides from Soil and Effects of Pesticides Residues on Plant

In the 1970s, there was use of organochlorine pesticides in China and the result was serious environmental pollution of soil and grain crops. Cai (1977) selected the cotton field where organochlorine pesticides had been largely used and contamination of the soil by DDT and BHC was the most serious, to measure the amount of these residues in the soil and edible parts of crops at harvest time. The results showed that (Table 9.15) DDT and BHC contents in the soil were 11.68 and 0.8 mg/kg, respectively. Among the nine crops studied, the amounts of DDT in tobacco and carrot were the highest (27.7 and 9.23 mg/kg, respectively). These two crops have a strong ability to accumulate DDT, followed by sweet potato and peanut, and the starch crops and cotton seeds were the weakest. The residue amount of BHC in the soil as well as in those of various crops were all lower than that of DDT, but the plant uptake ability of BHC was generally similar to that of DDT.

There are many kinds of crops and complicated rotation systems; the crop damage ac-

Table 9.15. Plant Uptakes of DDT and BHC from Soil[a]

	DDT		BHC	
Plant	Residues (mg/kg)	Uptake (%)	Residues (mg/kg)	Uptake (%)
Soil	11.68	–	0.84	–
Tobacco	27.70	237.15	1.34	159.52
Carrot	9.23	79.63	3.65	434.52
Sweet potato	1.19	10.19	0.11	13.10
Peanut	0.28	2.40	0.26	30.95
Corn	0.13	1.11	0.10	11.90
Wheat	0.11	0.94	0.13	15.48
Rice	0.06	0.51	0.04	4.76
Cotton seed	0.03	0.26	0.04	4.76

[a]Reprinted from Cai, 1977.

cidents caused by pesticide residues in the soil occur mainly in areas where crop rotation of wheat and rice was adopted in China. For example, chlorsulfuron and metsulfuron-methyl have been applied on a large scale in wheat fields and are safer for wheat-type crops, but can easily cause damage to some sensitive second crops because of their residues in the soil.

Su and Zhao (1988) showed that when the amount of chlorsulfuron applied to the soil was 10 mg/kg, 96% of it would be dissipated within 60 days, and the remainder degraded very slowly afterward, so that after one year the residue amount in the soil was still 0.1 to 0.2 μg/kg. If the applied amount was 10–30 g/ha, the residue in the 0–5 cm soil layer after one year was 0.08 to 0.16 μg/kg. This residual level still is effective on weed control and is not harmful to beets, but it still has some effect on the final production of soybean. Shan et al. (1995) studied the effect of chlorsulfuron on rice. Before the planting of rice, chlorsulfuron was well distributed into the cultivated soil layer in six different amounts: 0.375, 0.75, 1.5, 3.0, 6.0 g active ingredient/ha and zero for comparison. The results showed that when the application of chlorsulfuron was greater than 0.375 g a.i./ha, certain effects on the production of rice growth existed. If 0.375 g a.i./ha of the chemical would leave the safer residue dose, the allowed maximum chlorsulfuron residue in soil during the period of rice planting would be 0.17 μg/kg. In some dry crop regions, serious damage to second crops such as maize are caused by using methsulfuron-methyl and chlorsulfuron in wheat fields without following the rules. If methsulfuron-methyl and chlorsulfuron are properly used in the wheat fields of interplanting regions, the safety of the second crops can be ensured. Cai and Jiang (1995) applied 15 g a.i./ha of chlorsulfuron to the soil of southern Jiangsu Province. The residue of chlorsulfuron in the soil was 0.22 μg/kg before the planting of rice. If dry crops such as corn were planted at this time, there would be danger to the crop. Rice is very sensitive to chlorsulfuron; the seedling roots and growth will be affected when the residue of chlorsulfuron in the soil is over 0.1 μg/kg. Since the safe coefficient of the second rice crop is much lower, when applying chlorsulfuron to the rice and wheat rotation fields, it is necessary to strictly control the dosage to reduce the harm of pesticide residues to the second crops.

Effects of Pesticide Residues on Soil Organisms

The earthworm is a commonly seen soil organism which accelerates the transformation of materials, improves the soil environment, and enhances the fertility of soil. To protect

Table 9.16. Toxicity of Pesticides to Earthworms

Pesticide	LC_{50} [a](mg/kg)		Reference
	1 week	2 weeks	
Monocrotophos	0.31	0.18	Zhang et al., 1986
Carbendazim		4.27	Zhang et al., 1986
Carbofuran	25.7	12.9	Zhang et al., 1986
Lindane	106.0	70.9	Zhang et al., 1986
Parathion-methyl	79.9	74.5	Zhang et al., 1986
Fenvalerate	8.8		Zhang et al., 1986
Dimehypo	12.1		Zhang et al., 1986
Jiajiyiliulin	23.3	23.3	Cai et al., 1993
Midinyanglin	29.4	26.9	Cai et al., 1993
Kecaoan	91.7	84.5	Cai et al., 1993
Danjiami	917.0	753.0	Cai et al., 1993

[a]Results with the standard soil.

earthworms from the harm of pesticides, their safety must be taken into account during the development and application of any pesticide.

The acute toxicity is generally regarded as the index for estimating the effect of pesticides on the earthworm (Cai, 1987). One of the methods for measuring the acute toxicity to earthworms is directly taking the representative soil sample from the field as the experimental material. As toxicity values of pesticides to earthworms will be different in different soils and the results cannot be compared with each other, Zhang et al. (1985) used quartz sand, bentonite, peat, horse waste, lime, etc., to make the standard soil for simulating the basic properties of natural soil to determine the acute toxicity.

From Table 9.16, it can be seen that the sequence of half-lethal dose (LC_{50}) of the earthworm measured from the standard soil for the 11 experimental pesticides are monocrotophs, carbendazim, carbofuran, lindane, parathion-methyl, fenvalerate, dimehypo, jiajiyiliulin, midinyanglin, kecaoan, and danjiami. Among them, the toxicity of monocrotophs and carbendazim obviously increase with the lengthening of the experiment time, while that of lindane and danjiami have little increase. Others have no obvious temporal effects. The poisoning symptoms of the earthworm harmed by different pesticides are clearly different. In the case of carbendazim poisoning, the symptoms are body relaxing, ring belt swelling, and grain-sized blisters appearing on the head. The earthworm is in a paralyzed state and dies when the blisters are broken. Earthworms poisoned by monocrotophs, carbofuran, parathion-methyl, and fenvalerate are rapidly rolled up and have twisted bodies. The rolling will intensify through light or physical stimulation. With the lengthening of time, the earthworm's body color turns pale, the body breaks apart, and it dies. Earthworms poisoned by lindane indicate body relaxation, pale color, and paralysis before dying.

Zhang et al. (1986) did the comparison for the toxicity of carbofuran to the earthworm in three major soil types of China as well as in the standard soil. The experimental results showed that the toxicity of carbofuran to earthworms for the three natural soils is between 6.9 to 19.2 mg/kg (Table 9.17). The maximum value is 2.8 times the minimum value, while the LC_{50} value in standard soil is 12.9 mg/kg, equivalent to the average value of the three soil types.

In addition to the acute harm, there exists the chronic harm, the enrichment of pesticides inside the earthworm's body may further result in harm to the earthworm or pass through the food chain to affect the whole continental ecosystem.

Li et al. (1985b) studied the BHC residues in the bodies of earthworms in different crop fields. They found that the BHC residues were between 0.32 to 1.94 mg/kg (Table

Table 9.17. Toxicity of Carbofuran to Earthworms in Different Soils[a]

Soil Type	pH	O.M. (g/kg)	CEC (cmol/kg)	Texture	LC_{50}(mg/kg)		Toxicity Ratio[b]
					1 week	2 weeks	
Red soil	6.85	23.1	7.74	Heavy loam	17.2	12.7	0.98
Black soil	6.58	36.7	34.12	Light clay	27.2	19.2	1.49
Drab soil	7.95	19.3	12.38	Light loam	11.3	6.9	0.53
Standard soil	7.00	24.2	5.60	Tight sand	25.7	12.9	1.00

[a]Reprinted from Zhang et al., 1986.

[b]Refers to comparative toxicity, which is equal to the ratio of toxicity at 2 weeks of carbofuran in a soil to that in the standard soil.

Table 9.18. Residues of BHC in Soil and Earthworm and the Accumulation Factors of Earthworms[a]

Field Type	Residue in Soil (mg/kg)		Residue in Earthworm (mg/kg)		Accumulation Factors
	Range	Average	Range	Average	
Cotton	0.12–4.98	1.94	1.44–11.51	5.79	2.98
Wheat	0.49–0.57	0.53	1.09– 4.37	2.73	5.15
Corn	0.16–0.96	0.43	0.34– 4.03	1.67	3.88
Orchard	0.08–0.47	0.28	0.61– 0.63	0.62	2.21
Vegetable	0.04–0.59	0.23	0.27– 2.53	0.97	4.18
Grass	0.04–0.61	0.32	0.31– 2.20	1.05	3.28

[a]Reprinted from Li et al., 1985b.

9.18) in crop field soils, the main BHC residue was β-BHC, and the sequence was β-BHC (0.61 mg/kg) > α-BHC (0.04 mg/kg)> δ-BHC (0.03 mg/kg) > γ-BHC (0.02 mg/kg). The average content inside the earthworm's body was between 0.62 to 5.79 mg/kg. The isomer content of each BHC in an earthworm was similar to the soil components. The concentration coefficient of earthworms for BHC in each type of crop field was 2.21 to 5.15 mg/kg, with the average being 3.61 mg/kg. This residual level had no obvious effect on the normal growth of earthworms themselves. Nevertheless, earthworms are the food of birds and small animals, and the harmful residues might pass through the food chain to affect some biological species that are sensitive to BHC. The cumulative ability of different kinds of earthworms to a pesticides is different. Measuring results showed that the cumulative ability of *Allolbophore caligionsa* is the strongest, followed by *Eisenia foetiela*, with that of *Pheretima guillelmi* being the weakest. The difference between the maximum and the minimum cumulative abilities of different types of earthworms might be as great as three times.

In addition to earthworms, there exist various microorganisms in soil. Pesticide residues in soil will have an effect on these microorganisms. The effect can be expressed by the change of soil respiration. Soil respiration is the process of releasing CO_2 during the life activities of soil microorganisms. If the respiration process of a soil becomes weak after applying a pesticide, it means that the general activities of soil microorganisms are restricted.

Cai et al. (1986) measured the respiration intensity of soil microorganisms as follows: put the tested soil and a small cup that contains the base solution for collecting the CO_2 released during the respiratory process of the soil microorganisms in a 1-L bottle; seal, and leave to cultivate. Change the base solution at a certain period and measure the CO_2 quantity released at different time intervals. The black soil taken from northeast China

Table 9.19. Effects of Pesticides on Respiration of Soil Microorganisms

	Dosage (mg/kg)				
Pesticide	**1**	**10**	**100**	**1000**	**Reference**
Parathion-methyl	A	A[a]	A		Cai et al., 1986
Carbofuran	A	A	A		Cai et al., 1986
Lindane	A	A	A		Cai et al., 1986
Carbendazim	A	A	C		Cai et al., 1986
Shachongdan		A	C	C	Chen et al., 1987
PCP-Na			A	C	Chen et al., 1987
Paraquat			A	A	Chen et al., 1987
Trifluralin			A	B	Chen et al., 1987
Molinate			A	A	Chen et al., 1987
Butachlor			C	C	Chen et al., 1987
Jiajiyiliulin		C	C	C	Cai et al., 1992
Midinyanglin		A	A	B	Cai et al., 1992
Kecaoan		A	A	C	Cai et al., 1992
Danjiami		A	A	A	Cai et al., 1992

[a]A: No effect; B: Have effect at the initial period of experiment, inhibiting ratio>10%; C: Have inhibiting effect in the entire experimental period, the total inhibiting ratio >10%.

and paddy soil from Taihu Lake were used for measuring the effect of pesticides on the respiratory intensity of soil microorganisms. The concentrations used in the experiment for each pesticide were 1, 10, 100, 1000 mg/kg, and the experimental periods were 12 and 30 days. The experimental results are summarized in Table 9.19. Based on the experimental results, the effect of pesticides on the respiration of soil microorganisms could be divided into three types: (1) within the testing concentration range and the entire testing period, the six pesticides, namely, parathion-methyl, carbofuran, lindane, paraquat, molinate, and danjiami, generally have no obvious effect on the respiratory intensity of soil microorganisms in both soil types; (2) trifluralin and midinyanglin at 1000-mg/kg concentration have a certain effect on soil respiration in the first 1–2 days of the experimental period, but then return to normal—both pesticides have no effect during the entire experimental process when used in lower concentration; (3) carbendanzim, shachongdan, butachlor, PCP-Na, and jiajiyiliulin have certain effects on soil respiration intensity during the whole experimental process when treated at 100 mg/kg concentration. Among these pesticides, jiajiyiliulin has the most obvious restraining effect, which occurs when applied in 10-mg/kg concentration.

Most of the pesticides have certain restraining effects on the respiration process of soil microorganisms initially, followed by quick recovery. Among the pesticides studied, kecaoan is the most typical. When applied at 1000-mg/kg concentration, the restraining rate may be as high as 80% in the first few days, but then changes to a promoting effect 10 days later, and the respiratory intensity increases by 60%.

In the entire crop rotation system, the leguminous plants play an important role in enhancing the soil fertility, the main contribution of which is the nitrogen fixation effect of root nodule bacteria. Researchers in Northeastern China Agricultural College (1989) studied the effect on the formation of soybean root nodules and the activity of nitrogenase when using trifluralin and metribuzim as weed killers. Potted planting experimental results showed that for the use rate of 48% trifluralin at 1.0 kg/ha and 1.5 kg/ha, the root nodules of soybean have been reduced 41.5% and 85.4%, respectively. Compared with that in the CK, the activity of nitrogenase in root nodules have been reduced 18.2% and 56.9%, respectively. The experimental results of applying 0.5 kg/ha, 1.0 kg/ha, and 1.5

kg/ha of metribuzim showed that the root nodules of soybean have been reduced 56.0%, 72.3%, and 85.7%, respectively, while the nitrogenase activity of root nodules have been reduced 14.4%, 32.2%, and 56.2%, respectively. Trifluralin and metribuzim are the main herbicides used in soybean fields now in China; the above two group experiments indicated that their use in serious weed problem regions has obvious effects on increasing soybean production. Nevertheless, the harmful effects mentioned above cannot be neglected. Therefore, selecting a pesticide with good weeding effects and at the same time not affecting the growth of root nodules will achieve better ecological and economic benefits.

Endotrophic mycorrhiza (VAM) is one of the beneficial fungi in the soil that can be in symbiosis with many plants. Some plants cannot normally grow without VAM in the soil. Cheng (1985) studied the effect of the fungicides metalaxyl, aliette, folpet, PCNB, and piothiocarb on kidney beans and clover that are infected by VAM. He measured the VAM infection rate of roots one month after pesticides were applied. The infection rates for the two crops without using pesticides were 34.4% and 35.4%, respectively. The effect of PCNB on VAM was the greatest. Continually using the chemical for two years, the infection rate of kidney bean and clover decreased to 0 and 3.3%, respectively, but the production of these two crops was also reduced 49% and 74%, respectively. The infection rate for the crops treated by the other four chemicals also somewhat decreased.

Zeng and Wang (1986) studied the effect of BHC, fenvalerate, deltamethrin, permethrin, isocarbofos, and fenamiphos on the combined nitrogen fixation, *Azospirillum brasilense*, that widely exists in parts of plant roots. Experimental results showed that when the content of BHC and permethrin in the soil was greater than 20 mg/kg, the activities of *Azospirillum brasilense* would be seriously restrained. If the amount of isocarbofos and fenamiphos was lower than 20 mg/kg in the soil, the activities of nitrogen fixation were stimulated. The opposite was true when the amount was greater than 20 mg/kg, and intensified with the increase of chemical concentration. When the amount of deltamethrin and fenvalerate was within 200 mg/kg, the nitrogenase activity increased with the increase of chemical concentration. The study showed that the effects of different kinds of pesticides on nitrogen fixation are quite different.

GROUNDWATER CONTAMINATION BY PESTICIDES IN CHINA

Factors Affecting Groundwater Contamination by Pesticides

Groundwater contamination from field-applied pesticides is almost entirely unexpected, because there is such a long distance for pesticides to go through before reaching groundwater. Since several incidents (Holden, 1986) of groundwater contamination resulting from field application of pesticides had been confirmed in New York, U.S., in the late 1970s, the interest in groundwater residue and contamination by pesticides has increased greatly. The early findings had led to monitoring for other pesticides, and as a result of this, over 70 kinds of pesticides or their degradation products were detected in groundwater in 37 states in the U.S. (Hallberg, 1989).

In the late 1970s and the early 1980s, problems of groundwater contamination by pesticides were found in China. Cai (1977) reported that BHC residue was detected in shallow and deep groundwater in Taihu Lake District, Jiangsu Province; the average residue contents were 1.02 and 0.59 μg/L, respectively. Lou and Li (1984) reported BHC residue in groundwater in Sichuan Province.

Many factors can influence the leaching of pesticides in soil. What may influence the residue of pesticides in soil may affect the potential for groundwater contamination. Generally, factors which favor the leaching of pesticides in soil may increase the potentiality for groundwater contamination by pesticides. Zhu et al. (1994a) studied the leaching properties of aldicarb in soil under different conditions in Jiangsu Province, and concluded that the leaching depth of aldicarb in soil was closely related to the application rate, soil texture, the volume of irrigation water or rainfall and the period between the irrigation or rainfall, and the pesticide application date. In short, the results indicated that the higher the sand content of the soil, the larger the application rate of pesticide and the volume of irrigation water or rainfall, the shorter the time between the irrigation or rainfall and the pesticide application, the deeper is the pesticide leaching in the soil.

Characterization and Identification of Potentially Leachable Pesticides and Areas Vulnerable to Groundwater Contamination by Pesticides

The research on groundwater contamination by pesticides is new in China. The national state of groundwater contamination by pesticides is not clear and many studies are now underway or need to be carried out.

Cai et al. (1995) studied the potentiality of groundwater contamination by aldicarb in Jiangsu province. According to many previous findings (Cai et al., 1990, 1993; Shan et al., 1994b; Zhu et al., 1994b), they concluded that the potentiality was mainly decided by five factors, i.e., the soil texture (sand content), content of organic matter, soil pH, the total volume of irrigation water, and rainfall in the three months after pesticide application, and the depth from the soil surface to the groundwater table. Each factor has a contributing weight to the potentiality; namely 0.9, 0.7, 0.9, 1.0, and 1.0, respectively. Because the value of each factor differs in different sites, a relative weight was given (Table 9.20). The actual potentiality to a specific site can be calculated from the following formula:

$$PGWC = [(0.9W_1) \times (0.7W_2) \times (0.9W_3)\ (\times (1.0W_4)]^{1/4} + 1.0W_5 \qquad (9.4)$$

where *PGWC* is the potentiality of groundwater contamination, and W_1, W_2, W_3, W_4, W_5 represent the relative weight within the factors of soil texture, content of organic matter, soil pH, the total volume of irrigation water and rainfall in the three months after pesticide application, and the depth from the soil surface to the groundwater table, respectively. The higher the value of *PGWC*, the larger the groundwater contamination potential by aldicarb. Areas where their *PGWC* values are more than 1.60 are named highly vulnerable contaminated area, meaning that groundwater in these areas can easily be contaminated by aldicarb. The use of aldicarb therefore should be banned in these areas. Where PGWC values are lower than 1.40 are designated as nonvulnerable to contamination, meaning that groundwater in these areas generally will not be contaminated by the use of aldicarb. PGWC values between 1.40 to 1.60 are named moderately vulnerable areas, meaning that the use of aldicarb may cause groundwater contamination, and so its use should be restrained. Based on the differentiated standards above, the vulnerability to groundwater contamination by aldicarb in Jiangsu province was mapped, which showed that 18.9% were highly vulnerable contaminated areas, where the depth of groundwater is lower than 1.0 m or sand content of soil is more than 95%; 11.1% were moderately-vulnerable contaminated areas, where the depth of groundwater is between 1.0 and 3.0 m and sand content of soil is between 75% and 85%, or the depth of groundwater is between 3.0 and 5.0 m and sand content of soil is between 85% and 95%; 63.5% were nonvulnerable contaminated areas, where the depth of groundwater is greater than 5.0 m, and the remaining 6.5% were river areas.

Table 9.20. Factors and Their Weights to the Contribution of Potentiality of Groundwater Contamination by Aldicarb[a]

Factor and Its Range of Value		Fraction
Sand content (%)	$[W_1]$	
<50		0.40
50–70		0.50
70–85		0.60
85–95		0.80
>95		1.00
Organic matter content (%)	$[W_2]$	
>3.0		0.50
2.5–3.0		0.65
2.0–2.5		0.80
1.5–2.0		0.90
<1.5		1.00
Soil pH	$[W_3]$	
>7.5		0.60
6.5–7.5		0.80
5.5–6.5		0.90
<5.5		1.00
Total water volume in three months after aldicarb applied (mm)	$[W_4]$	
<450		0.50
450–475		0.60
475–500		0.70
500–525		0.80
525–550		0.90
>550		1.00
Depth to groundwater (m)	$[W_5]$	
>5		0.20
3–5		0.40
1–3		0.70
<1		1.00

[a]Reprinted from Cai et al., 1995.

REMEDIATION STRATEGIES FOR SOIL AND GROUNDWATER CONTAMINATION BY PESTICIDES IN CHINA

General Practices for Preventing and Controlling Soil and Groundwater Contamination by Pesticides

According to statistical data, the area in China where diseases, pests, and weeds are found are up to 150 million ha, and the areas where pesticides are used annually are up to 130 million ha (Lang, 1992). Modern agricultural practices have resulted in large increases in food production, feed grain, and domestic animal husbandry. The large increase in the use of pesticides and the prevalence of using irrigation water over the last 30 years have made agriculture a very significant nonpoint source of soil and groundwater contamination. Before 1983, the agricultural environment in China was seriously contaminated due to the extensive use of the organochlorine pesticides BHC and DDT. The national average residues of these chemicals in the soil were 0.742 and 0.419 mg/kg, respectively, in 1980, and were still 0.038 and 0.054 mg/kg, respectively, in 1992. It would take decades to eradicate such levels of contamination.

Nonpoint sources of contamination are more difficult to control than point sources, and often in these cases, the goal would be to minimize the effect of such contamination rather than to seek its total prevention (Zhu and Cai, 1993; Zhu, 1994). The threat of soil and groundwater contamination by pesticides could be reduced by (a) closely controlling their application rates and time, (b) selecting suitable pesticide formulation, (c) using easily biodegradable pesticides, (d) using pesticides as part of IMP in order to reduce their use rate, (e) adopting biological control of pests, (f) selecting suitable pesticide application, (g) controlling irrigation time and volume, and (h) controlling the pumping volume of groundwater, etc.

Proposed Strategies for the Protection of Soil and Groundwater

The resources of soil and groundwater play an important role in preserving the existence as well as the development of China and its people. How to effectively protect the environmental resources and increase the production of food to meet the needs of the Chinese people has great practical significance. Because pesticides are so widely and extensively used in China, and because of the difficulty and uncertainty of remedial actions on contaminated soil and groundwater, the best method for solving or controlling contamination is to prevent it. Proposed strategies for the protection of soil and groundwater are as follows:

1. Strengthen the registration and scientific management of pesticides, and perfect the legal system, laws, and regulations of pesticides.
2. Formulate and perfect every standard for pesticides in different kinds of media, strengthen monitoring and supervision organizations and their work, and rapidly solve every contamination occurrence.
3. Develop new kinds of pesticides with highly effective, low-toxic and low-residual properties; improve pesticide formulations and application equipment and methods.
4. Popularize the integrated approach to pest management and the application of biodegradable pesticides in agriculture.
5. Carry on active propaganda and education on the scientific use of pesticides, and prevent the abuse of pesticide usage.
6. Enhance the national groundwater management programs, characterize the national groundwater resources, and implement the registration regulations for groundwater use.
7. Strengthen the fundamental research works on pesticide behavior in the environment, ban or restrict the application of some pesticides which have the potential to contaminate the soil or groundwater.

SUMMARY AND CONCLUSIONS

Pesticides have been widely and extensively used in China, and will continue to be used in the future. The application of pesticides had resulted in greatly increased production of food and feed grain, and benefited domestic animal husbandry. It will continue to play its important role in guaranteeing the harvests of agricultural crops. This is especially important to China because of its very large population.

However, the heavy use of pesticides had contaminated, and may continue to result in the contamination of the ecosystem, bringing hazards to human health and life. How to manage well and effectively use pesticides, so as to reach the goal of being both able to control the harmful diseases, pests, and weeds and yet not contaminate the ecosystem is of utmost importance to mankind.

Pesticides behave in many different ways in soil and groundwater, depending on their properties and related environmental conditions. Although the behavior of pesticides in the soil is very complex and can be influenced by many factors, it can be summarized as adsorption, degradation, movement, volatilization, plant uptake, and residue. Related mechanisms, processes, and factors that affect their transformation are briefly discussed in different sectors of the text.

There are many organizations where researchers are engaged in behavior, residue, toxicity, environmental effects, and safety evaluation of pesticides, etc. Of these, Nanjing Agricultural University and Anhui Agricultural University are proficient at environmental behavior research of pesticides, especially in photodegradation mechanisms; Zhejiang Agricultural University is proficient at residue studies; the Tea Research Institute of Chinese Academy of Agricultural Sciences is skilled in effects and residues of pesticides in tea; the Northeastern Agricultural University is skilled in the effect of pesticides on microorganisms; the Institute of Occupational Medicine of the Chinese Academy of Preventive Medicine is expert in toxicity studies; and the Nanjing Research Institute of Environmental Sciences of the Chinese NEPA is expert in all aspects of the above, especially in environmental effects and safety evaluation of pesticides.

ACKNOWLEDGMENTS

The authors wish to acknowledge the assistance of Prof. DeFang Fan (Zhejiang Agricultural University), Prof. Zu Yi Chen (Nanjing Agricultural University), Prof Yue Yong De (Anhui Agricultural University) in the preparation for the text, and Prof. Jue Yang (Hohai University) for her useful comments and critical review of the text.

REFERENCES

Cai, D.J. Soil Contamination by BHC and DDT in China and Its Prevention (unpublished report, 1977).

Cai, D.J. The Standard of Safety Evaluation for Pesticides on Non-target Organisms. *Rural Eco-Environ.* 3(2), pp. 12–16, 1987.

Cai, D.J. Volatilization of Pesticides, in *Encyclopaedia of Chinese Agriculture, Pesticide Volume,* 1st ed., X.L. Han, Ed., Agriculture Publishing House, Peking, 1993a.

Cai, D.J. Soil Adsorption of Pesticides, in *Encyclopaedia of Chinese Agriculture, Pesticide Volume,* 1st ed., X.L. Han, Ed., Agriculture Publishing House, Peking, 1993b.

Cai, D.J., X.L. Jiang, and Y.Q. Cai. Effect and Evaluation of Pesticides on Activity of Soil Microorganisms. *Rural Eco-Environ.* 2(2), pp. 9–14, 1986.

Cai, D.J. Safety Evaluation of Herbicides on Eco-Environment (unpublished report, 1989).

Cai, D.J., P.Z. Yang, and J.L. Wang. The Behavior and Hazard of Organochlorine Pesticides in Environment. *Chinese J. Ecol.*, 3(1), pp. 12–17, 1983.

Cai, D.J. and G.C. Tang. Pesticide Usage in China (unpublished report, 1992).

Cai, D.J., F. Xiang, X.M. Jiang, Z.L.Zhu, X.M.Hua, and Z.K.Dai. Influence of Aldicarb on Groundwater Quality. *Acta Scientiae Circumstantiae.*10(4), pp. 482–487, 1990.

Cai, D.J., F. Xiang, X.M. Jiang, Z.L.Zhu, X.M.Hua, and Z.K.Dai. Fate of Aldicarb in the Vadose Zone Beneath a Cotton Field. *J Contam. Hydrol.* 14, pp. 129–142, 1993.

Cai, D.J., Z.L. Zhu, Z.J. Shan, X.M. Hua, and X.M. Jiang. Prediction of Vulnerable Areas and Division of Groundwater Contamination by Aldicarb. *Adv. Environ Sci.,* 3(1), pp. 11–37, 1995.

Cai, L. and M.Y. Jiang. Residue and Harness Studies on Chlorsulfuron in Soil. *Rural Eco-Environ.*, 11(2), pp. 39–42, 1995.

Cai, Y.Q. and D.J. Cai. Evaluation of the Toxicity and Safety of Four Pesticides Such as Isofenphos-methyl to Earthworm. *Pest. Sci. Admin.,* 2(3), pp. 23–27, 1993.

Cai, Y.Q., S.L. Wang, and D.J. Cai. Effects of Isofenphos-methyl and Three Other Pesticides on

Soil Respiration and Their Evaluation. *Rural Eco-Environ.*, 8(3), pp. 36–40, 1992.
Chen, Z.Y., Y. Shi, and S.L. Huang. *Bound Residues of ^{14}C-Chlorotoluron in Soil.* Atomic Publishing House, Peking, China, 1992.
Chen, Z.M. A Comparative Study of Thin Film Photodegradation Rates for 35 Pesticides. *Acta Scientiae Circumstantiae*, 5(1), pp. 48–54, 1985.
Cheng, G.S. Ecological Effect of Pesticide on Microbe in the Soil-The Influence of Fungicides on VA Mycorrhiza. *China Environ. Sci.*, 5(1), pp. 21–25, 1985.
Fan, D.F. and X.S. Huang. *Pesticide Contamination and Its Prevention.* Science Publishing House, Peking, China, 1982, pp. 48–57.
Hallberg, G.R. Pesticides Pollution of Groundwater in the Humid United States. *Agric. Ecosys. Environ.*, 26, pp. 299–367, 1989.
Han, X.L. and F. Li. *Metabolization, Degradation and Toxicology of Pesticide.* Chemical Industry Publishing House, Peking, China, 1991.
Holden, P.W. *Pesticides and Groundwater Quality: Issues and Problems in Four States.* National Academy Press, Washington, DC, 1986.
Hua, X.M. and Z.L. Shan. The Production and Application of Pesticides and Factors Analysis of Their Pollution in Environment in China. *Adv. Environ. Sci.*, 4(2), pp. 33–45, 1996.
Huang, H.K. *An Introduction to Water Contamination and Its Prevention.* Peking Agricultural University Press, Peking, China, 1991.
Huang, X. Effect of Soil Adsorption on the Dissipation of Carbofuran. *Pesticide,* 26(5), pp. 32–33, 1987.
Jiang, X.L. and D.J. Cai. Photolysis of Pesticides in Aqueous Solution and Soil Surface. *Rural Eco-Environ.*, 3(1):, pp. 16–19, 1987a.
Jiang, X.L. and D.J. Cai. Volatilization of Pesticides from Water and Soil Surfaces. *China Environ. Sci.*, 10(3), pp. 171–176, 1990.
Jiang, X.M. and D.J. Cai. Adsorption and Desorption of Pesticides in Soil. *Rural Eco-Environ.*, 3(4), pp. 11–14, 1987b.
Jiang, X.L., Y. Jin, and D.J. Cai. Research on the Residual Dynamics and Effects of the Herbicide Chlorotoluron in Wheat Field on the Successive Crops. *Agro-Environ. Prot.*, 11(6), pp. 252–255, 1992.
Kong, L.R., H. Hong, R.W. Xu, and W. Jin. The Photodegradation and Volatilization of Dimehypo. *Environ. Chem.*, 10(4), pp. 31–35, 1991.
Lang, Y.Z. *Pesticide and Ecology.* Zhongnan Industry University Press, Hunan, China, 1992.
Li, W.M., J.W. Yao, and Z.Q. Gong. *Pesticide and Environment.* Chemical Industry Publishing House, Peking, China, 1985a.
Li, Z.X., S.Z. Huang, and Q.Y. Bai. Studies on Biological Concentration of Technical Benzene Hexachloride(BHC) in Terrestrial-Food Chains. *Acta Agricultufae Universitatis Zhejiangensis,* 11(2), pp. 151–158, 1985b.
Lou, G.L. and M.L. Li. Pesticide Residue in Different Rice Field Soils. *Agro-Environ. Prot.* 3(3), pp. 8–11, 1984.
Northeastern China Agricultural College. *Effects of Herbicide on the Formation of Soybean Nodule and Azotase Activity* (unpublished report, 1989).
Qian, C.F. Pesticide Residue, in *Encyclopaedia of Chinese Agriculture, Pesticide Volume,* 1st ed., X.L. Han, Ed., Agriculture Publishing House, Peking, 1993.
Shan, Z.J. *Bound Residue of Pesticide.* Nanjing, China, (unpublished report, 1994).
Shan, Z.J., A.Y. Zhang, and D.J. Cai. Injured Dose of Chlorsulfuron Rice. *Rural Eco-Environ.*, 11(1), pp. 36–39, 1995.
Shan, Z.J., Z.L. Zhu, X.H. Hua, and D.J. Cai. Mobility of Three Pesticides in Soils. *Rural Eco-Environ.*, 10(4), pp. 30–33, 1994a.
Shan, Z.J., Z.L. Zhu, X.H. Hua, X.M. Jiang, and D.J. Cai. Influence of Alachlor on Groundwater. *Acta Scientiae Circumstantiae*, 14(4), pp. 72–78, 1994b.
South China Agricultural University. *Chemical Protection of Plant.* Agriculture Publishing House, Peking, China, 1990.

Su, D.S. and D.F. Fan. Degradation and Movement of Fenpropathrin in Soils. *Acta Scientiae Circumstantiae*, 9(4), pp. 446–452, 1989.

Su, S.Q. and C.S. Zhao. Dynamics of Chlorsulfuron in Soil. *J. Weed Sci.*, 2(2), pp. 17–22, 1988.

Wan, H.B. Photostability of Organophosphorus Pesticides and Its Relation with Their Molecular Structure. *Acta Scientiae Circumstantiae*, 11(4), pp. 468–474, 1991.

Wang, J.H., L.Z. Zhang, and G.M. Dai. Adsorption of Some Pesticides in Soil. *Environ. Chem.*, 8(5), pp. 21–27, 1989.

Wang, Z.H., X.Z. An, and C.H. Li. A Mathematical Approach to the Degradation Law of Pesticides. *Agro-Environ. Prot.*, 11(6), pp. 283–285, 1992a.

Wang, Z.H., C.G. Qiao, and S.M. Liang. A New Method of the Quantitative Prediction of Pesticide Residues. *China Environ. Sci.*, 12(5), pp. 393–396, 1992b.

Yang, D.W., S.X. Yang, and H.H. Mo. Transport of Pesticide in Unsaturated Soil and Its Influence Factors. *Acta Pedological Sinica*, 29(4), pp. 383–391, 1992.

Yang, P.Z., A.Y. Zhang, and D.J. Cai. The Degradation of Pesticide in Soil. *Rural Eco-Environ.*, 3(3), pp. 15–18, 1987.

Yu, Z.D. and D.M. Chen. Pesticides and Their Environmental Impacts. *Environ. Chem.*, 5(6), pp. 19–24, 1986.

Yue, Y.D.. Factors Affecting the Photodegradation of Ethylenethiourea in Soil. *Acta Scientiae Circumstantiae*, 9(3), pp. 338–345, 1989.

Yue, Y.D. and R.M. Hua. Studies on Photosensitive-degradation of Pyrithriod Insecticides. *Acta Scientiae Circumstantiae*, 12(4), pp. 466–472, 1992.

Zeng, K.R. and Z.F. Wang. Influence of Pesticide on Nitrogenase and Hydrogen Uptake Hydrogenase Activities of *Azospirillum Brasilense. Acta Scientiae Circumstantiae*, 6(3), pp. 357–363, 1986.

Zhang, A.Y. and D.J. Cai. Effects of Herbicide on the Microbial Activity, Nitrification and Ammonification in Soil. *Rural Eco-Environ.*, 6(3), pp. 62–67, 1990.

Zhang, A.Y., Z.J. Shan, and D.J. Cai. The Mobility of Four Pesticides in Soil TLC System. *Rural Eco-Environ.*, 10(2), pp. 29–32, 1994.

Zhang, L.C. Bound Residue of Deltamethrin. *Environ. Chem.*, 5(2), pp. 19–27, 1986.

Zhang, R.W., Z.X. Li, Q.Y. Bai, and D.J. Cai. Application of a Standardized Method for Determing Acute Toxicity of Some Pesticides in Earthworms. *Acta Scientiae Circumstantiae*, 5(3), pp. 327–334, 1985.

Zhang, R.W., Z.X. Li, Q.Y. Bai, and D.J. Cai. Toxicity and Evaluation of Pesticides on Earthworms. *Rural Eco-Environ.*, 2(2), pp. 14–19, 1986.

Zheng Z. Microbial Degradation of Pesticides. *Environ. Sci.*, 11(2), pp. 68–72, 1990.

Zhou, J. A Primary Discussion on the Harmfulness of the Non-point Pollution Source in the Agriculture Areas. *Agro-Environ. Prot.*, 9(1), pp. 22–25, 1990.

Zhu, Z.L. The Safe and Effective Use of Pesticide, Prevention of Contamination of Groundwater. *Pest. Sci. Admin.*, 13(3), pp. 34–37, 1994.

Zhu, Z.L. and D.J. Cai. Residue, Toxicology of Aldicarb and Its Effects on Ecosystem. *Rural Eco-Environ.*, 9(2), pp. 50–53, 1993.

Zhu, Z.L., Z.J. Shan, X.M. Hua, and D.J. Cai. Analysis on the Factors Affecting Aldicarb Ground Water Contamination. *Rural Eco-Environ.*, 10(1), pp. 25–28, 1994a.

Zhu, Z.L., F. Xiang, X.M. Jiang, X.M. Hua, and D.J. Cai. Dynamic Simulation of TEMIK Residues and Its Movement in Soil. *Environ. Pollut. Control*, 16(6), pp. 1–5, 1994b.

APPENDIX: CHEMICAL NAMES OF PESTICIDES MENTIONED IN CHAPTER 9

Common Name	Chemical Name
Alachlor	2-chloro-2′,6′-diethyl-N-(methoxymethyl)acetanilide
Aldicarb	2-methyl-2-(methylthio)propionaldhyde O-methylcarbamoylxime
Aldrin	1,2,3,4,10,10-hexachloro-1,4,4a,5,8,8a-hexahydro-endo-1,4-exo-5,8-dimethano-naphthalene
Aliette	aluminium tri-ethylphosphonate
Amitrole	3-amino-s-triazole
Asozine	methylarsinic sulphide
Atrazine	2-chloro-4-(ethylamino)-6-(isopropylamino)-s-triazine
Azinphos-methyl	3-dimethoxyphosphinothioylthiomethyl-1,2,3-benzotriazin-4(3H)-one
Azocyclotin	tri(cyclohexyl)-1H-1,2,4-triazol-1-yltin
Bensulfuron	methyl 2-[[[[[(4,6-dimethoxy pyrimidin-2-yl)amino] carbonyl]amino]sulfonyl] methyl] benzoate
Benzoximate	3-chloro-α–ethoxyimino-2,6-dimethoxybenzyl benzoate
α-BHC	α-1,2,3,4,5,6-hexachlorocyclohexane
β-BHC	β-1,2,3,4,5,6-hexachlorocyclohexane
δ-BHC	δ-1,2,3,4,5,6-hexachlorocyclohexane
Bifenthrin	2-methylbiphenyl-3-ylmethyl-(Z)-(1RS,3RS)-3-(2-chloro-3,3,3-trifluoroprop-1-enyl)-2,2-=dimethylcyclopropanecarboxylate
BPMC	2-sec-butylphenyl methylcarbamate
Buprofezin	2-tert-butylimino-3-isopropyl-5-phenyl-1,3,5-thiadiazinan-4-one
Butachlor	N-butoxymethyl-2-chloro-2′,6′-diethylacetanilide
Carbaryl	1-naphthyl methylcarbamate
Carbendazim	methyl benzimidazol-2-ylcarbamate
Carbofuran	2,3-dihydro-2,2-dimethylbenzofuran-7-yl methylcarbamate
Carbosulfan	2,3-dihydro-2,2-dimethylbenzofuran-7-yl(dibutylaminothio) methylcarbamate
Cartap	1,3-di(carbamoylthio)-2-dimethylaminopropane
Chlordimeforn	N′-(4-chloro-o-tolyl)-N,N-dimethyl-formamidine
Chlorfenvinphos	2-chloro-1-(2,4-dichlorophenyl)vinyl diethyl phosphate
Chlorfluazuron	1-[3,5-dichloro-4-(3-chloro-5-trifluoromethyl-2-pyridyloxy) phenyl]-3-(2,6-difluorobenzoyl)urea
Chlorothalonil	tetrachloroisophthalonitrile
Chlorpyrifos	O,O-diethyl O-3,5,6-trichloro-2-pyridyl phosphorothioate
Chlorsulfuron	2-chloro-N-[(4-methoxy-6-methyl-1,3,5-trizin-2-yl)amino carbonyl] beneze sulphonamide
Chlortoluron	3-(3-chloro-p-tolyl)-1,1-dimethylurea
Clofentezine	3,6-bis(2-chlorophenyl)-1,2,4,5-tetrazine
Copper sulfate	cupric sulfate
Cyfluthrin	(RS)-α-cyano-4-fluoro-3-phenoxybenzyl(1RS)-cis-trans-3-(2,2-dichlorovinyl)-2,2-=dimethylcyclopropanecarboxylate
Cypermethrin	(±)-α-cyano-(3-phenoxy)benzyl-(1RS)-cis-trans-2,2-dimethyl-3-(2,2-dichlorovinyl) cyclopropanecarboxylate
2,4-D	(2,4-dichlorophenoxy)acetic acid
2,4-DB	4-(2,4-dichlorophenoxy)butyric acid
Dalapon	2,2-dichloropropionic acid
DDE	1,1-dichloro-2,2-bis(p-chlorophenyl)ethylene
DDT	1,1,1-trichloro-2,2-bis(p-chlorophenyl)ethane
Deltamethrin	α-cyano-phenoxybenzyl(1R,3R)-3-(2,2-dibromoethenyl)-2,2-dimethyl cyclopropanecarboxylate
Diazinon	O,O-diethyl-O-(2-isopropyl-6-methyl-4-pyrimidinyl)phosphorothioate
Dibromochloropropane	1,2-dibromo-3-chloropropane
Dichlorovos	2,2-dichlorovinyl dimethyl phosphate
Diclofop-methyl	methyl(RS)-2-[4-(2,4-dichlorophenoxy)phenoxy]propionate
Dicofol	2,2,2-trichloro-1,1-bis(4-chlorophenyl)ethanol
Dieldrin	1,2,3,4,10,10-hexachloro-exo-6,7-epoxy-1,4,4a,5,6,7,8,8a-octahydro-1,4-endo-exo-5,8-dimethanonaphthalene
Diflubenzuron	1-(4-chlorophenyl)-3-(2,6-difluorobenzoyl)urea
Dimehypo	2-dimethylamino-1,3-di(sodium thiosulphate)propane

Common Name	Chemical Name
Dimethazon	2-(2-chlorophenyl)methyl-4,4-dimethyl-3-isoxazolidinone
Dimethoate	O,O-dimethyl-S-(N-methylcarbamoylmethyl)-phosphorodithioate
Dinoseb	2-sec-butyl-4,6-dinitrophenol
Diquat	9,10-dihydro-8a,10a-diazoniaphenanthrene
Disulfoton	O,O-diethyl-S-2-ethylthioethyl phosphorodithioate
Diuron	3-(3,4-dichlorophenyl)-1,1-dimethylurea
DNOC	2-methyl-4,6-dinitrophenol
EPN	O-ethyl-O-4-nitrophenyl phenylphosphonothioate
EPTC	S-ethyl dipropylthiocarbamate
Esfenvalerate	(S)-α-cycno-3-phenoxybenzyl-(S)-2-(4-chlorophenyl)-3-methylbutyrate
Ethion	O,O,O′,O′-tetraethyl-S,S′-methylene di(phosphorodithioate)
Ethofenprox	2-(4-ethyloxyphenyl)-2-methylpropyl-3-phenyloxybenzyl ether
Ethylene dibromide	1,2-dibromoethane
Ethylenethiourea	degradation product of the ethylene bisdithiocarbamate
Ethylmercury chloride	ethylmercury chloride
Fenamiphos	O-ethyl-O-(3-methyl-4-methylthiophenyl)-isopropylamino phosphate
Fenitrothion	O,O-dimethyl-O-4-nitro-m-tolyl phosphorothioate
Fenothiocarb	S-4-phenoxybutyl dimethylthiocarbamate
Fenpropathrin	(RS)-α-cycno-3-phenoxybenzyl-2,2,3,3-tetramethylcyclopropanecarboxylate
Fenpyroximate	tertbutyl-(E)-4-[(1,3-dimethyl)-5-phenoxypyrazol-1H-4-yl] methyleneamino-oxy)-p-toluate
Fenthion	O,O-dimethyl-O-[4-(methylthio)-m-tolyl] phosphorothioate
Fenuron	1,1-dimethyl-3-phenylurea
Fenvalerate	(RS)-α–cycno-3-phenoxybenzyl-(RS)-2-(4-chlorophenyl)-3-methylbutyrate
Fluazifop-butyl	butyl-2-[4-(5-trifluoromethyl-2-pyridinyl)oxyphenoxy] propionate
Flufenoxuron	1-[4-(2-chloro-2,2,2-trifluoro-p-tolyloxy)-2-fluorophenyl]-3-(2,6-difluorobenzoyl)urea
Fluvalinate	(RS)-α-cycno-3-phenoxybenzyl N-(2-chloro-α,α,α-trifluoro-p-tolyl-valinate
Folpet	N-trichloromethylthiophthalimide
Fonofos	O-ethyl S-phenyl(RS)-ethylphosphonodithioate
Haloxyfop	2-[4-(3-chloro-5-(trifluoromethyl-2-pyridinyl)oxy) phenoxy]propanoic acid methyl ester
Hexythiazox	trans-5-(4-chlorophenyl)-N-cyclohexyl-4-methyl-2-oxothiazolidone-3-carboxylamide
IBP	S-benzyl O,O-diisopropyl phosphorothioate
Isazophos	O-5-chloro-1-isopropyl-1H-1,2,4-triazol-3-yl O,O-diethyl phosphorothioate
Isocarbofos	O-methyl-O-(o-carboisopropoxyphenyl) phosphoroamidothioate
Kitazine	S-benzyl O,O-diethyl phosphorothioate
Lead arsenate	lead arsenate
Leptophos	O-(4-bromo-2,5-dichlorophenyl)-O-methyl phenylphosphorothioate
Lindane	γ-1,2,3,4,5,6-hexachlorocyclohexane
Linuron	3-(3,4-dichlorophenyl)-1-methoxy-1-methylurea
Malathion	O,O-dimethyl-S-(1,2-dicarbenthoxyethyl) phosphorodithioate
Maleic hydrazide	6-hydroxy-2H-pyridazin-3-one
MCPA	2-methyl-4-chlorophenoxyacetic acid
Metalaxyl	methyl N-((2,6-dimethyl phenyl)-N-(2′-methoxyacetyl)-DL-alaninate
Methomyl	1-(methylthio)ethylideneamino methylcarbamate
Methothrin	4-(methoxymethyl)benyl-(1RS)-cis-trans-2,2-dimethyl-3-(2-methyl-1-propenyl) cyclopropanecarboxylate
Methoxychlor	2,2-bis(p-methoxyphenyl)-1,1,1-trichloroethane
Methsulfuron-methyl	methyl 2-[3-(4-methoxy-6-methyl-1,3,5-triazin-2-yl) ureidosulphonyl]benzoate
Metobromuron	3-(p-bromophenyl)-1-methoxy-1-methylurea
Metolachlor	2-ethyl-6-methyl-N-(1′-methyl-2′-methoxyethyl) chloroacetanilide
Metribuzin	4-amino-6-tert-butyl-4,5-dihydro-3-methyl-thio-1,2,4-triazine-5-one
Mevinphos	2-methoxy-carbonyl-1-methylvinyl dimethylphosphate
Molinate	S-ethyl hexahydro-1H-azepine-1-carbothioate
Monocrotophos	dimethyl-1-methyl-3-(methylamino)-3-oxo-1-propenylphosphate
Monuron	3-(4-chlorophenyl)-1,1-dimethylurea
MTMC	m-tolyl-N-methylcarbamate

Common Name	Chemical Name
Neburon	1-butyl-3-(3,4-dichlorophenyl)-1-methylurea
Omethoate	O,O-dimethyl S-(N-methylcarbamoylmethyl)phorothioate
Oxamyl	N,N-dimethyl-2-methylcarbamoyloxyimino-2-(methylthio)acetamide
Paraquat	1,1-dimethyl-4,4′-bipyridylium ion
Parathion	O,O-diethyl-O-p-nitrophenyl phosphorothioate
Parathion-methyl	O,O-dimethyl-O-p-nitrophenyl phosphorothioate
PCNB(Guintozene)	pentachloronitrobeneze
PCP	pentachlorophenol
PCP-Na	sodium pentachlorophenate
Pendimethalin	N-(1-ethylpropyl)-2,6-dinitro-3,4-xylidine
Permethrin	3-phenoxybenzyl)methyl cis-trans(±)-3-(2,2-dichloroethenyl)2,2-dimethylcyclopropanecarboxylate
Phenothrin	3-phenoxybenzyl(±)-cis-trans-chrysanthemate
Phorate	O,O-diethyl-S-(ethylthio)-methyl phosphorodithioate
Phosalone	6-chloro-3-diethoxyphosphinothioylthiomethyl-1,3-benzoxazol-2(3H)-one
Phosmet	N-(dimethoxyphosphinothioylthiomethyl)phthalimide
Phoxim	2-(diethoxyphosphinothioyloxyimino)-2-phenylacetonitrile
Pirimicarb	2-dimethylamino-5,6-dimethylpyrimidin-4-yl dimethylcarbamate
Prometryne	2,4-bis(isopropylamino)-6-methylamino-1,3,5-triazine
Propanil	N-(3,4-dichlorophenyl)propionamide
Propazine	2-chloro-4,6-bis(isopropylamino)-1,3,5-triazine
Propoxur	2-isopropoxyphenyl methylcarbamate
Propylene dichloride	1,2-dichloropropane
Prothiocarb	S-ethyl-N-(3-dimethylamino-propyl)thiocarbamate
Pursuit	(±)5-ethyl-2-(4-isopropyl-4-methyl-5-oxo-2-imidazolin-2-yl)nicotinic acid
Pyrazosulfuron-ethyl	ethyl 5-[3-(4,6-dimethyl pyrimidin-2-yl)ureidosulfonyl]-1-methylpyrazole-4-carboxylate
Pyridaphenthion	2,3-dihydro-3-oxo-2-phenyl-6-pyridazinyl diethyl phosphorothionate
Quinalphos	O,O-diethyl O-quinoxalin-2-yl phosphorothioate
Quinclorac	3,7-dichloro-quinoline acid-8
Quizalofop-ethyl	ethyl-2-[4-(6-chloro-2-quinoxalinyloxy)phenoxy] propanoate
Resmethrin	5-benzyl-3-furylmethyl-(1RS)-cis-trans-chrysanthemate
S-5439	3-phenoxybenzyl-(RS)-2-(4-chlorophenyl)-3-methylbutyrate
Sebufos	S,S-di-sec-butyl O-ethyl phosphorodithioate
Sethoxydim	(±)2-[1-(ethoxyimino)butyl]-5-[2-(ethylthio)propyl]-3-hydroxy-2-cyclohexen-1-one
Simazine	2-chloro-4,6-bis(ethylamino)-1,3,5-triazine
Simetryne	2,4-bis(ethylamino)-6-methylthio-1,3,5-triazine
Sodium Fluoride	sodium fluoride
Sulphur	sulfur
Tebufenpyrad	N-(4-tert-butylbenzyl)-4-chloro-3-ethyl-1-methyl-pyrazole-5-carboxamide
Tetramethrin	cyclohex-1-ene-1,2-dicarboximidomethyl(()-cis-trans-chrysanthemate
Thiophanate-methyl	dimethyl 4,4′-(o-phenylene)bis(3-thioallophanate)
Toxaphene	octachlorocamphene
Triallate	S-(2,3,3-trichloroally)diisopropylthiocarbamate
Tribenuron-methyl	methyl-2-[3-(4-methoxyl-6-methyl-1,3,5-triazin-2-yl)-3-methylureidosulphonyl] benzoate
Trichlorphon	O,O-dimethyl-(2,2,2-trichloro-1-hydroxyethyl) phosphonate
Trietazine	2-chloro-4-diethylamino-6-ethylamino-1,3,5-triazine
Trifluralin	α,α,α-trifluoro-2,6-dinitro-N,N-dipropyl-p-toluidine
Vinolozolin	(RS)-3-(3,5-dichlorophenyl)-5-methyl-5-vinyl-1,3-oxazolidine-2,4=dione
XMC	3,5-xylyl-N-methylcarbamate
Danjiami[a]	2,4-dimethyl-phenyl-imino-methyl-methylaminechloride
Jiajiyiliulin[a]	N-isopropyl-O-methyl-O-[(2-isopropoxylcarbonyl) phenyl]phosphorothioate
Kecaoan[a]	N-ethoxymethyl-2-chloro-2′-ethylacetanilide
Midinyanglin[a]	O,O-diethyl-O-[2-methoxyl-4-methyl-6-pyrimidinyl] phosphate
Mieyouniao[a]	o-chlorophenylcarbonyl-N-p-chlorophenyl-urea
Shachongdan[a]	2-dimethylamino-1-(sodium thiosulphate)-3-(thiosulphatic acid)-propane

[a]These pesticides were developed by China, and there are no English common and/or trade names.

CHAPTER 10

PERSPECTIVES OF ENVIRONMENTAL POLLUTION IN DENSELY POPULATED AREAS: THE CASE OF HONG KONG

M.S. Yang and M. H. Wong

INTRODUCTION

Hong Kong is situated at the Pearl River Estuary of Guangdong Province, south of China. Over the past 95 years, the city has grown from being a fishing village to one of the most important financial centers in the world. The population increased from a few thousand to over six million. The city handles a huge volume of trade and, until recently, has been considered the only link the Western World has with China. Concurrent with such a remarkable transformation, a substantial amount of environmental problems unique to the city have arisen. In particular, problems associated with waste disposal have captured the majority of the attention over the past 20 years. This chapter reviews the problems that the population and its associated activities have imposed on the Hong Kong environment, as well as the remedial and control measures that are in place. Particular attention is directed toward groundwater quality and land use.

Geographical Setting of Hong Kong

Situated at the southern tip of China, Hong Kong is a territory composed of islands. Besides Hong Kong Island itself, the largest one is Lantau Island. In 1898, the Kowloon peninsula and its adjacent New Territories were incorporated into the territory. Together with continuous reclamation activity throughout the years, the total land area of Hong Kong is 1,078 km^2 (Fig. 10.1).

Hong Kong is built on granite rock overlain by thin layers of alluvial soil. The upland hilly terrain, which covers the New Territories and the middle of the two largest islands, descends rapidly to valley floors and low-lying flood plains (Chiu and So, 1986). There are about 600 rivers and streams which drain the hills. The majority of the rivers have a high flow during the summer monsoon season, which begins in April and ends in September, when high pressure systems develop over the South China Sea, bringing large amounts of precipitation (Poon, 1989). During the remaining of the year, the river volume is decreased markedly. These rivers drain to the open sea.

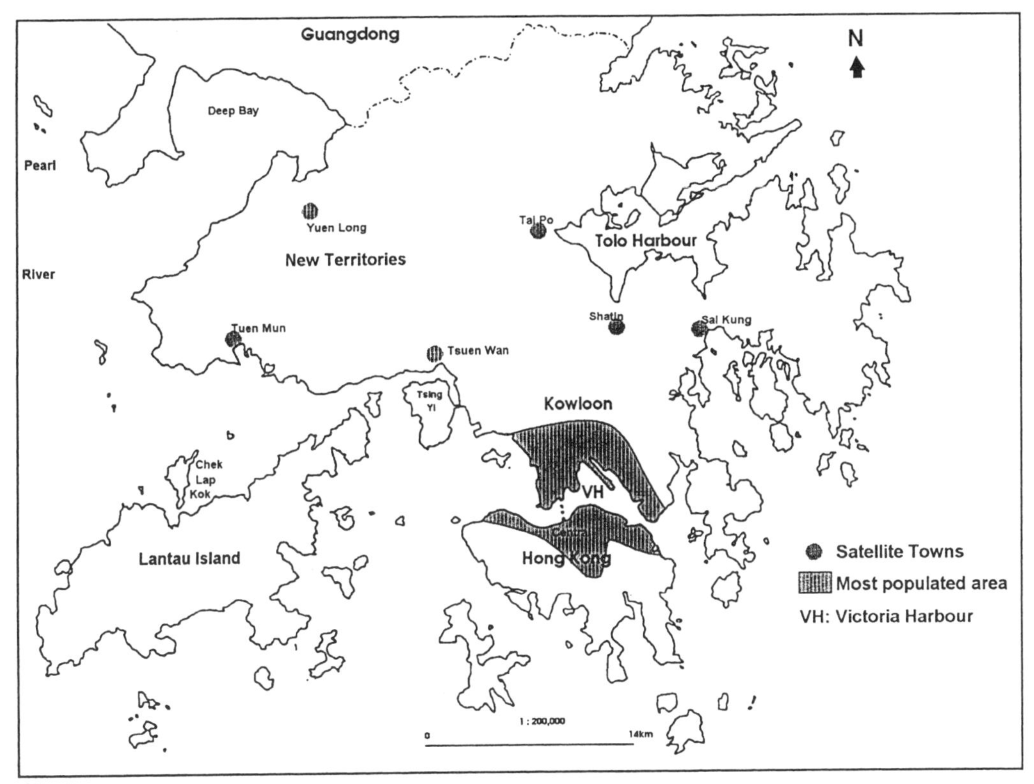

Figure 10.1. A map of Hong Kong indicating the most populated areas and the new town development (Source: Hong Kong Environmental Protection Department, Environment HK, 1990).

Population Growth and Distribution

Since 1898, the population of Hong Kong has grown enormously. About three-quarters of a million people entered Hong Kong in 1949 and 1950 due to the change of government in China. The population reached 3.1 million in 1967, and exceeded 6 million in 1993. The annual growth rate of the population averaged 1.1% over the past 10 years (1984–93). There has been a stable death rate at about 5 per 1,000 and a declining birth rate over the 10-year period (from 16 to 12 per 1,000), so the rate of natural increase has dropped steadily (from 11 to 6.8 per 1,000, respectively). The further increase in population has been due to immigration.

Even with the reclamation of coastal areas, which created some 18 km^2 of land for housing development, Hong Kong is still one of the most densely populated places in the world. The average density was 5,700/km^2 in 1993. The majority of the population are located in the urban coastal area around Central in Hong Kong island, and the southern tip of the Kowloon peninsula (Fig. 10.1), where the density exceeds 26,000/km^2.

The uneven distribution and high degree of overcrowding in these urban areas have long been recognized. The government's decentralization scheme helped to redistribute the population to several satellite towns such as Yuen Long, Shatin, and Tai Po in the New Territories. As a consequence, there has been a continuous urban growth in these areas (Fig. 10.1), with massive redevelopment and resettlement projects since 1961 (Lo, 1986).

The demographic pressure associated with rapid urbanization has caused the replacement of much needed fertile agricultural land by the expanding new towns. In addition to the fact that 25% of the Hong Kong population is housed on land reclaimed from the sea (Boyden et al., 1981), there has been a steady leveling of the hills around the city for various developments. Scattered villages with a lower population density can still be found in the hilly terrain, however.

Economic Growth

Prior to 1950, there were extensive agricultural activities in the hills, with rice being the most important crop. The granite base rock in the territory yielded little groundwater for irrigation. Rice paddies were designed as steps along the hillside to catch the down flow of surface runoff. Currently, many of the rice paddy-fields have been leveled and redeveloped into new towns, with little agricultural activity remaining except for growing cash crops, such as vegetables.

Although Hong Kong has a shortage of land and water for industrial growth, and lacks raw materials and sources of power, the city has been able to attract many industries. Manufacturing of textiles and plastics, and food processing have been the major industries. These have been gradually replaced by light industries such as electronics and optical equipment. The estimated gross domestic product* (GDP) per capita increased by approximately 13% every year between 1986 and 1994 (Census and Statistics Department, Hong Kong, 1996). More recently, a large number of manufacturing industries have been moved from Hong Kong to the Pearl River Delta due to lower cost in labor and raw materials. Manufacturers in Hong Kong have been moving away from labor-intensive production into making high value-added products that can compete on quality. The Hong Kong Productivity Council is striving to make Hong Kong a regional leader in the management, service, financial, and technology sector.

MAJOR SOURCES OF ENVIRONMENTAL POLLUTANTS AND CONTROL STRATEGIES

Having to cope with a variety of environmental problems generated by the rapid increase in population, the Hong Kong Government set up the Environmental Protection Agency (EPA) in 1982. In 1986, the agency, together with special units from the Departments of Agriculture and Fisheries, Labor, Marine, Engineering Development and Municipal Services, were consolidated into the Hong Kong Environmental Protection Department (HKEPD). The major function of the Department was to develop procedures for monitoring environmental pollution and enforce matters pertaining to environmental protection.

Domestic Sewage

The ever-increasing volume of domestic sewage was one of the foremost problems faced by the EPD. A substantial amount of the sewerage in Hong Kong was disposed into the river system and carried out to sea. The survey conducted in 1988 showed that two-thirds of the rivers were polluted by livestock waste, domestic sewage, and indus-

---

*Gross domestic product per capita represents the sum of the value added of resident producing units; e.g., factories, shops, service organization calculated at current market price.

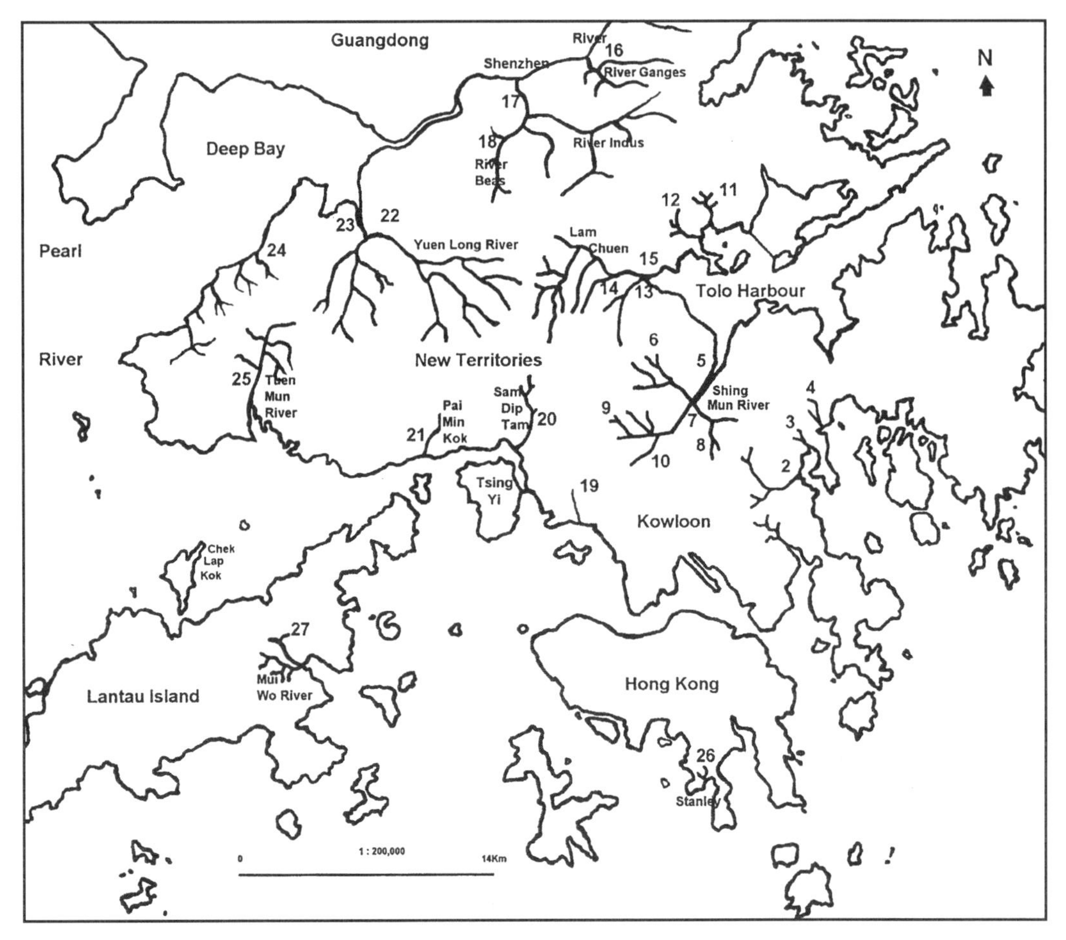

Figure 10.2. The major river systems of Hong Kong. The names of the rivers are listed in numbers which correspond to that in Table 10.1. (Source: Hong Kong Environmental Protection Department, River Quality of Hong Kong, 1993).

trial effluent. Consequently, the quality of the coastal waters has deteriorated. Since 1973, the water quality within the typhoon shelter in Victoria Harbour, which is situated between the most populated areas in Hong Kong (Fig. 10.1), has continued to decline. The dissolved oxygen (DO) in various stations ranged between 10 to 60% saturation and the *E. coli* count was above 2,000/100 mL (HKEPD, 1989). Although DO continued to improve between 1992 and 1995, there has been little change in *E. coli* content (HKEPD, 1996b). In the Deep Bay area which receives drainage from Yuen Long and Tuen Mun Rivers, the BOD_5 was reported to be 2.4 mg/L and *E. coli* count >1,281/100 mL. (HKEPD, 1989). The values were not improved in 1995 (HKEPD, 1996b)

Areas around the Tolo Harbour (Fig. 10.2), an enclosed body of water with limited mixing with the open sea, experienced the most rapid urban growth during the 1980s. The harbor receives drainage from the Shing Mun River and the Lam Chuen River, which pass through the Shatin and Tai Po new towns, respectively (Fig. 10.2). In 1982, one year after the establishment of the EPA, the Tolo Harbour was designated as the first Water Control Zone (WCZ) within which the quality of effluent was controlled. A sewage treatment plant was built along the Shing Mun River to collect and treat domes-

Table 10.1. The Changes in River Water Quality Between 1984 and 1992 in Different Monitoring Stations Around Hong Kong[a]

Rivers/Streams[b]		Change in Water Quality[c]
1.	Tseng Lan Shu Stream	F to G
2.	Ho Chung	VB to B
3.	Tai Chung Hau	B to F
4.	Sha Kok Mei	G to G
5.	Shing Mun Main Channel	F to F
6.	Fo Tan Nullah (Shing Mun)	B to F
7.	Siu Lik Yuen (Shing Mun)	F to G
8.	Kwun Yam Shan (Shing Mun)	E to G
9.	Tai Wai (Shing Mun)	B to G
10.	Tin Sum (Shing Mun)	G to G
11.	Shan Liu Stream	G to G
12.	Tung Tze Stream	G to G
13.	Tai Po Dau Stream	B to G
14.	Tai Po River	F to G
15.	Lam Tsuen	B to F
16.	River Ganges	VB to VB
17.	River Indus	VB to VB
18.	River Beas	VB to VB
19.	Kau Wah Keng Stream	F to F
20.	Sam Dip Tam Stream	F to F
21.	Pai Min Kok (Angler's) Stream	B to G
22.	Kam Tin River	VB to VB
23.	Yuen Long Creek	VB to VB
24.	Deep Bay River	VB to VB
25.	Tuen Mun River	VB to VB
26.	Stanley Stream	B to B
27.	Mui Wo River	B to G

[a]Hong Kong Environmental Protection Department, River Quality of Hong Kong, 1993.
[b]Locations of the rivers are indicated in Figure 10.2.
[c]The ranking of water quality is according to the Dutch Water Quality Index:

No. of Points Awarded	DO (5 Saturation)	BOD_5 (mg/L)	NH_4^+–N (mg/L)
1	91–110	<3	<0.5
2	71–90 or 111–120	3.1–6.0	0.5–1.0
3	51–70 or 121–130	6.1–9.0	1.1–2.0
4	31–50	9.1–15.0	2.1–5.0
5	<30 or >130	>15	>5.0

Total Points Awarded	Water Quality Condition
3.0–4.5	E
4.6–7.5	G
7.6–10.5	F
10.6–13.5	B
13.6–15	VB

tic effluent from the Shatin new town. The raw sewage is treated by screening, sedimentation, alum precipitation, and aerobic digestion before being discharged into the harbor, or transported through underground pipes to the Kai Tak Nullah. The water quality of some of the rivers that flow into the Tolo Harbour has improved significantly. In 1992, the quality of water in Fo Tan Nullah, and rivers that pass through Siu Lik Yuen and Tai

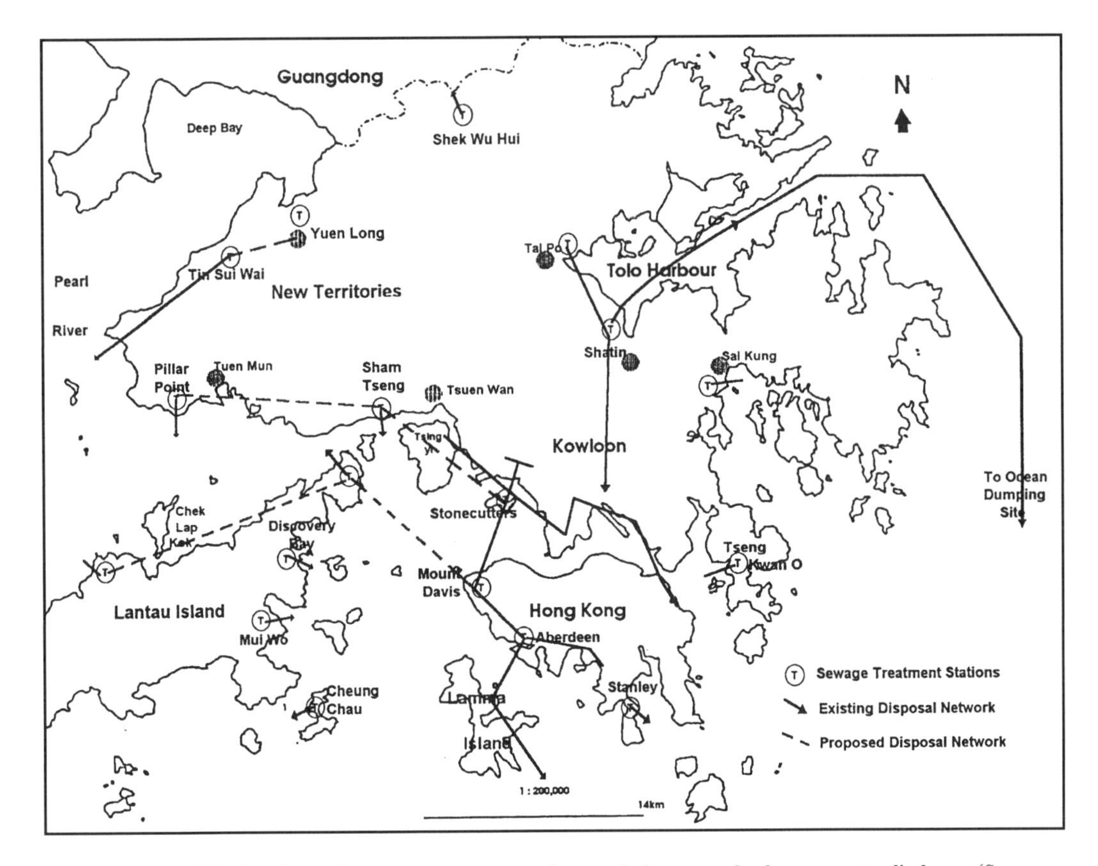

Figure 10.3. The locations of sewage treatment stations and the network of wastewater discharge (Source: Hong Kong Environmental Protection Department, Environment HK, 1990).

Wai (Fig. 10.2) gained significant improvement (Table 10.1). Since 1993, there was a significant reduction in *E. coli* counts recorded in various monitoring sites within the Tolo Harbour (HKEPD, 1996b).

The EPD divided the coastal waters of Hong Kong into 10 water control zones and the water quality is monitored continuously. It also drew up a Sewerage Master Plan and the Strategic Sewage Disposal Plan in which the entire territory was divided into 16 areas. Each area will have a sewage treatment plant built to receive domestic effluent (Fig. 10.3). After initial screening to remove large fragments, the water will be collected on a catchment by catchment basis, channeled through a network of deep tunnels and disposed through a submarine outfall that extends 30 km out into the South China Sea. Along selected strategic locations, sewerage would undergo a second stage of chemically enhanced treatment by addition of lime to reduce bacteria and heavy metals, and possibly a biological treatment to remove organic nutrients prior to disposal into the ocean (HKEPD, 1995). The project is expected to be in full operation in the early 2000s.

Livestock Waste

To control the disposal of livestock waste, the government declared urban Kowloon and the whole of Hong Kong Island to be the Livestock Waste Prohibition Zones in which

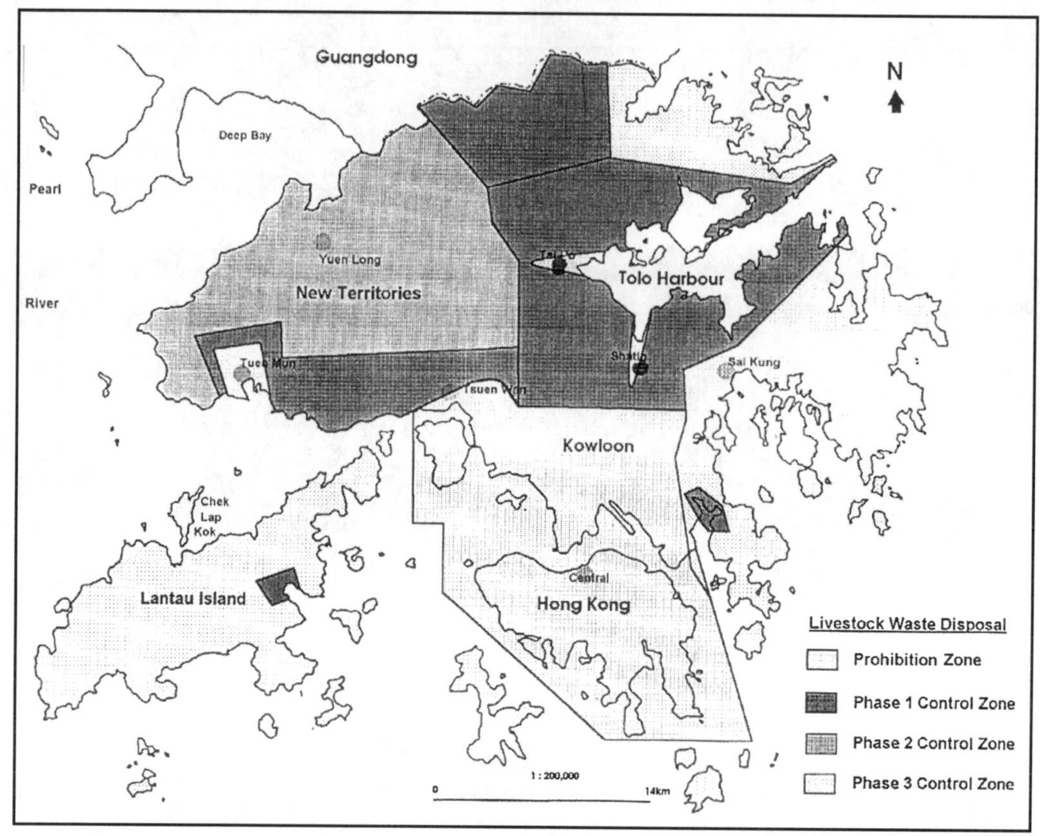

Figure 10.4. Livestock waste prohibition and control zones in Hong Kong (Source: Hong Kong Environmental Protection Department, Environment HK, 1990).

no disposal of livestock waste is permitted. The remaining areas were subsequently listed as the Livestock Waste Control Zones in which a permit is needed for disposing of livestock waste into open drainage (Fig. 10.4).

The institution of the Livestock Waste Prohibition Zones practically stopped all farming activities in urban Hong Kong and Kowloon. For those farmers in the New Territories who continue to survive within the Livestock Waste Control Zone, licenses must be obtained before disposal can be made into open drainage. Furthermore, farmers were advised to institute new waste management technologies; e.g., installing their own preliminary sewage treatment facility, the "dry muck out," or the "pig-on-litter" system in which a special bedding material, comprising sawdust and bacteria, is spread over the floor of the sty to promote degradation of pig manure. Complaints have been voiced by farmers who claimed that the investment for such high technology environmental protection devices are too great to warrant the existence of the industry. Illegal disposal of livestock waste might still be preferred by small-scale farmers. Although river water quality has been significantly improved within the Livestock Waste Prohibiting Zones, the water quality of rivers which pass through the Livestock Waste Control Zones have not changed. The quality of the River Ganges and the River Indus, which drain the northern New Territories into the Shenzhen River, and the Kam Tin River and Yuen Long Creek which drain into the Deep Bay (Fig. 10.2), continue to be graded as "Very Bad" (Table 10.1).

Pesticides

Only 5% of the Hong Kong land area is now being used for agriculture. Most of the arable land is located in the Yuen Long Plain northwest of the New Territories. Due to the limited agricultural activities, the use of pesticides in Hong Kong is low. It has been estimated that about 150 tonnes are used each year (Wu, 1988). Insecticides account for two-thirds of the total pestidices used and organophosphates are most common (Morton, 1989).

Despite the limited use of organophosphates, samples taken from human breast milk showed that the levels of DDT, DDE, and dieldrin to be 11.67, 2.17, and 10.24 ng/g lipid dry weight, respectively. These levels are significantly higher than those reported in other parts of the world (Phillips, 1989). A high level of DDT was also recorded in local biota. In green-lipped mussels collected from 15 locations around Hong Kong, DDT level ranged between 50–2,043 ng/g dry weight (Phillips, 1985). DDT has now been banned due to its potential toxicity. Presently, there has been no organized program to continue monitoring this substance.

Industrial and Chemical Wastes

The industrial sector in Hong Kong produces a large amount of hazardous chemical wastes. The electroplating and dye industries disposed of a large amount of toxic metals into the water courses, endangering aquatic life forms. Polychlorinated biphenyls (PCBs) are mainly used in transformers and capacitors. The two power companies in Hong Kong are the largest PCB users. It is estimated that over 3,300 PCB capacitors are still in use or in storage (Morton, 1989). A high PCB level (60 to 1,904 ng/g dry weight) has been detected in human milk (Phillips, 1989). Analysis of sludge collected from the Shatin Sewage Treatment Plant showed that the PCB content in sludge reached 240.46 ng/g wet weight (Yang et al., 1993). Fish (tilapia) fed with pellets containing 30% of the sludge showed a significant increase in PCBs in the flesh.

In April 1993, the first integrated chemical waste treatment center in Southeast Asia started its operation in the Tsing Yi Island near Tsuen Wan. Through incineration, encasement in cement, and oil/water separation, the plant can process up to 100,000 tonnes of various chemical wastes per year. The plant should reduce the loading of toxic metals and hazardous chemicals in the water system.

Municipal Wastes

Hong Kong produces an average of 20,000 tonnes of solid waste daily by the activities of 1.6 million domestic households, 280,000 commercial and manufacturing businesses, and over 400 construction sites. A network of waste collection centers was constructed and the collected wastes are being dumped into landfill sites throughout the territories.

There are nine completed and four operating landfills in Hong Kong (Table 10.2). The long-term strategy of solid waste treatment is to dispose of all municipal waste into three strategic landfills: South East New Territories (SENT) Landfill, West New Territories (WENT) Landfill, and North East New Territories (NENT) Landfill (Fig. 10.5). Constructions of these three sites are carried out using the state-of-the-art technology completed with polyethylene liners that trap the leachate for further processing, and pipelines for collecting the landfill gas. The WENT Landfill started its operation in November 1993 and was estimated to have a capacity of 57,000,000 m^3 and a life span of 25 to 30 years.

Table 10.2. Site History and Details of the 13 Landfill Sites in Hong Kong[a,b]

	Area (ha)	Quantity of Waste (Mt)	Opening Date	Closing Date	Predevelopment Terrain	Type of Site
Ngau Tam Mei	20.0	0.03	12/73	3/75	Shrubland	Valley fill
Ngau Chi Wan	13.5	0.734	1/76	12/77	Shrubland	Valley fill
Gin Drinker's Bay	29.0	3.54	6/73	2/79	Marine and reclaim	Built-up platform/ Marine fill
Ma Tso Lung	2.0	0.182	7/76	2/79	Agriculture	Valley/Built-up platform
Sai Tso Wan	14.0	1.59	3/77	2/80	Terrace	Valley/Built-up platform
Ma Yau Tong West	6.6	0.595	2/79	2/81	Terrace	Valley fill
Siu Lang Shui	11.7	1.19	11/78	12/83	Shrubland	Valley fill
Ma Yau Tong Central	5.8	1.05	2/81	4/86	Quarry	Quarry fill
Jordan Valley	6.5	–	4/86	before 1988	–	Quarry fill
Shuen Wan	48.0	–	6/74	ongoing	Marine and reclaim	Marine fill
Junk Bay Stage I	60.0	–	3/80	ongoing	Marine and reclaim	Marine fill
Junk Bay Stage II/III	35.0	–	–	ongoing	Marine and reclaim	Marine fill
Pillar Point Valley	34.0	–	8/83	ongoing	–	Valley fill

[a]Chan et al. 1996.
[b]Locations of landfill sites are indicated in Figure 10.5.

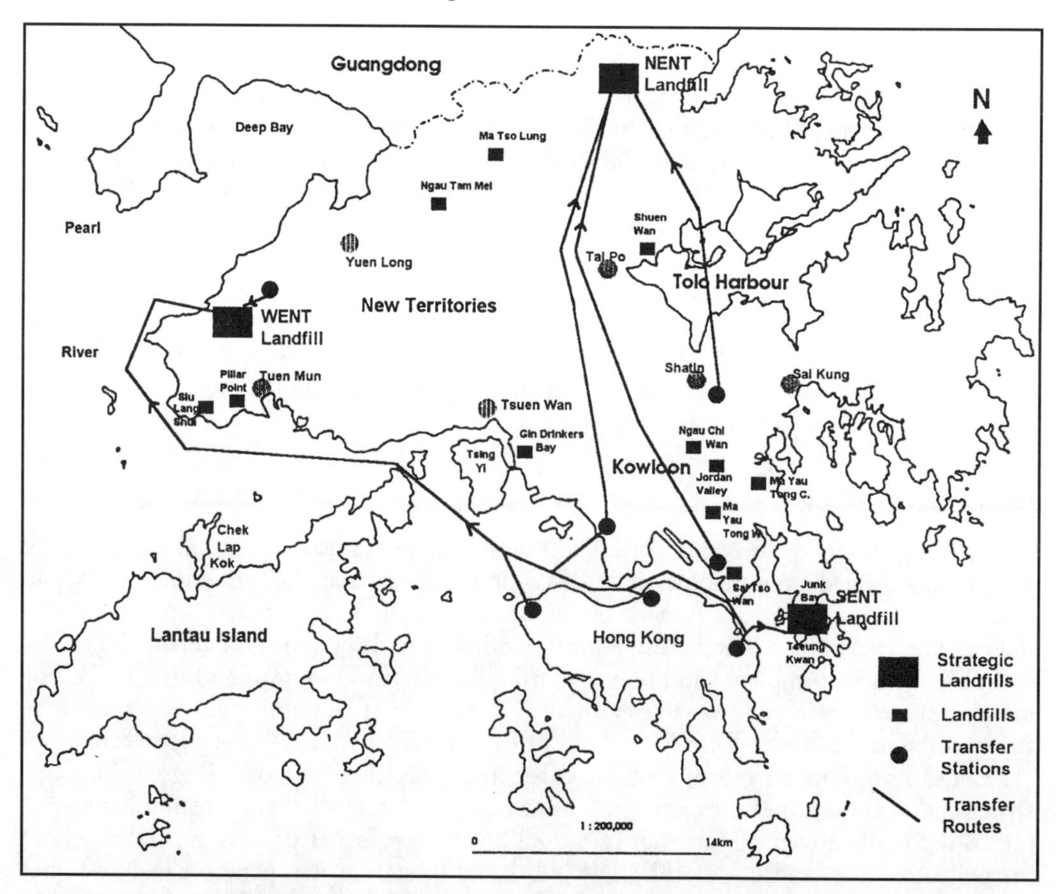

Figure 10.5. The locations of landfill sites, municipal waste transfer stations, and routes in Hong Kong (Source: Hong Kong Environmental Protection Department, Environment HK, 1990).

EFFECTS OF POLLUTION ON GROUNDWATER

The granite base of Hong Kong provides an elevated water table, but water can still flow through crests and crannies between rocks down the hillside. Paddy fields were constructed by barricading the runoff, creating wet puddles particularly fit for rice farming. While such hillside steps have been removed for high-rise construction, there is practically no groundwater reservoir left except those which remain trapped between rocks. Currently, the major domestic water supply in Hong Kong is from reservoirs. Only some villages in New Territories or on remote islands still rely on well water for domestic use. Thus, research on groundwater quality control and remediation has been very limited. The majority of the work aims to monitor the change in water quality in order to prevent health hazard.

The groundwater in Hong Kong can easily be infiltrated by surface runoffs from rivers and streams. A survey conducted by the EPD (HKEPD, 1988) showed that the water quality of wells reflected the local pollution status: those located in areas with intensive agricultural activity, livestock waste disposal sites and sewage outlets tend to be much more polluted. Figure 10.6 shows the locations of wells being monitored and the results of total coliform counts. In most of the sites studied, the total coliform content reached

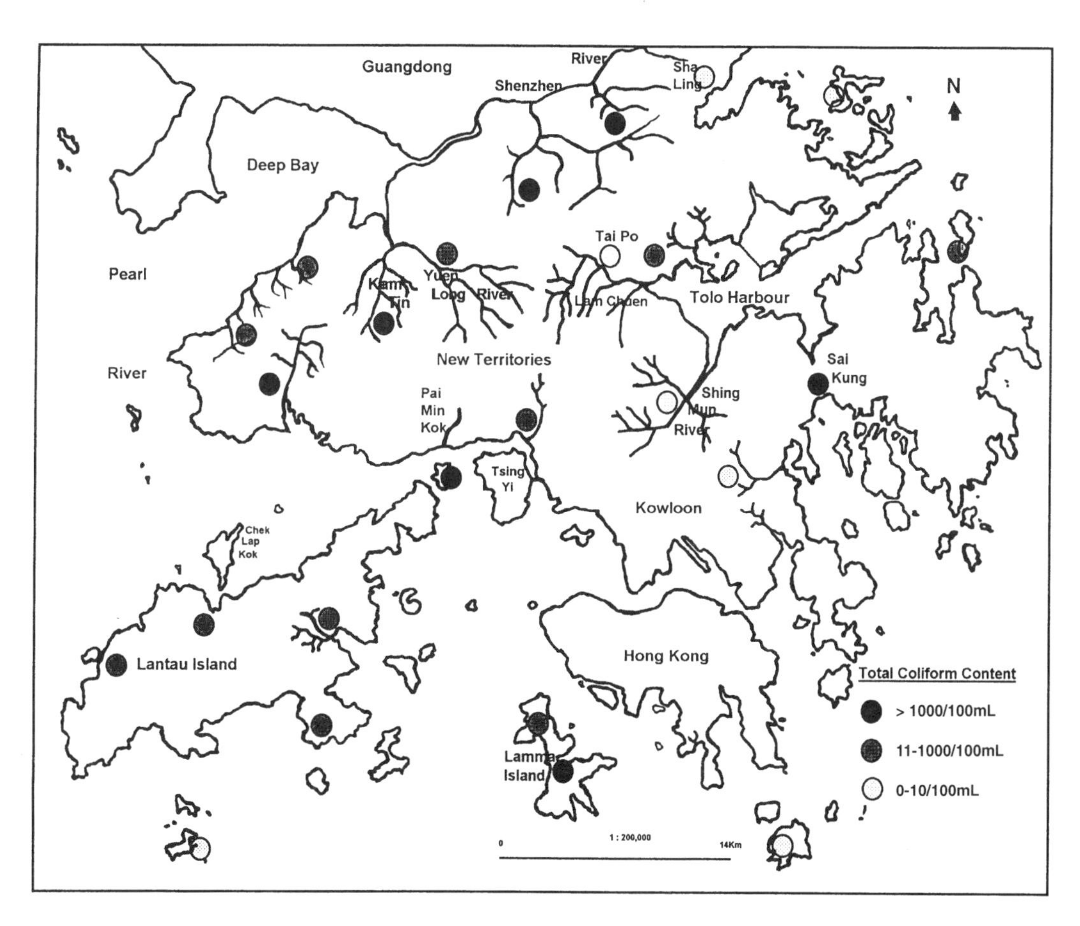

Figure 10.6. The locations of wells monitored by the EPD (Source: Hong Kong Environmental Protection Department, River Quality of Hong Kong, 1988).

Table 10.3. Chemical Contents of Well Waters in Hong Kong[a,b]

Area[c] (No. of Wells Sampled)	DO (mg/L)	BOD_5 (mg/L)	NH_4^--N (mg/L)	NO_3^--N (mg/L)	Total P (mg/L)
Tai Po (9)	3.15	0.74	0.026	0.43	0.05
	0.63–7.62	0.09–8.75	0.011–5.3	0.09–8.5	<0.01–2.7
Sai Kung (10)	3.92	1.10	0.012	0.81	0.07
	1.29–8.05	0.13–2.53	<0.01–5.9	0.2–7.8	<0.01–2.5
North New	2.32	0.89	0.053	0.75	0.01
Territories (23)	0.18–8.62	0.12–3.68	<0.01–18	0.002–20	<0.01–2.3
Kam Tin (15)	3.16	0.53	0.028	4.40	0.05
	0.24–8.58	0.04–4.84	0.028–8.3	0.19–16	<0.01–0.32
Tuen Mun (11)	2.04	0.85	0.019	2.00	0.05
	0.37–8.75	0.33–3.84	<0.01–8.3	0.064–8.7	<0.01–0.19
Lamma Island (14)	7.82	0.72	0.024	0.48	0.01
	3.77–9.69	0.08–3.26	<0.01–0.11	0.02–9.6	<0.01–0.2
Ma Wan (7)	4.83	1.42	0.025	2.50	0.03
	1.42–6.45	0.62–2.6	0.012–2.9	0.04–15	<0.01–0.18

[a]HKEPD, River Quality of Hong Kong, 1988.
[b]Data represents annual medians and ranges of monthly samples.
[c]Locations of the monitoring areas are indicated in Figure 10.6.

1,000/100 mL, well exceeding the World Health Organization (WHO) bacteriological standard for drinking (0 per 100 mL). Organic pollution was also noted in some wells indicated by a high 5-day biochemical oxygen demand (BOD_5, 8.7 mg/L), as well as ammoniacal-nitrogen (NH_4^+-N, 18 mg/L), nitrate-nitrogen (NO_3^--N, 20 mg/L) and total phosphate-phosphorous (total P, 2.7 mg/L) (Table 10.3).

The well waters in three locations, namely Pai Min Kok, Lam Tsuen Valley, and Sha Ling had been selected for long-term monitoring by the EPD, at monthly intervals, since 1988. Figure 10.7 shows the changes of water quality from 1990 to 1993 of well water collected from the three locations.

Pai Min Kok

Although the DO contents were similar (5.5 to 5.8 mg/L) throughout the monitoring period, the contents of BOD_5, NO_3^--N, NH_4^+-N, total P, and *E. coli* increased steadily during the first three years, with significant reductions in all the contents in 1993 (Fig. 10.7). The wells in Pai Min Kok are located at the Angler's River catchment, and the area was designated as the Livestock Waste Control Zone in 1988 (Fig. 10.3). However, as the number of residents increased around the area, the wells have become a sink for all the domestic activities, resulting in the continuous deterioration of water quality, until the tighter control has been enforced recently.

Lam Tsuen Valley

The second site is located in the catchment of the Lam Tsuen River. A similar trend, except NO_3^--N contents, indicating the steady increase of all parameters was observed throughout the years. The levels of NH_4^+-N and total P were higher than Pai Min Kok (Fig. 10.7) due to the higher agricultural activities of the region. In addition, the residents in this area use one of the wells for fish culture, resulting in a significant increase in *E. coli* count (>1,000/100 mL) in the 1993 record.

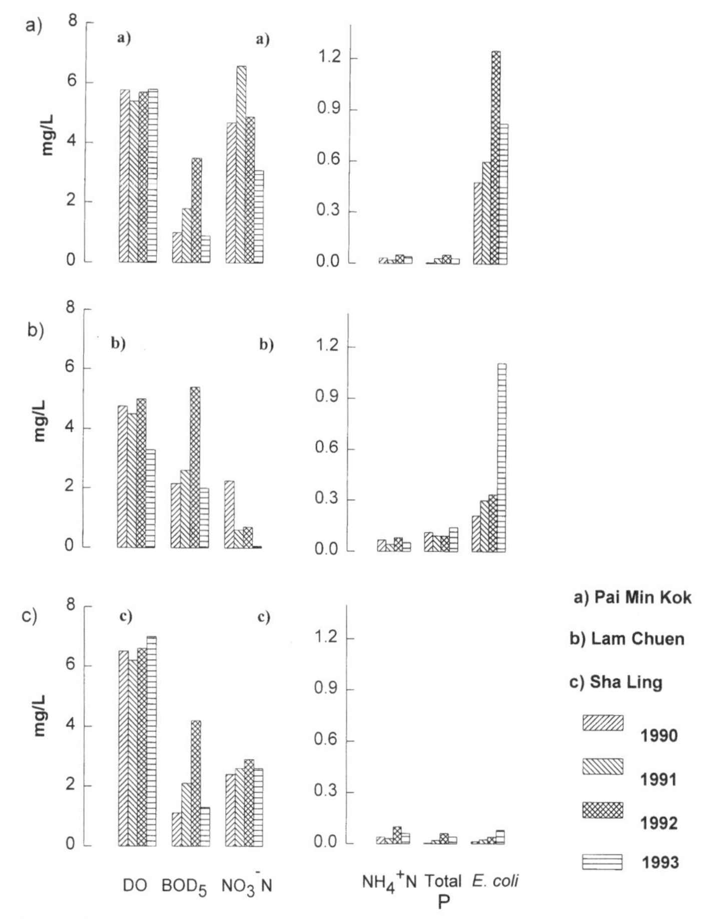

Figure 10.7. The changes of different parameters in groundwater in three different locations between 1990–1993. Units for all parameters are expressed as mg/L except E. coli, which is expressed as 103 counts/100 mL (Source: Hong Kong Environmental Protection Department, River Quality of Hong Kong, 1988).

Sha Ling

Being situated at the remote area in the Northern Territories, Sha Ling is relatively free of pollutants. The well water contained the highest levels of DO (dissolved oxygen) and the lowest *E. coli* contents among the three sites. However, most parameters also showed a similar trend where there was a steady increase throughout the years (Fig. 10.7). The water had a rather high level of nitrate content (530 mg/L had been recorded), which was due to the agricultural effluent discharged into the area. A livestock waste composting plant was built near the area and the groundwater was retrieved to monitor its environmental effect.

The results of the long-term monitoring showed that the well water quality reflected closely the river quality of the catchment areas where the wells are located. Implementation of the livestock waste disposal seemed to improve the water quality. Continuous monitoring of river quality replaced well water sampling. The quality of rivers near Sha Ling were graded excellent (HKEPD, 1996a), and at stations along the Lam Tsuen River and the Angler's Stream, were graded fair to good (HKEPD, 1996a).

CONTROL AND REMEDIATION OF DEGRADED SOIL AND LAND

The origin of soil pollution associated with a city like Hong Kong ranges from isolated incidences, such as effluent from factories or mines contaminating agriculture lands, to pollution resulting from traffic and municipal waste.

A study on the effluent discharged from a small scrap plastics recycling factory, situated at an elevation of 25 m above sea level in the New Territories, showed poor planning by placing an industrial establishment within an agricultural area. The recycling plant used acid to remove metals from the plastic surfaces. The spent acid was collected by another factory for metal recovery, whereas the rinse water was discharged into nearby fish ponds, killing the fish. The runoff affected vegetable cultivation on the downhill slope of the factory. The levels of Ni and Cu were high in both soil (385 and 194 mg/kg, respectively), and vegetables (30 and 40 mg/kg, respectively), posing health hazards to the consumers (Chui et al., 1988).

Emission of sulfur dioxide by an acid works factory located at Ching Lung Tau, a remote part of the New Territories, had also caused damage to the surrounding environment. In addition to causing chlorosis in plants, the large amount of sulfur dioxide emitted was deposited onto soil and vegetation as sulfuric acid after rainfall, resulting in severe erosion of the topsoil, followed by the death of plants (Wong, 1978).

A comprehensive study on the environmental impact of the iron ore tailings located at Ma On Shan (near Shatin) related to soil and plant has been made (Wong, 1981). The semibare ground was sparsely covered with grasses due to the lack of plant nutrients, poor texture, and fairly high contents of calcium, magnesium, iron, manganese, and copper (Wong and Tam, 1977). The tailings were subsequently removed and the area has been developed into a large housing estate.

Atmospheric pollution from automobile exhaust emission was one of the major sources of heavy metal contamination in roadside dust and soil. Being an urban center, Hong Kong has a high density of vehicular traffic. Lead (Pb) contamination of soil and roadside vegetation and street dust was detected (Lau and Wong, 1982; Tam et al., 1987). Other metals, such as Mn, Zn, Fe, Cu, and Cd, present in vehicles, are also discharged into the environment as a result of wear and tear, contaminating the roadside ecosystem. The level of Pb in roadside dust ranges between 107 to 915 mg/kg dry weight. Soil Pb level was reported to be 40 mg/kg dry weight, and Pb in the leaves of *Bauhina variegata* collected from the roadside was 10 mg/kg dry weight. It has been demonstrated that the water-extracts of roadside soil and dust can inhibit seed germination and root growth of vegetables (Wong et al., 1984). Moreover, the roadside populations of *Cynodon dactylon* and *Elensine indica* (both grass species) were more tolerant to elevated levels of Pb, compared with those collected from the control site without any vehicular traffic (Wong and Lau, 1985). The high levels of metals can also pose a serious public health hazard when they are detected in roadside dust at recreational parks where the elderly and children rest and play (Tam et al., 1987). With the gradual reduction of Pb content of gasoline from 0.84 g/L in 1980 to 0.4, 0.25, and 0.15 g/L in 1983, 1985, and 1987, respectively, the problem with Pb pollution has been somewhat alleviated (Chan et al., 1989).

A comprehensive assessment of trace metal distribution and contamination in surface soils of Hong Kong has been conducted (Chen et al., 1997). From results of cluster analysis, and comparisons among soil types and areas, it is clearly shown that increases in trace metal concentrations in soils were generally extensive and obvious in urban and orchard soils, and less so in vegetable soils, while rural and forest soils were subjected to the least impact of anthropogenic sources of trace metals. However, some of the forest soils also contained elevated levels of As, Cu, and Pb. Urban soils in Hong Kong were heavily polluted by Pb from gasoline combustion. The metals in both orchard and vegetable soils were from applications of pesticides, animal manures, and fertilizers. In general, trace metal pollution in soils of the industrial areas and Pb pollution in the soils of the commercial and residential areas were still obvious.

Recent interest in studying soil quality has centered around revegetation of completed landfill sites. Among the completed landfill sites in Hong Kong, the Gin Drinker's Bay landfill occupies the largest area (29 ha). The site originally consisted of an open body of water with Pillar Island at the mouth of the bay (Fig. 10.5). Dumping commenced in 1955. Around 1960, a rock mound bund, which connected the northern part of Pillar Island to the mainland, was constructed. From then until 1967, the former bay area was progressively filled by refuse. Open dumping was practiced until 1973, when the Public Works Department took over and replaced it with sanitary landfill method, which included liners to drain landfill leachate and gas extraction system to remove landfill gas. The area was eventually closed for dumping in February 1979, at which stage a final convering of inert fill approximately 1.5 m thick was laid on the area. The original plan for the site development was to convert it into a multirecreational park by 1991. However, due to the problem of landfill gas, the park is still closed to the public (Pugh and Choi, 1993).

The composition of the landfill gas at the Gin Drinker's Bay was monitored and the results showed that the methane level was as high as 46 % (v/v), ethane 1.96% (v/v), CO_2 18% (v/v), and O_2 0.23% (v/v), (Wong and Yu, 1989a). The high CH_4 contents reflect the active decomposition of waste by the anaerobic bacteria. The problem of leachate seepage occurs in confined areas, and usually at the edge of a site (Wong and Yu, 1989b). The leachate from Gin Drinker's Bay had high values of conductivity (2.5 to 11.8 mS/cm), chemical oxygen demand (147–1590 mg/L), total nitrogen (137–1,060 mg/L), and ammoniacal-nitrogen (65–883 mg/L). There are also fairly high levels of soluble iron (1.01 to 3.45 mg/L) and manganese (0.05 to 0.64 mg/L) (Cheung et al., 1993; Chu et al., 1993). Although high concentrations of these substances exert harmful effects on plant growth, it has been demonstrated that the diluted leachate can serve as irrigation water for crop growth (Wong and Leung, 1989). Recirculation of landfill leachate back to the landfills can effectively lower the phytotoxic substances (Wong et al., 1990).

The 13 landfill sites occupied a total land area of more than 250 ha. Revegetation of such degraded land can significantly improve the aesthetic value of the city, provide recreation for the general public, stabilize the slope, and maintain moisture content of the soil (Chan et al., 1996). Continuous effort has been made to study the growth of legume plants on landfills, as their associated nitrogen-fixing bacteria are able to fix the atmospheric nitrogen in the landfill soils, where nitrogen is commonly deficient (Chan et al., 1991).

FUTURE PERSPECTIVES

The recent "Open and Reform" policy of China has prompted a tremendous economic growth in the South China region. Situated adjacent to Hong Kong, Guangdong has become a newly formed economic powerhouse. With the assistance of the professionals in management and the well-developed communication facilities, Hong Kong's companies benefit from relatively cheap labor by moving its factories to Guangdong Province, while workers in Guangdong benefit from a relatively high wage. Large volumes of investment from Hong Kong and its neighboring city, Macao, has established a light and processing industry attracting a large influx of labor from different parts of China to the Pearl River Delta. The industrial and commercial growth in Guangdong was 25% in 1991 compared with 11% in China as a whole and 4% in Hong Kong (Emily, 1992).

The water network of the Pearl River consists of 100 main channels totaling a length of some 1,600 km with a density of 0.13 km/km^2. The yearly runoff of the delta area reaches 20 billion m^3 and enters the sea through eight estuaries. The tremendous increase in industrial establishment and population, together with an uncontrolled disposal of waste into the Pearl River eventually influenced the water quality of Hong Kong. Furthermore, Hong Kong has been purchasing freshwater from China mainland to meet the domestic need. Water is pumped from the East River through pipelines to reservoirs in Hong Kong. Deterioration of the river quality poses serious health risks that warrant administrative concern.

SUMMARY AND CONCLUSIONS

Like other countries in the world, the technological advancement over the past few centuries has improved the standard of living in Hong Kong, but at the same time generated a vast amount of by-products. The various environmental problems have been aggravated by a limited space with the ever-increasing number of inhabitants. The stabilization of China in recent years, together with the "Open and Reform" policy, produce yet another dimension of environmental problems faced by Hong Kong. Being situated at the mouth of the Pearl River, the quality of Hong Kong's coastal waters has been affected by the increasing industrial activities of southern China. In addition to the institution of new legislature by the Hong Kong Environmental Protection Department to prevent pollution, it is essential that the Department should work more closely with the Guangdong Environmental Bureau in tackling this regional problem.

ACKNOWLEDGEMENT

The authors thank the former director of the HKEPD, Dr. Stewart Reed for his consent to the use of materials published by the Department.

REFERENCES

Boyden, S., S. Millar, K. Newcombe, and B. O'Neill. *The Ecology of a City and Its People. The Case of Hong Kong.* Australian National Univ. Press, London, 1981.

Census and Statistics Department. Hong Kong Annual Digest of Statistics. Government Printer, Hong Kong, 1996.

Chan, G.Y.S., V.W.D. Chui, and M.H. Wong. Lead Concentration in Hong Kong Roadside Dust after Reduction of Lead Level in Petrol. *Biomed. Environ.* Sci. 2, pp. 131–140, 1989.

Chan, G.Y.S., M.H. Wong, and B.A. Whitton. Effects of Landfill Gas on Subtropical Woody Plants. *Environ. Management* 15, pp. 411–431, 1991.

Chan, G.Y.S., M.H. Wong, and B.A. Whitton. Effects of Landfill Factors on Tree Cover: A Field Survey at 13 Landfill Sites in Hong Kong. *Land Contam. Reclam.* 4, pp. 115–128, 1996.

Chen, T.B., J.W.C. Wong, H.Y. Zhou, and M.H. Wong. Assessment of Trace Metal Distribution and Contamination in Surface Soils of Hong Kong. *Environ. Pollut.* 96, pp. 61–68, 1997.

Cheung, L.C., L.M. Chu, and M.H. Wong. Toxic Effect of Landfill Leachate on Microalgae. *Soil, Air Water Pollut.* 69, pp. 337–349, 1993.

Chiu, T.N. and C.L. So. *A Geography of Hong Kong,* 2nd ed. Oxford University Press, Oxford, 1986.

Chu, L.M., K.C. Cheung and M.H. Wong. Variation in the Chemical Properties of Landfill Leachate. *Environ. Management.*18, pp. 105–117, 1993.

Chui, V.W.D., G.Y.S., Chan, Y.H. Cheung, and M.H. Wong. Contamination of Soil and Plant by Recycling of Scrap Plastics. *Environ. International.* 14, pp. 525–529, 1988.

Emily, P. Paying the Way to Prosperity. *U.S. News and World Report.* July, pp. 55–56, 1992.

Hong Kong Environmental Protection Department. River Quality of Hong Kong. Government Printer, Hong Kong, 1988.

Hong Kong Environmental Protection Department. Marine Quality of Hong Kong. Government Printer, Hong Kong, 1989.

Hong Kong Environmental Protection Department. Environment Hong Kong. Government Printer, Hong Kong, 1990.
Hong Kong Environmental Protection Department. River Quality of Hong Kong. Government Printer, Hong Kong, 1993.
Hong Kong Environmental Protection Department. Environment Hong Kong. Government Printer, Hong Kong, 1995.
Hong Kong Environmental Protection Department. River Quality of Hong Kong. Government Printer, Hong Kong, 1996a.
Hong Kong Environmental Protection Department. Environment Hong Kong. Government Printer, Hong Kong, 1996b.
Lau, W.M. and M.H. Wong. An Ecological Survey of Lead Contents in Roadside Dusts and Soil in Hong Kong. *Environ. Res.* 28, pp. 39–54, 1982.
Lo, C.P. The Population: A Spatial Analysis. In: *A Geography of Hong Kong*. Chiu, T.N. and So, C.L., Eds. Oxford University Press, Hong Kong, 1986, pp. 148–184.
Morton, B. Pollution of the Coastal Waters of Hong Kong. *Mar. Poll. Bull.* 20, pp. 310–318, 1989.
Phillips, D.J.H. Organochlorines and Trace Metals in Green-Lipped Mussels *Perna viridis* from Hong Kong Waters: A Test of Indicator Ability. *Mar. Biol.* 1, pp. 1–15, 1985.
Phillips, D.J.H. Trace Metals and Organochlorines in the Coastal Water of Hong Kong. *Mar. Poll. Bull.* 7, pp. 219–327, 1989.
Poon, H.T. A Shallow Water Wave Model for Hong Kong Waters. Technical Note (local) No. 47. Royal Observatory, Hong Kong, 1989.
Pugh, M.P. and C.C. Choi. Restoration of Urban Landfills — The Hong Kong Experience. In *Technical Meeting of the Institution of Water and Environmental Management*, Hong Kong Branch, March 30, Hong Kong, No. 2, 1993.
Tam, N.F.Y., W.K. Liu, M.H. Wong, and Y.S. Wong. Heavy Metal Pollution in Roadside Urban Parks and Gardens in Hong Kong. *Sci. Total Environ.* 59, pp. 325–328, 1987.
Wong, M.H. and N.F.Y. Tam. Soil and Vegetation Contamination by Iron-Ore Tailings. *Environ, Pollut.* 14, pp. 241–254, 1977.
Wong, M.H. An Ecological Survey of the Effect of Sulfur Dioxide Emitted from an Acid Work Factory. *Bull. Environ. Contam. Toxicol.* 19, pp. 715–723, 1978.
Wong, M.H. Environmental Impacts of Iron Ore Tailings—The Case of Tolo Harbour, Hong Kong. *Environ. Management.* 5, pp. 135–145, 1981.
Wong, M.H., L.C. Cheung, and W.C. Wong. Effects of Roadside Dust on Seed Germination and Root Growth of *Brassica chinensis* and *B. parachinensis. Sci. Total Environ.* 33, pp. 87–102, 1984.
Wong, M.H. and W.M. Lau. Root Growth of *Cynodon dactylon* and *Eleusine indica* Collected from Motorways at Different Concentrations of Lead. *Environ. Research.* 36, pp. 257–267, 1985.
Wong, M.H. and C.K. Leung. Landfill Leachate as Irrigation Water for Tree and Vegetable Crops. *Waste Management and Res.* 7, pp. 311–324, 1989.
Wong, M.H., M.M. Li, C.K. Leung, and C.Y. Lan. Decontamination of Landfill Leachate by Soils with Different Textures. *Biomed. Environ.* Sci. 3, pp. 429–442, 1990.
Wong, M.H. and C.T. Yu. Monitoring of Gin Drinker's Bay Landfill, Hong Kong. I. Landfill Gas on Top of Landfill. *Environ. Management.* 13, pp. 743–752, 1989a.
Wong, M.H. and C.T. Yu. Monitoring of Gin Drinker's Bay Landfill, Hong Kong. II. Gas Content, Soil Properties and Vegetation Performance on Side Slope. *Environ. Management.* 13, pp. 753–762, 1989b.
Wu, R.S.S. Marine Pollution in Hong Kong: A Review. *Asian Mar. Biol.* 5, pp. 1–23, 1988.
Yang, M.S., W.F. Fong, T.C. Chan, V.W.D. Chiu, and M.H. Wong. Feasibility Studies on the Use of Sewage Sludge as Supplementary Feed for Rearing Tilapia. II. PCBs of the Treated Fish and Biochemical Response in the Fish Liver. *Environ. Technol.* 14, pp. 1163–1169, 1993.

CHAPTER 11

ENVIRONMENTAL IMPACTS AND MANAGEMENT STRATEGIES OF TRACE METALS IN SOIL AND GROUNDWATER IN THE REPUBLIC OF KOREA

J.E. Yang, Y.K. Kim, J.H. Kim, and Y.H. Park

INTRODUCTION

Recently in the Republic of Korea, the conservation of soil and groundwater quality has been stimulated by recognizing that these natural resources are critically important components of the earth's biosphere and are directly related to human health. Soil plays an important role, not only in the production of food and fiber, but also in the maintenance of environmental quality. Groundwater is increasingly relied upon as a source for drinking water and many other industrial uses. Thus, soil and groundwater contamination by toxic chemicals, such as trace metals, is an environmental concern in the Republic of Korea.

For the last several decades, trace metal concentrations in some soil and groundwater have increased due to heavy industrialization. Sources of trace metals causing soil and groundwater contamination in Korea are mostly derived, directly or indirectly, from mining sites, industrial or domestic wastewater, solid wastes, and sewage sludges. Also, the smeltering industry and the combustion of fossil fuel constitute a significant contribution to the trace metal contamination in the soil and groundwater (Kim, 1993).

Previous work has shown that the levels of trace metals in soil and groundwater have increased in the last several decades; however, current concentrations of these trace metals are, except in a few extreme cases, lower than the regulatory levels (Kim, 1993; Lim, 1994). Critical impacts of trace metals on human health, such as Minamata and Itai-Itai diseases as reported in Japan (Salomons and Forstner, 1984), have not been documented in Korea.

The Republic of Korea is one of the few countries which is endowed with ample clean freshwater. However, with rapid industrialization and urbanization since the 1960s, many rivers and streams have become polluted (Ministry of Environment, 1992). This development has depreciated the quality of surface water and, as a result, groundwater is increasingly relied upon as a source of water for drinking and industrial uses in Korea. Contrary to a great concern about the groundwater quality, however, little is known about the extent of nationwide groundwater contamination, specifically by trace metals.

In Korea, realizing the importance of preventing soil and groundwater from trace metal contamination, a number of pollution control measures are being taken to conserve, prevent, and remediate the soil and groundwater contamination by trace metals. This chapter will be devoted to the discussion of the present status, impacts, management strategies, and perspectives of trace metals in soil and groundwater in Korea.

PRESENT STATUS AND ENVIRONMENTAL IMPACTS OF TRACE METALS IN KOREAN SOILS

Sources of Trace Metal Contamination in Soils

Any metal can be considered a pollutant when it exists in excess concentrations in the wrong place. Human activities often mobilize and redistribute the metals such that they can cause adverse effects. Certain trace metals may accumulate in soils to concentrations that are either phytotoxic or pose a health hazard to domestic animals and humans.

Trace metals of environmental concern in Korea, in terms of the degree of phytotoxicity or undesirable entry into the food chain, are predominantly As, Cd, Cr^{6+}, Cu, Hg, Pb, and Zn. Table 11.1 summarizes the major sources of metals and their pollution pathways into soils. Metal accumulation in soils is closely connected to the specific local sources such as discharges from smelters, metal-based industries, chemical manufacturing industries, active, inactive, and abandoned mining sites, and irrigation water.

Recently, the industrial districts in rural areas have been subsidized by the Korean government, and the numbers of industries in these districts are increasing (Ministry of Environment, 1993a). Metals from the solid wastes and wastewaters of these subsidized industries are potential sources of soil contamination. Numbers of inactive or abandoned mines have been increasing due to the depressed mining industries over the last two decades (Ministry of Environment, 1993a). The probability for the influx of metals into soils and streams from these mining sites is high. The Ministry of Environment is aware of these potential environmental problems and continues to make all possible efforts to

Table 11.1. Selected Hazardous Metal Sources and Major Contamination Pathways in Korean Soil[a]

Metals	Pollution Source Industries	Pollution Pathways
As	Mining, agrochemicals	Irrigation water
Cd	Mining	Irrigation water
	Smelter	Smoke
	Electroplating/finishing	Suspended particles
Cr	Mining, smelter, electroplating/finishing, and leather	Irrigation water
Cu	Mining	Irrigation water
	Smelter	Smoke
	Agrochemicals	Suspended particles
Hg	Chemical manufacturing, agrochemicals, battery, mining, metallurgy, paints, dyes, pulp, and paper	Irrigation water
Ni	Electroplating/finishing	Irrigation water
Pb	Automobile exhaust, mining, metallurgy, smelter, and agrochemicals	Irrigation water, smoke, and suspended particles
Zn	Mining and metallurgy	Irrigation water
	Smelters	Suspended particles

[a]Kim, 1989.

Table 11.2. Natural Abundance of Trace Metals in Paddy and Upland Soils

Soils	No. of Samples	As	Cd	Cr	Cu	Hg	Ni	Pb	Zn	Crops Grown[d]
		mg/kg								
Paddy	407[a]	c	0.130	c	4.15	c	c	4.67	3.95	
	330[b]	0.560	0.140	0.49	4.00	0.09	0.82	5.38	4.36	
	Japan	8.500	0.410	c	9.10	c	c	6.70	15.20	
Upland field	105	0.616	0.159	0.28	4.00	0.089	c	5.49	9.94	Barley
	56	0.256	0.149	0.27	2.05	0.089	c	5.49	9.94	Soybean
	51	0.431	0.160	0.29	2.15	0.091	c	3.34	9.71	Corn
	420	c	0.128	0.57	3.78	c	c	1.93	16.23	Vegetables
	325	c	0.216	0.63	3.59	c	c	1.81	24.61	Fruits
	Average	0.434	0.162	0.41	3.11	0.089	c	2.96	13.73	

[a]Topsoils (0~15 cm) were collected at harvesting periods from the unpolluted areas (Kim et al., 1982).
[b]Topsoils (0~15 cm) were collected during 1987-1988 from the unpolluted areas (Rhu et al., 1988).
[c]Not analyzed.
[d]Topsoils (0~15 cm) were collected from the fields where these crops were grown (Kim et al., 1990).

conserve the soils from pollution sources, by enacting the Soil Environment Conservation Law in 1995; expanding the soil quality monitoring networks, and mandatory metal analysis for soil quality and its impact assessment; and setting up the standard for quality assessment of soils in agricultural and industrial areas.

Distribution of Trace Metals in Soils

The natural abundance of trace metals in Korean paddy and upland field soils collected from the unpolluted areas is shown in Table 11.2. Similar metal contents exist between these two soils in general; however, Zn contents in the upland field soils for vegetables and fruits are greater than those in paddy soils, possibly because of the use of fertilizers. Natural contents of trace metals in Korean paddy soils are generally lower than those found in Japan (Kim et al., 1982; Rhu et al., 1988). This difference is possibly due to the difference in soil parent materials between the two countries, where granite is the major parent material in Korea and volcanic ash in Japan.

Of the total land area in Korea (about 9.93 million ha), 65% is forest area and 21% is cultivated land. A significant feature of farmland use in Korea is the abundance of paddy soils, comprising 63% of the total cultivated areas. A large majority (74%) of these paddy fields are irrigated for rice growth. As shown in Table 11.1, irrigation is a major pathway for metal contamination in soil. Not surprisingly, concentrations of metals in these irrigated paddy soils, especially near the industrialized areas, are usually greater than those in other areas.

Average concentrations of metals in major crops grown in uncontaminated soils are shown in Table 11.3. These data have been used as the natural background concentration of metals in crops. Concentrations of Cu and Zn, which are essential plant nutrients, were greater than those of the nonessential nutrients (i.e., Cd, Cr, Pb, Hg, and As). Zinc concentrations in all crops were the highest among the metals surveyed. Concentrations of As and Cd were similar between rice and upland crops, but those of Cu, Pb, and Zn were higher in upland crops than in rice. Those of Cd, Pb, and Zn tended to be higher in vegetables, whereas Cu was higher in corn.

Metal concentrations in soils and crops are subject to change due to different environ-

Table 11.3. Natural Contents of Metals in Crops Grown in the Unpolluted Paddy and Upland Field Soils

Crops	No. of Samples	As	Cd	Cr	Cu	Hg	Pb	Zn	References
		mg/kg							
Brown rice	407	n.a.[a]	0.05	n.a.	3.31	n.a.	0.44	20.55	Kim et al., 1982
	50	0.09	0.06	n.a.	2.31	n.a.	0.43	16.56	Rhu et al., 1988
Barley grain	108	0.115	0.050	0.21	4.49	0.044	0.54	19.35	Kim et al., 1990
Soybean grain	56	0.044	0.098	0.47	7.87	0.051	0.99	37.53	Kim et al., 1990
Corn grain	51	0.081	0.050	0.12	10.04	0.053	1.03	22.99	Kim et al., 1990
Vegetables	420	n.a.	0.149	0.57	5.50	n.a.	2.63	50.79	Kim et al., 1992
Fruits	374	n.a.	0.074	1.59	3.79	n.a.	2.50	20.70	Kim et al., 1993
	Average	0.080	0.084	0.59	6.34	0.049	1.14	30.26	

[a]n.a. : Not analyzed.

mental situations. Table 11.4 represents the metal distributions in the upland field soils from various parent material origins. Concentrations varied with the type of metals. For example, the concentrations of Cd, Zn, and As were greater in soils derived from basalt, porphyry, and limestone, respectively, whereas those of Cu and Pb were higher in alluvial soil. Metal concentrations were generally on the order of Zn > Pb > Cu > As > Cd and were similar to the natural abundance levels described in Table 11.2.

More than 60% of farming land in Korea is used for rice production. Table 11.5 shows examples of metal distributions in paddy soils and rice as influenced by the metal pollution source. Metal concentrations varied with sampling locations and pollution sources, but those in soils and rice were much higher than the natural occurrence levels (Tables 11.2 and 11. 3), indicating that the soil contamination is arising from water or air pollution.

Kim (1993) reported the damage to rice from mining wastes, which unfortunately were also partly used as the irrigation source. The pH and sulfate concentrations of the irrigated soil were 3.6 and 1860 mg/kg, and the respective values of the paddy water were 2.8 and 2850 mg/L. The pH of the irrigation water was 2.8. These results demonstrate that careful monitoring of the degree of contamination in the soils may be the simplest way of reducing human exposure to toxic chemicals.

Table 11.4. Natural Contents of Heavy Metals in Topsoils (0~15 cm) Derived from Different Parent Materials from the Unpolluted Upland Fields[a, b]

Parent Materials	Sample No.	As	Cd	Cu	Pb	Zn
		mg/kg				
Granite	47	0.273	0.031	2.62	3.01	8.12
Gneiss	23	0.031	0.158	2.55	3.55	6.97
Limestone	20	1.510	0.226	2.84	4.05	7.17
Porphyry	9	0.335	0.142	3.74	4.71	12.39
Schist	8	0.432	0.083	3.08	4.48	11.68
Granite gneiss	6	0.345	0.141	2.96	3.43	8.59
Basalt	5	0.398	0.320	1.34	1.25	4.55
Alluvials	19	0.494	0.128	4.05	5.58	4.58

[a]Kim, 1993.
[b]All values are the average over the numbers of samples specified.

Table 11.5. Average Metal Contents in Paddy Soil (0~15 cm) and Brown Rice Collected from the Areas Influenced by Different Pollution Sources

Samples	Location	Pollution Sources	Samples	As	Cd	Cu	Pb	Zn	References
				mg/kg					
Soil	(12)[a]	Pb & Zn mining waste	796	–[b]	1.11	8.54	27.40	30.30	Kim, 1990
Rice	(12)	Pb & Zn mining waste	679	–	0.22	4.07	0.58	4.94	Kim, 1990
Soil	Siheung	Zn mining	6	–	7.68	–	–	938.00	Yoo & Lee, 1990
	Changwon		8	–	1.25	–	–	60.90	
	Seongju		12	–	1.53	–	–	105.80	
	Uljin		9	–	2.16	–	–	104.60	
	Chilgok		10	–	1.79	–	–	17.80	
Rice	Siheung	Zn mining	6	–	0.87	–	–	34.70	Yoo & Lee, 1990
	Changwon		8	–	1.57	–	–	29.61	
	Seongju		12	–	0.55	–	–	28.66	
	Uljin		9	–	0.43	–	–	29.33	
	Chilgok		10	–	1.11	–	–	24.11	
Soil	Ulju	As mining	50	14.32	–	–	–	–	Kim et al., 1982
	Bongwha		50	3.54	–	–	–	–	
	Yangsan		10	2.69	–	–	–	–	
	Youngdong	As smeltering	56	3.57	–	–	–	–	Kim et al., 1982
	Chengsong		51	1.70	–	–	–	–	
Soil	Janghang	Metal smeltering	30	–	1.92	106.28	123.11	120.13	Kim & Yang, 1985
Rice	Janghang	Metal smeltering	30	–	0.71	4.44	6.33	30.34	Kim & Yang, 1985
Soil	Mankyeong River	Municipal & industrial wastes	30	–	0.68	–	–	86.40	Kim et al., 1994
Rice	Mankyeong River	Municipal & industrial wastes	30	–	0.14	–	–	26.07	Kim et al., 1994

[a]Samples from 12 different locations.
[b]Not analyzed.

Table 11.6. Metal Contents in the Sources of Irrigation Water, and Paddy Soils Using These Waters for Rice Growth[a]

Irrigation Water Sources	Sources of Water						Paddy Soil					
	pH	COD	Cd	Cu	Pb	Zn	As	Cd	Cr	Cu	Pb	Zn
		mg/L					mg/kg					
General irrigation water	7.4	10.1	0.01	0.003	0.016	0.031	0.088	0.17	0.35	4.11	4.47	4.30
Agricultural and industrial district drainage	7.5	44.7	0.01	0.003	0.014	0.094	0.159	0.18	0.30	3.71	5.18	4.94
Industry complex drainage	7.4	52.4	0.03	0.005	0.023	0.100	0.377	0.27	0.64	8.47	6.37	13.27
Municipal sewage	7.4	35.2	0.01	0.004	0.016	0.040	0.299	0.23	0.44	6.38	6.55	10.18
Limestone mine drainage	3.1	13.8	0.07	0.004	0.009	0.118	0.771	0.13	0.41	5.61	4.76	3.62
Livestock wastewater	7.4	63.3	0.03	0.011	0.059	0.207	0.023	0.08	0.55	4.50	2.87	3.64

[a]Yoo, 1990.

Yoo (1990) also reported the influence of mining wastes on the chemical characteristics of the soils (Table 11.6). The pH values of the water influenced by mining waste were the lowest, as compared to other sources. Metal contents and chemical oxygen demands (COD) in wastes of industries, mines, city, and livestock were much higher than those in noncontaminated irrigation water. As expected, metal concentrations in soils irrigated with industrial wastewater were greater than those in unpolluted soils.

Better management of wastewater before being discharged into the environment is still required to prevent soil from pollution. The Ministry of Environment requires the mandatory installment of wastewater treatment facilities at industrial plants. All expense is borne by the producers of pollutants, based on the principle of "Polluters Pay." However, many industrial plants are still in the stage of either installing or planning to install the mandatory treatment facilities (Ministry of Environment, 1993a).

Soil contamination is closely related with air pollution in Korea. Table 11.7 shows the metal distribution in paddy soils and brown rice collected from smelter-influenced areas,

Table 11.7. Heavy Metal Contents in Paddy Soils and Brown Rice as Influenced by the Distance from the Smelters[a]

Distance from Smelters	No. of Samples	Paddy Soil				Brown Rice			
		As	Cd	Cu	Pb	As	Cd	Cu	Pb
		mg/kg							
< 1 km	19	29.7	1.24	111.4	279.4	0.17	0.444	8.12	0.66
1–2 km	10	12.5	0.74	76.4	87.1	0.11	0.192	6.87	0.96
2–3 km	16	8.5	0.38	31.0	39.4	0.15	0.075	5.19	0.56
3–4 km	22	4.4	0.31	28.3	25.5	0.15	0.069	5.83	0.54
4–8 km	6	3.1	0.22	16.1	17.6	0.29	0.133	4.21	0.47
8–12 km	9	1.4	0.20	8.9	12.3	0.19	0.072	4.43	0.54
12–16 km	9	1.0	0.15	7.3	10.6	0.16	0.066	5.35	0.60
16–20 km	7	0.9	0.12	5.8	9.7	0.16	0.074	4.79	0.55
Natural abundance		0.56	0.13	4.1	5.0	0.09	2.810	2.81	0.43

[a]Kim, 1993.

Table 11.8. Specific Harmful Substances and Requirements for Designation as Agricultural Land Soil Pollution[a]

	Threshold of Danger Levels		Corrective Action Levels	
Substrate	Agricultural Area[b]	Factory/ Industrial Area[c]	Agricultural Area	Factory/ Industrial Area
Cd	1.5	12	4	30
Cu	50	20	125	500
As	6	20	15	50
Hg	4	16	10	40
Pb	100	400	300	1,000
Cr^{6+}	4	12	10	30

[a]Ministry of Environment, 1996.
[b]Agricultural area indicates paddy, upland farm, etc.
[c]Factory/Industrial Area indicates factorial and industrial sites, abandoned metalliferous mines, etc.

according to the distance from the contaminant source. The closer to the contaminant source, the greater the concentrations of As, Cd, Cu, and Pb detected in both soils and rice. Metal concentrations were higher than the natural occurrences even at sampling sites 20 km from the contaminant source. The Soil Environment Conservation Law designates the pollution standard for concentrations of As, Cd, and Cu (Table 11.8). The standards are based on threshold values for limited crop growth and are used as both the maximum permissible levels of such metals in agricultural soil and as the critical values for assessing the environmental impact on the soil. Concentrations of As within 1 km from smelters exceed the maximum permissible levels and those of Cu were close to the limit. Concentration of Cd in rice was relatively higher than the natural occurrence, but lower than the critical concentration. Kim (1993) reported distributions of As, Cd, Cu, and Pb in soil samples of different horizons and observed the highest concentrations of these metals in the top soil (0–15 cm). However, the concentrations in the lower horizons were still relatively high and are indicative of the potential influence of contamination sources. These results clearly demonstrate the necessity of controlling the pollutants at the emission source.

The Ministry of Environment (1994), based on the Basic Environmental Policy Law, constructed monitoring networks for measuring heavy metal concentrations in the soil at 522 locations in Korea. The numbers of sites have increased to 780 since 1996. The pH and concentrations of As, Cd, Cu, Hg, Pb, and Zn have been monitored every year since 1987. Figures 11.1 and 11.2 and Table 11.9 show selected data for metal distributions in Korean soils. Metal distributions in soils from the agricultural lands were similar to those of natural contents. Metal concentration in soils are fairly constant with time; however, concentrations of some metals such as Cu, Hg, and Pb tended to decrease with time. Metal concentrations in soils, collected from the mining areas, the waste landfill areas, and the lower reaches of a stream that was influenced by the type of the industrial wastes discharged into it, were generally greater than those in soils from farming lands. Metals in soils from these farm areas also tended to decrease with time. In addition, lead concentrations in soils from the roadside were higher than those in soils from other areas, but decreased with time. This distribution pattern is probably related to the use of unleaded gasoline in motor vehicles. Concentrations of metals in soils from residential areas were lower than other industrial areas but were also similar to the natural occurrence. The average metal concentrations at the area shown in Figures 11.1 and 11.2 and Table 11.9 were far lower than the critical concentrations that cause phytotoxicity.

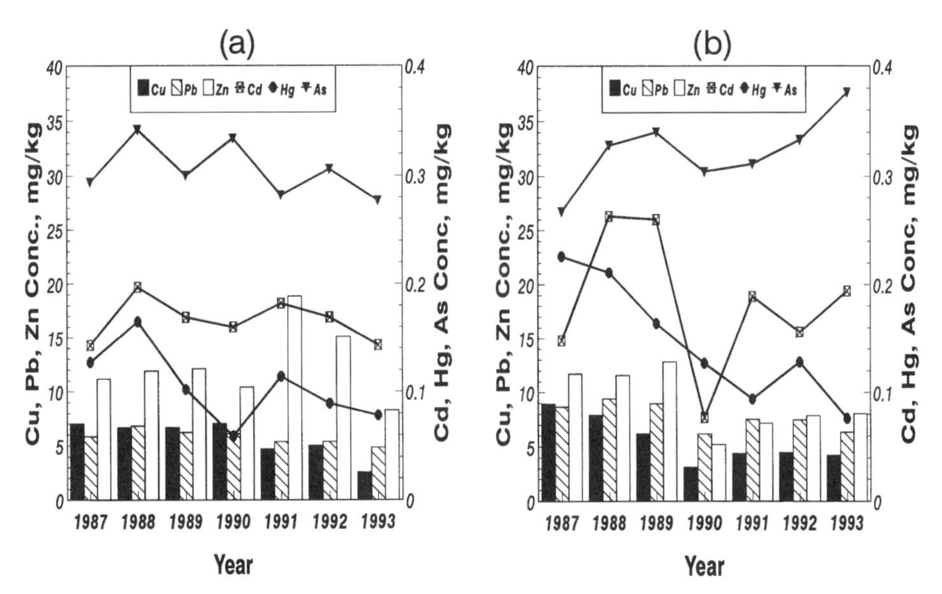

Figure 11.1. The yearly changes of trace metals in soils measured from the nationwide soil monitoring networks. Soils were taken from (a) the crop cultivating lands and (b) the downstream areas of the industrial wastewater discharges (Ministry of Environment, 1994).

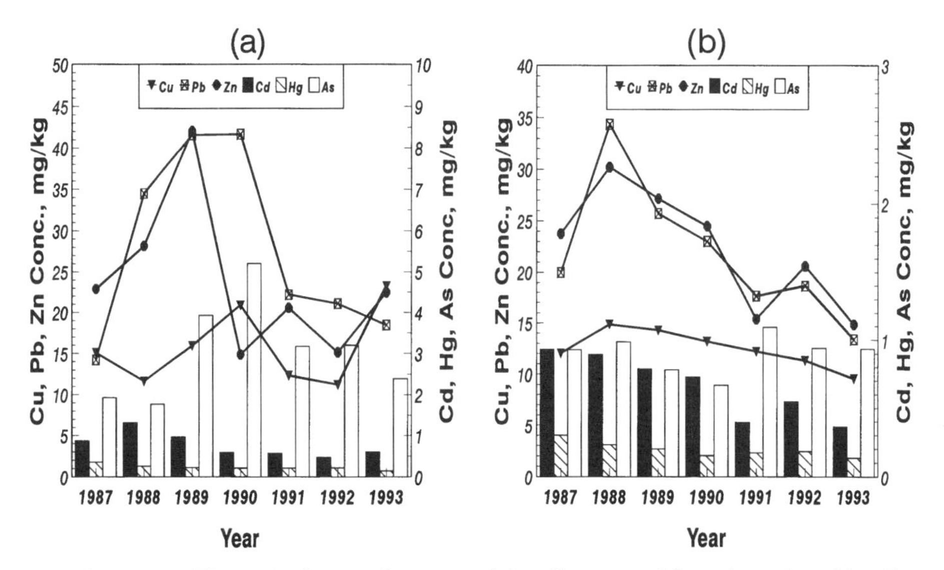

Figure 11.2. The yearly changes of trace metals in soils measured from the nationwide soil monitoring networks. Soils were taken from (a) the smeltering areas and (b) the metal-mining areas (Ministry of Environment, 1994).

These results indicate that metal contents in Korean soils were influenced by several contamination sources, such as wastewater and mining discharges. However, the average values of the metal concentrations were similar to the natural abundances and appeared to be safe for crop cultivation, except for a few contaminated areas. Measurements of metal contents in soils from the monitoring networks are continuing to assess soil quality and environmental safety.

Table 11.9. Ranges and Averages (Shown in Parentheses) of Metal Contents in Soils (mg/kg) Collected from Areas Subject to the Different Pollution Sources[a]

Sources	Sites	Year	pH	As	Cd	Cu	Hg	Pb	Zn
Irrigation water	Cultivated land	1987	4.7–6.8 (5.8)	ND-0.72 (0.23)	ND-0.45 (0.16)	1.83–11.45(4.54)	0.01–0.43(0.12)	2.57–18.71(8.10)	1.47–14.48(6.43)
		1989	5.0–7.7 (6.0)	ND-1.78 (0.32)	ND-0.46 (0.17)	1.27–11.81(4.98)	ND-1.97 (0.12)	1.75–19.31(8.57)	1.52–92.26(7.49)
		1991	3.2–6.9 (5.9)	ND-0.67 (0.22)	ND-0.45 (0.17)	0.67–11.92(3.75)	ND-0.33 (0.08)	ND–32.76(6.68)	1.26–62.80(9.37)
		1993	4.2–7.3 (5.6)	ND-0.97 (0.33)	0.03–0.43(0.17)	0.13–10.29(3.02)	ND-0.19 (0.06)	ND–15.32(4.03)	0.02–20.62(4.54)
Wastes	Domestic wastes landfill	1987	4.9–8.1 (6.3)	ND-0.57 (0.46)	0.01–3.61(0.19)	0.47–30.21(5.20)	ND-1.12 (0.15)	1.03–17.69(7.42)	1.44–25.66(9.62)
		1989	4.5–8.8 (6.3)	ND-2.28 (0.46)	ND-0.50 (0.19)	ND-26.23(4.62)	ND-1.93 (0.11)	ND–31.95(7.71)	ND –121.32(12.47)
		1991	3.9–8.3 (5.8)	0.01–1.80(0.33)	ND-0.49 (0.18)	ND-15.50(4.22)	ND-0.37 (0.10)	ND–26.92(7.21)	ND –54.43(7.05)
		1993	3.9–7.9 (5.7)	ND-2.01 (0.40)	0.02–0.71(0.16)	0.09–28.36(3.64)	0.01–0.45(0.08)	ND–29.37(5.59)	0.32–77.77(7.04)
	Industrial wastes landfill	1987	5.6–7.9 (6.2)	0.20–0.42(0.34)	0.19–0.89(0.42)	0.46–9.57(6.80)	0.01–0.38(0.20)	ND–17.40(10.81)	2.42–44.25(17.80)
		1989	5.4–8.6 (6.8)	ND-1.39 (0.52)	ND-0.56 (0.27)	1.98–22.00(9.10)	ND-0.45 (0.20)	2.71–50.00(15.66)	5.42–78.86(23.58)
		1991	4.3–9.0 (6.1)	0.04–1.12(0.51)	ND-0.41 (0.18)	0.81–11.50(3.80)	0.01–0.40(0.12)	2.22–9.09(6.12)	1.37–25.68(7.86)
		1993	4.5–7.7 (5.9)	0.02–1.66(0.55)	0.02–0.35(0.14)	0.98–12.18(3.46)	ND-0.17 (0.08)	0.31–7.78(3.76)	0.95–34.37(5.31)
Atmosphere	Roadside areas	1987	4.8–6.6 (5.8)	ND-0.43 (0.26)	2.24–12.18(0.17)	ND-0.35(5.01)	0.05–0.33(0.13)	1.84–23.19(9.88)	1.83–23.19(10.67)
		1989	4.7–7.0 (5.4)	ND-1.11 (0.31)	ND-0.39 (0.16)	1.91–37.08(5.79)	ND-1.29 (0.13)	2.50–32.44(9.02)	2.61–147.80(12.14)
		1991	5.0–7.1 (5.7)	0.01–1.67(0.38)	ND-0.56 (0.20)	0.72–12.79(4.88)	0.02–0.27(0.10)	ND–27.86(6.15)	1.12–30.32(8.68)
		1993	4.8–6.6 (5.5)	0.04–1.34(0.39)	0.02–0.34(0.16)	0.78–8.15(4.05)	ND-0.20 (0.07)	0.27–28.66(5.73)	0.80–40.87(6.65)
Residence	Resort and park areas	1987	4.9–7.3 (6.4)	0.04–0.29(0.21)	0.02–2.23(0.36)	0.86–11.51(4.04)	0.01–0.12(0.05)	0.88–19.54(5.52)	3.26–21.25(9.77)
		1989	4.7–7.5 (6.3)	ND-1.63 (0.25)	ND-0.34 (0.11)	0.54–18.57(5.21)	ND-0.21 (0.04)	0.57–15.47(6.33)	0.75–64.88(13.03)
		1991	4.4–8.5 (6.0)	0.04–0.93(0.31)	ND-0.47 (0.14)	0.91–5.88(2.18)	ND-0.14 (0.05)	1.12–14.58(6.81)	1.37–32.92(6.14)
		1993	4.4–8.5 (6.0)	0.05–0.91(0.31)	ND-0.37 (0.09)	0.26–10.86(2.36)	ND-0.50 (0.05)	1.30–13.13(4.79)	1.44–36.21(0.32)
Critical level[b]				15	25	125	40–50	400–500	150–500

[a] Data from the monitoring networks of soil quality measurements by Ministry of Environment, Korea (1994).
[b] Critical level causing phytotoxicity (Ministry of Environment, Korea, 1994).

Impacts of Trace Metal Contamination in Soil

Research concerning metal speciation, transformation, and bioavailability in Korean soils has been much more limited than that on metal distribution. Because it is well known that metal speciation and transformation in the environment are directly connected to its bioavailability and biotoxicity, more vigorous research activity in these fields is necessary in Korea to assess the impact of trace metals on the soil-plant system.

The presence of trace metals in metal-contaminated soils affects chemical and biological characteristics of the soil (Im, 1995; Lee, 1995; Yang and Skogley, 1989, 1990; Yang et al., 1995). Contamination with Cu or Cd changed the ionic distributions between soil and solution, resulting in temporary increases of cations such as Ca, Mg, and K in the soil solution and decreases in nutrient buffering capacity, due to the adsorption of the metal on the soil particles. Magnesium was a more responsive cation than Ca for changing the ionic distribution induced by Cu or Cd additions (Lee, 1995; Yang et al., 1995). These changes were considered to degrade the function of the soil to supply nutrients to plant roots over the long term. Most of the metals discharged into the soil were later adsorbed by the soil, and low concentrations of the metals remained in the soil solution. Concentration of Cd in the soil solution was higher than that of Cu (Yang and Skogley, 1989, 1990). The primary species of Cu, Ca, Mg, and K in soil solutions, as predicted by the GEOCHEM modeling program (Parker et al., 1987), were free ionic forms (> 99 molar %) and the remaining Cu was complexed with Cl-, SO_4^{2-}, and OH-, depending on the soil pH. In the case of Cd, Cd^{2+} and $CdCl^+$ forms were its primary species in the soil solution, followed by the secondary species such as $CdSO_4$, $CdHPO_4$, and $CdCl_2$ complexes. More than 73 molar % of Cd in soil solution existed as free Cd species at concentrations <100 mg/kg additions. Increasing Cd additions increased the Cd concentration in the soil solution, but Cd species complexed with Cl increased in compensation for the decreases of the free Cd species (Lee, 1995; Yang et al., 1995). Additions of trace metals such as Cu, Cd, Ni, and Zn decreased soil urease activities. The order of metals that reduced 50% of the urease activity was Cu > Zn > Cd > Ni (Im, 1995).

Table 11.10 summarizes the impact of trace metals on growth and yield of crops. Generally small amounts of metals, as compared to the treatment levels, were translocated into the plants; most of the metals were found in roots. Trace metals caused decreases in several growth parameters such as germination, height, leaf growth, root growth, and number of tillers. Nutrient uptake and crop yield also decreased with increased concentration of trace metals. Decreases in some of the physiological parameters such as chlorophyll content and transpiration rate were also reported.

Intake of trace metals from foods plays a significant role in assessing the risk to human health. The major food consumption in Korea is through the food crops such as rice, vegetables, and fruits. The estimated average metal intake per capita per day is estimated in Table 11.11 from the available data on the average trace metal concentrations in rice, vegetables, and fruits, as shown in Table 11.3, and the average nationwide food intake. The average nationwide intake of rice, vegetables (including both fresh and processed products), and fruits was 318.6, 281.0, and 68.7 g per capita per day in 1990 (Ministry of Health and Social Affairs, 1990). Average total food consumption per day was 1048 g, composed of 850 g of vegetative food intake (81.1%) and 198 g of animal food intake (18.9%). The ratio of vegetative to animal food intake tends to decrease with time. Average intake of trace metals per day from rice was estimated to range from 0.019 g for Cd to 5.913 g for Zn. Because the permissible levels of the average dietary intake (ADI) of trace metals from the specified food consumption have not been established in Korea, those intakes from foods are compared with the ADI for a healthy adult of Japan having

Table 11.10. Summary for the Impacts of Trace Metals on Plant Growth

Metals	Plants	Impacts	References
As	Rice (*Oryza sativa* L.)	Yield decrease; higher [As] in root; sterilized grain increase; root growth retardation; abnormal cells in roots; decrease in transpiration, height, tillers, and leaf growth	Lee et al., 1986, 1987, 1988
Cd, Cu Ni Zn	Raddish (*Raphanus sativus*) Cabbage (*Brassica campestris*)	Germination inhibited; yield decrease; leaf and root growth retarded	Moon et al., 1990
Cd	Rice (*Oryza sativa* L.)	Decrease in height, tillering, and yield	Kim et al., 1983
Cd, Zn	Corn (*Zea mays* L.)	Decrease in dry weight, height, peroxidase activity, chlorophyll and nutrient uptake	Lee and Kim, 1985
Cr	Cabbage (*Brassica campestris*)	Chlorosis and necrosis; root growth retardation; decrease in nutrient uptake and yield	Jeong, 1977
As, Cd, Cu, Pb, Zn	Corn (*Zea mays* L.)	Poor growth; decrease in height, ear length and yields	Lee et al., 1994

Table 11.11. Estimation of the Average Intake of Trace Metals Per Capita Per Day in Korea from Rice, Vegetables, and Fruits

Food Group	Metal	Mean Concentration [a] (mg/kg)	Mean Intake[b] (mg/kg)	ADI[c] (mg)	[ADI × (food/TFCX)][d] (mg)
Rice	As	0.09	0.029	0.1–0.3	0.03–0.09
	Cd	0.06	0.019	0.6	0.182
	Cu	2.81	0.895	2–5	0.61–1.52
	Pb	0.44	0.140	0.3–0.5	0.09–0.15
	Zn	18.58	5.913	1–5	0.30–1.52
Vegetables	Cd	0.15	0.042	0.6	0.16
	Cr	0.57	0.160	0.05	0.013
	Cu	5.50	1.546	2–5	0.53–1.34
	Pb	2.63	0.739	0.3–0.5	0.08–0.13
	Zn	50.79	14.272	1–5	0.27–1.34
Fruits	Cd	0.07	0.005	0.6	0.03
	Cr	1.59	0.109	0.05	0.003
	Cu	3.79	0.261	2–5	0.13–0.33
	Pb	2.50	0.172	0.3–0.5	0.02–0.03
	Zn	20.70	1.424	1–5	0.07–0.33

[a]Average trace metal concentrations adapted from data in Table 11.3.
[b]Mean intakes of metals from the specified foods: [the nationwide average food intake (g) per capita per day] * [average concentration of metals in food]. The nationwide average food intakes (g) of rice, vegetables and fruits in 1990 were 318.6, 281, and 68.8 g per day, respectively (Ministry of Health and Social Affairs, 1990).
[c]Average Dietary Intake of metals per adult (70 kg) per day in Japan (Yoshikawa, 1972).
[d]ADI of metals contributed by rice, vegetables, and fruits as relative to total food consumption (TFC), which was 1,048 g per capita (Ministry of Health and Social Affairs, 1990).

a similar pattern of food consumption. Intakes of As, Cd, Cu, and Pb from rice were lower than the ADI in Japan. However, Zn intake from rice was estimated to be higher than the ADI. Metal intake from vegetables and fruits was higher than that from rice, possibly due to the higher metal concentrations in those crops. These estimations from the exemplified foods do not absolutely evaluate the impact of the metals on human health, but they may suggest the importance of metal concentrations in crops in assessing risk to human health. This approach demonstrates the necessity of more careful monitoring for metals in soils to control metal translocation from the soil to the food chain.

MANAGEMENT STRATEGIES FOR TRACE METALS IN KOREAN SOILS

Designation of Specific Harmful Substances for Soil Pollution Assessment Criteria

Trace metals of Cd, Cu, As, Hg, Pb, and Cr^{6+}, which are likely to cause soil contamination and restricted crop growth, were designated as specific harmful substances under Article 14 and 16 of the Soil Environment Conservation Law, Ministry of Environment (1996) (Table 11.8). According to this law, standards of the six metalliferous ions are applicable only to the agricultural and factory/industrial areas. Thus, expanding species of harmful metalliferous substances and covering the whole national land, including residential and recreational areas, need to be established to conserve the quality of the soil, ecosystem, and human health.

Monitoring Trace Metals in Soils

The Ministry of Environment installed a nationwide monitoring network of soil quality assessment for trace metals in 1987, according to Article 15 of the Basic Environmental Policy Law. The number of monitoring locations of the network has been increased. In 1996, the monitoring network surveys metal concentrations in soils from 780 locations. Each location is classified by five contamination pathways and 16 contamination sources, as specified in Table 11.12. The monitoring network covers soil samples from agricultural upland fields, paddy fields, urban residences, etc. Six Regional Environmental Offices are in charge of soil sampling and analyses for pH, As, Cd, Cu, Hg, Pb, and Zn. Analysis of Cr^{6+} was added in 1996. The selected results are reported in Figures 11.1, 11.2, and Table 11.9. The objectives of installing the monitoring network are to (1) observe the changes of metal concentration in soils, (2) use these data for soil contamination control countermeasures, and (3) prevent soil contamination in advance. The analytical results have been reported annually by the Ministry of Environment. Expansion of the monitoring network is planned to enhance its efficiency to 10,000 locations by the year 2005.

Designation of the Soil Pollution Policy Area

Based on the analytical results on metal concentrations from the monitoring network, the Ministry of Environment may designate areas, corresponding to the requirements for the designation (Table 11.8), as agricultural soil pollution policy areas for the more intensive but detailed survey of metals. Metal mining and smeltering areas are cases in point. The National Institute of Environmental Research is in charge of metal analyses in soils, irrigation water, sediments, slags, and brown rice.

Metal concentrations in soils from the metal mining areas are much higher than those in other areas. Mining industries in Korea are declining, and the corresponding number

Table 11.12. Monitoring Networks for the Soil Quality Assessment in South Korea That Were Installed from 1987[a]

Major Pollution Pathways	Soils[b]	Survey Locations[c]	Classification of Pollution Sources and Site Selection Criteria
Agriculture	P; U; O	80	• Exclusively use the agricultural irrigation water: only affected by fertilizers and pesticides
		20	• Areas cultivating mainly fruits and vegetables
Water	P	42	• Riverside areas of potential intensive pollution: use the river water of BOD exceeding 10 mg/L as an irrigation source
	P	40	• Downstream of industrial wastewater: use the river water receiving the industrial wastewater as an irrigation source
	P; U; O	40	• Riverbed areas: flooded areas during the heavy rainfall in the summer
	P	20	• Areas using the special water as an irrigation source during the dry season: use the livestock wastewater and dye, paper mill, and leather manufacturing industries wastewater
Atmosphere	P	30	• Roadside areas: highway areas influenced by automobile emissions
	P; U	16	• Smeltering areas: influenced by the emission gas and suspended particles from smelters
Wastes	P; U; O	84	• General waste landfill areas influenced by leachate from landfill
	P; U; O	12	• Special waste landfill areas influenced by leachate from landfill
	P	50	• Metal mining areas influenced by mining waste, slags, etc.
	P; U; O	20	• Near the human waste treatment plants influenced by leachate from plants
Residence	O	24	• Areas surveyed for the residents' health
	O	20	• Resort and park areas: national parks, provincial parks, resorts, etc.
	O	12	• Playground soils in the large apartment complexes
	O	12	• Soils in the golf courses

[a]Ministry of Environment, 1994.
[b]P = paddy soil; U = upland soil; O = other general soils except for paddy and field soils.
[c]Five sampling sites per location. Survey agency = six regional Environment Offices; survey frequency = 261 out of total 522 locations per year, thus each location is surveyed every other year.

of inactive and abandoned mines is increasing. Therefore, metal contaminations of soils in these areas are becoming increasingly serious. The Ministries of Environment, Agriculture and Forestry, and Energy and Resources, and Provincial Governments provide guidance, inspection, subsidies, and other assistance; make an attempt to prevent soil contamination and crop damage; and establish the soil pollution control measures in these areas.

Establishment of Soil Pollution Policy Projects

After the designation of a policy area, the Provincial Governor and Mayor are authorized to design the soil pollution policy projects for prevention and removal of soil pollution, and the rationalized use of contaminated agricultural land. When a soil is judged to be contaminated by a specific harmful metal (Table 11.8), with the approval of the Minister of Environment and help from the Minister of Agriculture and Forestry, the Governor and Mayor can limit the crop growth in these contaminated soils, and can collect and remove the agricultural products from these soils. The cost is to be placed on the person responsible for the contamination.

The Governor and Mayor implement the remedial work in the soil pollution policy areas with physical, chemical, and biological remedial methods. Physical methods are soil covering/dressing, soil removing/dressing, reversing, soil layer mixing, isolation of contaminated soils, and burial of the contaminated soil into an impermeable layer. Chemical methods are to precipitate metals to form insoluble hydroxide, carbonate, or phosphate compounds by increasing soil pH with additions of silicate, calcium carbonate, or calcium hydroxide, etc., or to make metals less soluble by promoting reducing conditions in a soil by adding organic materials such as compost, grasses, etc., or S-containing compounds. Biological methods are to grow specific plants that uptake these metals from a contaminated soil. Ornamental or wild plants such as *Celtis sinensis, Euonymus japonica,* and *Buxus microphylla,* rather than an esculent plant, are commonly used in phytoremediation.

Regulatory Strategies for Soil Pollution Prevention

Laws relating to conservation and control measures for soil pollution are the Soil Environment Conservation Law, the Basic Environmental Policy Law, the Water Quality Conservation Law, the Air Quality Conservation Law, and the Mining Safety Law, etc. For preventing agricultural and forest soil pollution, a Provincial Governor or Mayor, after designating a policy area, may set a stricter standard than effluent water standards stipulated in accordance with the Water Quality Conservation Law.

River water quality standards to be used as a source of irrigation water for rice growth are pH 6 to 8.5, BOD < 8 mg/L, COD < 8 mg/L, dissolved oxygen > 2 mg/L, suspended solid < 100 mg/L, EC < 1 dS/m, and total nitrogen < 5 mg/L. Permissible wastewater discharge standards for irrigation water are pH 5–9, Cr < 2 mg/L, Fe < 10 mg/L, Zn < 5 mg/L, Cu < 3 mg/L, Cd < 0.1 mg/L, Hg < 0.005 mg/L, As < 0.5 mg/L, Pb < 1 mg/L, Cr^{6+} < 0.5 mg/L, and soluble Mn < 10 mg/L (Ministry of Environment, 1993a).

Based on the inspection of wastewater discharge facilities, the Minister of Environment may take the administrative measures against polluters, such as ordering improvement and repair, temporary suspension, license withdrawal, accusation, transfer, and others. Also, the Minister of Environment may impose the pollution charges to the producers.

Soil Environment Conservation Law

In recent years, there have been increasing occurrences of soil contamination having metals discharging from the mining, urban, and industrial areas. It is believed that this increasing trend has been partly triggered by recent urban development, as well as waste disposal problems that are now becoming increasingly serious. As the awareness of metal impacts on the ecosystem and human health is increasing, stricter measures to control soil pollution are necessary.

A separate law dealing with soil pollution had not been enacted in Korea by 1994, and rather few articles were included in the Water Quality Conservation Law. Under this condition, proper control and management of soil pollution had been limited. The Soil Environment Conservation Law was promulgated in 1994 by the Ministry of Environment and was effective from 1995, in which soils of agricultural fields, forests, and urban areas will be included for application of the pollution standard. The Ministry of Environment will issue the soil environmental quality standards for additional toxic substances which may affect crop growth and human health. Based on the continuous measurements of pollutants in soils through the monitoring networks, the soil pollution policy areas have been designated, and effective control measures have been established. The cost of the soil pollution policy projects is to be placed on any person responsible for such pollution. A stricter administrative measure will be taken against anyone who causes such pollution.

TRACE METALS IN GROUNDWATER

Water entering groundwater zones comes mainly from atmospheric precipitation, which is generally acidic. Groundwaters contain a variety of chemical species and most of the ionic species in groundwater are derived from the surrounding rocks and solids. The constituents of groundwater depend on the type of minerals, their solubility, and time of contact.

While the composition of groundwater is determined primarily by its contact with soils, minerals, and rocks, nonpoint surface pollution is often the primary source of groundwater contamination. Contamination of groundwater by metal ions may occur as a result of municipal or industrial landfills, disposal of liquid mining wastes or mining tailings, or excessive use of agricultural chemicals. Increased usage of groundwater resources and an increase of inputs of nonpoint surface pollutants into groundwater zones may cause contamination and general deterioration of groundwater quality.

After the occurrence of several recent incidents caused by the pollution of drinking water sources, the government conducted an investigation on 774 groundwater samples throughout Korea in 1993. Seventeen percent of the samples contained levels in excess of the maximum contaminant level (Ministry of Environment, 1993b). In this study, the samples were collected at about 30 m in depth. Most of the analyses were focused on primarily conventional parameters and hazardous organic substances. A summary of the currently available analytical data with respect to trace metals in the groundwater is provided in Table 11.13. In addition, the results of the groundwater analyses on samples in selected areas of Seoul are presented in Table 11.14, in which only trace metals are included (Chung, 1993).

As a result of the industrial expansion and rapid population growth during the last two decades in Korea, securing landfills to dispose of industrial and municipal solid wastes has become inevitable. Most of the currently opened or closed solid waste landfills are not properly designed, and they are without liners and leachate collection systems. Consequently, landfill leachates generated by rainfall infiltration can severely contaminate large portions of adjacent aquifers. Although the definite impacts of the leachates on groundwater quality near landfill sites have not yet been quantified, a strong indication of potential groundwater contamination has been reported based on the detection of some trace metals from the leachate collected from the Nanjido Landfill site, which is the largest landfill site in the Seoul metropolitan area. Concentration of Cd in the leachate ranged from traces to 0.58 mg/L. Maximum levels for Cr^{6+} and Hg were 0.062 mg/L and 0.012

Table 11.13. Trace Metal Concentrations in Groundwater from the Selected Areas in South Korea

Metals	Concentrations (μg/L)	References
As	1.6–6.8	Chung, 1993
Cd	0.2–450	Chung, 1993; Ministry of Environment, 1993a
Cr	0.7–17	Chung, 1993
Cu	0.6–35	Chung, 1993
Hg	0.8–0.4	Chung, 1993
Mn	1.0–52	Chung, 1993
Pb	1.0–232	Chung, 1993; Ministry of Environment, 1993a
Zn	3.0–880	Chung, 1993

Table 11.14. Trace Metal Concentrations in Groundwater of the Selected Sampling Locations in the City of Seoul

Location (District)	Groundwater Use	Concentrations (μg/L)				
		Cd	Cu	Mn	Pb	Zn
Yangchun-ku	Car wash	2.8	21.5	113.9	12.3	ND
Mapo-ku	School swimming pool	1.6	7.7	67.2	4.4	ND
Seodaemun-ku	School water supply	0.8	3.1	ND[a]	8.3	74.5
Kangnam-ku	Portable water supply	0.8	3.8	6.4	ND	ND
Kwanak-ku	Cooling water	1.2	10.0	85.5	16.2	23.8

[a]ND = Not detected.

Table 11.15. Trace Metal Concentrations in Groundwater from the Vicinity of Mining Areas[a]

Mining Industries and Location (Ore: Province)	Concentrations (μg/L)								
	Cr	Mn	Co	Ni	Cu	Zn	Cd	Ba	Pb
Yoochun (Gold: Chungnam)	240	1231	155	89	964	4403	19	19724	5834
Dalseong (Fe & Mn: Kyungbuk)	1	9288	165	36	9977	4401	62	9	0
Haman (Copper: Kyungnam)	0	1	2	1	17	9	0	8	0
Kwimyung (Gold: Kyungnam)	0	129	0	3	11	2636	5	222	0
Sangdong (Tungsten: Kangwon)	0	11	0	1	1	7	1	7	0

[a]Park, 1995.

mg/L, respectively.

Appreciable amounts of metals can be detected in areas rich in ore deposits. The mining industries also create groundwater pollution by discharging the by-products from other minings, including Fe, Cu, Zn, and Pb. Mining has been a declining industry, and many of the mines are left unmanaged, thereby creating a serious environmental problem such as surface water and groundwater contamination. Table 11.15 shows the extent of groundwater contamination in five temporarily or permanently closed mining areas around the nation (Park, 1995). Groundwaters from Yoochun and Dalseong mines show substantially higher metal concentrations than the naturally occurring levels. The Yoochun mine had been mining gold, whereas iron and manganese were the primary ores at the Dalseong mine.

MANAGEMENT STRATEGIES OF TRACE METALS IN GROUNDWATER

Natural levels of trace metals are commonly below the levels prescribed by the drinking water standards. As a result, these metals have received little attention in spite of their potential toxicity. Most of the metals in natural waters precipitate and have relatively low solubility. Mobility of metals and their solubility depends on the presence of organic acids, CO_2 levels, and acid precipitation. Therefore, detection of metals above the naturally occurring levels may indicate the contamination by artificial (point and nonpoint) sources. From this point of view, proper management of the surface pollution, particularly the industrial wastes, should be essential to prevent groundwater contamination by heavy metals.

The currently available information on trace metals and their concentrations in groundwater in Korea is not sufficient to develop any definite groundwater management strategies. Due to this limitation, we introduce here some preventive approaches for water resources management to be administered by the government.

Most water supplies in Korea come from reservoirs, which provide drinking water for 18 million people, and also water for industries in the metropolitan area and the central region. Various control measures to protect the water quality of the reservoirs are now being developed and implemented. However, there has not been a nationwide groundwater data collection program, and even in some previous groundwater studies, certain conventional pollutants were the primary focus. Only recently, the extent of heavy metals contamination in groundwater has begun to receive public attention.

Although there are a number of environmental statutes in Korea, there has been no explicit, comprehensive national legislative mandate to protect groundwater from contamination. In recognition of the increasing importance of groundwater as a drinking water source, the government enacted the Groundwater Conservation Law, which was the first law enacted to manage the groundwater resources and prevent groundwater from contamination. This legislation passed in 1993 and became effective in August, 1994. It establishes a national policy on groundwater protection, development, and management. Some government policies for environmental protection with respect to water resources management are as follows (Ministry of Environment, 1992).

Industrial Wastewater Management

Throughout the country, approximately 728 million tonnes of wastewater are discharged daily from 11,200 point sources. The government has installed and operated treatment facilities since 1984 to ensure the efficient treatment of highly contaminated industrial wastewater at six major industrial sites. While the government subsidized the installation of these facilities, the industries themselves are responsible for operational costs.

Waste Management

In Korea, about 94% of municipal wastes and 35% of industrial wastes are disposed of in solid waste disposal sites. Most of the disposal sites are not classified as sanitary landfills, and these can be considered as eventual point sources of groundwater contamination. In the Nanjido Landfill site described above, a preliminary plan to install a barrier around the landfill to prevent leachate from reaching groundwater and nearby surface waters is being developed. Thirty-four new large-scale landfill sites will be established in Korea.

Toxic Chemicals Management

The Toxic Chemicals Control Act which was enacted in 1990 provides for the regulation of manufacture, import, and use of 427 chemicals designated as toxic chemicals.

SUMMARY AND CONCLUSIONS

Following the heavy industrialization and urbanization in Korea during the last several decades, soil and groundwater contamination by trace metals is becoming one of the major environmental concerns because these natural resources may impact human health through the food chain. Sources of trace metals causing soil and groundwater contamination are closely related to industrial and human activities, such as smeltering, wastewater from mining and manufacturing industries, solid wastes, etc. Trace metal concentrations in soil and groundwater that have been influenced by these pollution sources have been reported to exceed the pollution control standards, even though these phenomena are regional. However, critical impacts of trace metals on human health due to soil and groundwater pollution have not yet been reported in Korea.

Trace metals such as As, Cd, Cr^{6+}, Cu, Hg, Pb, and Zn are mainly considered for soil and groundwater pollution from the viewpoints of phytotoxicity or undesirable entrance into the food chain. Trace metal levels in arable soils were similar to their natural abundance, which is relatively safe for crop growth. Metals caused an adverse effect on nutrient availability properties of the soil by changing the ionic speciation and distribution in soil solution and decreasing the buffer capacity and urease activity. Where soils were subject to metal contamination, however, metal contents in crops were high enough to inhibit nutrient uptake and crop growth, resulting in yield decreases. Metal contents in major crops grown in unpolluted areas were lower than the pollution standards. Groundwater is increasingly important as a source of drinking water and industrial uses in Korea. Trace metal concentration in groundwater is in general similar to the natural abundance and lower than the pollution standard due to the geochemical or chemical characteristics of trace metals. However, data on trace metal concentrations in groundwater and its impacts are limited. A more detailed database is necessary to describe the current status and impact of trace metals in groundwater.

Various control measures for soil and groundwater pollution have been made for the preservation of soil and groundwater quality, through the Soil Environment Conservation Law, the Basic Environmental Policy Law, Water Quality Conservation Law, Groundwater Conservation Law and Mining Safety Law, etc. Soil and groundwater pollution standards were established for trace metals, and permissible concentrations of metals were designated. The nationwide monitoring network for trace metal measurements in soil has been in operation since 1987 to monitor metal contents and establish soil pollution countermeasures. A comprehensive nationwide program to prevent groundwater resources from contamination and to monitor the groundwater quality, however, has not yet been implemented in Korea. Administrative measures, such as to improve and repair, issue temporary suspensions, order license withdrawals, or transfers, and the pollution charges are subject to any person responsible for pollution. Remedial projects for soil and groundwater pollution prevention and removal have been made through the physical, chemical, and biological methods. For proper control and management of soil and groundwater pollution, the Soil Environment Conservation Law and the Groundwater Conservation Law were promulgated and enacted.

Close monitoring of the metal pollution sources in soil and groundwater may be the simplest way of keeping human exposure to metals at a low level. The Ministries of Environment, Agriculture and Forestry, and Commerce & Resources, and Provincial Governments shall endeavor to provide guidance, inspection, subsidies, and other assistance, toward the prevention of soil and groundwater pollution.

REFERENCES

Chung, Y. Groundwater Quality Standards and Establishment of Standards in Groundwater Contamination. A report submitted to Korea Environmental Science Council, Seoul, Korea, 1993.

Im, I.G. Influence of heavy metals on soil urease activity. M.S. thesis. Kangwon National University, Chuncheon, Korea, 1995.

Jeong, Y.H. Study on the Effects of Heavy Elements in Agricultural Crops and the Control Measures. I. Effect of Chromium on Chinese Cabbage and the Control Measures. *J. Korean Soil Sci. Fert.* 10, pp. 205–210, 1977.

Kim, B.Y. K.S. Kim, J.K. Cho, M.H. Lee, S.K. Kim, Y.S. Park, and B.J. Kim. Survey on Heavy Metals in Paddy Soil and Brown Rice in Korea. *Res. Rept. Office Rural Development* 24, pp. 51–57, 1982.

Kim, B.Y., K.H. So, K.S. Kim, J.K. Cho, I.H. Cho, and K.D. Woo. Survey on the Natural Heavy Metal Contents in Upland Soil and Crop Grains in Korea. *Res. Rept. Rural Development Agency*. 32, pp. 57–68, 1990.

Kim, B.Y. Soil Pollution and Improvement Countermeasures. *Agric. Indust. Technique*. 7, pp. 135–143, 1990.

Kim, B.Y. Soil Pollution Status and its Improvement Countermeasures, in Soil Management for Sustainable Agric. *Kor. Soc. Soil Fert.*, Suwon, Korea, 1993, pp. 68–98.

Kim, B.Y., K.H. So, K.S. Kim, K.D. Woo, and S.H. Woo. Survey on the Natural Heavy Metal Contents in Vegetables and Upland Soil in Korea. *Res. Rept. Rural Development Agency*. 34, pp. 56–70, 1992.

Kim, B.Y., K.S. Kim, C.S. Lee, and S.H. Woo. Survey on the Natural Heavy Metal Contents in Fruits and Orchard Soils in Korea. *Rural Development Agency. J. Agri. Sci.* 35, pp. 280–290, 1993.

Kim, J.J. Soil Pollution, in *Agricultural Environmental Chemistry*. K. Han et al., Eds., Dong-Hwa Technol. Pub. Co., Seoul, 1989, pp. 169–214.

Kim, K.S., B.Y. Kim, and Y.S. Park. Effect of Various Cadmium Compounds on the Growth and Cadmium Uptake of Paddy Rice. *Korean J. Environ. Agric.* 2, pp. 6–12, 1983.

Kim, S.J. and H.S. Yang. Studies on the Heavy Metals in Paddy Rice and Soils in Jang–Hang Smelter. *J. Korean Soil Sci. Fert.* 18, pp. 336–347, 1985.

Kim, S.J., S.H. Baek, U.S. Kim, K.W. Yoon, K.H. Moon, and G.W. Kang. Variation of Cadmium and Zinc Content in Rice and Soil of the Mangyeong River Area. *Korean J. Environ. Agric.* 13, pp. 142–150, 1994.

Lee, J.P., N.K. Park, and B.J. Kim. Influence of Heavy Metal Contents in Soils Near Old Zinc-Mining Sites on the Growth of Corn. *Korean J. Environ. Agric.* 13, pp. 241–250, 1994

Lee, K.W. Heavy Metal-Induced Changes of Ionic Distribution and Species Between Soil and Soil Solution. M.S. thesis. Kangwon National University. Chuncheon, Korea, 1995.

Lee, M.H. and B.Y. Kim. The Effect of Cd and Zn Elements Applied to Soil on the Growth and Their Uptake of Corn Plant. *Korean J. Environ. Agric.* 4, pp. 11–17, 1985.

Lee, M.H., S.K. Lim, and B.Y. Kim. Behaviors of Arsenic in Paddy Soils and Effects of Absorbed Arsenic on Physiological and Ecological Characteristics of Rice Plant. II. Effect of as Treatment on the Growth and as Uptake of Rice Plant. *Korean J. Environ. Agric.* 5, pp. 95–100, 1986.

Lee, M.H., S.K. Lim, and B.Y. Kim. Behaviors of Arsenic in Paddy Soils and Effects of Absorbed Arsenic on Physiological and Ecological Characteristics of Rice Plant. IV. Effect of as Content in Water Culture on Transpiration, Stomatal Resistance, Temperature and Humidity in the Leaves of Rice Plant. *Korean J. Environ. Agric.* 6, pp. 39–45, 1987.

Lee, M.H., S.K. Lim, Y.D. Park, and S.H. Lee. Behaviors of Arsenic in Paddy Soils and Effects of Absorbed Arsenic on Physiological and Ecological Characteristics of Rice Plant. V. Effect of Arsenic Added to Soil on Ecological Characteristics of the Rice Plant. *Korean J. Environ. Agric.* 7, pp. 21–25, 1988.

Lim, S.K. Survey on the Establishment of the Soil Quality Standard. Korea Council Environ. Sci.. Seoul, 1994.

Ministry of Environment. National Report of the Republic of Korea to UNCED, Kwachon, Korea, 1992.

Ministry of Environment. Korea Environmental Yearbook. Vol. 6. Kwachon, Korea, 1993a.

Ministry of Environment. Current Status of Environment, Kwachon, Korea, 1993b.

Ministry of Environment. Guidelines for the Soil Monitoring Networks. Government Printing No. 12000–67630–66–9, Kwachon, Korea, 1994.

Ministry of Environment. Environmental Protection in Korea. Kwachon, Korea, 1996.

Ministry of Health and Social Affairs. National Nutrition Survey Report, Kwachon, Korea, 1990.

Moon, Y.H., Y.H. Kim, and H.S. Ryang. Effects of Heavy Metals Cr, Ni, Cd, Cu, Zn on Growth of Radish and Chinese Cabbage in Soils. *Korean J. Environ. Agric.* 9, pp. 113–119, 1990.

Park, Y.H. A Study on the Current Status of Pollution and Management Strategy in Inactive or Abandoned Mining Areas. Korea Environmental Technology Development Institute, Seoul, Korea, 1995.

Parker, D.R, L.W. Zelazny, and T.B. Kinraide. Improvement to the Program GEOCHEM. *Soil Sci. Soc. Am. J.* 51, pp. 488–491, 1987.

Rhu, H.I., Y.S. Suh, S.H. Jun, M.H.Lee, S.J. Yu, S.N. Hur, and S.Y. Kim. A Study on the Natural Content of Heavy Metals in Paddy Soil and Brown Rice in Korea. *Nat'l Inst. Environ. Res.* pp.1–77, 1988.

Salomons, W. and U. Forstner. *Metals in the Hydrocycle.* Springer–Verlag, Berlin, 1984.

Yang, J.E. and E.O. Skogley. Influence of Copper or Cadmium on Soil K Availability Properties. *Soil Sci. Soc. Am. J.* 53, pp. 1019–1023, 1989.

Yang, J.E. and E.O. Skogley. Effects of Copper or Cadmium on Potassium Adsorption and Buffering Capacity. *Soil Sci. Soc. Am. J.* 54, pp. 739–744, 1990.

Yang, J.E., K.W. Lee, J.J. Kim, and H. Lim. Changes of Chemical Species in Soil Solution Induced by Heavy Metals. *Korean J. Environ. Agric.* 14, pp. 263–271, 1995.

Yoo, S.H. Investigation of the Agricultural Environment Pollution Status. *Kor. Rural Development Agency*, pp. 3–44, 1990.

Yoo, S.H. and C.Y. Lee. Contents of Cd and Zn in Paddy Soil and Brown Rice in Zn Mining Areas. *Kor. Nat'l Acad. Sci. J.* 19, pp. 255–266, 1990.

Yoshikawa, H. Metal and Its Effect on Human Health. *Pollution and Measures.* 8, pp. 525–534, 1972.

CHAPTER 12

TRANSPORT, RESIDUES, AND TOXICOLOGICAL PROBLEMS OF AGROCHEMICALS IN AGROECOSYSTEMS AND A REMEDIATION PLAN IN THE REPUBLIC OF KOREA

K.S. Lee and B.H. Song

INTRODUCTION

In 1976 Korea had achieved self-sufficiency in the supply of rice, our major cereal. Although this was the result of the new hybrid breeding of *Japonica* × *Indica*, we also remember the development of cultivation methods for that strain. One of those methods was the use of pesticides for protection against some pests and diseases such as rice stem borer, hoppers, rice blast, rice sheath blight, etc.

Under Korean field conditions approximately 4,310 pests, diseases, and weeds have been reported, with 290 in paddy fields and 4,042 in upland areas. To control this problem, natural pyrethrin was introduced into Korea as the first pesticide in 1906, but until 1945, there were very few new synthetic organochloro insecticides applied, and then only in the case of experimental farming.

Initially, BHC and DDT were introduced to the Korean people as hygienic chemicals for control of lice, fleas, bedbugs, etc. Korean farmers soon realized the effect of these organochloro insecticides, and applied them to their crops. Since the late 1950s, consumption amounts of pesticides have been continually increasing. From the mid-1960s, many different kinds of pesticides have been formulated and syntheses of them were started in the early 1970s. In 1949 a total of 23 kinds of pesticides were recorded, and increased to 534 by 1993. In the late 1960s pesticide residue analytical methodology studies were started on crops. At the same time, the evaluation of the residue of organochloro insecticides was also begun by monitoring market samples. In 1970, the U.N.-Korea Plant Protection Programme, a United Nations Development Programme (UNDP) Project, was started by the Office of Rural Development (ORD), which then changed to Rural Development Administration (RDA) of Ministry of Agriculture and Forest (MAF) to carry out the monitoring survey of field crops on a national scale. In addition, since the mid-1970s, residue levels in soil have been surveyed, although the survey for river water was not frequently conducted up until the 1980s. There is still no monitoring survey of residues for groundwater.

After the early 1970s, MAF of Korea has been mainly recommending the safe use of pesticides and safe intervals for preharvest. The tolerances (or Maximum Residue Limit [MRL]) for foods were established by the Korean Environmental Protection Agency (KEPA), which was founded in 1980. Furthermore, regulation of pesticide residues has been enforced under the Soil Conservation Law, which is established by KEPA since 1996 (KEPA recently became the Ministry of Environment).

CHANGES OF THE PATTERNS IN PESTICIDE USE

In Korea, modern pesticide has been used by the people since the end of World War II. At the beginning, BHC and DDT were used for the improvement of hygiene and for agricultural purposes. Farmers, realizing the effect of pesticides, had increased the amount of pesticide applications, which resulted in high yields of production. Average application amounts per 10 a (= 0.1 ha, area unit in Korea) were over 10 kg in the 1990s, making Korea one of the major pesticide application countries. It is therefore essential to discuss the changes of pesticide application patterns with reference to amounts, formulations, and application uses in this chapter.

Change of Pesticide Application Amounts During 1963–1993

After World War II only small quantities of imported synthetic pesticides were used, although some plant insecticides such as natural pyrethrin and nicotine sulfate had been manufactured by Korean pesticide companies. The Korean War (1950) devastated our agriculture and industries. Therefore, until 1953, pesticide application amounts were insignificant. Since that time more than 20 pesticide manufacturing companies have been built, and many different kinds of synthesized pesticides were formulated using imported ingredients.

In 1949, 26 kinds of pesticides, totaling 1,029 MT (metric tons) of the actual amounts were used (Table 12.1). Usage was increased to 30 kinds totaling 14,114 MT in 1955. Over the years this trend continued, so that 120 kinds (1,783 MT) (a.i.: active ingredient) were used in 1960, 346 kinds (1,392 MT) in 1965, and 148 kinds (3,719 MT) in 1970. Many new pesticides were introduced successively in the 1970s, reaching over 200 kinds by 1979, and over 300 kinds in 1985. In 1988, 416 kinds were recorded and as many as 501 kinds in 1992. In proportion to the increase in the different kinds of pesticides, the amounts of pesticides also increased. It was 8,619 MT (a.i.) in 1975, 11,309 MT in 1978, 16,132 MT in 1980, and 21,322 MT in 1986. Thereafter the increase rate stabilized slightly to reach 25,999 MT in 1993.

In 1958 the proportion of pesticides by use was 82.1% of fungicides, 15.9% of insecticides, and only 3.0% of herbicides. The use of insecticides increased very quickly by 1963 so that it was reported that 49.2% of fungicides, 48.9% of insecticides, and 1.2% of herbicides were used. After insecticide had become the major pesticide in Korean agriculture, 64.7% of total pesticides was insecticide in 1968. At the same time the use of herbicide was increasing. In 1973, total pesticide use consisted of 15.5% fungicide, 64.5% insecticide, and 17.4% herbicide. Herbicide use exceeded fungicide in that year. A similar use pattern was shown in 1978 as 19.3% of fungicide, 59.7% of insecticide, and 19.5% of herbicide were used. But the use of fungicides and herbicides relatively increased in 1983 as indicated by the following statistics: 25.4% of fungicide, 42.8% of insecticide, and 25.1% of herbicide. Fungicide use increased to the extent that, since 1988, it exceeded the use of

Table 12.1. Changes of Registration Numbers and Consumption Amounts of Pesticides

Year	Pesticides (Number of Item) Fungicide	Insecticide	Fungicide and Insecticide	Herbicide	Growth Regulator and Others	Total	(MT)	Remark
1949	8	10	0	0	8	26	1,029	Formulation (MT)
1955	6	20	0	1	3	30	4,114	Formulation (MT)
1960	19	90	1	1	9	120	1,783	a.i.[a](MT)
1965	61	209	5	30	41	346	1,392	a.i.(MT)
1970	37	78	4	20	9	148	3,719	a.i.(MT)
1975	38	76	7	20	13	154	8,619	a.i.(MT)
1978	52	92	8	30	14	196	11,309	a.i.(MT)
1979	64	90	8	30	15	207	14,454	a.i.(MT)
1980	71	104	8	35	15	233	16,132	a.i.(MT)
1986	111	137	13	59	18	338	21,322	a.i.(MT)
1988	137	173	18	69	19	416	21,967	a.i.(MT)
1992	179	185	12	100	25	501	26,718	a.i.(MT)
1993	187	201	10	107	25	530	26,999	a.i.(MT)

[a]Active ingredient.

insecticide. In 1988, the use was 37.1% of fungicide, 33.0% of insecticide, and 20.9% of herbicide. In 1993 fungicide was used to an even greater extent, showing 39.0% of fungicide, 29.4% of insecticide, and 20.3% of herbicide.

The changes of pesticide use patterns were closely related to the changes in cultivation conditions and understanding of the effects of pesticide applications by farmers. Therefore, one of the reasons for a decrease in the use of insecticides would be the pesticide residue problem, as well as the development of more potent pesticides. Other factors that contribute to the decrease include the use of plant growth regulators, chemicals for stored grains and surfactants. In Table 12.2, changes of consuming amounts by different pesticide groups are listed.

Changes of Pesticide Formulations

Different types of pesticides are used in the agricultural environment. Generally, granule (G) and dust (D) forms are applied more than emulsifiable concentrate (EC) and wettable powder WP) or solution (S) to the same unit area. Although using EC usually means smaller amounts of the pesticide are required, farmers have preferred the use of G and D as they are more easily applied.

In 1949, about 730 MT of sulfur powder was used as a fungicide, and 15 MT of derris powder and 33 MT of natural pyrethrin powder as insecticides. DDT and BHC dust were also used widely from 1953 on. Totals of 96 MT (12.3%) of EC, 244 MT (31.3%) of D, and 439 MT (56.4%) of WP were used in 1958. In 1963, it was 650 MT (44.6%) of EC, 108 MT (7.4%) of D, and 699 MT (48.0%) of WP. It was clear that the use of EC had increased rapidly, whereas that of D had quickly decreased. Since 1965, granule type pesticides had been introduced to the Korean market, although only 6 MT were applied in that year. Nevertheless, use of G had increased very quickly because of the farmer's preference for this form of pesticide.

In 1968 1,238 MT (61.7%) of EC, 108 MT (5.4%) of D, 24 MT (1.2%) of G, and 616

Table 12.2. Changes of Application Amounts of Different Pesticide Groups

MT					
Fungicide	**Insecticide**	**Herbicide**	**Others**	**Total**	**Year**
1958	640	123	16	—	779
1963	717	713	18	9	1,457
1968	415	1,300	274	19	2,008
1970	767	1,735	1,122	95	3,719
1973	1,044	4,344	1,170	171	6,729
1978	2,184	6,755	2,204	166	11,309
1980	5,448	6,407	3,374	903	16,132
1983	3,959	6,681	3,912	1,052	15,604
1987	8,384	8,069	4,666	2,110	23,229
1990	7,778	9,332	5,509	2,463	25,082
1993	10,143	7,642	5,270	2,944	25,999

MT (30.7%) of WP were used. In 1973, G usage was increased to 20.2%. However, 6,170 MT (39.5%) of EC, 80.5 MT (5.2%) of D, 4,229 MT (27.1%) of G, and 3,721 (23.8%) of WP were used in 1978, indicating that use of D had continued to decrease while use of G continued to increase. This trend was maintained throughout the 1980s. In 1988, 36.8% of EC, 1.4% of D, 25.1% of G, and 32.8% of WP were used. These numbers show that WP was more widely used than EC, while use of D decreased steadily.

In 1993, 8,704 MT (33.5%) of EC, 212 MT (0.8%) of D, 5,086 MT (19.6%) of G, and 10,307 MT (39.6%) of WP were used. The proportional use of D and G was decreasing. This decreased use of pesticide per unit area would be helpful to the agricultural environment. Furthermore, this shows that the majority of Korean pesticide companies have been developing new formulations that are effective as well as environmentally attractive.

Pesticide Application in Different Land Uses and Unit Areas

Land use patterns in Korea could be divided into two main categories, namely paddy fields and upland cropping areas. In 1958 the use of agrochemicals for paddy were 82 MT and 697 MT for upland crops. These figure equate to 0.07 kg/ha and 0.86 kg/ha, respectively.

Until 1968, more pesticides were applied to upland cultivation than to the paddy fields. This trend was reversed to show that pesticides used for paddy were 1,455 MT, which exceeded the 988 MT used on upland crops. Since 1970, amounts used have been increasing rapidly, and more than 3,000 MT of a.i. was recorded in 1970 and 5,000 MT in 1973. In 1978 more than 10,000 MT was used and increased to over 20,000 MT in 1987, after which the increasing rate of pesticide usage stabilized. It should be noted that, since 1988, horticultural use of pesticide has exceeded that used for paddy, and this trend should continue in the future. Table 12.3 shows changes of pesticide consumption by different crops.

The average consumption rate of pesticides was 0.39 kg/ha in 1958 and has gradually increased year by year: 0.9 kg/ha was recorded in 1963, 0.86 kg/ha in 1968, and then 1.06 kg/ha in 1969. Thereafter, the application rate began to increase more quickly, so that 2.93 kg/ha was used in 1973 and 5.01 kg/ha in 1978. In 1988 9.72 kg/ha were used, and finally the 10 kg/ha milestone was broken by the use of 10.78 kg/ha in 1990. Also, use of

Table 12.3. Changes of Pesticide Consumption by Different Crops Active Ingredient (MT)

Year	Rice	Horticulture	Others	Total
1958	82	697	—	779
1963	386	1,062	9	1,457
1968	962	1,027	19	2,008
1969	1,455	988	26	2,469
1970	2,546	1,078	95	3,719
1973	3,091	3,467	171	6,729
1978	6,143	5,000	166	11,309
1982	7,505	5,838	1,084	14,427
1986	9,813	9,529	1,980	21,322
1987	10,863	10,256	2,110	23,229
1988	9,728	10,259	1,980	21,967
1991	11,665	12,939	2,872	27,476
1992	10,539	13,153	3,026	26,718
1993	7,920	15,135	2,944	25,999

pesticides on paddy fields had increased gradually to reach the peak use of 9.65 kg/ha in 1991, and then decreased slightly to 9.11 kg/ha in 1992. On the other hand, horticultural use of pesticide also had increased year by year to exceed over 10 kg/ha in 1986, and reached 14.91 kg/ha in 1992. The trends of increasing horticultural pesticide consumption per unit area should be continued by the expansion of cultivation in covered structures and increased production of high quality vegetables and fruits.

PESTICIDE RESIDUES IN AGRICULTURAL ENVIRONMENTS

Pesticides used for controlling pests, diseases, and weeds directly penetrate into the soil, or are adsorbed and translocated through the soil into crops to remain as residues. These residual components could be harmful to the human body through the food chain. In this section, an evaluation is made of the residue levels in crops and soils in the Korean agricultural environments, as well as in the irrigation waters, lakes, and rivers. A similar evaluation is also done on soil that has been cultivated under man-made structures.

Residue Levels in Cultivated Soil

Monitoring surveys of organochloro pesticide residues in crops began in the late 1960s at the National Health Institute (NHI) and Radio Radiation Research Institute in Agriculture (RRIA) in Korea. During the early stages of the surveys, 10–30 basket samples collected from the markets were taken, and later were increased to several hundred basket samples after the UNDP project of ORD was begun. Monitoring surveys for basket samples are still performed by NHI of Ministry of Health and Welfare (MHW). This system is now continued by the Health and Environment Institute of local government and regional extension branches of ORD. Also, in cooperation with MAF, the Korean Agricultural Chemicals Industrial Association (KACIA) has carried out pesticide monitoring surveys on basket and field samples for large quantities of consumption crops or for studies pertaining to toxicological problems. Five-year intervals, results of the monitoring surveys from the late 1960s to 1990, are shown in Tables 12.4, 12.5, and 12.6.

The residue level of total-BHC (see Appendix) was at a peak in the late 1960s when

Table 12.4. Residue Levels of Organochloro Pesticides in Crops

Organochlorine Pesticides		Period					References
		1966–70	1971–75	1976–80	1981–85	1986–90	
		(mg/kg)					
BHC	Maximum	1.020	0.444	0.706	0.800	0.061	Lee, S.R. (1993)
	Mean	0.073	0.013	0.021	0.025	0.003	
DDT	Maximum	0.700	2.217	0.134	0.374	0.123	Park, C.K. and
	Mean	0.075	0.017	0.007	0.016	0.003	Lee, Y.D. (1987)
Heptachlor	Maximum	0.160	0.552	0.065	0.068	0.096	
& epoxide	Mean	0.004	0.010	0.004	0.005	O.O1Z	
Drins	Range	ND-0.520	ND-0.027	ND-0.061	ND-0.054	ND-0.049	
Others	Maximum			0.007	0.512	4.540	
	Mean			0.001	0.057	0.016	

Table 12.5. Residue Levels of Organophosphorus Pesticides in Crops

Organochlorine Pesticides		Period					References
		1966–70	1971–75	1976–80	1981–85	1986–90	
		(mg/kg)					
Chlorpyrifos	Maximum				ND	0.168	Lee, S.R. (1993)
	Mean				ND	0.003	
Diazinon	Maximum		0.190	0.584	0.267	0.443	Park C.K. and Lee, Y.D. (1987)
	Mean		0.011	0.006	0.007	0.007	
Dimethoate	Maximum				0.102	0.122	
	Mean				0.004	0.002	
EPN	Maximum			0.032	0.344	0.023	
	Mean			0.006	0.009	0.007	
MEP	Maximum		0.086	0.268	0.163	0.453	
	Mean		0.002	0.002	0.006	0.005	
MPP	Maximum		0.067	1.077	0.400	0.097	
	Mean		0.007	0.016	0.016	0.001	
IBP	Maximum		ND	2.275	0.180	0.070	
	Mean		ND	0.040	0.008	0.003	
Malathion	Maximum		0.059	0.235	0.038	0.102	
	Mean		0.001	0.010	0.011	0.009	
Parathion	Maximum	0.210			0.091	0.245	
	Mean	0.003			0.005	0.004	
PAP	Maximum		0.219	0.003	0.063	0.181	
	Mean		0.003	ND	0.003	0.003	
Others	Maximum				0.273	0.212	
	Mean				0.090	0.002	

that chemical had been used widely. After the prohibition of BHC in 1979 and lindane(r-isomer) prohibition in 1982, residue level of BHC gradually reduced until the late 1980s.

DDT shown in Table 12.4 included p,p′-DDT, DDE, DDD and all of o,p′-isomers. This insecticide showed a residue level of 0.075 mg/kg in the late 1960s. After prohibition of the use of DDT in 1972, the residue level of this chemical was lowered. Although the average DDT residue level remained at 0.016 mg/kg until the early 1980s, it was reduced to 0.003 mg/kg by the end of the 1980s. A similar reduction was found with BHC residue levels.

Table 12.6. Residue Levels of Carbamate Pesticides in Crops

Carbamate Pesticides		Period				References
		1976–80	1981–85	1986–87	1988–90	
		(mg/kg)				
NAC	Maximum	ND	0.720	0.845	0.858	
	Mean	ND	0.042	0.024	0.021	Lee, S.R. (1993)
BPMC	Maximum	ND	0.902	0.511	0.280	Park, C.K. and
	Mean	ND	0.031	0.007	0.054	Lee, Y.D. (1987)
MIPC	Maximum	ND	0.574	0.234	0.072	
	Mean	ND	0.031	0.012	0.001	
Carbofuran	Maximum	ND	ND	ND	ND	
	Mean	ND	ND	ND	ND	

Heptachlor had been used in tobacco fields throughout the 1970s, and so it was possible to detect a high level of its residues in the following crops such as potato and Chinese cabbage. Captafol residue level was 4.54 mg/kg during 1986–1990. For this reason, use of this chemical was banned from 1993.

The average residue level of organophosphorus insecticides showed almost less than 0.01 mg/kg. Diazinon which was introduced during the late 1960s and widely used in the 1970s was showing an average of 0.006 to 0.007 mg/kg of residue during the late 1970s to early 1980s, and MPP also averaged over 0.01 mg/kg. In the late 1980s, no chemical exceeded a residue level of 0.01 mg/kg, indicating a close relation with a decrease in the use of organophosphorus insecticides.

IBP, the only organophoshrous fungicide, had relatively higher residue levels in brown rice grain, but became much lower in late 1980s in proportion to the decrease of consumption.

The carbamate insecticides were widely used on rice, vegetables, and fruits. Table 12.6 shows the residue levels in brown rice grains. NAC and MIPC showed a similar pattern with the amounts of use, and BPMC was used most for control of rice hoppers during 1988–1990. No residue was found to be from carbofuran. More than 5,000 crop samples from baskets and fields had been collected and analyzed for residue annually by governmental institutes in the 1990s.

Residue levels of organochloro insecticides in soils are summarized in Table 12.7. The average residue levels in paddy field soils decreased rapidly year by year, and a much similar pattern was found in upland soils, with the exception of that in Cheju Island. Generally the higher residue levels shown in upland soils were due to application amounts. In Cheju Island more BHC and heptachlor were used in crop fields than other organochloro insecticides. The results of soil surveys show that the levels of residues are related to that of application amounts and type of chemical used.

In orchard soil, α-endosulfan showed 0.14 mg/kg of residue level. This figure was higher than that found in paddy and vegetable field soils. Also, p,p′-DDT and p,p′-DDD were 16–270 and 9–90 times higher than those of paddy and vegetable field soils, respectively.

Kethane and chlorbenzilate, organochloro acaricides, were detected in citrus orchard soil of Cheju Island. These chemicals should be included in future monitoring surveys due to the possibility of widespread contamination.

Table 12.7. Residue Levels of Organochloro Pesticides in Korean Soil

Soil	No. of Samples	Residue Levels (mg/kg) α-BHC	γ-BHC	PCNB	H. Chlor	Aldrin	H. Epoxide	α-Endo-sulfan	Dieldrin	p.p-DDD	p.p-DDT	Kelthane	Akar	References
	95	0.002-0.085 (0.012)[a]	0.002-0.036 (0.008)	—	ND-0.028 (0.002)	ND-0.088 (0.005)	ND-0.025 (0.001)	ND-0.069 (0.002)	ND-0.045 (0.010)	ND-0.048 (0.007)	ND-0.270 (0.016)	—	—	Park et al. (1982)
Paddy	33	t-0.007 (0.003)	ND-0.004 (t)	—	ND-t (t)	ND-0.007 (.001)	ND-0.001 (t)	—	ND-0.007 (t)	ND-0.024 (0.002)	ND-0.065 (0.006)	—	—	Suh et al. (1984)
	108	t-0.008 (0.002)	t-0.005 (0.001)	,	t-0.001 (t)	—	t-0.008 (0.001)	—	t-0.005 (0.001)	t-0.004 (0.002)	t-0.008 (0.002)	—	—	Choi et al. (1987)
	59	0.003-0.049 (0.009)	0.001-0.034 (0.009)	ND-0.011 (0.004)	ND-0.017 (0.003)	ND-0.017 (0.003)	ND-0.310 (0.012)	ND-0.033 (0.003)	ND-0.110 (0.010)	ND-0.062 (0.005)	ND-0.200 (0.020)	—	—	Park et al. (1983)
Upland	30	t-0.005 (0.002)	t-0.013 (0.001)	—	ND-0.011 (0.004)	ND-0.005 (t)	ND-0.050 (0.014)	—	ND-0.032 (0.017)	ND-0.220 (0.020)	ND-0.152 (0.033)	—	—	Suh et al. (1982)
	30	ND-0.967 (0.391)	ND-0.590 (0.218)	—	ND-0.819 (0.163)	—	ND-0.150 (0.034)	—	—	—	—	—	—	Lee (1981)
Orchard	43	t-0.025 (0.008	t-0.040 (0.010)	—	ND-0.062 (0.012)	ND-0.340 (0.094)	ND-0.110 (0.008)	ND-1.70 (0.140)	ND-3.30 (0.246)	ND-1.400 (0.180)	ND-3.200 (0.540)			Park et al. (1982)
	150	—	—	—	—	—	—	—	—	—	—	t-1.359 (0.251)	t-0.925 (0.120)	Lee (1980)

[a]( ) Average.

Pesticide Residues in Cultivated Soils Sheltered by Man-Made Structures

In 1970, Korea had only 320 ha of cultivated areas that were sheltered by structures. Since the late 1970s, the area had expanded rapidly to 16,500 ha in 1985, 23,680 ha in 1990, and 59,888 ha in 1993. The plastic (vinyl) greenhouse has become the most popular pattern of sheltered cultivation. However, more recently, farmers are showing greater interest in glass houses. The environmental conditions inside the plastic house is of high humidity and temperature, which are more susceptible to the breakout of diseases than in a field condition.

One research paper recommends that reentering the plastic houses should not be allowed until 3 hours after treatment by dichlorvos (Korea Research Institute of Chemical Technology, 1985). More research is needed in order to solve this toxicological problem. Residue levels of organochloro pesticides for cultivated soils in sheltered structures were evaluated in 1981 and 1982. Also, 26 pesticides including 5 organochloro pesticides were investigated in 1991. Table 12.8 shows that the residue levels of -BHC, -BHC, dieldrin, and p,p′-DDT were reduced to 1/28, 1/8, 1/51, and 1/7, respectively, in the 10 years from 1981 to 1991.

Other detected pesticides, besides the organochloro pesticides, were in the majority fungicides. They were procymidone, vinclozolin, and chlorothalonil in strawberry fields, procymidone, ethroprophos, captafol, and vinclozolin in green pepper fields, and chlorothalonil, procymidone, vinclozoline, dichlofluanid, and BPMC in cucumber fields (Choi and Kim, 1991).

Residues of organochloro insecticides should continue to be less of a problem in cultivated soils sheltered by structure in the future. However, evaluation of fungicide residues should be continued together with the establishment of some policies to decrease the residues of these chemicals. Futhermore, monitoring surveys on crop lands should be enlarged, and changes of pesticide residues on the point source should also be periodically monitored.

Pesticide Residues in Aquatic Environments

Paddy rice is the major crop in Korea. Pesticides applied to the paddy fields have contaminated different aquatic environments such as streams, rivers, and lakes through irrigation waters. It is possible that mud and sediments of river estuaries and the coast areas could also be polluted by residual pesticides.

Table 12.9 shows the residue levels of organochloro insecticides in different water samples. Relatively high residue levels of organochloro pesticides were detected in water taken from the Kwangyang Gulf river estuaries in 1974. For example, 13.0 μg/kg of aldrin and 16.9 μg/kg of dieldrin residue were detected. These residues were higher than those of other chemicals. From 1982 to 1983, slight traces of α-, γ-BHC, and heptachlor epoxides were detected while other river or irrigation water samples showed less than 0.003 μg/kg of organochloro insecticides. Fifteen sampling sites of six major rivers in Korea were investigated in April and August of 1991. However, no organochloro pesticide residues were found in any of these water samples in the soils.

Residue levels of organophosphorus pesticides in river water are shown in Table 12.10. While diazinon and IBP were detected in all samples, they did not show any special residual pattern. Comparison of these residue levels with LC50 of carp was 1/6,500–1/320,000 in diazinon and 1/990–1/200,000 in IBP, respectively. Also, residues of other detected organophosphrous insecticides were only found in very low levels. In addition, when the maximum detected residue level was compared with the standard of drinking water, 20 μg/kg of diazinon, 0.16 μg/kg of fenitrothion, 0.17 μg/kg of

Table 12.8. Residue Levels of Organochlorine Pesticides in Sheltered Soils

Year	No. of Samples	Residue Levels (mg/kg)										References
		α-BHC	γ-BHC	PCNB	H. Chlor	Aldrin	H. Epoxide	α-Endo-sulfan	Dieldrin	p.p-DDD	p.p-DDT	
		0.002-0.047 (0.014)[a]	0.001-0.057 (0.013)	ND-0.040 (0.002)	ND-0.075 (0.011)	ND-0.079 (0.014)	ND-0.120 (0.026)	ND-0.075 (0.003)	ND-0.240 (0.036)	ND-0.310 (0.030)	ND-1.100 (0.120)	Park et al. (1982)
1982	59	t-0.050 (0.002)		ND-0.100 (0.010)					ND–0.090 (0.013)	ND-1.500 (0.104)	ND-1.600 (0.102)	Suh et al. (1984)
1991	90	0.0002–0.0012 (0.0005)	0.0007–0.0021 (0.0016)			ND	-	~	ND-0.0018 (0.0007)		0.009-0.033 (0.016)	Choi et al. (1991)

[a]( ) Average.

Table 12.9. Residue Levels of Organochloro Pesticides in Water

		Residue levels (μg/kg)							
Year	**Sites**	**α-BHC**	**γ-BHC**	**H. Chlor**	**H. Epoxide**	**Aldrin**	**Dieldrin**	**Endrin**	**References**
1974–1975	Bay	1.0–55.0 (8.9)	0.6–18.0 (5.8)	0.7–7.6 (2.6)	0.0–5.3 (2.4)	1.0–47.6 (13.0)	5.0–36.4 (16.9)	0–20.0 (3.4)	Lee et al. (1976)
1981	Lake	ND-0 15 (0.033)	t–0.015 (0.007)	— —	ND-0.003 (t)	—	—	—	Park et al. (1982)
1982	River	ND-0.035 (0.009)	ND-0.038 (0.003)	ND-0.060 (0.010)	ND-0.110 (0.01)		—	ND- t (t)	Lee et al. (1983)
1982–	Irrigation	ND-0.008 (0.002)	ND-0.01 (0.001)	ND-0.001 (t)	ND-0.004 (t)	ND-0.004 (t)	ND-0.002 (t)	—	Lee et al. (1985)
1991	River	ND	ND	ND	ND	ND	ND	ND	A.R.I. (1991)

Table 12.10. Residue Levels of Organophosphorous Pesticides in River Water

	Residue Levels (mg/kg)						
Year	**Diazinon**	**Fenitrothion**	**IBP**	**H. Malathion**	**Parathion**	**Phenthoate**	**References**
1982.8	0.16–2.0 (0.46)	ND-0.16 (0.04)	2.0–11.0 (4.3)	ND-0.17 (0.05)	ND-0.24 (0.11)	ND-1.3 (0.22)	
1982	ND-0.07 (0.01)	ND	ND-0.53 (0.10)	ND	ND	ND	Park et al. (1984)
1983.6	0.08–0.67 (0.16)	ND	0.05–0.32 (0.15)	ND	ND	ND-1.2	
1983	t–0.63 (0.046)	t–0.04 (t)	t–1.38 (0.086)	—	—	—	Lee et al. (1984)
1991	ND-0.4	ND-t	ND-0.83	ND	ND	ND	A.R.I. (1991)
1992	0.2–1.5	—	0.05–2.21	—	—	—	Choi et al. (1992)

malathion, and 0.24 μg/kg of parathion were the same as 1/10, 1/200, 1/1,470 and 1/200 of tolerance of drinking water, respectively.

In 1991, river water was tested for approximately 36 pesticides. 0.09 μg/kg of chlorpyrifos was detected in the Kum river. Procymidone and euparean were determined to be 100% and 53% of frequencies, respectively. Out of 26 chemicals tested in 1992, 9 were detected to be diazinon. IBP, α-, β-endosulfan, BPMC, carbofuran, alachlor, and butachlor were also apparent.

EVALUATION OF THE EFFECTS OF PESTICIDE RESIDUES ON AGRICULTURAL ENVIRONMENT

Pesticides applied to the earth could have an effect on the soil microflora and physicochemical properties of soil. Furthermore, if residues remain for a long period of time in the soil environment, remedial strategies, and preservation should be provided. This section therefore discusses the fate of some pesticides in soil.

Direct Effects of Pesticide Residues on Soils

As pesticides are introduced into a soil environment, it is generally found that the activities of microorganisms are primarily affected and secondarily, as a result of the lasting effects of the pesticide residues, the metabolic pathways of soil nutrients also are affected. Hence, a change of microorganism numbers, enzyme activities, and chemical properties of soils are considered in this section.

Changes of Microorganism Numbers and Enzyme Activities in the Soil

Hong and Choi (1969) reported that BHC inhibited some soil bacteria such as *Bacillus subtillis*, *B. sphaericus*, and *B. cerius* more than DDT did. Han and Kim (1974) treated soil with six different fungicides; namely, ferbam, kasugamycin, diamonium ethylenebis dithiocarbamate, maneb, IBP, and blasticidin and also four different insecticides such as fenthion, diazinon, γ-BHC, and carbaryl. They noted that five fungicides inhibited growth of soil bacteria except for ferbam, which showed an increase in bacteria growth from 15 days after treatment. They reported that all insecticides inhibited bacterial growth at the early stage after application. From the eighth day after treatment, however, the bacteria recovered and its number grew to more than those in the control sample. Actinomycetes showed a much similar pattern to that of bacteria. Fungi had a different pattern from that of bacteria and actinomycetes. After two days of fungicide treatments, bacteria and actinomycetes showed higher numbers than normal, but quickly decreased back to the minimum numbers on the fifth day. Fungi numbers were similar to those in the control samples in all treatments. In the case of insecticide treatments, the numbers of fungi were also generally very similar to those in the control samples.

Han (1975) said that phosvel inhibited the number of bacteria up to 4 days after treatment, and thereafter the bacteria recovered gradually. Fungi increased up till 4 days after treatment, then decreased quickly until 4 weeks after treatment. The number of actinomycetes were similar to those in the control experiments.

Kim (1975) published a report that polyoxin reduced the number of bacteria up to 3 weeks after treatment, and up to 4 weeks in the case of actinomycetes. He also pointed out that the number of fungi was affected up to 5 weeks after treatment.

Kim and Jeong (1976) reported the effects of butachlor treatment on soil microflora under paddy soil condition. They said that the number of bacteria reached the maximum in 35 days after treatment and thereafter decreased slowly. Actinomycetes showed periodical fluctuation patterns at weekly intervals. The authors mentioned that the fungi number in control samples increased for a week and then decreased gradually. They also pointed out that actinomycetes showed a similar periodical fluctuation pattern by heptachlor treatment. Kim and Han (1977) published a report that isoprothiolane inhibited the numbers of bacteria and actinomycetes during the early stage of treatment, but the numbers increased in the course of time. The fungi numbers also increased in the early days of treatment.

Kim (1978) applied four kinds of herbicides; namely, nitrofen, CNP, avirosan, and mamet, to soil and counted the number of microflora. Within 14 days the number of bacteria grew to more than those in the control samples, then became lower from 14 to 28 days. Actinomycetes were lower in numbers between 14 and 28 days after treatment.

Fungi number decreased up till one week after treatment and then gradually increased as the fungi recovered.

Ryu et al. (1980) applied three major pesticides (butachlor, carbofuran, and IBP) to paddy rice grown under submerged soil condition. Twenty days after the application, the bacteria, actinomycetes, and fungi reached the maximum count. The addition of organic matter (rice straw) showed an increase of their numbers, but an application of high concentration of the pesticides showed the opposite pattern. The number of nitrite-forming bacteria peaked 50 days after treatment, whereas the nitrate-forming bacteria peaked in 10-35 days. Applying a high concentration of IBP to the samples increased the number of nitrate-forming bacteria. Nitrate-reducing bacteria count was the highest 10 days after treatment, and then gradually decreased. After 10 days of chemical treatment the number of denitrification bacteria was reduced to the lowest level and it took 50 days before their number reverted to the same as that in the control sample. In this instance the addition of rice straws helped decrease the number of denitrification bacteria.

Kang (1978) used bentazon to find out about the change of soil microflora. He said that the total number of bacteria was influenced by the concentration of the chemicals used and total fungi had increased for 2 weeks. He also mentioned that compared with the control sample, the number of nitrite-forming bacteria was 1.3 times higher in 1 week, 3.6 times in 2 weeks, and 2.6 times in 4 weeks after treatment at 50-mg/kg application. Song et al. (1981) presented their findings that replications of IBP treatment had little effect on bacteria and fungi in soil.

Lee et al. (1982) reported that the number of fungi was the lowest 15 days after treatment, and that there was hardly any influence on bacteria and actinomycetes. Rho and Baik (1981) treated isoprothiolane, acephate, and butachlor with a recommended dose and also one that was 10 times higher to both submerged soils and upland soils, and counted the number of microorganisms 7 days after incubation. They discovered that isoprothiolane slightly hindered the growth of fungi in both submerged soils and in upland soils; a high dose of acephate decreased the number of fungi in paddy soil condition, while butachlor showed no harmful effect to the growth of bacteria, actinomycetes, and fungi.

Kim et al. (1987) examined the effects of fungicides (isoprocarb and captan) and insecticides (carbofuran, propoxar, and acephate) to changes of soil microflora. They found that all chemicals decreased the number of fungi during the early days of treatment, but that the fungi recovered in due course. They also discovered that fungicides inhibited the fungi more than the insecticides did. Actinomycetes growth was slightly hindered by carbofuran and also by two fungicides up to 30 days after treatment, then recovered so that approximately 60 days after treatment, their number was becoming equal to that in the control sample. They also found that bacteria was reduced to the minimum number 6 days after application, but reverted to the control level 60 days after treatment.

Kim et al. (1988) tested four different herbicides such as dicamba, CNP, linuron, and simetryne. They ascertained that fungi growth apparently became inhibited from 2 to 10 days, and gradually regained the original number as time went by, although it was still lower than those in the control experiment even 60 days after treatment. The treated chemicals were not hazardous to actinomycetes, and the bacteria count was comparable to those in the control sample although a little lower in number.

Lee et al. (1988) investigated the fate of nitrification microorganisms in soils after being treated with carbofuran, acephate, MIPC, and simetryne. Ammonium-forming bacteria decreased in all chemical treatments. Among them, acephate affected little damage and MIPC caused some inhibition up to 60 days after treatment. While nitrite-forming bacteria showed a lower inhibition rate than ammonium-forming bacteria, it was also found to be hindered by herbicide application. Carbofuran and simetryne caused more inhibitory effects to the growth of nitrite-forming bacteria. The authors also mentioned that the number of denitrification bacteria was lower than those in the control sample at 10 days after treatment, and that nitrate-reducing bacteria had no effect by tested chemicals. Kim et al. (1989) stated that soil bacteria increased until day 2; thereafter the number remained constant up till 10 days after treatment. Fungi increased within the first 4 days of diazinon treatment under paddy soil condition and thereafter remained constant throughout the test period.

Lee (1989) said that under submerged soil conditions the number of bacteria increased during the 10 days after treatment with diazinon and chlorpyrifos. The growth rate doubled in the case of diazinon treatment. Among other bacteria, pseudomonas showed a similar growth rate up to 5 days after treatment, but increased especially rapidly with diazinon treatment. The number of fungi increased till day 5 and then slowly decreased. Treating the samples with three times higher concentration resulted in a higher number of fungi than using the normal dose of diazinon and chlorpyrifos treatment. Activities of the soil enzymes are much influenced by the fate of soil microorganisms.

The following are the results of some researchers who investigated these phenomena. Hong and Choi (1969) reported that amylase activities were increased when treated with BHC and DDT at 50 mg/kg concentration, and decreased in proportion to increasing the concentration of the two chemicals. They also mentioned that saccharase activities were the highest in 200 mg/kg of BHC and 100 mg/kg of DDT treatments. They isolated *Bacillus substillis*, *B. sphaerius,* and *B. cerius* from soils and tested the effects of the two chemicals on amylase and saccharase activities. The activities of those two enzymes were highest when 50–100 mg/kg concentrations of BHC and DDT were used.

Roh and Baik (1981) reported about the dehydrogenase activities in the soils which had been treated with isothiolane, acephate, and butachlor. Their activities were steadily decreased by the chemical treatments in the upland aerated condition. In submerged soils, dehydrogenase activities increased 4.7 times in control samples, 5.2 times in isoprothiolane-treated samples, 5.5 times in acephate treated samples, and 4.9 times in butachlor treated samples at 12 days after treatment.

Hong and Cho (1979) compared urease activities in the soils when some herbicides with nitrogen heteroatom were applied to urea-treated and nontreated soils. Both control and chemical treated samples showed the highest enzyme activities 5 days after treatment in urea-nontreated soils, especially as they maintained their activities in proportion to the concentrations of asulam. Compared with the control, all chemical treatments resulted in lower urease activities in urea-treated soils, and reached the highest activities 6 weeks after treatment. They also mentioned that application of high concentrations of dimetametryne and linuron sustained urease activities for a relatively long period in urea treated soils. Ryu et al. (1980) also reported that the dehydrogenase activities in control soils were much the same as in those treated with recommended doses of butachlor, carbofuran, and IBP. It was also proved that high concentrations of the chemicals inhibited enzyme activities. Catalase activities were also greater when the applied dose of chemicals was increased. Lee and Lee (1983) published a report that urease activities in paddy soils were increased by 2,4-D and oxadiazon treatments. Kim et al. (1987) studied the activities of some soil enzymes when treated with three insecticides and two fungicides. Polygalacturonase activities were little affected by pesticide treatment, yet about 60% of inhibition was observed at 10 days after captan and isoprocarb were applied. Dehydrogenase activities were affected by the inhibition effects of pesticide for 10 days after treatment, and then increased gradually to control level at 30 days after treatment. Also they reported that phosphatase activities decreased during the early days of all chemical treatments, but returned to control level from 20 days after treatment. Cellulase activities were generally not influenced by pesticide treatments except when acephate was used, which showed a 15% inhibition on the first day after treatment.

Kim et al. (1988) investigated the effects of four different kinds of herbicides on some soil enzyme activities. They said dehydrogenase activities were lower in chemical treated samples than those in control up till 10 days after treatment, after which they recovered and even surpassed the control level after 20 days. Activities of phosphatase and protease were not affected by herbicide treatments, while urease activities were inhibited for 2 days initially, recovered 6 days after treatment, and then showed a lower activity level than the control for quite a long time. β-Glucosidase activities were influenced by chemical treatments, such as using CNP and linuron, which increased slightly the activities of this enzyme. The activities of cellulase and polygalacturonase were not affected by herbicides. Choi et al. (1990) investigated changes of activities for monooxygenase, α- and β-esterase when diazinon and chlorpyrifos were introduced to soils under submerged conditions. They found that monooxygenase activities appeared only 12 hours after diazinon treatment, but 3 days in the case of chlorpyrifos treatment. The α-esterase activities were 10 times higher than those of β-esterase; they were high during the early stage of chlorpyrifos treatment and also 5–8 days after diazinon treatment.

Effects on Chemical Properties of Soils

From the various research reports mentioned above, we can see that many pesticides affect the growth of soil microorganisms and soil enzyme activities. As a result they also affected some of the chemical properties of soils such as the pH, Eh, and nitritification processes.

Oh (1973) tested for changes of chemical properties in three replication treatments of five different herbicides including CNP, lorox, 2,4-D, lasso, and PCP. Lorox treatment showed some different results like decrease in pH, increase of exchangeable and hydrolytic acids and exchangeable potassium, and a decrease of exchangeable sodium and calcium and organic matter contents. He noted that the other herbicides showed an increase of soil organic matter contents.

Han and Kim (1974) reported that soil pH was increased by pesticide treatment with time, and had remained steadily higher since 15 days after treatment. Use of maneb, IBP, and blasticydin resulted in higher pH than control even 2 days after treatment. The Eh value was not affected within 2 days after treatment, but oxidative condition occurred at 5 days and reductive condition at 15 days after treatment. Among the chemicals, oxidative condition occurred 8 days after treatment with kasugamin and blasticydin. They found that the concentration of NH_4-N was increased rapidly within 2 days, then decreased quickly and showed the lowest value on day 5, after which it increased very slowly from the 22nd day after treatment. Generally, all treatments resulted in lower levels of NH_4-N than control, and more inhibition effects of NH_4-N forming at high dose application than at the recommended dose. They also mentioned that concentrations of NO_2-N and NO_3-N increased during the early days of incubation, but was not apparent over the later stage of incubation. Chemical treatments showed lower NO_2-N and NO_3-N than non-treatment, and the higher concentration of chemicals showed the lower NO_2-N and NO_3-N. They noted that treatments with IBP, ferbam, and four insecticides such as baycid, diazinon, γ-BHC, and sevin showed an upward tendency of NO_2-N and NO_3-N 15 days after incubation, after which they gradually disappeared.

Han (1975) applied phosvel to submerged soil, as a result of which an increase of pH was apparent within 7 days after treatment. The pH level also remained steady in the range of 7.5 to 7.6 up to 6 weeks after treatment. On the other hand, Eh decreased quickly after 4 days, and then gradually rose. A reduction in value was shown throughout incubation. He observed that NH_4-N showed the highest concentration at day 4, then continuously decreased until 28 days after treatment. Also, NO_3-N increased rapidly 7 days after incubation, but disappeared at 21 days. The higher concentration of phosvel produced more NO_3-N, while NO_2-N disappeared at 14 days of incubation.

Kim (1975) reported that polyoxin treatment increased soil pH but lowered soil Eh in comparison with nontreated soil. He also found that NH_4-N peaked 4 days after treatment, then decreased slowly until 28 days, after which it again increased. NO_2-N increased gradually and showed the highest value 14 days after treatment. Kim (1976) also reported that the pH and Eh of soils were not affected by six herbicides. He concluded that nitrofen, benthiocarb + simetryne and propanil didn't influence the production of NH_4-N from urea, while butachlor and perfluidon inhibited production of NH_4-N during the early days of treatment, but then soon recovered to the control level. His investigations found that inhibition of nitrification was not apparent in all treatments with the exception of propanil. An eight times higher concentration of propanil treatment showed an accumulation of NH_4-N and the inhibition of NO_2-N and NO_3-N production.

Kim and Jeong (1976) observed that by increasing its concentration, butachlor reduced the formation of NH_4-N. NO_2-N and NO_3-N also were reduced by butachlor treatment; however, they both disappeared 6 weeks after treatment.

Lim et al. (1977) investigated the degradation of urea and change of chemical states for nitrogen when 2,4-D, butachlor, and nitrofen as herbicides, fthalide, neoasozine, and phenazin as fungicides, and chlorofenvinphos, diazinon, fenitrothion, and bux as insecticides were applied to soil. They found that degradation of urea was not affected at 100 mg/kg concentration by all fungicides and herbicides, while it apparently was inhibited at 200 mg/kg except for fthalide and neoasozine. A 500 mg/kg treatment of insecticides also clearly hindered NH_4-N formation. The formation of NH_4-N increased under submerged soil conditions, especially showing a notable increase 2 weeks after treatment. At the same time, about 20% of inorganic nitrogen disappeared within the same period. Among the chemicals used, fungicide had a greater effect than the others. Kim and Han (1977) pointed out that isoprothiolane had little effect on soil pH and that Eh was lower than control from 28 days after application in submerged soils. NH_4-N increased during the early stage of incubation, then decreased rapidly until 21 days, only to increase again after 28 days. NO_2-N and NO_3-N disappeared completely from all soil samples after about 2 weeks. As these results indicate, isoprothiolane cannot be hazardous to the soil environment. Kang (1978) mentioned that bentazon inhibited nitrification in relation to concentration, and Kim (1978) observed that the formation of NH_4-N was affected by herbicides such as nitrofen, CNP, avirosan, and mamet. He noticed that NH_4-N formation was inhibited for 3 weeks, and that the higher chemical concentration was the greater inhibition. NO_3-N formation was also dependent on the chemical concentration, even though it was not detected until 3 weeks after incubation.

Ryu et al. (1980) reported that butachlor, carbofuran, and IBP applied to soils under submerged conditions increased soil pH until 15 days, then maintained the pH level at 6.3 throughout the incubation period. Eh was not affected by herbicides and insecticides, yet with a high dose rate of fungicides it decreased remarkably. They also observed that high application dosage of IBP produced more NH_4-N and yet showed no significant effects on exchangeable calcium, magnesium, and potassium.

Ryu et al. (1983) reported that a higher application rate of CNP resulted in a higher soil pH and a lower soil Eh. Also, the formation of NH_4-N was affected by similar conditions. They further pointed out that the high NH_4-N concentration was determined from the soil with high organic matter content when the soil maintained a reductive condition. On the other hand, NO_3-N formation was delayed by CNP treatment. Inhibition of nitrification was also confirmed by a high dose application of CNP.

Lee and Lee (1983) observed that glyphosate, 2,4-D, and oxadiazon increased the formation of NH4-N and NO3-N, while their production was inhibited approximately 30% by paraquat and 17% by butachlor. Also, they found that the inorganification rate of urea as a result of terbutryne and butachlor treatments was 0.7% and 0.35%, respectively.

Mitigation of Pesticide Residues in Soils

Pesticide residues disappear through the degradation and metabolic processes of soil microorganisms, runoff from the soil surface, leaching by soil water, diffusion into the air, photolysis and translocation into crops. Some research findings on the decrease of residual components in the soils are summarized as follows.

Translocation of Pesticide Residues into Crops

It is very rare to find any sign of direct damage to crops by pesticide residues. Since the late 1970s, poor hop growth was found in Kangwon Province where more than 40% of Korean hop were cultivated. The cause was attributed to be damage to the crop from heptachlor applications resulting in the closure of more than 10% of hop farms in 1980. A field survey during 1981–1982 revealed that heptachlor and heptachlor epoxide were the major causes of hop damage. Some time later it was verified that heptachlor epoxide was more toxic to hop roots, and that it had a minimum toxic level of 0.009 mg/kg in soil (Park and Hwang, 1982). These results were substantiated by Han et al. in 1993.

Ryang et al. (1991) planted winter Chinese cabbage, radish, spinach, onion, and garlic and then selected the suitable herbicides for each crop. Herbicides were applied to the soil in one area at recommended dose and also at double strength to another area. After harvesting, seeds were taken and examined and compared to establish the degree of persistence of herbicides in the soils. Upon the completion of a cropping period of 200–240 days, alachlor, triflurarin, and prometryn showed no significant harmful effects to the following crops. Recommended doses of pendimethalin, metalachlor, linuron, and methabenzthizuron produced the same nonharmful results. Doubling the application amount of these chemicals also caused very little damage to Italian ryegrass. The application of 150–300 g(a.i.)/10 a of napropamide to the soil hindered slightly the growth of Italian ryegrass, rice, and barley, whereas the use of 75 g/10 a of nitralin caused notable damage to the same crops. Sesame, green perilla, and spinach showed little growth inhibition when these two herbicides were used. Ryang et al. (1991) experimented with summer crops such as potato, carrot, corn, sesame, soybean and watermelon. The crops were seeded, and recommended herbicides for each crop were applied to the soils in a double dose rate. On completion of a cropping period of 100–120 days, the crops were planted and the toxic effects of residual herbicides evaluated by observing the growth status of the crops. They found no hazardous effect from using alachlor, trifluralin, ethalfluralin, metribuzin, and prometryn. Using recommended doses of pendimethalin, metolachlor, linuron, methabenzthiazuron, and simazine also showed no significant damage to the crops. However, all of the above chemicals caused slight damage to Italian ryegrass. If twice the recommended dose were used, 300 g/10 a of nepropamide affected the growth of Italian ryegrass and barley even at 140 days after treatment, while no damage to radish and Chinese cabbage was observed. They also found that nitralin inhibited the growth of

Gramineae such as rice and barley at an application rate of 150 g/10 a. Hence, the uptake of soil residues by different crops should be known.

Park (1975) cultivated radish and Chinese cabbage in soils which contained 0.003 mg/kg of heptachlor and 0.002 mg/kg of heptachlor epoxide, and then evaluated the residue levels in the crops. He found 0.003 mg/kg of these chemical residues in the radish leaf, 0.005 mg/kg in the radish root, and 0.007 mg/kg in the Chinese cabbage.

Lee (1981) tested and found that Chinese cabbage absorbed 0.002 mg/kg of α-BHC from 0.22 mg/kg of soil, only a trace amount of β-BHC from 0.16 mg/kg of soil, 0.002 mg/kg of heptachlor from 0.29 mg/kg of soil and 0.001 mg/kg of heptachlor epoxide from 0.06 mg/kg of soil. He also reported garlic uptake of the chemicals at 0.002 mg/kg of α-BHC from 0.82 mg/kg of soil and 0.001 mg/kg of heptachlor from 0.02 mg/kg of soil. Furthermore, Lee et al. (1992) conducted an experiment to establish the limit of pesticide residues in soils. Ethoprophos was used in the experiment to determine its residue levels left in the soils, by evaluating its levels found in lettuce, brown rice, and rice straws and comparing them to the maximum residue limits or tolerance levels for these crops. They explained that 0.02 mg/kg of residues found in lettuce, which is the FAO/WHO tolerance level for ethoprophos, was the same as 12 mg/kg of residues in the soil, and 0.02 mg/kg of residues in brown rice, which is acceptable to the EPA (U.S.) standards, coincided with 40 mg/kg of residues in the soil. Out of every 2.0 mg/kg ethoprophos residues in the soils, 0.02 mg/kg of it was absorbed by rice straws, indicating a high rate of absorbency for the chemical.

Lee et al. (1988) applied ^{14}C-bentazone to soybean and radish growing soil. Soybeans absorbed 42.66 ± 1.19% of total radio activity into the roots, and 2.76 ± 0.13% into the sheaths. Also, radish absorbed 16.64 ± 0.02% into the roots and 4.85 0.03% into the sheaths. Lee et al.(1989) applied ^{14}C -bentazon to paddy soil and found that within 42 days of application, rice plants absorbed 25.8 to 29.4% into the root area and 12.6 to 12.7% into the stems and leaves. Lee et al. (1989) carried out a similar experiment using ^{14}C -carbofuran. Rice plants absorbed 6.52 ± 0.58% into the root zone and 20.45 ± 3.73% into the sheaths. We need more information about the translocation of soil residues in edible parts of crops because the absorbtion rate varies with chemicals and crops.

Behavior of Pesticides in Soils and Waters

To enable the safe use of pesticide it is essential to research and establish the absorption rate into crops, leaching rate into soil, and so forth. Some researchers have carried this out in controlled environments, while others were conducted under normal field conditions

Chung et al. (1976) mentioned that diazinon residues were found in equal amounts from the upper and lower levels of the crops when granules were applied into the soil. Park and Je (1983) noted that the degradation rate of BPMC was higher in upland soil condition than in submerged condition, while carbofuran showed the opposite result. They also pointed out that the increase of organic matter contents caused the retardation of pesticide degradation, and that high temperatures together with high concentration of organic matter stimulated the degradation of both chemicals.

Oh et al. (1984) sprayed isoprothiolane and chlorpyriphosmethyl (EC) on rice plants and examined the residues at harvesting time. Only 0.19 to 0.99% of the chemical residues were found in rice straws and 0.01 to 0.48% in brown rice, with the majority of the remainder being contained in the soil.

Hong and Hong (1984) compared the releasing velocity of carbofuran technical and sand-coating granules into water. Technical granules released 93% of carbofuran in 12 hr, while the sand-coating granule reached their maximum releasing velocity after 5 days.

They also observed that 3-hydroxy carbofuran and 3-ketocarbofuran showed shorter persistence in submerged soil conditions than in that of upland soils.

Park and Lee (1985) applied isoprothiolane granules (12%) to paddy soils and evaluated the residue levels in water and rice plants. They explained that the residue levels were at a maximum within 2–5 days in water, and 3–8 days in rice plants.

Lee et al. (1986) treated paddy soils with ^{14}C-carbofuran and detected 6.2% of total radioactivity from rice plants after 56 days, and that the majority remained in the soils including 46% of nonextractable residues. Park (1986) applied IBP (17%) granules to paddy soils and detected residues from water and rice plants with time. The residues progressively decreased and could not be detected after 8-15 days, and the maximum residue level was found on day 1 in rice plants.

Park also experimented with BPMC EC (50%) in the same way. He reported that BPMC residues were at the maximum level 3 hr after it was put into the water, soil, and rice plants, and could not be detected after 15 days in soil and 3–5 days in water. Park and Oh (1986) applied ^{14}C-carbofuran to paddy soil and investigated the distribution rate of the chemical in rice plants, soil, volatile components, and CO_2. They observed 0.6% of the residues in rice plants and 87.0% in soil only 1 hr after treatment; 14.3% in rice plants, 67.0% in soil, 3.8% in volatile components, and 0.5% in CO_2 at 3 days after treatment, and 28.1%, 43.0%, 7.6%, and 0.6%, respectively, at 9 days after treatment. Ma et al. (1987) reported that pretilachlor degraded faster in silty loam soil with high organic matter and clay than in sandy loam. High soil temperature and submerged soil conditions also stimulated the degradation. Lee et al. (1988) tested ^{14}C-BPMC under submerged soil conditions, and observed rapid degradation of this chemical. They also mentioned that nonextractable residues increased with time, and attained a level of 22% at 72 days after treatment. They also suggested that BPMC should be easy to move and leach in the soil because of its low (<1.0%) absorption and distribution coefficients.

Moon (1990) observed that submerged soil condition and high temperature(35°C) stimulated the degradation of fenitrothion, and that a higher concentration (30 mg/kg) of the chemical retarded degradation more than a lower concentration (10 mg/kg) did. At the same time, he said that the addition of complex fertilizer, IBP, and butachlor to the soil did not significantly influence the degradation of fenitrothion.

Lim and Bong (1992) presented their findings on soil adsorption of paraquat and alachlor. They found that paraquat showed a high regression with clay contents and alachlor with organic matter contents in the soil. Lee and Oh (1993) applied carbofuran, bentazon, and TCBA labeled ^{14}C-radio to soil and observed the mobilities of these chemicals in the soil. They found 92% of ^{14}C-carbofuran and 96% of ^{14}C-bentazon leached into the soil, while ^{14}C-TCBA showed no migration at all.

Song and Lee (1993) used humic materials such as humin, humic acid, and fulvic acid to monitor the behavior of diazinon in submerged soils. They compared the adsorption rate of diazinon by using different kinds and concentrations of humic materials. It was found that 12.4% of applied diazinon was recovered from humin fraction, 10.4% in humic acid fraction, and 11.86% in fulvic acid fraction, respectively. They said the higher contents of organic materials resulted in the higher adsorption rates of diazinon, yet the higher concentrations of diazinon did not reduce a higher adsorption rate.

Moon et al. (1993) investigated the soil mobility of ethoprophos, and found almost all components remaining in 0–2 cm of soil depth. They also found that a little portion of the ethoprophos moved into 2–4 cm of soil depth, and only 0.2 to 0.5% of the chemical was detected in 4–6 cm of soil depth during 15–30 days after treatment. Kim et al. (1993) estimated the possibilities of water contamination through leaching of procymidone and ethoprophos into the soils and waters. They examined the adsorption and distribution coefficiencies, mobile lengths, and leaching amounts of the chemicals. The adsorption and distribution coefficiencies based on organic carbon (Koc) were 88–100 for procymidone and 362–449 for ethoprophos, and Koc for humic acid was 594 for procymidone and 326 for ethoprophos. From these results they reasoned that ethoprophos could have a higher leaching ability than procymidone. Kim and Han (1994) conducted water culture experiments with young rice plants to establish the adsorption pattern of ^{14}C-bensulfuron methyl. They noted that 10.8% of total ratio activity was found in rice plants 12 hours after treatment, and increased to 11.7% 48 hours later. They reported that the addition of butachlor did not influence the adsorption of ^{14}C-bensulfuron methyl, but quinchlorac inhibited its adsorption at the rate of 1.0 mg/kg. A deficiency of nitrogen and an excess of sulfur in the soil also inhibited the adsorption of chemicals.

Acceleration of Pesticide Degradation

Research into the mitigation of pesticide residues in soils has been going on since the late 1970s and experiments into the addition of organic matter and lime treatment have been carried out by some researchers. Some active trials were also conducted using microorganisms.

Oh et al. (1979) applied crushed rice straw at levels of 1.0, 3.0, and 5.0% to nitrofen treated soil, 0.1, 0.5, and 1.0% for simazine and 0.1, 0.5, 1.0, and 3.5% for butachlor-treated soil, and analyzed the residues by time course. At 1.0% addition level, nitrofen showed no clear effect, while half-lives were shortened by 5 days in submerged condition and 3 days in upland condition for butachlor treatment, also 90 days in upland condition for simazine application. They also investigated the effects of lime treatment, and ascertained a 3- to 5-day shortening of half-lives in butachlor and 8–15 days in nitrofen treatment. Lee et al. (1979) treated 2% of rice straw in soil and investigated to the half-lives of diazinon, carbofuran, and BHC isomers. All chemicals used during the experiment resulted in a shortening of 2.4 days in diazinon, 10 days in BHC isomers, and 7.4 days in carbofuran.

Park et al. (1993) compared the effects of zeolite, weathered porphyry, rice straw powder, compost, and lime on iprobenfos- and parathion-treated soils. Two weeks after applying 10 mg/kg each of the chemicals they observed that zeolite, weathered porphyry, and rice straw powder treatment resulted in a higher degradation rate than nontreatment in iprobenfos, while no significant effect was found in parathion. Six weeks after application of 50 mg/kg of each chemical, iprobenfos was degraded 96.3% in control, but lime and rice straw treatment resulted in 93.1% and 82.1% of degradation, respectively. The result also showed that parathion was degraded 97.8% in rice straw treatment and 87.3% in compost treatment. These degradation rates were much higher than that of control, which was 61.9% under submerged conditions.

Some researchers compared the biological degradation rates by using sterilized soils. Oh et al. (1979) reported butachlor was postponed four times the half-life, and that of nitrofen was six times longer in sterilized soils. Lee et al. (1979) also mentioned the retardation of half-lives of diazinon and carbofuran as 2.5 times and 2.2 times, respectively, in sterilized soil conditions. Song et al. (1981) observed that the half-life of IBP was 12 days in nonsterilized soil and 37 days in sterilized conditions.

Park and Je (1983) noted that the degradation rate of carbofuran was delayed 2 times in sterilized soil. Hong and Kim (1984) reported that the degradation of dinoseb in sterilized soil resulted in 16 days longer of half-life. Park and Oh (1986) experimented with ^{14}C-carbofuran and observed the increase of organic phase extracts and the decrease of nonextractable fractions in sterilized conditions. Choi and Lee (1987) also observed that the degradation of diazinon and chlorpyriphos were delayed by three times of half-life in sterilized soil. Kim et al. (1989) found that diazinon showed a 95% of degradation rate in nonsterilized soil after 7 days' incubation at 30°C, while more than 30% of diazinon was detected after 10 days' incubation in sterilized soil. Moon (1990) said that fenitrothion showed about 7 days of half-life in nonsterilized soil, but that almost no degradation was observed in sterilized soil under submerged conditions. Kim and Choi (1992) found that bifenthrin showed 85.1 days of half-life in nonsterilized soil but only 11% of degradation was observed in sterilized conditions. The degradation of cyhalothrin was 54.6 days of half-life in nonsterilized soil, similar to that of bifenthrin in sterilized soil.

As previously stated, many researchers have emphasized that pesticides are degraded by soil microorganisms, Hence, efforts were made to find a powerful microorganism that could be used for treatment of pesticide contaminated soil.

Lee et al. (1984) identified three strains of fungi each from PCNB- and endosulfan-treated soil, out of which they selected two strains of Fusarium for PCNB degradation. They reported that the superior strain shortened the half-life of PCNB from 42.4 days to 31.9 days. Kim et al. (1987) identified a powerful strain of fungi for PCB's degradation and investigated the degradation rate in media. They found that 45% of alachlor 1016 was degraded 9 days after treatment, but in the case of alachlor 1254, γ-BHC, and p,p′-DDT, the degradation was less than 20%. Kim et al. (1991) selected eight strains of microorganism for parathion degradation, and observed a Bacillus strain degraded parathion to 98% in 50 mg/kg level 7 days after incubation at 35°C. Park et al. (1992) confirmed the effect of the selected Bacillus strain for parathion degradation. They observed that the degradation rates were 1.5 to 1.7 times higher than those in noninoculated conditions. At the same time, that strain did not affect the degradation of ethoprophos and fenitrothion. Lee et al. (1991) identified eight different strains of fungi from 20 orchard soils and two paraquat replicative treated soils. Among these they selected 3–4 strains and observed the degradation rate of paraquat at 10 mg/kg of treatment. The result showed that the selected strains effected a degradation rate twice faster than control.

Fate of Some Pesticides in the Soil

Degradation and metabolism of pesticides can be considered from two different views. One is the half-life of pesticides in Korean soil condition, and the other is the metabolic pathway.

Table 12.10 shows half-lives of some pesticides carried out in experiments in Korea. Half-life (DT50) and DT90 were compulsory data that must be registered for the Korean market, and recently the authorities have demanded that the degradation data under Korean conditions must also be registered. Hence, much data for half-life in soils has already been submitted to the government in order to get permission for registration.

Synthetic pyrethroids, bifenthrin, and cyhalothrin showed over 120 days of DT50 in submerged soil, but were much shorter under upland soil condition. Flufenoxuron showed 1.8 times longer in sandy loam than in loam soil. Among herbicides, DT50 of esprocarb was 59 days and 2.3 days in molinate, and 28–38 days in butachlor, one of the major herbicides for paddy rice. IBP showed 12 days of half-life, 15–18 days in IB, and 52 days in isoprothiolane.

Not only had there been little study into the metabolism of each chemical, but also the majority of them were carried out using ^{14}C-labeled pesticides.

Park and Je (1983) investigated the degradation patterns of pesticides and studied the metabolites using ^{14}C-labeled carbofuran under submerged soil conditions. They found 88% of unchanged carbofuran, 1.7% of 3-hydroxy benzofuranol, 1.6% of 3-hydroxy car-

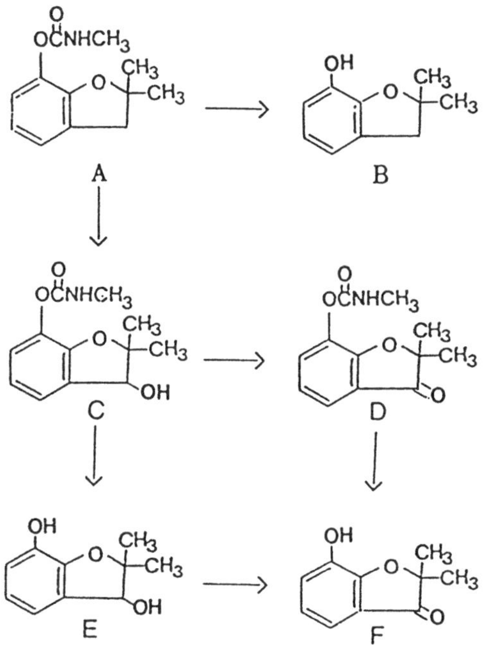

Figure 12.1. Degradation pathway of 3-^{14}C-carbofuran in paddy soil (Lee et al., 1986).
A: Carbofuran
B: Carbofuran phenol (19.1%)
C: 3-Hydroxycarbofuran (0.7%)
D: 3-Ketocarbofuran (1.1%)
E: 3-Hydroxycarbofuran phenol (0.6%)
F: 3-Ketocarbofuran phenol (4.3%)

bofuran, 1.3% of 3-ketobenzofuranol, 0.1% of benzofuranol and a trace amount of 3-ketocarbofuran at 24 days after application. They also hydrolyzed water-soluble extract of soil using HCl and cellulase, and confirmed that carbofuran and 5 other metabolites were existing in aglycone form, and identified 7-benzofuranol and 3-hydroxy carbofuran as the major components of aglycone. They also found much similar metabolites in rice plants and concluded that metabolism of carbofuran was initiated by phase I reaction and changed to conjugate forms through phase II reaction with relative rapidity. Lee et al. (1986) applied ^{14}C-carbofuran and identified the metabolites. They found 19.1% of carbofuran phenol, 4.3% of 3-keto carbofuran phenol, 1.1% of 3-ketocarbofuran, 0.7% of 3-hydroxy carbofuran, and 0.6% of 3-hydroxy carbofuran phenol at 56 days after application. Lee et al. (1993) examined the mobility of ^{14}C-carbofuran and its metabolites using a soil column. They found 3-ketocarbofuran phenol as a major metabolite, as well as 3-hydroxy carbofuran and 3-hydroxycarbofuran phenol in leachate.

Figure 12.1 shows the degradation pathway of 3-^{14}C-carbofuran in paddy soil (Lee et al., 1986).

Choi et al. (1987) investigated metabolism of diazinon and chlorpyrifos under submerged soil condition. They found diazoxon, 2-isopropy-6-methyl-pyrimidine-4-one, O,O- diethyl-O-[2(1-hydroxy-1,1-dimethyl)6-methyl] pyridmidinyl phosphorothioate, O,O- diethyl phosphorothioate, and sulfotep as metabolites of diazinon. They only identified O,O-diethyl phosphorothioate as a metabolite of chlorpyrifos. Figure 12.2 shows the degradation pathway of diazinon in paddy soil (Choi et al., 1987).

Lee et al. (1984) compared metabolism of alachlor in submerged soil conditions and the inoculation of degrading microorganisms on media. They found 1-formyl-2,3-dihydro-7-ethyl indole, 2,6-diethy aniline, 2,6-diethylacetanilide, 2,6-diethyl-N-(methoxy methyl)acetanilide, 2-hydroxy-2′,6′-diethyl-N-(methoxy methyl) acetanilide, and 3 unidentified metabolites in submerged soil. They also observed 25% of 2-hydroxy-2′,6′-diethyl-N-(methoxy methyl) acetanilide in *Streptomyces lavendulae* Ru 3340-8 inocula-

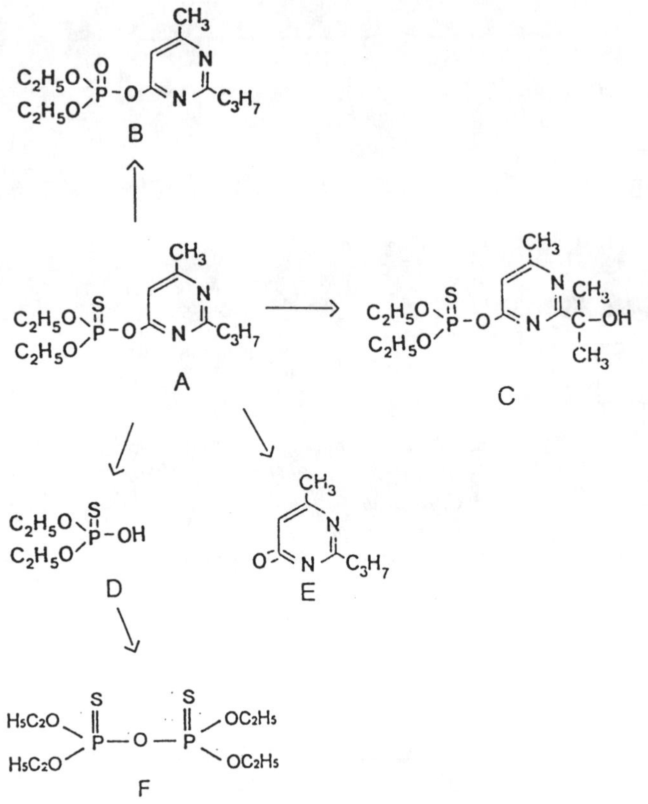

Figure 12.2. Degradation pathway of diazinon in paddy soil (Choi et al., 1987).
A: Diazinon
B: Diazoxon
C: Phosphorothioate (0,0-diethyl-0-[2(1-hydroxy-1,1,-dimethyl)6-methyl]pyrimidinyl, 0,0,-diethyl phosphorothioate
D: 0,0-Diethyl phosphorothioate
E: 2-Isopropyl-6-methyl-pyrimidine-4-one
F: Sulfotep

tion. Figure 12.3 shows the degradation pathway of ring ^{14} C-alachlor paddy soil (Lee et al., 1984).

Lee et al. (1986) examined the metabolism of alachlor under upland soil condition. They found 8-ethyl-2-hydroxy-N-(methoxy methyl)-1,2,3,4-tetrahydroquinoline and N-hydroxyl-2, 3-dihydro-7-ethulindol, which are different from the metabolites of submerged soil condition. Also they did not find 2,6-diethyl-N-(methoxy methyl) acetanilide, a major metabolite in submerged condition. They concluded that 2-hydroxy-2′,6;-diethyl -N-(methoxy methyl) acetanilide was one of the major metabolites in both soil conditions, and suggested that alachlor metabolized through different routes in submerged (anaerobic, reductive condition) soil and upland (aerobic, oxidative condition) soils. Kim et al. (1991) explained that bifenthrin metabolized by the typical pyrethoid degradation process and the hydrolysis of ester linkage, and that 2-methylbiphenyl-3-yl methanol was a hydrolysis product. Chang et al. (1993), who conducted an experiment to study the degradation of bentazon in soil incubation and culturing media with degrading microorganism, also suggested the metabolic process. Lee et al. (1993) reported that bentazone hydrolyzed first and formed anthranic acid isopropylamide and H_2SO_4, before producing anthranilic acid through deamination.

PROTECTION STRATEGIES FOR AGRICULTURAL ECOSYSTEMS

Fatal accidents during spraying of pesticide and insecticides to eradicate killfish (*Oryzias latipes*), loach (*Misgurnus an guillicaudatus*), and MCP and locust (*Oxya chinensis*) for

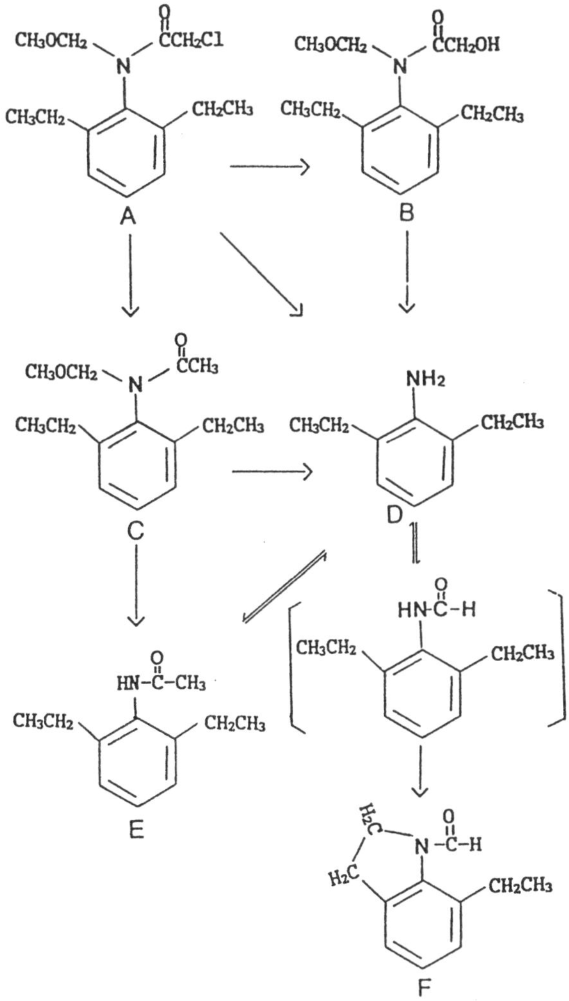

Figure 12.3. Degradation pathway of ring ^{14}C-alachlor in paddy soil (Lee et al., 1984).
A: Alachlor
B: 2-Hydroxy-2′, 6′,-diethyl-N-(methoxymethyl)acetanilide
C: 2, 6-Diethyl-N-(methoxy methyl)acetanilide
D: 2. 6-Diethylaniline
E: 2, 6-Diethylacetanilide
F: 1-Formyl-2,3-dihydro-7-ethyl indole

mosana shiraki (*O. velox*) by insecticides for rice grasshopper control in paddy fields have created adverse public opinion for pesticide applications. This situation has quite naturally initiated a dispute on the problems of pesticide residues. The Korean Government therefore has made every effort to diminish the pesticide residue in crops by establishing the *Pesticide Safe Use Guideline and Tolerance.* Nevertheless, consumer organizations still want to see safer crops, and many farmers have been favoring and changing to organic farming methods.

Regulation of Residual Pesticides

Generally speaking, the regulatory methods pertaining to pesticide residues are divided mainly into two ways. One is the *Pesticide Safe Use Guideline* which is a restriction of spraying frequencies and the last application day prior to harvesting. The other is the tolerance level which is the limit of residue level for agroproducts. The former has been applied in U.K., France, Belgium, and Poland as well as Korea.

From the early 1970s, experiments for establishing the *Pesticide Safe Use Guideline* were carried out by using 384 out of the 584 registered pesticides. They included 114 for paddy pesticides and 267 for upland and horticultural use. More than six applications of the pesticide were carried out ranging 75 to 3 days before harvesting of the crops. Samples were collected, prepared, analyzed, and the results were compared with the known tolerance of maximum residue limit (MRL). The final application day was determined by selecting the day when the residue level reached just under the known maximum residue limit.

Since tolerance levels were not established in Korea until the 1970s, the tolerance levels of foreign countries were used as a guideline. The *National Tolerance Guidelines*, enacted in the 1980s, fit both national and foreign standards. It meant the establishment of tolerance levels or MRL for each crop, which practice is popular in the U.S., Canada, Australia, the Netherlands, and Germany.

In Korea, the KEPA established tolerance levels for 21 pesticides in 1981 and 30 in 1987 based on the *Water Quality Law*. Since then, MHW have set up tolerance limits for 16 pesticides on 28 crops on August 26, 1988 based on the *Food Hygiene Act*. These guidelines became effective from September 1, 1989 (Notification 88-60 of MHW). The basic rules for establishing the tolerance limits were as follow:

1. Covering the large consumption crops and residual pesticides.
2. ADI (acceptable daily intake) follows FAO/WHO guidelines.
3. Food factor based on average food consumption amounts for the most recent 5 years, taking into consideration the *Survey of National Nutrition* (1987), and *Food Demand and Supply Record* (1985).
4. Average body weight is 50 kg per capita (now it changes to 55 kg).
5. Considering the tolerance levels of Korea, FAO/WHO, USA, Japan, and Australia, and also pesticide safe use guideline.

In keeping with these rules, MHW reestablished tolerance limits for 16 chemicals on 51 crops, including those imported from foreign countries to take effect from January 1, 1991 (Notification 90-85 of MHW). The maximum limits of five pesticides on 32 crops were also added (Notification 91-88 of MHW) and put into operation from January 1, 1993. Furthermore, tolerance limits enforced from January 1, 1995 on 105 pesticides for the control of residues on crops produced in Korea, as well as on imported foods, are rapidly on the increase. Korea has now emerged, like Japan, as a country that is using both pesticide safe use of MAF and tolerance limits of MHW. Table 12.11 shows the tolerance limits of some pesticide in/on crops in Korea.

Some Standard Methods for Mitigation of Pesticide Residues

Tests for pesticide residual patterns in agricultural ecosystems have included residual features in soil and water, but not in crops. The tests were carried out by following standard experimental methods for pesticides that were officially announced by RDA, MAF [Agrochemicals Research Institute (A.R.I.), R.D.A. 1994].

The following are some examples of establishing pesticide safe use guidelines for rice and Chinese cabbage in accordance with the experimental method for pesticides by RDA. Relevant chemicals should be applied to corresponding crops between 2 to 6 times during the 3-75 growth days. Examples in Table 12.12 showed that for both malathion and chlorofenvinphos EC application, 75-60 days before harvest, two applications of the chemicals were the basal treatment. For the preharvest intervals of 75-60-30, 75-60-15, 75-60-7, and 75-60-3 days, the frequencies were for three applications; for intervals of 75-60-30-15, 75-60-15-7, 75-60-7-3 days, it required four applications; for intervals of

Table 12.11. Half-Life (DT_{50}) of Some Pesticides in Korean Soil

Pesticides	Soil	Half-Life(day)	Reference
Insecticides			
Bifenthrin	Upland	61–85	Kim et al. (1991)
	Paddy	128	
BPMC	Paddy (L)	5	Park and Je (1983)
Carbofuran	Paddy (SCL)	8–12	Lee et al. (1986)
Cyhalothrin	Upland	3–5	Kim et al. (1991)
	Paddy	120	
Diazinon	Paddy (SiL)	2.2	Choi et al. (1987)
Chlorpyrifos	Paddy (SiL)	10.8	Choi et al. (1987)
Ethoprophos	Upland (CL)	15	Lee et al. (1992)
Fenitrothion	Upland (L)	3.2	Moon and Ryang (1990)
	Paddy (L)	1.6	
Flufenoxuron	Upland (L)	51	Unpublished
	Upland (SL)	97	
Fungicides			
Chlorothalonil	Upland (SL)	9	Lee et al. (1990)
Diethofencarb	Upland (L)	35	Unpublished
	Upland (SL)	33	
Dodine	Upland (L)	14	Unpublished
	Upland (SL)	12	
IB	Paddy (SCL)	15–18	Park (1986)
IBP	Paddy (SiL)	12	Park (1986)
Isoprothiolane	Paddy (SL)	52–63	Park and Lee (1985)
Metalaxyl	Paddy (L)	23	Unpublished
	Paddy (SL)	17	
Herbicides			
Butachlor	Paddy (SL)	28–38	Moon and Ryang (1990)
Esprocarb	Paddy (SL)	45	Unpublished
	Paddy (L)	59	
Hexazinon	Upland (SL)	19	Unpublished
	Upland (L)	18	
Metobromoron	Upland (L)	37	Unpublished
	Upland (SL)	36	
Molinate	Paddy (S)	0.3	Unpublished
	Paddy (SL)	0.6	
Nitrofen	Paddy (SL)	33–35	Unpublished
Paraquat	Upland (L)	151	Unpublished
	Paddy (SL)	177	

75-60-30-15-7 and 75-60-30-15-3 days, it was for five applications and for intervals of 75-60-30-15-7-3 days, it was six applications, with all samples being harvested on the same day. The samples were analyzed to evaluate the residue levels, which were then compared to the maximum residue limit to determine the safe application days before harvest.

Experimental results of four applications of malathion EC on brown rice and chlorofenvinphos EC on Chinese cabbage are shown in Table 12.13. From the degradation

Table 12.12. Tolerance Limits of Some Pesticides in/on Crops in Korea

Pesticides	Tolerance (mg/kg)	Crops
DDT (banned)	0.1	Sesame, Lemon, Green pepper, Wheat, Almond, Orange, Grapefruit, Pumpkin, Hops
	0.2	Rice, Barley, Maize, Common bean, Potato, Sweet potato, Chinese cabbage, Cabbage, Spinach, Welsh onion, Radish, Beet radish, Onion, Red pepper, Cucumber, Eggplant, Tomato, Strawberry, Apple, Pear, Citrus fruits, Persimmon, Grape, Garlic, Soybean, Peanut, Bracken, Watermelon, Melon, Apricot, Prune, Papaya, Mango, Cherry, Pineapple, Banana, Dried grape, Avocado, Lettuce, Mushroom, Asparagus, Peach, Lettuce
BHC (banned)	0.1	Wheat
	0.2	Rice, Barley, Common bean, Potato, Sweet potato, Chinese cabbage, Cabbage, Lettuce, Spinach, Welsh onion, Radish, Beet radish, Red pepper, Cucumber, Eggplant, Tomato, Strawberry, Apple, Citrus fruits, Peach, Persimmon, Grape, Garlic, Soybean, Lemon, Green pepper, Watermelon, Apricot, Prune, Mango, Cherry, Orange, Grapefruit, Pear, Pineapple, Avocado, Pumpkin, Celery, Asparagus, Maize, Onion, Melon
Aldrin & Dieldrin	0.01	Rice, Barley, Common bean, Potato, Sweet potato, Chinese cabbage, Cabbage, Lettuce, Spinach, Welsh onion, Radish, Beet radish, Red pepper, Cucumber, Eggplant, Tomato, Strawberry, Apple, Pear, Citrus fruit, Peach, Persimmon, Garlic, Soybean, Peanut, Lemon, Green pepper, Watermelon, Apricot, Prune, Wheat, Papaya, Mango, Cherry Orange, Grapefruit, Pineapple, Banana, Dried grape, Avocado, Pumpkin Melon, Celery, Asparagus, Maize, Onion, Grape
Captafol (banned)	0.05	Peanut
	0.1	Pineapple
	0.5	Lemon, Orange, Grapefruit
	1.0	Red pepper, Cucumber, Grape, Soybean, Watermelon, Pumpkin
	2.0	Melon, Prune
	5.0	Apple, Pear, Peach, Apricot, Cherry
Diazinon	0.05	Wheat
	0.1	Rice, Common bean, Potato, Sweet potato, Chinese cabbage, Cabbage, Lettuce, Spinach, Welsh onion, Radish, Cucumber, Eggplant, Strawberry, Korean melon, Pear, Citrus fruits, Persimmon, Grape, Garlic, Crown daisy, Soybean, Peanut, Watermelon, Melon, Almond, Cherry, Banana, Pumpkin, Hops
	0.3	Tomato
	0.5	Red pepper, Apple, Bracken, Green pepper, Apricot, Papaya, Mango, Pineapple, Avocado, Mushroom, Celery, Asparagus, Prune
	0.7	Peach, Lemon, Kiwi, Orange, Grapefruit
Malathion	0.2	Crown daisy
	0.3	Rice
	0.5	Common bean, Potato, Chinese cabbage, Cabbage, Spinach, Radish, Beet radish, Red pepper, Cucumber, Tomato, Strawberry, Apple, Pear, Citrus fruits, Peach, Persimmon, Soybean, Peanut, Lemon, Green pepper, Watermelon, Melon, Apricot, Prune, Kiwi, Almond, Papaya, Mango, Cherry, Grapefruit, Pineapple, Dried grape, Avocado, Mushroom, Pumpkin, Hops, Asparagus, Orange
	1.0	Celery
	2.0	Wheat

Table 12.12. Continued

Pesticides	Tolerance (mg/kg)	Crops
Parathion	0.1	Rice, Potato, Sweet potato, Soy bean, Almond
	0.3	Barley, Common bean, Maize, Chinese cabbage, Cabbage, Spinach, Banana, Welsh onion, Radish, Beet radish, Onion, Red pepper, Cucumber, Tomato, Strawberry, Korean melon, Apple, Pear, Citrus fruits, Peach, Persimmon, Grape, Garlic, Crown daisy, Peanut, Lemon, Green pepper, Watermelon, Melon, Apricot, Kiwi, Wheat, Papaya, Cherry, Orange, Grapefruit, Pineapple, Avocado, Pumpkin, Celery, Hops, Asparagus
	0.5	Prune, Mango
	0.7	Bracken
Carbaryl	0.2	Potato
	0.5	Chinese cabbage, Cabbage, Radish, Red pepper, Citrus fruits, Peach, Grape, Pineapple, Banana, Pear
	1.0	Rice, Apple, Soybean, Lemon, Green pepper, Melon, Apricot, Prune, Almond, Cherry, Orange, Grapefruit, Celery, Kiwi, Asparagus, Pumpkin
	2.0	Peanut
	3.0	Wheat
Carbofuran	0.1	Banana, Mushroom, Celery, Asparagus
	0.2	Rice, Maize, Common bean, Wheat
	0.4	Melon
	0.5	Potato, Welsh onion, Beet radish, Cucumber, Garlic, Dried grape, Pumpkin, Hop, Peanut
Bentazone	0.2	Rice, Barley, Maize, Wheat, Rye, Sorghum, Oat, Buckwheat, Millet, Soybean, Mung bean, Adzuki bean, Kidney bean, Horse bean, Pea bean, Peanut, Common bean, Radish(root), Radish(leaf), Chinese cabbage, Cabbage, Kale, Crown daisy, Lettuce, Onion, Welsh onion, Spinach, Garlic, Asparagus, Beet radish, Celery, Tomato, Green pepper, Leek, Eggplant, Cucumber, Pumpkin, Ginger, Shiitake mushroom, Mushroom, Vegetables (others)
Cyperme hrin	0.05	Soybean, Peanut, Common bean, Potato, Sweet potato, Radish(root), Beet radish, Mushroom, Taro
	0.1	Onion
	0.2	Maize, Wheat, Sunflower seed, Sesame, Cottonseed, Seed oils
	0.5	Barley, Red pepper, Eggplant, Cucumber
	1.0	Rice, Rye, Sorghum, Oat, Buckwheat, Millet, Cabbage, Kale
	2.0	Lettuce, Tomato, Green pepper, Watermelon, Melon, Spinach, Almond, Citrus, Lemon, Grapefruit, Orange, Citrus fruits, Apple, Chinese quince, Peach, Apricot, Prune, Cherry, Strawberry, Grape, Persimmon, Banana, Papaya, Avocado, Pineapple, Mango, Fruits (others), Gingko nut, Chestnut, Green pepper, Nut, Dried fruit, Pear, Ume, Kiwi
	5.0	Radish (leaf), Chinese cabbage, Crown daisy, Welsh onion, Leek, Garlic, Asparagus, Celery, Pumpkin, Ginger, Shiitake mushroom, Vegetables (others)
Diflubenzuron	0.1	Soybean, Mushroom
	0.2	Cottonseed
	1.0	Cabbage, Red pepper, Citrus, Apple, Chinese quince, Peach, Apricot, Prune, Cherry, Strawberry, Grape, Persimmon, Banana, Kiwi, Papaya, Avocado, Fruits, Sunflower seed, Pineapple, Sesame, Seed oils, Gingko nut, Chestnut, Green pepper, Almond, Nut, Dried fruits, Mango Pear, Ume
	3.0	Lemon, Grapefruit, Orange, Citrus fruits

Table 12.12. Continued

Pesticides	Tolerance (mg/kg)	Crops
Fensulfothion	0.02	Soy bean, Banana, Cottonseed
	0.05	Peanut, Sweet potato, Pineapple
	0.1	Maize, Rye, Sorghum, Oat, Buckwheat, Potato, Radish(root), Tomato Onion, Millet
Aluminum phosphide	0.01	Mungbean, Azuki bean, Kidney bean, Other beans, Dried vegetable & Spices, Green pepper, Almond, Nut, Peanut
	0.1	Rice, Barley, Maize, Wheat, Rye, Sorghum, Oat, Buckwheat, Millet, Soy bean, Seed oils

equations it was found that malathion and chlorofenvinphos were degraded by the 0.1 mg/kg MRL of each chemical for each crop at 12 days and 16 days intervals before harvest, respectively. Pesticide safe use guidelines were created by this method which regulated the application frequency and the final application day before harvest. Table 12.14 shows the fixed results for pesticide safe use guideline and for chemicals in Table 12.13.

Figure 12.4 shows the comparison of degradation equation and MRL. Pesticide residue experiments in soils were conducted in both field and incubation conditions. With all pesticides used in Korean fields being the subject of investigation, some chemicals were found to have no hazardous effect to soil and could therefore be exempted from adhering to the guidelines.

The experiment must be performed with two soils which are different in soil texture, organic matter contents, and CEC, etc. The recommended dose should mainly be applied, while the replication plot with another application should follow at 7- to 10-day intervals in field treatment, must be included.

Using augers, 200-g each soil samples were collected from 4 sites of 10 cm top soil. Also, paddy soil was collected with water using a specially designed paddy soil collector.

Table 12.13. Experimental Results for Establishment of Pesticide Safe Use Guideline

		Rice		Chinese Cabbage			
		mg/kg					
Application Frequency	Preharvest Interval	Malathion EC (70%)	Paration G (5%)	Chlorfenvinphos EC (24%)	Phosmet D (3%)	Chlorpyrifos D (2%)	Chlorfenvinphos D (1.5%)
0	—	ND	ND	ND	ND	ND	ND
2	BT[a]	ND	0.050	ND	ND	ND	ND
3	BT-30	ND	0.069	t	0.040	0.026	ND
	BT-15	0.031	0.125	0.107	0.213	0.254	0.015
	BT- 7	0.292	0.137	0.385	0.440	1.344	0.040
	BT- 3	0.031		0.618	0.838	3.316	0.012
4	BT-30-15	0.067	0.136	0.382	0.220	0.392	0.041
	BT-15-7	0.102	0.157	0.411	0.649	1.472	0.043
	BT- 7- 3	0.360		0.604	1.892	3.672	0.103
5	BT-30-15-7	0.445	0.430	0.598	0.701	1.444	0.036
	BT-15- 7-3	0.461		0.898	2.048	4.124	0.138
6	BT-30-15-7-3	0.486		1.061	2.216	5.312	0.155
Tolerance		0.1	0.1	0.1	0.1	1.0	0.1

[a]BT : Basal Treatment (BT means applications twice at 75 and 60 days before harvest).

Table 12.14. Fixed Pesticides Safe Guideline Derived from Data in Table 12.13

Crops	Applied Target	Items	Applicable Frequencies (times)	Preharvest Interval (day)	MRL (mg/kg)
Rice	Aphid	Malathion EC 70%	4	12	0.1
	Soil insect	Parathion G 5%	3	22	0.1
Chinese cabbage	Leaf roller moths	Chlorpyrifos WP 25%	4	8	1.0
	Common cabbage worm	Chlorfenvinphos EC 24%	3	16	0.1
	Common cabbage worm	Phosmet D 3%	3	22	0.1
	Cabbage soil insect	Chlorpyrifos D 2%	4	10	1.0
	Soil insect	Chlorfenvinphos D 1.5%	5	7	0.1

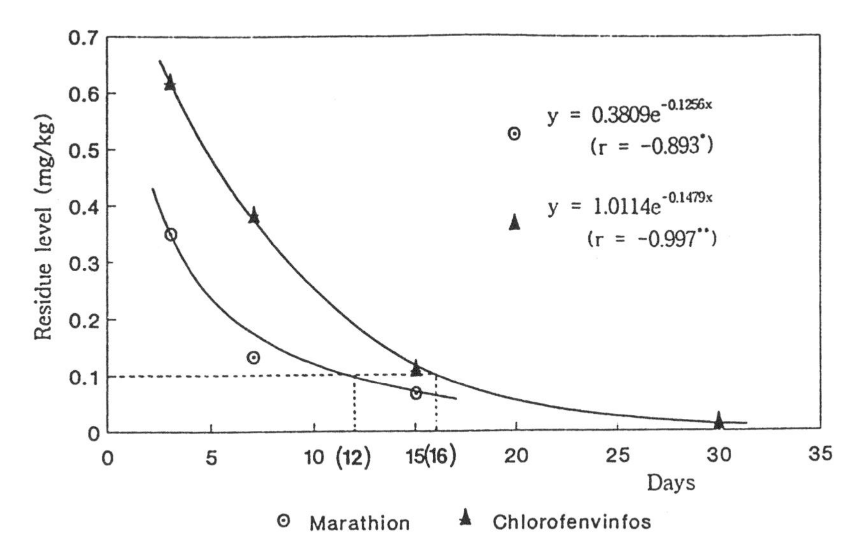

Figure 12.4. Comparison of degradation equation and maximum residue level.

All soil samples were collected more than six times including pretreatment and posttreatments. Collected soil was crushed and sieved to 2 mm, and analyzed in the proper way. The results were then expressed using DT50 and degradation curve on semilogarithmic paper. In the laboratory experiment, beaker, glass vial, and Erlenmeyer flask could be used for incubation higher than 1.0 cm of soil depth although more than 1 week of preincubation time is needed to conduct a proper experiment. Incubation was conducted at the set temperature range of 25–30°C. The 50–61% of the maximum water-holding capacity in upland condition and the over 1.0-cm water depth in paddy condition were maintained throughout the experimental period. Up until June 1994, 248 chemicals were investigated and Table 12.15 shows the half-life in soil by the standard experimental methods mentioned above. From Table 12.15 it can be found that more than 95% of tested chemicals showed less than 100 days of half-life. Furthermore, since December 30, 1991, the regulation for residue persistence in the soil has been enforced by reducing the period from 12

Table 12.15. Residue Persistence of Major Pesticides Used in Korea

	Maximum Half-Life in Soil (day)						
	<15	15–30	30–60	60–100	100–200	200<	Total
Fungicide	25	22	10	12	5	2	76
Insecticide	41	20	20	17	2	—	100
Herbicide	29	14	8	10	3	—	64
Plant growth regulator	5	2	—	1	—	—	8
Total	100	58	38	40	10	2	248

Table 12.16. Prohibited Pesticides in Korea Due to Residual Problems in the Soils

Year	Pesticide
1969	DDT, Phenyl mercury acetate(spray), Endrin, Aldrin
1970	Dieldrin
1976	Phenyl mercury acetate(seed treatment)
1979	BHC, Heptachlor
1982	Lindane
1990	Picloram, Carbophenothion

months to 6 months as the basis of half-life. Table 12.16 shows a list of pesticides that are prohibited for use in Korea because of their residual persistency in the soil.

The target pesticides for residual experiment in water showed that the chemicals might cause serious contamination of water. These chemicals were listed by the pesticide residue board of the Pesticide Management Committee (PMC). Chemicals with less than 2.0 mg/kg of LC50 for fish must be included in the target pesticide. Pot experiments without a flow of water could be carried out for this purpose. The soil sample was 2–3 cm in depth, and 3 cm of water was added to it. Then all floating organic matter was discarded from the surface. The sample was then left for 1–3 days with the water depth adjusted to 5.0 cm before the target chemical was applied.. Water samples were collected before treatment, then again at 1-3 hr after treatment, as well as 1, 3, 7, and 14 days after treatment with duplication. Soil samples were also analyzed, and recovery from both samples must exceed 70%

Action of Regulatory Organizations

In Korea, all matters connected with pesticides are primarily controlled by MAF and established Pesticide Management Laws such as Law No. 445 of 28 March 1957. This law has since been revised several times, and has been applied since August 8, 1996 as the Law No. 5153. The purpose of this law is for the improvement of quality, smooth distribution, and proper use of pesticides. Hence, it is used to enforce control of manufacture, imports, sales, and usage of the chemicals for safe production of crops and the preservation of the Korean agricultural environment.

This law includes some articles for Item Notification (Article 8), and Pesticide Safe Use Guideline (Article 19). At the same time the Enforcement Ordinance of this Law clarified the PMC (Article 11) and its function (Article 12). Chairmanship of PMC is performed by the Vice Director of RDA; included are 15 members consisting of officials

Table 12.17. Classification of Prohibited Pesticides in Korea and the Reasons for the Ban

Banned Reason	Items (Banned Year)
Acute toxicity (13)	Kayaphos EC('81), Kayaphos G('83), Disulfon EC('89), Parathion EC('90), Aldicarb G('91), Heptenophos EC('91), Chlorfenson + Dichlorvos EC('91), Pirimicarb + Dichlorvos EC('91), Vamidothion('91), Quinalphos + Thiometon EC('91), Dialifos EC('91), Chlorfenvinphos EC('91), Diazinon WP('91)
	Phenyl chlorophenyl dimethylformalin EC('77), Lindane G ('81), Maleic hydrazide LQ('83), Tetrachlomethane + Carbondisulfide F ('86),
Chronic toxicity (21)	Maneb WP('89), Chlorobenzilate EC('89), Metabenzthiazuron+Amitrole WP('89), BinapacrylEC('89), Sevin D ('89), Zineb WP ('90), Mancozeb FL('90), Mepronil+Captan WP ('90), 1,3-Dichloropropene+Chloropicrin F('91), Captafol WP('93), Captafol FL ('93), Penconazole + Captafol WP ('93), Polyoxin + Captafol WP ('93)
Infertility (3)	Nitrofen WP ('82), Nitrofen G ('82), ACN + NIP + MCPB G('82)
Teratogenesis (4)	2,4,5-TP L('84), Cyhexatin WP('89), Tetradifon + Cyhexatin WP('90), Benzoximate + Cyhexatin WP('90)
Total	58

of MAFF, KEPA, MHW, and RDA, specialists in pesticides, pesticide manufacturers, and representatives of consumer organizations.

Special sectional committees could be set up in PMC. At the present time, three of them are operating; namely, the Pesticide Residue Committee, Pesticide Toxicity Committee, and temporarily, the Risk/Benefit Committee.

All technical evaluations are conducted by the special sectional committees, and some pesticides which help resolve agricultural and environmental problems are reexamined by PMC. For example, all residual results are conducted by the standard experimental method (V-B) that have been overlooked by the Pesticide Residue Committee, and some problem pesticides are discussed in the general meeting of the PMC.

All prohibited pesticides except the residual ones specified by PMC are listed in Table 12.17.

MHW also deals with some matters concerning pesticides. The Food Hygiene Law includes some articles for pesticide regulation, such as the Establishment of Tolerance (Article 7), Inspection of Pesticide Residue (Article 13), Prohibition of Sales for Disqualified Goods(Article 15), and Disuse of Disqualified Good (Article 56). The Food Hygiene Committee carefully discusses the data submitted by MHW for establishment of pesticide tolerance. Therefore, MHW conducts monitoring surveys for several thousands of basket samples every year through NHI and the Health and Environment Institute of Regional Government. MHW also has regulated residue levels of some insecticides in drinking water based on the Food Hygiene Law. There were 70 μg/kg of carbaryl, 20 μg/kg of diazinon, 40 μg/kg of fenitrothion, 200 μg/kg of malathion, and 60 μg/kg of parathion, respectively.

Hence, pesticide residues in the soil were regulated by Ministry of Environment (NDE), former KEPA, through the Water Quality Law until 1994. From 1995 on, it will be replaced by the Soil Conservation Law, which also includes the regulation of production and use of pesticides.

Therefore, MAF could regulate the manufacture and use of residual pesticides, while MHW could also restrict the use of pesticides and forbid the use of crops which have been treated by a pesticide that has exceeded the tolerance limit. MOE could control discontinuance of production and use of residual pesticides based on the results of soil monitoring surveys. So a close relationship of those three organizations should establish the most reasonable way to regulate residual pesticides in the soil.

SUMMARY AND CONCLUSION

Change of pesticide use patterns and evaluation of pesticide residues in soil and water were considered in this chapter. The effects of pesticides on soil microorganisms, soil enzyme activities, and soil chemical properties such as pH, Eh, and nitrification were also discussed. In addition, translocation of pesticides into crops, behavior of pesticides in the soil, and mitigation strategies of pesticides in the soil were also examined. Finally, systems and standards for regulation of pesticide residues in the soil and crops, and the existing experimental methods and establishing processes of pesticide safe use guidelines were introduced. It should be possible to draw some conclusions from the above considerations. Use of pesticides in Korea rapidly increased during the late 1960s and the late 1980s, but has stabilized since then. Also, the pattern of use has changed to embrace a friendly relationship with the environment. The use of granules and dusts has been decreasing to some extent because of a decline in the use of pesticides in the paddy fields, but an increase of usage in orchard and field crops could be continued in the future, owing to an increase of cultivation areas under covered structures.

Monitoring surveys of crops are performed by RDA of MAF and NHI of MHW on several thousands of samples every year. However, monitoring surveys of soils have not been conducted as much. Residues of organochloro insecticides such as BHC, DDT, and drines are decreasing annually, so that it is now quite negligible in Korean cultivated soil. There remains a need for further investigations into cultivation soils from under covered structures, which method of production has been rapidly expanding in the region. Although residue levels of pesticides in the water were not high, the results of the monitoring surveys of irrigation waters, lakes, or rivers were insufficient to justify establishment of contamination levels. A search of the existing literature revealed no publication that covered pesticide residues in groundwaters. Hence, magnification of water samples including groundwater may be required; also, pesticide residue limits in drinking water must be reassessed to cover more and a greater variety of pesticides.

Residual components affected the number of microorganisms in the early stages of application, and then recovered in time to slightly lower than control level. The activities of soil enzymes showed a pattern much similar to that of the number of microorganisms. Pesticide residues had little influence on soil pH, but soil Eh affected by the activities of soil microorganisms were lowered during the early days after treatment. As a result of soil Eh changes, ammonification was hindered under reductive potential condition. However, these situations also recovered in the course of time to control level.

To establish mitigation methods for pesticide residues in the soil, translocation amounts of soil residue into crops were examined. It was found that the translocation rate into crops varied by chemicals and crops. Distribution, leaching, and mobility of the residues in soil and water were examined and it was established that those properties were also dependent on the chemical nature of pesticides. Data were obtained from some experiments carried out to investigate whether pesticide degradation was being accelerated by treating with organic matter, lime, and other soil conditioners, and it was apparent that addition of organic matter promoted the degradation rate in upland (aerated, oxidative) as well as submerged (anaerated, reductive) soil conditions. Although microorganisms identified from the soil sometimes showed a slightly higher effect of degradation, in practice it would be difficult to achieve. Therefore, more research on pesticide degradation would be needed in order to decrease pesticide residues in the soil.

Furthermore, examination of DT50 and DT90 ought to be continued to enable pesticide registration, and the metabolism studies should also be encouraged. The system for regulating pesticide residues in the soil and in the environment has a dual function : to set up pesticide safe use guidelines and to establish the maximum tolerance levels. The powerful Pesticide Management Law, Food Hygiene Law, and Soil Conservation Law must be enforced so as to eliminate misuse of pesticides. In conclusion, Korea is one of the strongest regulating countries for control of pesticide residue.

ACKNOWLEDGMENTS

The authors are grateful to Dr. J. W. Choi and Miss J. R. Cho for their help during preparation of this chapter.

REFERENCES

Agricultural Chemicals Industrial Association. Agrochemical Year Book, 1994.

Agricultural Chemicals Industrial Association. '94 Handbook of Pesticide Use, 1994.

Agricultural Chemicals Industrial Association. 20 Years of Agrochemical Industry, 1993.

Agrochemicals Research Institute. '91 Evaluation of Pesticide Residue Levels of River Water. Unpublished data, 1991.

Agrochemicals Research Institute. Evaluation of Pesticide Residues and Safety in/on Crops and Agricultural Ecosystems. 1–4. Unpublished data, 1994.

A.R.I., R.D.A. Pesticide Use and Safety, 1994.

A.R.I., R.D.A., and A.C.I.A. Standard Methods of Pesticide Experiments. (Notice-1994-1) 1994.

Chang, M.S., Y.T. Kim, Y.H. Moon and H.S. Ryang. Residual Amount of Herbicide Dimepiperate and Hydroxy Dimepiperate in Surface Water and Leachate. *Kor. J. Environ. Agric.*, 12:1, pp. 27–34, 1993.

Choi, J.H. and C.S. Kim Evaluation of Pesticide Residues in Cultivated Soil under Structure. Unpublished data, 1991.

Choi, J.H., J.K. Lee, and C.S. Kim. Evaluation on Pesticide Residue Levels of Irrigation Water in Paddy Field. Annual Research Report of ARI. RDA. 466–472, 1993.

Choi, J.H., C.S. Kim, and J.K. Lee. Evaluation of Pesticide Residue Levels in Irrigation Water. Annual Research Report of ARI. RDA, 1992.

Choi, J.W. and K.S. Lee. Degradation of Diazinon and Dursban in Submerged Soil, *J. Kor. Environ. Agric.*, 6:2, pp. 1–11, 1987.

Choi, J.W., K.S. Lee, J.K. Lee, and K.B. Noh. Residue Levels of Organochlorine Pesticides on Paddy Field Soils in the Chungnam Area. *Kor. J Environ. Agric.*, 6:2, pp. 12–21, 1987.

Choi, J.W., Y.H. Rhee, and K.S. Lee. Effect of Activities of Monooxygenase, α-, β-esterase on the Degradation of Diazinon and Dursban in Submerged Soil. *Kor. J. Environ. Agric.*, 9:2, pp. 97–103, 1990.

Chun, J.C. and K.W. Han. Effects of Mineral Nutrients and Mixed Herbicide on the Absorption and Translocation of Bensulfuron-Methyl in Rice, J. *Kor. Environ. Agric.*, 13:1, 60–65, 1994.

Han, D.S. Effects of phosvel on the soil microflora and change of inorganic matter. Kangwon University Theses Coll., 9, pp. 153–158, 1975.

Han, D.S. and J.J. Kim Effects of pesticides on the soil microflora and change of inorganic matter. Kangwon University Theses Coll., 8, pp. 89–104, 1974.

Han, D.S., C.K. Park, C.U. Son, and J.H. Hur. Studies on the Heptachlor-Caused Phytotoxicity at the Growing Stage of Hop and Hansam Vine. *Kor. J. Environ. Agric.*, 12:1, pp. 59–67, 1993.

Hong, J.U. and S.M. Cho. The Changes of the Activity of Nitrogen-Containing Herbicides in Soils; Part 1. Effects on the Urease Activity in Soils. *J. Kor. Agric. Chem. Soc.*, 22:41, pp. 217–220, 1979.

Hong, J.U. and J. Choi. Effects of DDT and BHC on the soil microflora and carbohydrase activity. Kyungpook University Theses Coll., 13, pp. 7–14. 1969.

Hong, J.U. and H. Kim. Degradation of Dinobuton in Soil and Solution, *J. Kor. Environ. Agric.*, 3:2, pp. 16–22, 1984.

Hong, M.K. and J.U. Hong. Release of Carbofuran from Granular Formulations in Water and Its Degradation Patterns in Soil. *Kor. J. Environ. Agric.*, 3:2, pp. 9–15, 1984.

Jeong, Y.H., S.S. Kuem, K.T. Cho, K.S. Lee, and Y.C. Hong. Studies on the Spatial Distribution of Pesticides in/on Rice Plants on Different Formulation, Annual Research Report. Ins. Agric.Sci. RDA, pp. 53–57, 1976.

Kang, K.Y. Effect of the Herbicide Bentazon on Nitrification, and on Numbers of Bacteria and Fungi in the Soil., *Kor. J. Agric. Chem. Soc.*, 11:2, pp. 81–83, 1978.

Kim, C.J., M.J. Oh, and J.S. Lee. Degradation of Organochlorinated Pollutants by Microorganism-Degradation of PCBs and Organochlorine Pesticides and Degradation Products. *J. Kor. Agric. Chem. Soc.*, 30:4, pp. 31–39, 1987.

Kim, C.S., J.H. Choi, and J.K. Lee. Fate of Pesticide in Soil. Annual Research Report of ARI. RDA. pp. 473–479, 1993.

Kim, C.S., J.H. Choi, Y.C. Choi, and Y.H. Jeong. Mitigation of Pesticide Residue in Agricultural Ecosystem. Annual Research Report of ARI. RDA. pp. 192–196, 1991.

Kim, E.H. and D.S. Han. Fluctuations of Microorganisms in Soil and Changes of Inorganic Nitrogens Treated Agricultural Medicines, Kangwon University, 11, pp. 133–137, 1977.

Kim, E.H. and D.S. Han. Fluctuations of microorganisms in soil and changes of inorganic nitrogens treated agricultural medicines. Kangweon University Theses Coll., 11, pp. 133–137, 1978.

Kim, J.E. and T.H. Choi. Behavior of Synthetic Pyrethroid Insecticide Bifenth-Rin in Soil Environment. II. Identification of Degradation Product and Leaching of Bifenthrin in Soil. *Kor. J. Environ. Agric.*Theses Coll, 11:2, pp. 125–132, 1992.

Kim, J.H., Y.H.Rhee, J.W. Choi, and K.S. Lee. Microbial Degradation of Diazinon in Submerged Soil, *J. Kor. Microbial.* 27:2, pp. 139–136, 1989.

Kim, J.J. The effect of rising and falling inorganic nitrogen and soil microorganism fluctuation in accordance with polyoxin. Kangwon University Theses Coll., 9, pp. 161–167, 1975.

Kim, J.J. Effect of herbicide on the soil microorganism and inorganic matters., Kangweon University Theses Coll,12, pp. 95–102, 1978.

Kim, J.J. and H.S. Jeong. The Influence of Herbicides on Soil Microflora—Influence of Butachlor, *J. Kor. Soc. Soil Sci. Fert.*, 9:1, pp. 25–31, 1976.

Kim, K.S., Y.W. Kim, J.A. Kim, and H.W. Kim. Effect of Herbicides on Microflora and Enzyme Activity in Soil, *J. Kor. Soc. Soil Sci. Fert.*, 21:1, pp. 61–71, 1988.

Kim, K.S., Y.W. Kim, M.C. Lee, and H.W. Kim. Effect of Pesticides on Microflora, Soil Respiration and Enzyme Activity in Soil, *J. Kor. Soc. Soil Sci. Fert.*, 21:1, pp. 375–385, 1987.

Kim, M.K. The Influence of Some Soil-Treated Herbicides on the Mineralization of Nitrogen Fertilizers in a Flooded Paddy Soil. *Kor. J. Pl. Prot.*, 15:4, pp. 205–214, 1976.

Korea Research Institute of Chemistry and Technology. A Study on Toxicological Research for Chemical Compounds (IV). pp. 323–349, 1985.

Lee, H.K. Effect of Rice Straw Amendment and Repeated Application of Diazinon on the Persistence of Diazinon in Submerged Soils. *J. Kor. Agric. Chem. Soc.*, 24:1, pp. 1–6, 1981.

Lee, H.K., Y.D. Lee, and Y.S. Park. Evaluation of Pesticide Residues of River Waters in 1983. Annual Research Report. ORD., 26:1, pp. 46–53, 1977.

Lee, H.K., Y.D. Lee, and Y.S. Park. Effect of Soil Microorganism to the Degradation of Pesticide, Annual Research Report. ORD., pp. 91–95, 1984.

Lee, H.K., Y.D. Lee, Y.S. Park, and Y.H. Shin. A Survey for Pesticide Residues in Major Rivers of Korea. *Kor. J. Environ. Agric.*, 2:2, pp. 83–89, 1983.

Lee, H.K., Y.S. Park, J.U. Hong, and N.S. Talekar. Persistence of Cyanofenphos on Chinese Cabbage. *Kor. J. Environ. Agric.*, 1:2, pp. 89–98, 1982.

Lee, H.K., Y.H. Joeng, B.Y. Oh, B.H. Song, and B.M. Lee. Study on Degradation of Insecticides in Soil. Annual Research Report of ARI. RDA. pp. 78–92, 1979.

Lee, J.K. Degradation of the Herbicide, Alachlor, by Soil Microorganisms, III. Degradation Under an Upland Soil Condition. *J. Kor. Agric. Chem. Soc.*, 29:2, pp. 182–189, 1986.

Lee, J.K. and J-C. Fournier. A Study on the Evolution of 3,4-DCA and TCAB in Some Selected Soil-Degradation of ^{14}C-3,4-DCA and ^{14}C-TCAB, *J. Kor. Agric. Chem. Sci.*, 21:2, pp. 71–79, 1978.

Lee, J.K. and K.S. Kyung. Change in the Non-Extractable Bound Residue of TCAB as a Function of Aging Period in Soil. *Kor. J. Environ. Agric.*, 10:2, pp. 149–157, 1991.

Lee, J.K. and K.S. Oh. Leaching Behavior of the Residues of Carbofuran, Bentazon, and TCAB in Soil. *Kor. J. Environ. Agric.*, 12:1, pp. 9–18. 1993.

Lee, J.K., K.R. Cho, K.S. Oh, and K.S. Kyung. Degradation of the Herbicide Bentazon by Soil Microorganisms. *Kor. J. Environ. Agric.*, 12:1, pp. 121–128, 1993.

Lee, J.K., Y.B. Ihm, Y.G. Cho, and K.S. Kyung. Microbial Degradation of the Persistent Pollutant TCAB: (II) Degradation of TCAB by Isolated Microorganisms. *J. Kor. Agric. Chem. Soc.*, 34:4, pp. 299–306, 1991.

Lee, J.K., K.S. Kyung, and F. Fohr, Bioavailability of Soil-Aged Residues of the Herbicide Bentazon to Rice Plants. *J. Kor. Agric. Chem. Soc.*, 32:4, pp. 393–400, 1989.

Lee, J.K., K.S. Kyung, and W.B. Wheeler. Uptake of the Fresh and Aged Residues of Carbofuran by Rice Plants from Soil. *Kor. J. Environ. Agric.*, 8:2, pp. 103–118, 1989.

Lee, J.K., J.H. Choi, and C.S. Kim. Establish of Safety Limits of Pesticide Residues in Soil. Annual Research Report of ARI. RDA. pp. 378–383, 1992.

Lee, K.B., I.K. Cho, J.H. Shim, and Y.T. Suh. Residue Determination of Chloro-Thalonil in Sesame and Soil. *Kor. J. Environ. Agric.*, 9:1, pp. 15–22, 1990.

Lee, K.B., Y.W. Kim, and K.S. Kim. Effect of Pesticides on Microorganisms Related to the Nitrogen Cycle in the Submerged Soil, *J. Kor. Soc. Soil Sci. Fert.*, 21:2, pp. 149–159, 1988.

Lee, K.S. On the Absorption Rate of Some Residual Organochloro Insecticide in Soil to the Certain Vegetables, Annual Research Theses, Cheju University, 12, pp. 67–70, 1981.

Lee, K.S. A Survey of Pesticide Residues of Citrus Fruits and Citrus Orchard Soil in Jeju Island. *J. Kor. Agric. Chem. Soc.*, 23:3, pp. 184–188, 1980.

Lee, K.S. Organochlorine Insecticide Residues of Field Soils and Vegetables in Cheju Island. *J. Kor. Agric. Chem. Soc.*, 24:3, pp. 155–160, 1981.

Lee, K.S. Degradation and Its Inhibition of Granular Type OP-Insecticides Using Paddy Field. Research Report of the Korean Foundation of Science and Technology (867-1502-001-3), 30–33, 1989.

Lee, K.S. and J.S. Lee. Transformation of Urea-N and Population Changes of Microorganism in Soil Treated with Several Herbicides, Chungnam University, 2, pp. 55–63, 1983.

Lee, S.R. *Food Safety and Toxicology*. Ewha Womans University Press, Seoul, 1993.

Lee, S.R., S.Y. Kang, C.K. Park, J.K. Lee, and C.S. Rho. A Survey on the Residues of Organochlorine Pesticides in Water, Mud and Clam Samples from the Kwangyang Bay, Korea. *J. Kor. Agric. Chem. Soc.*, 19:3, pp. 113–119, 1976.

Lee, Y.D., Y.K. Kim, K.H. Lee, and H.M. Park. Behavior of Carbofuran in Paddy Ecosystem. Annual Research Report of ARI. RDA. pp.78–86, 1986.

Lee, Y.D., H.K. Lee, and S.H. Lee. Effect of Soil Conditions to Acceleration on Pesticide Degradation. Annual Research Report of ARI. RDA. pp. 118–126, 1982.

Lee, Y.H., E.C. Hwang, and C.K. Park. Evaluation of Polychlorinated Biphenyls(PCBs) and Organochlorine Insecticide Residues in Irrigation Waters in the Periphery of Suwon. *Kor. J. Environ. Agric.*, 4:2, pp. 95–101, 1985.

Lim, S.K. and W.A. Bong. Studies on the Several Soil Factors Affecting on Alachlor and Paraquat Adsorption by Soils. *Kor. J. Environ. Agric.*, 11:2, pp. 101–108, 1992.

Lim, S.U., K.Y. Kang, and S.O. Park. Studies on the Behaviors of Urea in Soils; [Part 1] Effects of Some Pesticides on the Urea Decomposition and Nitrogen Transformation in Flooded Paddy Soil. *J. Kor. Agric. Chem. Soc.*, 20:1, pp. 58–65, 1977.

Ma, S.Y., Y.H. Moon, and H.S. Ryang. Movement of Herbicide Pritilachlor in Plants and Soil, *J. Kor. Agric. Chem. Sci.*, 30:4, pp. 351–356, 1987.

Moon, Y.H. Effects of Soil Environmental Conditions on the Decomposition Rate of Insecticide Fenitrothion in Flooded Soils. *Kor. J. Environ.* Agric., 9:1, pp. 1–8, 1990.

Moon, Y.H. and H.S. Ryang. Dissipation of Fenitrothion, IBP and Butachlor in Flooded Soil Under Outdoor Conditions. *Kor. J. Environ. Agric.*, 9:1, pp. 9–14, 1990.

Moon, Y.H., Y.T. Kim, Y.S. Kim, and S.K. Han. Simulation and Measurement of Degradation and Movement of Insecticide Ethoprophos in Soil. *Kor. J. Environ. Agric.*, 12:3, pp. 209–218, 1993.

Oh, B.Y., Y.K. Kim, and Y.S. Park. Effects of Pesticide Formulations on the Residues in Paddy Rice. *Kor. J. Environ. Agric.*, 3:2, pp. 1–8, 1984.

Oh, B.Y., Y.H. Jeong, B.H. Song, H.K. Lee, and B.M. Lee. Study on Degradation of Herbicides in Soil. Annual Research Report of IAS, RDA. pp. 93–110, 1979.

Oh, W.K. Studies on the Effects of Continuous Application of Herbicides on Chemical Nature of Upland Soils. *J. Kor. Soc. Soil Sci. Fert.*, 6:1, pp. 9–16, 1973.

Park, C.K. Studies on the Residues of Chlorinated Organic Insecticides. 3. Heptachlor Residues in Soil 15 Years after Yearly Treatment of the Soil Insecticide in a Tobacco Field. *J. Kor. Agric. Chem. Soc.*, 18:2, pp. 61–64, 1975.

Park, C.K., D.S. Han, and J.H. Hur. Organophosphorus Pesticide Residues in Major Environmental Components of Nakdong River. *Kor. J. Environ. Agric.*, 3:1, pp. 36–44, 1984.

Park, C.K. and E.C. Hwang. Evaluation of Polychlorinated Biphenyls and Organochlorine Insecticide Residues in Waters, Sediments and Crucian Carps in Soho lake. *Kor. J. Environ. Agric.*, 1:2, pp. 105–114, 1982.

Park, C.K. and K.S. Lee. Pesticide (VII-4), in *History of Agricultural Technology* (Edited by Publication Committee of History of Korea Agricultural Technology). pp. 632–652, 1983.

Park, C.K. Behavior of Pesticides in Paddy Ecosystem. Annual Research Report of IAS, RDA. pp. 87–92, 1986.

Park, C.K. and Y.T. Je. Degradation Patterns of BPMC and Carbofuran in Flooded Soil, *J. Kor. Environ. Agric.*, 2;2, pp. 65–72, 1983.

Park, C.K. and Y.D. Lee. Fate of Pesticide Under Paddy Field Condition. Annual Research Report of IAS, RDA. pp. 89–99, 1985.

Park, C.K. and Y.D. Lee. Trends of Pesticide Use and Residue in Korea. Memorial Symposium for 40 Anniversary of Kangwon National University (1987.6.11–12). pp. 59–83, 1987.

Park, C.K. and Y.S. Ma. Organochlorine Pesticide Residues in Agricultural Soils-1981. *Kor. J. Environ. Agric.*, 1:1, pp. 1–13, 1982.

Park, C.K. and S.R. Oh. Fate of C-14 Labelled Carbofuran in Paddy Plants and Soil. *Kor. J. Environ. Agric.*, 5:2, pp. 9–15, 1986.

Park, C.K. and N.D. Park. Analysis of Organochlorine Pesticide Residues in the Presence of Polychlorinated Biphenyls (PCBs). *J. Kor. Agric. Chem. Soc.*, 23:1, pp. 59–63, 1980.

Park, K.H., Y.C. Choi, J.H. Choi, and C.S. Kim. Selection of Microorganisms for Acceleration on Degradation of Pesticide Residues. Annual Research Report of IAS, RDA. pp. 384–391, 1992.

Park, K.H., Y.H. Jeong, J.H. Choi, and C.S. Kim. Mitigation of Pesticide Residues in Agricultural Ecosystem. Annual Research Report of IAS, RDA. pp. 480–484, 1993.

Park, K.Y., D.W. Ree, C.K. Park, and D.S. Han. Studies on the Effect of Heptachlor Residues in Soil on the Growth of Hop; 1.Phytotoxic Symptom of Heptachlor Residues in Hop. *Kor. J. Environ. Agric.*, 1:2, pp. 99–104, 1982.

Park, S.K., K.Y. Kim, J.W. Lee, Y.A. Shin, and E.H. Lee. Effect of Application of Woody Charred Materials on the Plant Growth and Chemical Properties of Soil in the Continuous Cropping Field of Red Pepper, *J. Kor. Environ. Agric.*, 12:1, pp. 1–8, 1993.

Roh, J.K. and O.R. Baik. Effects of Some Pesticides on Korean Paddy Soil Microorganisms. *J. Kor. Agric. Chem. Soc.*, 24:3, pp. 174–180, 1981.

Ryang, H.S., K.W. Han, Y.H. Moon, and Y.C. Choi. Effect of Transplanting Depths on Growth of Transplanted Rice by Dithiopyr., *Kor.J. Weed Sci.*, 3, pp. 174–177, 1991.

Ryang, H.S., Y.H. Moon, E.S. Choi, M.S. Jang, and J.H. Lee. Residual Activity and Effect of Soil Applied Herbicides on Succeeding Crops in Vegetable Field; Residual Activity and Effect of Soil Applied Herbicides on Succeeding Crops in Winter Crops., *Kor. J. Weed Sci.*, 11:1, pp. 32–49, 1991.

Ryang, H.S., Y.H. Moon, E.S. Choi, M.S. Jang, J.H. Lee, and Y.N. Chang. Residual Activity of Soil Applied Herbicides on Succeeding Crops in Vegetable Field., *Kor.J. Weed Sci.*, 11:2, pp. 117–121, 1991.

Ryang, H.S., Y.H. Moon, E.S. Choi, M.S. Jang, J.H. Lee, and Y.N. Chang. Residual Activity and Effect of Soil-Applied Herbicides on Succeeding Crops in Vegetable Fields; 4. Residual Amount of Herbicides Nitralin and Napropamide. *Kor. J. Environ. Agric.*, 10:2, pp. 113–118, 1991.

Ryang, H.S., Y.H. Moon, E.S. Choi, M.S. Jang, J.H. Lee, and Y.N. Chang. Residual Activity and Effect of Soil Applied Herbicides on Succeeding Crops in Vegetable Field; Residual Activity and Effect of Soil Applied Herbicides on Succeeding Crops in Summer Crops., *Kor. J. Weed Sci.*,11:1, pp. 50–59, 1991.

Ryu, J.C., M. Araragi, and H. Koga. The Influence of Pesticides on Some Chemical and Microbiological Properties Related to Soil Fertility; 1. Effects of Herbicide (CPN) on Some Soil Chemical Factors Concerning Nitrogen Mineralization. *J. Kor. Soc. Soil Sci. Fert.*, 16:4, pp. 372–381, 1983.

Ryu, J.C., B.H. Song, H.H. Lee, and J.H. Ha. Study on Effects of Pesticides Application to Chemicals Properties and Microflora in Soil. Annual Research Report of IAS. RDA. pp. 416–462, 1980.

Song, B.H., Y.H. Jeong, and Y.S. Park. Effect of Repeated Application of IBP on the Degradation of Pesticides in Flooded Soil. *Kor. J. Environ. Agric.*, 1:1, pp. 65–70, 1982.

Song, B.H., Y.S. Park, and Y.H. Jeong. Study on Pesticide Degradation in Soil. Annual Research of ARI, RDA, pp. 49–60, 1981.

Suh, Y.T., J.G. Im, and J.H. Shim. Evaluation of Organochlorine Pesticide Residues in the Mud Flat. *Kor. J. Environ. Agric.*, 5:2, pp. 113–118, 1986.

Suh, Y.T., R.D. Park, and J.H. Sim. Levels of Organochlorine Pesticides in the Cultivating Soils in the Suburbs of Gwangju-city, Jeollanam-Do.Kor. *J. Environ.Agric.*, 1:2, pp. 83–88, 1982.

Suh, Y.T., J.H. Shim, and R.D. Park. Evaluation of Organochlorine Pesticide Residues in Soil by Steam Distillation. *Kor. J. Environ. Agric.*, 3:2, pp. 23–29, 1984.

Yang, C.S. Effects of Pesticides on Soil Microflora—Changes in Soil Microflora by Application of Organochlorine Pesticides. *J. Kor. Soc. Soil Sci. Fert.*, 17:3, pp. 299–306, 1984.

Appendix: List of Pesticides Mentioned in Chapter 12

Common Name	Use	Chemical Name(IUPAC)
Acephate	insecticide	O,S dimethyl acetylphosphoramidothioate
Alachlor	herbicide	2-chloro-2′ ,6′-diethyl-N-methoxymethylacetanilide
Aldrin	insecticide	(1 R, 4S, 4αS, 5S, 8R, 8αR)-1, 2, 3, 4, 10, 10-hexachloro- 1,4, 4α,5,8,8α-hexahydro-1,4: 5,8-dimethanonaphthalene
Akar (chlorobenzilate)	acaricide	Ethyl 4,4i-dichlorobenzilate
Asulam	herbicide	Methyl sulfanilycarbamate
Avirosan	herbicide	Dimethametryn [N^2-(1, 2-dimethylpropyl)-N^4-ethyl-6-methylthio-1,3,5-triazine-2,4-diamine) +Piperophos] S-2-methylpiperidinocarbonymethyl-0, O-dipropyl phosphorodithioate
Baycid (Fenthion)	insecticide	0,0-dimethyl O-4-methylthio-m-tolyl phosphorothioate
Bensulfuron methyl	herbicide	α -(4, 6-dimethoxypyrimidin-2-ylcarbamoylsulfamoyl-0-toluic acid
Bentazone	herbicide	3-isopropyl-lH-2, 1 ,3-benzothiadiazin-4(3H)-one 2, 2-dioxide
Benthiocarb (Thiobencarb)	herbicide	S-4-chlorobenzyl diethylthiocarbamate
BHC	insecticide	1, 2, 3, 4, 5, 6-hexachlorohexane (mixed isomers)
Bifenthrin	insecticide	2-methylbiphenyl-3-ylmethyl(z)-(lRS)-cis-3-(2-chloro-3, 3,3-trifluoroprop-l -enyl) 2, 2-dimethylcyclopropane carboxylate
Blasticidin-S	antibiotic	1-(4-amino- 1, 2-dihydro-2-oxopyrimidin- 1-yl) -4- [(S) 3-amino-5- (l-methylguanidino) valeramido -1, 2, 3, 4-tetradeoxy-l5-D-erythro-hex-2-enopyranuronic acid
BPMC	insecticide	2-sec-buthylphenyl methyl carbamate
Butachlor	herbicide	N-butoxymethyl-2-chloro-2′, 6′-diethylacetanilide
Bux (Bufencarb)	insecticide	3-(1-methyl buthyl)phenyl methyl carbamate
Captapol	fungicide	N-(1,1, 2, 2-tetrachloroethylthio) cyclohex-4-ene-1, 2-di carboximide
Carbaryl	insecticide	1-naphthyl methyl carbamate
Carbofuran	insecticide	2 ,3-dihydro-2 ,2-dimethylbenzofuran-7-yl methylcarbamate
Chlorothalonil	fungicide	tetrachloroisophthalonitrile
Chlorpyrifos	insecticide	O, O-diethyl O-3, 5, 6-trichloro-2-pyridylphosphorothioate
Chloropyrifos	insecticide	O, O-dimethyl O-3, 5, 6-trichloro-2 pyridylphosphoromethyl thioate
CNP	herbicide	4-nitrophenyl 2,4,6-trichlorophenyl ether
Cyhalothrin	insecticide	(RS)-a-cyano-3-phenoxybenzyl(Z)-(lRS)-cis-3-(2-chloro-3, 3, 3 trifluoropropenyl)-2, 2-dimethylcyclopropane carboxylate
2,4-D	herbicide	(2,4-dichlorophenoxy) acetic acid
DDT	insecticide	1,1,1-trichloro-2, 2-bis (4-chlorophenyl) ethane
Diazinon	insecticide	O, O-diethyl O-2-isopropyl-6-methylpyrimidin-4-yl phosphorothioate 2,2 dichlorovinyl dimethyl phosphate
Dichlofluanid	fungicide	N-dichlorofluoromethylthio-N′ ,N″-dimethyl-N-phenyl sulfamide
Dichlorvos	insecticide	2,2-dichlorovinyl dimethyl phosphate
Dieldrin	insecticide	(lR,4S,4aS, 5R,6R, 7S, 8S, 8aR)-1, 2, 3 ,4,10,10-hexachloro-1, 4, 4a, 5 ,6, 7, 8, 8a-octahydro-6, 7-epoxy-1,4 :5, 8-dimethanonophthalene
Dimepiperate	herbicide	S- 1-methyl- 1 -phenylethyl pipiridine- 1 -carbamate
Dimetametrynes	herbicide	N^2-(1 ,2-dimethylpropyl)-N^4-ethyl-6-methylthio-1,3,5 triazine-2 ,4-diamin)
Dinoseb	herbicide	2-sec-butyl-4, 6-dinitrophenol
Endosulfan	insecticide	(1,4,5,6,7,7-hexachlor-8,9, 10-trinorborn-5-en-2,3-yle nebismethylene) sulfite
Esprocarb	herbicide	S-benzyll,2-dimethylpropyl(ethyl)thiocarbamate
Ethalfluralin	herbicide	N-ethyl-α, α, α-trifluoro-N- (2-methylallyl)-2, 6-dinitro P-toluidine
Ethoprophos	insecticide	O-ethyl S, S-dipropyl phosphorodithioate
Fenitrothion	insecticide	O,O-dimethyl 0-4-nitro-m-tolyl phosphorothioate
Ferbam	fungicide	Iron tris(dimethyldithiocarbamate)

Appendix: (Continued)

Common name	Use	Chemical Name(IUPAC)
Flufenoxuron	insecticide	1- [4- (2-chloro- α, α, α-trifluoro-p-tolyloxy) -2-fluorophenyl] -3- (2, 6-difluorobenzoyl) urea
Glyphosate	herbicide	N-(phosphonomethyl) glycine
Heptachlor	insecticide	1,4, 5, 6, 7, 8, 8-heptachloro-3α,4, 7, 7 α-tetrahydro-4, 7 methanoindene
Iprobenfos	fungicide	S-benzyl O,O-di-isopropyl phosphorothioate
Iprothiolane	fungicide	Di-isopropyl 1, 3-dithiolane-2-ylidenemalonate
Kasugamycin	antibiotic	lL-1, 3, 4/2, 5, 6-i-deoxy-2, 3 ,4, 5, 6-pentahydroxycyclohexyl 2-amino-2, 3, 4, 6-tetradeoxy-4- (α-iminoglycino) -α -D-α rabino-hexopyranoside
Kelthane	acaricide	2, 2, 2-trichloro- 1 ,1 -bis (4-chlorophenyl) ethanol
Linuron	herbicide	3- (3, 4-dichlorophenyl) -1-methoxy-1-methylurea
Malathion	insecticide	Diethyl (dimethoxyphosphinothioylthio) succinate
Maneb	fungicide	manganese ethylenebis(dithio carbamate)
MCP	herbicide	4-chloro-O-tolyloxyacetic acid
Methabenzthiazuron	herbicide	1-(1 ,3-benzothiazol-2-yl)-1 ,3-dimethylurea
Metolachlor	herbicide	2-chloro-6′-ethyl-N-(2-methoxy-1-methylethyl)acet-o-toluidide
Metribuzin	herbicide	4-amino-6-tert-butyl-3-methylthio-1, 2,4-triazin-5 (4H) -one
MIPC	insecticide	2-isopropylphenyl methylcarbamate
Molinate	herbicide	S-ethylenezepane- 1 -carbothioate
MPP	insecticide	O, O-dimethylO-4-methylthio-m-tolylphosphorothioate
Oxadiazon	herbicide	5-tert-butyl-3-C2,4-dichloro-5-isopropyloxyphenyl)-1,3, 4-oxadiazol-2(3H)-one
Paraquat	herbicide	1,1′-dimethyl-4 ,4′-bipyridinium
Parathion	insecticide	O,O-diethyl 0-4-nitrophenyl phosphorothioate
PCNB	fungicide	Pentachloronitrobenzene
PCP	herbicide insecticide fungicide	Sodium pentachlorophenete
Pendimethalin	herbicide	N- (1-ethylpropyl)-2, 6-dinitro-3, 4-xylidine
Perfluidone	herbicide	1,1,1 trifluoro-2′-methyl-4′-(phenylsulfonyl)methane sulfonanilide
Phosvel	insecticide	0-4-bromo-2, 5-dichlorophenyl-O-methyl phenyl phosphonothilate
Polyoxins	antibiotic	PolyoxinB 5- (2-amino-5-o-carbamoyl-2-deoxy-L-xylonamido) -1, 5 dideoxy-1-(1 ,2,3,4-tetrahydro-5-hydroxymethyl-2.4-dioxopyrim-idin- 1 -yl) -β-D-allofurnuronic acid
Pretilachlor	herbicide	2-chloro-2′, 6′-diethyl-N- (2-propoxyethyl) acetanilide
Procymidone	fungicide	N- (3, 5-dichlorophenyl) -1, 2-dimethylcyclopropane- 1, 2 dicarbox-imide
Prometryne	herbicide	N^2, N^4-di-isopropyl-6-methylthio-1, 3, 5-triazine-2 ,4 diamine
Propanil	herbicide	3 ‘, 4′-dichloropropionanilide
Pthalide	fungicide	4, 5, 6, 7-tetrachlorophthalide
Quinclorac	herbicide	3,7-dichloroquinoline-8-carboxylic acid
Sanggamma-S		
Simazine	herbicide	6-chloro-N^2,N^4-diethyl-1,3,5-triazine-2,4 diamine
Simetryn	herbicide	N^2,N^4-diethyl-6-methylthio-1,3 ,5-triazine-2 ,4-diamine
TCBA	herbicide	2, 3, 6-trichlorobenzoic acid
Terbutryne	herbicide	N2-tert-butyl-N4-ethyl-6-methylthio 1, 3, 5-triazine-2, 4-diamine
Trifluralin	herbicide	α ,α, α-trifluoro-2, 6-dinitro-N, N-dipropyl-p-toluidine
Vinclozolin	fungicide	(R,S)-3-(3,5-dichlorophenyl)-5-methyl-5-vinyl-1,3-oxazolidine-2, 4-dione
Napropamide	herbicide	(R,S)-N,N-diethyl-2-(1-naphthyloxy) propionamide
Nitralin	herbicide	4-methyl sulfonyl-2,6-dinitro-N,N-dipropylaniline
Nitrofen	herbicide	2,4-dichlorophenyl-4-nitrophenyl ether

CHAPTER 13

TOXIC METALS AND AGROCHEMICALS IN SOILS IN MALAYSIA: CURRENT PROBLEMS AND MITIGATION PLANS

Y.M. Khanif, I.C. Fauziah, and J. Shamshuddin

INTRODUCTION

In recent years, serious environmental and natural resources degradation problems have captured the attention of and generated concern in the world community. Agriculture is viewed as a contributor to, as well as a victim of, many of these environmental problems. Until recently the foremost priority of the agricultural community in Malaysia and globally was to ensure and enhance the productivity and profitability of agriculture. This priority is being reexamined to include environmental considerations. This is because of the concern about the impact of agricultural activities on the ecological system.

Of the many natural resources involved in agriculture, soil is of foremost importance and is very vulnerable to ecological disturbances. Given the continuous increase in population along with rapid industrialization and public concern over the environment, management of the soil resource has an added challenge. For viable agriculture, soil management should sustain high productivity and have minimal risks to the ecosystem health.

Malaysia is located close to the equator. Thus, the climate is characterized by temperatures between 23° and 32°C with minor variation. The mean annual rainfall is about 2000 mm. The rainfall occurs throughout the year, with a slightly drier period in February and March, while higher rainfall occurs in November and December. Rates of biodegradation of most contaminants are maximal at the prevailing temperature range. Because of adequate supply of moisture from rainfall, major crops cultivated, except rice and vegetables are not irrigated. Thus, major crops such as oil palm, rubber, and cocoa are distributed throughout the country. These crops, however, are not grown in the northern part of Peninsular Malaysia because of insufficient rainfall for optimal production. Rice production is confined to the coastal plains and riverine areas, because of its dependence on irrigation.

Malaysian agriculture has undergone tremendous technological changes. It is envisaged that the trend will continue in the future. To cater to the growing demand for food and other agricultural products due to population increase, an increase in agricultural production is required. This can be achieved by increasing the land area for agricultural production or by increasing productivity. Both options have important environmental implications. Increasing productivity usually requires high production inputs, which include fertilizer and other agrochemicals. Some of the important problems associated with soil and the environment are toxic metal accumulation and pesticide pollution. This chapter attempts to discuss the problems associated with soil and the environment, and offers research and development strategies in anticipation of the future when Malaysia becomes a fully developed and highly industrialized nation.

FERTILIZER AND AGROCHEMICAL USAGE

Malaysia is an important producer of agricultural commodities globally. Currently, it is the world's largest producer of palm oil, besides being a leading producer of natural rubber and cocoa. Previously, agriculture was the mainstay of the Malaysian economy. Now, its prominence is being overshadowed by the rapid industrialization. To maintain its competitiveness, agriculture's productivity has to be increased. Malaysian soils are generally of low fertility, and the crops are subjected to serious disease, pest, and weed problems. Thus, the dependence on fertilizer and agrochemicals in the future will be inevitable.

Plantation crops (oil palm, rubber, and cocoa) occupy about four million hectares (or 80%) of the total cultivated land (Table 13.1). The rest of the cultivated areas are planted with rice and other cash crops of minor importance. The areas planted with fruits and other high value horticultural crops (data not shown) are becoming increasingly important. Although the area planted with crops, except for oil palm, has either stabilized or declined, the use of fertilizer and agricultural chemicals is expected to increase. This is because of the higher dependence on chemical herbicides due to a labor shortage and increased rate of fertilizer application for higher productivity (Table 13.2). This is especially true in the plantation sectors. The use of insecticides and fungicides is mainly important in the production of rice, vegetables, pepper, tobacco, and other high value crops.

Fertilizer

Given the low fertility status of tropical soils, optimum economic crop production depends heavily on fertilizer usage. Fertilizer consumption in Malaysia is high and continues to increase. The oil palm industry consumes almost 60% of the fertilizer applied (Tay et al., 1994). Rubber, cocoa, and rice account for about 30% and the rest of the crops account for about 10% of the fertilizer usage. The N, P, and K fertilizers are the important fertilizers applied. To most of the crops, N is the nutrient required in the greatest amount. However, for oil palm, the consumption of K exceeds N.

Malaysia imports most of the fertilizers consumed, except urea, which is manufactured locally. The types and the amount of fertilizer imported are presented in Table 13.3. The main sources of N are urea and ammonium sulfate. Phosphorus is applied as phosphate rock, and muriate of potash (MOP) is the principal source of K. The choice of the fertilizer depends on agronomic and economic factors.

Heavy metal contamination in fertilizers is confined mainly to phosphate rock. The levels of toxic metals in other fertilizers are not significant. High amounts of phosphate rock are being applied especially in the plantation crops. Continuous application of this P source can cause serious heavy metal accumulation in the soil and plants. The types and amount of heavy metals in phosphatic fertilizer depend on the origin of the phosphate

Table 13.1. Total Area Under Selected Crops : Malaysia (Ha)[a,b]

Year	Rubber	Oil Palm	Cocoa	Coconut	Pepper	Pineapple	Tobacco	Paddy	Coffee	Tea	Sugarcane
1980	1,999,300	1,023,306	138,500	354,500	13,295	10,729	13,243	716,873	10,882	2,428	12,705
1986	1,905,606	1,599,311	322,334	280,500	6,120	10,495	15,822	627,500	15,919	3,263	33,318
1987	1,874,600	1,672,875	343,972	320,592	9,704	7,295	12,314	644,800	15,969	3,130	20,489
1988	1,867,300	1,805,923	398,376	327,812	8,667	8,365	9,447	665,800	14,879	3,249	23,970
1989	1,849,000	1,951,256	409,444	314,645	9,500	7,985	12,311	661,600	14,000	3,141	23,000
1990	1,932,900	2,029,464	415,628	315,481	11,210	9,302	10,168	651,700	14,000	2,987	23,000
1991	1,823,000	2,094,028	400,200	315,925	11,230	9,211	14,953	656,000	14,000	3,056	23,000
1992e[d]	1,807,000	2,167,396	388,700	n.a.[c]	11,267	9,081	11,905	660,700	n.a.	2,600	n.a.

[a]Ministry of Primary Industry, 1993.
[b]Based on main crop equivalent area concept.
[c]n.a. = not available.
[d]e = estimate.

rock (Syers et al., 1986). The metals that appear to be the most harmful to plant and animal systems include Cd, Hg, Ni, Pb, and Zn (Jones and Jarvis, 1981). The typical heavy metal contaminant in the commonly used phosphate rock is presented in Table 13.4. The metal that has been given serious attention is Cd, because its concentration is fairly high in some phosphate rock. In the long term, the use of phosphate rock with high Cd concentration can be harmful, since Cd may build up in the soil and ultimately become a health hazard to humans through uptake by plants. The concentration of Cr in phosphate rock, although not being given much attention, can be substantial in some of the phosphate rock.

Table 13.2. N, P, K Nutrients Used Per Hectare in Malaysia, 1975–93 (kg/ha)[a]

Year	N, P, K Nutrients
1975	70
1980	105
1985	115
1990	129
1991e[b]	132
1992f[c]	147
1993f	156

[a]Vaes,1992.
[b] = estimated.
[c] = forecast.

Agrochemicals

Under tropical conditions, crop losses from weeds, pests, and diseases are high. Malaysian agriculture depends heavily on agrochemicals to maintain high productivity. The agrochemicals used are herbicides, insecticides, and fungicides (Table 13.5). Herbicides constitute about 80% of the agrochemicals consumed in Malaysia. The heavy dependence on herbicides, especially in the plantation sector, is due to a labor shortage. The shift from transplanting to direct seeding in rice production has created a serious weed problem and also increased herbicide usage. The use of insecticide ranks second to herbicide and it is mainly used in rice, vegetable and tobacco production. Consumption of fungicide is less important than that of herbicide and insecticide. It is mainly confined to vegetable, fruit, and ornamental plant production.

More than 37 types of herbicides are registered with the Malaysian Pesticide Board. The most popular chemicals used in both the plantations and small holdings are dichlorophenoxy acetic acid (2, 4-D), paraquat, and diuron, marketed under various trade names. The other commonly used herbicides are MCPA, 245-T, and monosodium methane arsonate (MSMA) (Mohamad and Ali, 1986).

Insecticides are mainly used in vegetables, rice, and tobacco, with a smaller portion being used in plantation crops. Numerous insecticides with different active ingredients are being used in Malaysia. The main insecticides are gamma-BHC, carbofuran, chlorpyrifos, cypermethrin, malathion, and endosulfan.

The use of fungicides is confined mainly to horticulture crops and ornamental plants and for treatment of common root, panel, and crown disease of rubber. Small quantities are used in vegetables and tobacco. The chemicals often used for control of fungal disease are copper oxychloride, maneb, and mancozeb.

Future Trends

The future consumption of fertilizers and agrochemicals in Malaysia is dictated by economic factors and environmental consciousness. Consumption of both fertilizer and agrochemicals in oil palm production is likely to increase. This is because palm oil is enjoying a very favorable price in the world market. Serious weed problems due to a labor shortage and the increased area for direct-seeded rice will leave the agriculture sector with no choice but to increase herbicide consumption. Serious efforts, however, are being made

Table 13.3. Volume and Value of Malaysian Imports of Fertilizers (Tonnes)[a]

Fertilizer	1989 Volume	1990 Volume	1991 Volume
Organic			
Guano	1,501	1,810	2,862
Other natural animal/vegetable fertilizer	24,051	19,797	275
Total	**25,552**	**21,607**	**3,137**
Nitrogenous			
Ammonium nitrate	12,520	15,705	22,097
Ammonium sulfate	209,073	301,930	54,849
Ammonium chloride	94,735	75,457	95,203
Urea	334,795	326,185	96,058
Calcium ammonium sulfate	101	244	7
Sodium nitrate	208	77	167
Fertilizers containing N, P, K	96,018	83,220	74,274
Fertilizers containing N & P	541	68	19
Other nitrogen fertilizers	1,948	994	276
Total	**749,939**	**803,880**	**842,950**
Phosphatic			
Basic slag	23	2	1,000
Superphosphate	15,571	15,146	14,422
Other phosphatic fertilizer	185	18,716	11,819
Ammonium phosphate	33,996	37,491	21,851
Rock phosphate (ground)	237,361	300,825	159,469
Rock phosphate (unground)	138,799	180,276	147,158
Fertilizer containing P & K	20	11	150
Total	**425,955**	**552,467**	**455,869**
Potassic			
Potassium sulfate	3,349	2,982	7,713
Other potassic fertilizers (mostly MOP)	631,985	866,016	826,465
Total	**635,334**	**868,998**	**834,178**
Miscellaneous			
Other fertilizers not containing N & P	9,244		34,995
Other fertilizers in lozenge and similar preparation form	5,649		1,653
Other fertilizers in packings < 10 kg		23,981	24,789
Total		**26,752**	**61,437**

[a]Tay et al., 1994.

to reduce the dependence on agrochemicals, especially in food crop production. There is an intense public pressure on minimizing the pesticide residues in agriculture produce. Integrated Pest Management (IPM) with minimum use of agrochemicals is being employed, and it is gaining momentum especially in vegetable production. A fully comprehensive study on the impact of agricultural activities on toxic metal and agrochemical pollution in the environment is lacking. Future consumption of fertilizer and agrochemicals in agriculture is not expected to change until substantial evidence of environmental pollution from fertilizers and agrochemicals is available.

Table 13.4. Heavy Metals in the Phosphate Rock Fertilizer Used in Malaysia (mg/kg)[a,b]

	Fe	Cd	Cu	Zn	Mn	Pb
China Rock Phosphate (CRP)	2781.52 (±0.50)	9.43 (±0.50)	16.33 (±1.26)	187.68 (±0.19)	631.84 (±0.6)	135.61 (±8.32)
Morrocoan Rock Phosphate (MRP)	1107.39 (±3.01)	19.07 (±2.35)	18.26 (±2.31)	149.41 (±0.88)	28.08 (0	19.14 (±8.32)
Jordan Rock Phosphate (JRP)	1204.06 (± 1.31)	8.20 (± 0.58)	17.87 (0)	170.81 (± 0.47)	21.92 (0)	31.61 (± 13.58)
Christmas Island Rock Phosphate (CIRP)	2960.87 (± 4.99)	21.79 (± 0.50)	54.47 (± 1.94)	299.79 (± 0.47)	448.03 (± 23.58)	52.41 (± 8.32)
North Carolina Rock Phosphate (NCRP)	2276.09 (± 3.69)	31.42 (± 2.14)	18.26 (± 1.47)	344.74 (± 0.77)	32.08 (± 0.62)	19.14 (± 8.32)

[a]Sivadasan, 1995.
[b]Results are mean of 4 replicates ( ) - std. deviation.

Table 13.5. Malaysia: Estimates of Agrochemical Market (Rm Million)[a]

	1989	1990	1991	1992	1993
Herbicides	262.1	261.3	230	210	200
Insecticides	39.3	42.8	40	41	39
Fungicides	13.1	14.6	13	13	13
Rodenticides	9.8	10.5	10	12	0
Total	324.3	329.2	293	276	262

[a]Malaysia Agriculture Chemical Assoc., Annual Report and Directory, 1992/93.

ECOLOGICAL PROBLEMS OF TOXIC METALS AND AGROCHEMICALS

Soil is a very specific component of the biosphere. It is not only a geochemical sink for contaminants, but also acts as a natural buffer controlling the transport of chemical elements and substances to the atmosphere, hydrosphere, and biota. The type of soil can affect the environment in terms of heavy metals and agrochemical contamination. Erosion and dispersion of soil can lead to production of dispersed clay in runoff waters. This clay is very readily transported, as its settling velocity is negligible and its capacity to carry heavy metals and agrochemicals to surface water is apparent. Beyond the role of the sed-

Table 13.6. Agricultural Activities in Malaysia in Relation to Location and Soil Types

Sedentary Soils	Alluvial Soils	Peat Soils
1. *In the highland* Located at Cameron Highland; classified as Inceptisols and Ultisols; used for tea, vegetable, flower cultivations. 2. *In the lowland* Distributed throughout the country; occupy an area approximately 70% of Malaysia; classified as Ultisols and Oxisols; characterized by low pH, low P, Ca and Mg, but high Al; used for oil palm, rubber, cocoa, fruit, and annual crop cultivations.	1. *Acid sulfate soils* Found mainly along the coastal plains of the west coast of Peninsular Malaysia; occupy an area approximately 500,000 ha; classified as Entisols and Inceptisols; mainly under mangrove forest; reclaimed for rice, oil palm and coconut cultivations. 2. *Sandy beach soils* Found mainly along the coastal plains of the east coast of Peninsular Malaysia; occupy an area approximately 140,000 ha; classified as Entisols and Spodosols; used for tobacco and fruit cultivations. 3. *Riverine alluvial soils* Found along the major rivers in the country; classified as Entisols, Inceptisols, Alfisols and Ultisols; used for rice, rubber, oil palm, fruit and annual crop cultivations.	Mainly found in Sarawak, Johore, Selangor and Pahang; occupy an area approximately 2.4 million ha; mainly classified as Histosols; Sago and nipah palm grow in peat swamp; oil palm, vegetables; pineapple, fruit and annual crops are grown on the drained peats.

iment itself as a pollutant impairing water quantity, heavy metals and pesticides that sorb to colloids are delivered to rivers and reservoirs, which are primary sources of potable water in Malaysia.

Extent and Chemistry of Malaysian Soils

The common soil orders found in Malaysia include Entisols, Inceptisols, Alfisols, Ultisols, Oxisols, Spodosols, and Histosols. The majority of them fall within the acid, highly weathered soils (ultisols and oxisols), and histosols. Table 13.6 summarizes the agricultural use in relation to location and soil types in Malaysia. About 200,000 ha of tin tailings (previously tin mining land) are present in Malaysia. The soils, sandy and poorly structured, are mainly located in the vicinity of Kinta Valley, Perak, and Klang Valley, Selangor. These soils are often used for intensive vegetable and fruit cultivation.

The chemical properties of the soils are related to their texture, mineralogy, and the degree of weathering, which in turn depend on the types of parent materials. Studies on Zn adsorption of selected Malaysian soils by Denamany (1992) showed that the adsorption conformed to the Langmuir adsorption isotherm (Table 13.7). The Kangkong, Selangor, and Jawa series, which are alluvial soils, generally have higher adsorption capacity than the upland soils. The adsorption maxima are related to the cation exchange capacity (CEC), organic matter, and clay content.

The Ultisols and Oxisols in Malaysia have been studied in detail recently. The clay

Table 13.7. Adsorption Maxima and Langmuir Constant of Zn Adsorption Fitted to Langmuir Equation[a]

Soil Series	Texture	Organic Matter (%)	Adsorption Maxima (b) (mg kg^{-1})	Langmuir Constant (K) (μg kg^{-1})
Bungor	Sandy clay	1.17	53.82	0.267
Jerangau	Sandy clay	2.32	102.16	0.190
Padang Besar	Clay	1.81	109.61	0.131
Apek	Sandy clay	1.44	129.45	0.218
Chuping	Loam	0.96	215.11	0.164
Selangor	Clay	1.30	1598.13	0.182
Jawa	Clay	1.98	2423.54	0.016
Kangkong	Clay	2.65	2809.15	0.30

[a]Denamy, 1992.

Table 13.8. Mineralogical Composition of the B Horizon of Typical Malaysian Ultisols and Oxisols[a]

Soil Series	% Minerals in Clay Fraction			
	Kaolinite	Gibbsite	Goethite	Mica
Ultisols				
Serdang	77.8	1.6	3.9	2
Rengam		12.3	8.0	1
Oxisols				
Segamat	35.7	1.5	10.9	0
Kuantan	41.1	32.0	24.4	0
Prang	55.4	2.9	10.5	0
Sg. Mas	32.7	0.8	23.8	0

[a]Tessens and Shamshuddin, 1983.

fraction of the soils is dominated by variable-charge minerals such as kaolinite, gibbsite, and goethite (Table 13.8). In general, kaolinite is very high in the Ultisols, whereas in the Oxisols, besides kaolinite, gibbsite and goethite are also present in significant amounts (Tessens and Shamshuddin, 1983). Occasionally, mica is present in some of the Ultisols.

Charges in the Ultisols and Oxisols are very low. At the soil pH, the Ultisols are usually net negative-charged (Table 13.9). The values are altered by changing pH, resulting from agricultural activities. Our record shows that under normal circumstances the soil pH of Malaysian Ultisols and Oxisols is between 4 to 5. At low soil pH and due to the presence of sesquioxides, these soils have high capacity to fix As.

Under oxidizing and acid environment, and below pH 5, Cd^{2+}, Co^{2+}, Ni^{2+}, Zn^{2+}, Cu^{2+}, Cr^{3+}, and Hg^{2+} are moderately mobile in soils, and this mobility increases with decreasing pH (Kabata-Pendias and Pendias, 1992). These positively charged cations can be adsorbed to the negatively charged sites on the soil minerals listed in Tables 13.8 and 13.10. On the other hand, anions from fertilizers and pesticides can be fixed onto the positive-charged sites of the same minerals. Because of the presence of low negatively charged surface, the heavy metals (especially Cd) will be mobile and can be easily leached (Sposito and Page, 1984).

Acid sulfate soils develop from oxidation of pyrite (FeS_2) when exposed to the atmosphere as a consequence of drainage. When the soils are allowed to undergo free oxidation under aerobic conditions, Al, Mn, Fe, Ca, Mg, K, Na, and S are released into the

Table 13.9. Negative, Positive, and Net Charges in the B Horizon of Typical Malaysian Ultisols and Oxisols[a]

	Charge in the Soil (cmolc/kg)		
Soil Series	**Negative**	**Positive**	**Net**
Ultisols			
Serdang	1.33	0.82	–0.51
Rengam	2.08	1.26	–0.49
Oxisols			
Segamat	2.71	3.53	+0.82
Kuantan	2.68	4.10	+1.42
Prang	1.77	3.40	+1.62
Sg. Mas	1.44	3.65	+2.24

[a]Tessen and Shamshuddin, 1983.

soil solutions from the breakdown of clay minerals (Shamshuddin and Auxtero, 1991). Acid sulfate soils are characterized by low pH (< 3.5) and high extractable Al, which affect crop growth. Jarosite, natrojarosite, and alunite are formed as a result of the pyrite oxidation.

Nickel, Co, Mn, Cu, Pb, and Zn partially replace Fe in the pyrite structure (Deer et al., 1966). Likewise, pyrite-S can be partially replaced by As to form arsenopyrite (Read, 1962). Our recent study (unpublished) indicated that jarosite, natrojarosite, and alunite found in Malaysian acid sulfate soils contained variable amounts of the above metals, which may have originated from pyrite. Furthermore, As and Cd are found in soil solutions and natural waters from drainage canals near acid sulfate soils (Hasniah, 1994).

It is, therefore, imperative to be cautious in developing acid sulfate soils for agricultural or industrial purposes. Care should be taken not to allow excessive oxidation of pyrite in the soils. If it does, the soils can release toxic amounts of Al, and also other heavy metals to the soil environment; consequently plants and aquatic life in the surrounding areas are adversely affected.

The sandy soils are widely used for tobacco, vegetable, and fruit cultivation. The sand content of these soils is generally more than 90%; thus the CECs are extremely low. The condition favors high rates of leaching and mobility of the heavy metals. Earlier reports have indicated the presence of pesticides and NO_3^- in the groundwater below these soils (Sharma et al., 1993). Under these conditions, heavy metal contamination of groundwater can also be expected to occur.

Although no specific study in Malaysia has been conducted to evaluate the effect of soil properties on pesticides and toxic metals contamination, based on their properties, the Oxisols, Ultisols and especially the sandy soils are more susceptible to contamination. This condition can result in contamination of the groundwater as well as accumulation in plant tissue.

Levels of Heavy Metals and Agrochemicals

Heavy Metals

Soil composition varies widely, and it reflects the nature of the parent material. The principal factors determining these variations are the selective incorporation of particular elements in specific minerals during igneous rock crystallization, the relative rates of weathering, and the modes of formation of sedimentary rocks (Mitchell, 1964). In the absence

Table 13.10. Total Micronutrient Content of Soils Developed over Igneous and High Grade Metamorphic Rocks[a]

Soil Series	Classification	Parent Material	Heavy Metal Concentration					
			Mn (ppm)	Cu (ppm)	Zn (ppm)	Co (ppm)	Ni (ppm)	Fe_2O_3 (%)
Tampin	Clayey, kaolinitic, isohyperthermic, Orthoxic, Tropodult	Granite	40–58	21–36	60–130	30–106	1.4–4.5	1.5–1.6
Kulai	Clayey, kaolinitic, isohyperthermic, Orthoxic, Tropodult	Rhyolite	177–210	24–56	60–97	30–75	6.2–18.0	3.2–6.4
Segamat	Clayey, Oxidic, isohyperthermic, Tropeptic Hapludox	Andesite	420–935	102–146	168–243	65–125	3.8–10.8	17.7–18.6
Kuantan	Clayey, Oxidic, isohyperthermic, Haplic Acrudox	Basalt	290–430	172–208	157–168	50–110	39.5–64.2	19.3–20.6
Prang	Clayey, Oxidic, isohyperthermic, Haplic Acrudox	Schist	600–700	80–96	211–228	25–75	21.5–34.0	20.5–21.0

[a]Lee et al., 1992.

of pollution, total metal contents are related to soil parent material, organic matter content, soil texture, and soil depth (Urea and Berrow, 1982). The background levels of Fe, Cu, Zn, and Mn in Malaysian soils have been reported by Lee et al. (1992). In most soils the order of abundance is Fe>Mn>Zn>Cu. The concentration of Zn and Cu are generally low. The levels of these metals are related to the soil parent materials. The background levels of these metals in some selected soils are presented in Table 13.10. Analyses of the background levels of the toxic metals such as Pb, Cd, Cr, and As are not available.

There are few reports on the extent of metal contamination on agricultural lands in Malaysia. An example of such a report is on Cd levels in soil and plant samples due to long-term P fertilization (IFDC, 1992). At a site where a total of 4165 kg P_2O_5/ha, as Christmas Island phosphate rock, was applied during 1959 to 1976, the total Cd in soil samples taken in 1992 was 1.26 mg/kg. In comparison, the amount of total Cd in unfertilized plots was 1.13 mg/kg soil. Extractable Cd in both groups was only 20 to 70 µg/kg soil. Cadmium content of rubber leaves from fertilized areas was 108 µg/kg and from the unfertilized plots was 37 µg/kg. Samples of the underbrush from both areas also indicated a content of 37 µg/kg.

In another long-term application of P fertilizers, North Carolina rock phosphate (NCRP) has been applied to corn grown on Tebok soil series (an Ultisol), at a rate of 0, 25, 100, and 200 kg P/ha during each crop cycle for a period of three years (Shaharani, 1995). During these three years, corn was grown in six cycles. Results showed that at the different rates of NCRP applied throughout the six corn cycles, the amount of Cd accumulated in the soil profile was not significantly higher than the control. The concentration of Cd was less than 1.5 mg/kg for all the treatments. The distribution of Cd was rather uniform throughout the 30-cm soil depth. Also, from the results, the amount of Cd taken up during the latter cycle (6th cycle) was the highest, compared to the earlier corn cycles, which indicates the influence of residual NCRP on Cd uptake.

In Malaysia, many of the data collected on levels of heavy metals are from sediments. The reconnaissance survey of coastal stream sediment has been of great value for highlighting geochemical anomalies important to agriculture and for showing regional patterns of soil pollution. A study on the distribution of total mercury in the water and sediment of the Klang estuary has also been conducted (Law and Singh, 1987). The mean level in water was 1.69 µg/L, with a range of 0.10 to 6.50 µg/L, while in the sediment, the mean was 0.20 mg/kg of wet sediment with a range of 0.03 to 0.40 mg/kg of wet sediment.

Under urban condition the levels of heavy metals in soils have been shown to be elevated (Ramlan and Badri, 1989). In this study, the level of Pb was reported to be 2466 µg/g with 98% due to anthropogenic activities. The levels of Zn, Cu, and Cd were generally low, but the major portions were made up of the nonresistant fraction, which is usually thought of as having originated from anthropogenic activities.

Pesticides

Research has shown that contamination of soil, water, and fish in a rice ecosystem may arise as a result of using pesticide for pest control (Cheah, 1988). Some of these pesticides, e.g., lindane and endosulfan, are persistent in the environment. Alpha-HUT and gamma-HCH have been detected in the soil and water samples in a rice agroenvironment.

The presence of 2, 4-D and paraquat in groundwater has been reported (Cheah et al., 1996). Paraquat and 2, 4-D are two of the most commonly used herbicides in Malaysia. Both have broad spectrum activity and are nonselective. Paraquat is a contact herbicide while 2, 4-D is a systematic herbicide. Field studies on the soil repeatedly treated with paraquat in oil palm growing areas have shown the presence of paraquat residues. Levels of paraquat were found to range from undetectable levels (< 0.05 mg/kg) to 8.40 mg/kg.

The analyses of pesticide residues in vegetables from various parts of Peninsular Malaysia were started in 1984 to get a general impression of the pesticide residue in vegetables. Initially 196 vegetable samples were analyzed for residues of the four main groups of pesticides, that is, organochlorines, organophosphorus esters, synthetic pyrethroids, and dithiocarbamates. From this work it was concluded that residue problems were most likely to occur for insecticides in the organophosphorus ester groups and fungicides of the dithiocarbamate group. Since then, the monitoring of pesticide residues in vegetables has been carried out regularly, especially on vegetables grown in the highlands, and the emphasis has been on these two groups of pesticide.

As of September 1989, 3799 analyses have been carried out, including analyses on samples from Sabah and Sarawak (East Malaysia). Out of the 696 samples analyzed for residues of insecticides in the organophosphorus ester group, 154 were found to have residues more than the allowed limits. Out of the 2551 samples analyzed for dithiocarbamate residues, 546 were found to exceed the permissible limits. In organochlorine and synthetic groups, out of 552 analyses, only nine results exceeded the maximum residue limits (MRL).

Impact of Toxic Metals and Agrochemicals on Soil and the Environment

Surface and Groundwater

Until recently, not much attention has been directed toward environmental problems, particularly contamination of surface and groundwater as a result of the continuous use of pesticides. A preliminary study conducted on water samples from the main reservoir in Cameron Highlands revealed the presence of diazinon residues. The level ranged from 0.00091 to 0.0037 mg/L (Cheah and Lum, 1993). Low levels of profenophos residues (0.0001 mg/L) were detected in one of the water samples.

A recent broadscale survey in Kelantan (a state in the northeast part of Peninsular Malaysia) investigating pesticide residues in surface and groundwater showed the presence of pesticide residues as contaminants (Sharma et al., 1993). The organochlorines detected were lindane, aldrin, dieldrin, and DDT. Aldrin was detected in 42 samples. Out of that, eight samples exceeded the maximum level of 0.03 µg/L for the Malaysian standard. Also, residues of insecticides such as DDT, heptachlor, and lindane have been reported in Malaysian rivers (Tan, 1992).

The result of the survey in Kelantan also showed the presence of paraquat residues in groundwater. Residue levels detected were low and should not be of any toxicological concern. However, the presence of paraquat residues was rather unexpected, considering its properties of strong adsorption to soil colloidal particles. The presence of these herbicide residues in groundwater could have arisen because of preferential flow, which refers to the rapid transport of water and solute through a small portion of the soil volume which is receiving input over its entire inlet boundary (Guth and Mani, 1991). The heavy metal contents of groundwater collected from rubber and oil palm plantations were generally low (Khairiah Jusoh, 1993).

Human and Animal Health Effects

Pesticide residues in vegetable produce have become a serious issue. Insecticide residues were first reported in local newspapers after more than 100 people were poisoned in Singapore after eating methamidophos-contaminated *Brassica alboglabra* (*The New Straits Times*, 1988). Studies carried out have shown that organophosphorus insecticide residues have been detected in the vegetables. Seven out of ten *Brassica alboglabra* samples taken on a single day in December 1988 were found to contain methamidophos residues exceeding the maximum residue limits (MRL) of 1 mg/kg. A few samples contained more than 30 mg/kg. Extremely high methamidophos residues, about 100 mg/kg, were also detected in some samples. Recently, consumption of methamidophos-treated sweet corn also resulted in a few human poisoning cases in Johore (southern part of Peninsular Malaysia). The problem of excessive pesticide residues in harvested vegetable farm produce is now of common public concern.

In the 1960s, sodium arsenite was used to kill off weeds and old rubber trees in plantations. In 1974, 770 cases of death of livestock due to poisoning (mostly due to sodium arsenite) were reported (Reddy, 1977); it was eventually banned in late 1976 (Umakanthan, 1983).

In a study on pesticide element's in blood serum (Wong, 1980), the level of DDT (an organochlorine compound) detected in Malaysian paddy farmers, rubber estate workers, and the public was much higher than those in similar categories in the United States. From the study, it was found that the total mean DDT concentration in Malaysian paddy farmers was 0.11 ppm, the level for rubber estate workers was 0.09 ppm, and the level for the public was 0.066 ppm.

Pesticides washed into rivers are rapidly adsorbed by sediment, and eventually absorbed by plankton, algae, aquatic invertebrates, aquatic vegetation, and fish. Some organisms pick up the pesticides directly from the water. Fish can accumulate pesticides in their bodies. Sometimes these concentrations prove fatal to the fish (Ooi and Lo, 1990). Fish is a major source of protein among Malaysian farmers. The concentration of pesticide residues in Malaysian paddy field fish has been determined, and based on their findings on dieldrin, chlordane, BHC and aldrin residues, researchers conclude that farmers are at risk if they consume a relatively large daily ration of the pesticide-contaminated fish (Mcier et al., 1983).

Laboratory and field studies were also conducted by Ooi and Lo (1990) to determine the acute toxicity of 11 herbicides common to paddy field fish in Malaysia. The studies indicated that herbicide bensulfuron, 2, 4-D, metsulfuron, and quinclorac have low toxicity (LC50 > 10 ppm) to sepat siam (*Trichogaster pectoralis*) and catfish (*Claria batrachu*), while butachlor, molinate, oxadiazon, pretilachlor, propanil, and thiobencarb are moderately toxic (LC50 0.5 to 10 ppm), but fenoxaprop is highly toxic (LC50 < 0.5 ppm). However, field studies showed bensulfuron, 2, 4-D, fenoxaprop, oxadiazon, and propanil at double the effective rates for weed control were safe to the fish.

Effects on Soil Biota

Beneficial soil organisms such as earthworms may also be killed by pesticides. Pollution of soils by heavy metals and organochlorine compounds may inhibit microbial enzyme activity and decrease the diversity of populations of soil flora and fauna. A study was conducted whereby the effects of the herbicides paraquat and alachlor on microbial populations in peat were investigated (Ismail and Mohd. Azib, 1991). Paraquat and alachor at 250 mg/kg caused a reduction of about 75% and 78% in bacterial populations, respectively. Both herbicides also reduced soil fungi population, again with alachor being relatively more effective compared to paraquat.

MITIGATION PLANS

Regulation and Legislation

Malaysia has comprehensive legislation for pollution control and environmental preservation. The Department of Environment (DOE) is the agency responsible for formulating and enforcing regulations related to the environment. The Environment Quality Act 1974 contains 13 sets of regulations addressing a wide range of problems, including palm oil waste, raw natural rubber, sewage and industrial effluents, schedule of wastes, and provision for environmental impact assessment. The approach to environmental protection places more emphasis on preventive than curative measures. Since 1987, proposals for all major development projects are required to include Environmental Impact Assessment before they can be considered for approval.

The DOE maintains an up-to-date inventory of toxic and hazardous wastes. The disposal and treatment of these wastes are closely monitored. An integrated waste treatment and disposal facility is under construction and will be in operation in the near future.

Comprehensive monitoring of air and water quality (rivers and marine) has been carried out by DOE for several years, but not soil quality. Currently, there is no comprehensive nationwide program to check, monitor, and conduct research on the extent of soil contamination, whether due to heavy metals, pesticides, or other persistent organics. There are, however, research activities carried out by individuals that are rather isolated and not coordinated. A comprehensive soil quality monitoring system needs to be established, with standard sampling methods and analytical procedures. Also, information on soil loading capacity and inventory of contaminants is required.

Research Needs

A comprehensive study on the contamination profile of pesticides and heavy metals in the environment, especially soils and water sources, is necessary. This information, which is currently sparse and limited, is important for the formulation of environmental management strategy. As a first step, the accumulation of baseline information or the setting up of an organized database on the extent and nature of environmental contamination should be developed. Also, development of reliable, scientifically sound indicators of ecological change is necessary. These indicators of environmental change should be monitored intensively, often, and over long periods of time.

The toxic effect of a metal is determined more by its form than by its concentration. Currently in Malaysia, no emphasis has been placed on heavy metal speciation studies.

Until recently, most research on pesticide residues has been focused primarily on agricultural produce, especially edible crops, and commodities for export. Data of pesticide degradation trial are required in order to establish preharvest intervals and maximum permitted residue level for pesticides in food. Currently, Malaysia is adopting limits set by the Codex Alimentarius Commission, the World Trade Organization's intergovernmental panel that develops global standards to protect health and the environment, whereby data are mostly generated from trials carried out under temperate climatic conditions. Data from tropical countries are required to ensure the preharvest interval stipulated and that the MRL established can be used under tropical condition. Also, recognition of the need for further research in the environmental fates and effect of pesticides has gained momentum only in recent years, especially after a number of alarming global incidents adversely affecting the environment.

Consideration for remediation strategies should be similarly and simultaneously developed according to the specific nature of the contamination and the ecosystem components involved. Equal emphasis should be given to preventive and remedial measures. Bioremediation should be given priority as a safe and cost-effective approach, but systems that work in tandem with physical and other methods should not be ignored. Environment-friendly pesticides and fertilizers with appropriate application technology and formulation should be considered. Research on natural products and alternative control methods should also be given emphasis, such as biological control, resistant crops, cultural practices, sterilization, physical control methods, and integrated pest control.

SUMMARY AND CONCLUSIONS

Agriculture has been the mainstay of Malaysia's economy for a long time. Inputs of fertilizers and agrochemicals are expected to continue to play an important role in the agricultural sector. In Malaysia, it is currently believed that the magnitude of soil and environmental degradation because of agricultural activities (nonpoint source pollution) is relatively mild as compared to industrial activities (point source pollution). Data collected on levels of heavy metals in stream or coastal sediments can normally be traced back to effluent discharged from industrial sectors. Another reason for this belief might be due to lack of research or inadequate documentation on the extent of soil contamination because of agricultural activities. A more comprehensive baseline data of heavy metals, other than micronutrients in soils, is needed as a measure of the extent of contamination. Also for heavy metals, more research is needed in identifying sources of the contamination for understanding the fate and transport of these metals in the environment, especially the soil system. As for pesticides, research is needed in identifying the nature and mechanisms of the transformation processes to establish the disappearance rate and the factors affecting it, to identify the degradation products, and also to predict persistence. Finally, for both these contaminants, ways of controlling them in the environment and remediation of sites polluted by them need to be investigated.

REFERENCES

Cheah, U.B. Status of Pesticide Residue Problems in Agricultural Produce. *Malaysia Agric. Chem. Assoc. Newsletter*, 1988.

Cheah, U.B. and K.Y. Lum. Pesticide Contamination in the Environment, in *Waste Management in Malaysia: Current Status and Prospects for Bioremediation*. B.G. Yeoh, K.S. Chee, S.M. Phang, Z. Isa, A. Idris, and M. Mohamed, Eds. Ministry of Science, Technology and the Environment, Malaysia, 1993, pp. 169–176.

Cheah, U.B., M.L. Sharma, B.Y. Aminuddin, C.H. Mahmuddan, and M. Mohd Zain. Pesticide Residues in Water Resources of the Agricultural Plain of Kelantan, in *Agricultural Impact on Groundwater Quality*. B.Y. Aminuddin, M.L. Sharma and I.R. Willet, Eds., ACIAR Proceedings No. 61, Australia, 1996, pp. 22–28.

Deer, W.A., R.A. Howie, and J. Zussman. *An Introduction to the Rock Forming Minerals.* Longmans, London, 1966.

Denamany, G. Adsorption and Exchange of Zn in Some Cocoa Growing Soils, in *Secondary and Micro Nutrients in Malaysian Agriculture*, H.A.H. Sharifudd, P.Vimala, and H. Aminuddin, Eds. MSSS, Kuala Lumpur, Malaysia, 1992, pp. 105–103.

Guth, J.A. and J.T. Mani. Experimental Methods for Measuring Movement of Pesticides in Soil. Paper presented at the British Crop Protection Council Symposium 25–27th March, 1991, University of Warwick, Coventry, U.K., 1991.

Hasniah, M. Release of acidity and metals during acid sulfate weathering. B.Agric. Sc. Thesis. Faculty of Agriculture, Universiti Pertanian Malaysia, Serdang, Malaysia, p. 43, 1994.

International Fertilizer Development Center (IFDC). Annual Report. 1992. Muscle Shoals, Alabama, U.S., 1992.

Ismail, S. and S. Mohd. Azib. Effects of Weedicides on Soil Microbial Activities, in Zakri, K.S. Chee, S.M. Phang, Z. Isa, A. Idris and M. Mohamed, Eds. *Research Priorities for the Advance of Science and Strategies in Agriculture, Industries and Medicine.* First Universiti Kebangsaan Malaysia Symposium on Intensification of Research of Priority Areas. 1991, pp. 362–367.

Jones, L.H.P. and S.C. Jarvis. The Fate of Heavy Metals, in *The Chemistry of Soil Processes*. D.J. Greenland and M.H.B. Hayes, Eds., John Wiley & Sons, New York, 1981, pp. 593–620.

Kabata-Pendias, A. and H. Pendias. *Trace Elements in Soils and Plants*. 2nd ed. CRC Press, Boca Raton, FL 1992.

Khairiah, J. Recycling of agricultural waste water: Input and availability of heavy metals. Ph.D. Dissertation, Universiti Kebangsaan Malaysia, Bangi Malaysia. p. 391, 1993.

Law, A.T. and A. Singh. Distribution of Mercury in the Kelang Estuary, *Pertanika* 10(2), pp. 175–181, 1987.

Lee, S.C., J.S. Lim, and O. Wahid. Micronutrients Status in Major Soils in Peninsular Malaysia, in *Secondary and Micronutrients in Malaysian Agriculture*. H.A.H. Sharifuddin, P. Vimala and H. Aminuddin, Eds. MSSS, Kuala Lumpur, Malaysia. 1992, pp. 131–148.

Malaysian Agricultural Chemical Association (MACA). MACA reports and directories 1992/93. MACA. Kuala Lumpur, Malaysia, 1993.

Mcier, P.G., D.C. Fook, and K.F. Lagler. Organochlorine Pesticide Residues in Rice Paddies in Malaysia, *Bull. Environ. Contam. Toxicol*. 30, pp. 351–357, 1981.

Ministry of Primary Industry. Statistics on commodities. Kuala Lumpur, Malaysia, 1993.

Mitchell, R.L. Trace Elements in Soil, in *Chemistry of the Soil*, F.E. Bear, Ed. van Nostrand Reinhold, New York, 1964, pp. 320–368.

Mohamad, R. and S.M. Ali. The Status of Herbicide in Malaysia, in *Symposium in Weed Science*. J.V. Pancho, S.S. Sastroutomo, and S. Tjitro Semito, Eds. Semeo-Biotrop, Bogor, Indonesia, 1986, pp. 69–76.

Ooi, G.G. and N.P. Lo. Toxicity of Herbicides to Malaysian Rice Field Fish. Paper presented at 3rd Inter. Conf. on Plant Protection in the Tropics, 20–23 March, Genting Highlands, Malaysia, 1990.

Ramlan, M.N. and M.A. Badri. Heavy Metals in Tropical City Street Dust and Roadside Soils: A Case of Kuala Lumpur, Malaysia. *Environ. Technol. Lett*., 10, pp. 435–444, 1989.

Read, H.H. *Rutley's Elements of Mineralogy*. 25th ed. Thomas Murphy and Co., London, 1962.

Reddy, D.B. Future Trends of Pesticides Use in the World with Particular Reference to Asia and Pacific, in *Pesticide Management in South East Asia*, Bogor, Indonesia, Biotrop Special Pub. No.7, 1977.

Shaharani, M.S. Contamination of agricultural lands from long-term applications of P fertilizers. B. Agric. Sci. Thesis. Faculty of Agriculture, Universiti Pertanian Malaysia, Serdang, Malaysia, 1995.

Shamshuddin, J. and E.A. Auxtero. Laboratory Studies on the Release of Cations and Anions in Acid Sulfate Soils from Pulau Lumut, Selangor. *Pertanika*, 14 (3), pp. 257–264, 1991.

Sharma, M.L., B.Y. Aminuddin, A.R. Ahmad, C.H. Mohammad, U.B.Cheah, M. Zulkifli, M.M. Zain, and K.Y. Lum. Impact of Agricultural Practices on Groundwater Quality in Kelantan Plain, Malaysia : Survey Results. CSIRO Report No. 93/3, 1993

Sivadasan, R. Phosphate rock: Morphology, mineralogy and solubility of trace elements. B. Agric. Sc. Thesis. Faculty of Agriculture, Universiti Pertanian Malaysia, Serdang, Malaysia. p. 50, 1995.

Sposito, G. and A.L. Page. Cycling of Metal Ions in the Soil Environment, in *Metal Ions in Biological Systems*, H. Sigel, Ed. Marcel Dekker, New York, 1984, pp. 287–332.

Syers, J.K., A.D. MacKay, M.W. Brown, and L.D. Curie. Chemical and Physical Characteristics of Phosphate Rock Materials of Varying Reactivity. *J. Food Agric*. 37, pp. 1057–1064, 1986.

Tan, G.H. Environmental Monitoring of Organochloride Pesticide in Selected Malaysian Rivers: An Assessment. *International Conf. on Pesticide in Perspectives*, Kuala Lumpur, Malaysia, 1992.

Tay, T.H., B.L. Kho, and K.C. Chang. Malaysia Agricultural Directory and Index 1993/94. Agriquest, Petaling Jaya, Malaysia, 1994.

Tessen, E. and J. Shamshuddin. *Quantitative Relationship Between Mineralogy and Properties of Tropical Soils*. UPM Press, Selangor, Malaysia, 1983.

The New Straits Times. Singapore Destroys Johore Kailan. 9 Dec., 1988.

Umakanthan, G. Pesticides in Our Environment—Boon or Bane? Paper presented at the

Malaysian Plant Protection Society (MAPPS), Panel discussion on pesticides, Kuala Lumpur, March, 1983.

Ure, A.M. and M.L. Berrow, The Chemical Constituents in Soils, in *Environmental Chemistry* H.J.M. Bowen, Ed. Royal Chemical Society, London, 1982.

Vaes, A.G. Fertilizer Scenario in ASEAN Region: Present and Future Situation, in *Fertilizer Usage in the Tropics*. B. Aziz, Chief Ed. Malaysian Society and Soil Science. Kuala Lumpur Malaysia, 1992, pp. 16–26.

Wong, K.K. Impact of Pesticide Usage—A Case Study of Organochlorine Compound Levels in the Blood Serum of Selected Malaysian Population Groups, Consumer Assoc. Penang (CAP) Seminar on Economics, Development and the Consumer (unpublished), 1980.

CHAPTER 14

STATUS OF CADMIUM, LEAD, AND SELENIUM IN THE SOILS OF SELECTED AFRICAN COUNTRIES AND PERSPECTIVES OF THEIR EFFECTS ON HUMAN AND ENVIRONMENTAL HEALTH

B. Waiyaki

INTRODUCTION

A heavy metal is one having a density at least five times greater than that of water. Although metals have many physical properties in common, their chemical reactivity is quite different and their toxic effect on biological systems is even more diverse.

For life to be sustained, many of the heavy metals are essential, even those that are found in extremely low concentrations in nature. For any metal to be essential to sustain life, its concentration level in the human body must be within a given optimum range, beyond which a metal becomes toxic. The toxicity of a metal depends on its route of administration and the chemical compound with which it is bound. Thus, when a given metal is combined with an organic compound, this may either increase or decrease its toxic effects on human cells. For example, the combination of a metal with sulfur to form a sulfide results in a less toxic compound than the corresponding hydroxide or oxide. This is because the sulfide is less soluble in body fluids than the corresponding hydroxide or oxide.

Three main kinds of sources of heavy metals in the environment are known to exist. The most obvious is the process of extraction and purification—mining, smelting, and refining. The second source is fossil fuels such as coal or oil, which release metals when they are burned. Cadmium, lead, mercury, nickel, and copper are all present in these fuels, and considerable amounts enter the air, or are deposited in ash, from them. A third source is industry as a result of the production and use of metal-containing products. The chemical and plastic industries are among the major development activities which act as sources of heavy metals in the environment.

One source of mercury is water-based latex paints, to which mercury compounds are added to inhibit the growth of bacteria and molds in the interior of houses. These paints are known to release inorganic mercury into environment; another source is skin-lightening soaps and creams that penetrate the human skin or enter the human body through inhalation, which may result in high mercury concentration levels, particularly in newborn babies. Other recorded effects of mercury on humans include acute toxicity in workers, which may result in chest pains, dyspnea, cough, and other ailments. Effects on the central nervous system following occupational exposure have been known to lead to psychotic reactions such as delirium, hallucinations, excitability, insomnia, and excessive shyness.

Manganese is an essential element for man and animals and thus occurs in body cells. Over 90% of its global production is by the steel industry (the rest is used in the production of batteries, chemical fertilizers, dyes, paint dryers, etc.). Retarded growth of hair and nails, mild dermatitis, and pigment changes on the hair and beard (as well as some weight loss) are some of the symptoms that are caused by manganese deficiency in humans. On the other hand, exposure to manganese dusts is a health hazard in the mining industry and may result in chronic poisoning. Chronic poisoning appears in stages. Examples of symptom progression are initially insomnia, hypersexuality, and headache; speech disturbance, eventual muteness, and an increased tone of facial muscles in the second stage; onto the establishment of chronic poisoning whose symptoms include marked rigidity of the body, muscle pain, disturbances of the libido, and other symptoms.

Nickel, another heavy metal, can find its way into the human body indirectly, e.g., through food which has been handled, processed, or cooked by utensils containing large quantities of nickel. The element can also enter the human body through smoking, but there is no agreement on the chemical nature of the element in tobacco smoke and its health significance. Nickel carbonyl is the most acutely toxic nickel compound. In its immediate toxic effect, this compound causes frontal headache, nausea, vomiting, insomnia, and irritability. Delayed pulmonary symptoms that follow the immediate effect include chest pains, dry coughing, dyspnea, visual disturbances, sweating, and weakness. Other pathways through which nickel enters the human body are consumption of plants and animals that contain compounds of the metal. Other heavy metals—cadmium, lead, tin, vanadium, molybdenum, cobalt and chromium—also pose occupational hazards to man.

Today, increasing emphasis is being placed on the carcinogenic, mutagenic, and teratogenic effects of metals and numerous experiments have been conducted on laboratory test animals to date. For example, chromium, nickel, lead, and cadmium are all proven or suspected causes of certain cancers associated with industrial processes. Large doses of cadmium and nickel are teratogenic in animals, but this effect is not well established in man.

It must be mentioned that the foregoing have only been examples of the sources and effects of specific heavy metals in man and are by far not conclusive in themselves, nor does it dwell on areas of concern such as fish, plants, and microorganisms. Cadmium, lead, and selenium are, however, discussed in this chapter in some detail, but once again, only a few important issues of relevance to human and environment health can be raised.

SOIL CHARACTERISTICS IN FIVE SELECTED AFRICAN COUNTRIES (ETHIOPIA, GHANA, MALAWI, SIERRA LEONE, AND TANZANIA)

The range of variation in Ethiopian soils is very wide, although heavy-textured soils predominate. Most of the soils show strong to moderate acidity. High organic carbon (OC) content and cation exchange capacity (CEC) are characteristic, but the electrical conductivity and $CaCO_3$-equivalent values are usually low.

On an international basis, the N, and, especially, K contents of Ethiopian soils are on the high side, but P contents are on the low side, while Ca and Mg correspond to the average international level. In all cases the variations are substantial. Boron (B) and Cu levels are relatively low in general in Ethiopian soils, while wide variations characterize Fe, Mn, Mo, and Zn levels (Sillanpaa, 1982).

Ghana soils are mostly coarse textured and the average texture index (I, 1–30) is one of the lowest among the selected countries. Other typical characteristics of the soils are their strong to moderate acidity, medium OC content, low CEC, low electrical conductivity, and low $CaCO_3$-equivalent.

With the exception of medium total N contents, the soils of Ghana are generally poor in macro nutrients. The status of six micronutrients (B, Cu, Fe, Mu, Mo, and Zn) varies from low (Cu and Mo) through low-medium (Zn) and medium (B, Fe) to high (Mn).

The soils of Malawi are mostly coarse to medium-textured and strongly to moderately acidic, with medium OC content and electrical conductivity and low CEC and $CaCO_3$-equivalent.

Relatively high P, low to medium N and K, and low Ca and Mg contents are typical of most Malawi soils. Fe and Zn usually show normal values. The contents of B and Cu are low, those of Mo lower than in most other countries in Africa, while the Mn contents are among the highest (Sillanpaa, 1982).

The majority of soils in Sierra Leone are coarse textured and, with a few exceptions, have a high acidity content. In spite of a generally high OC content, the CEC of soils remains relatively low due to the coarse texture. Low electrical conductivity and $CaCO_3$-equivalent are typical of Sierra Leone soils.

The Tanzanian soils vary widely in texture, but coarse and medium textures predominate. Wide variations are characteristic also for pH, OC content and CEC, but soils with moderate acidity, medium OC content, and low to medium CEC are most common. The electrical conductivity and $CaCO_3$-equivalent values are usually low. Wide variations characterize also the contents of all macronutrients but on the average, the N, Mg, and Ca contents of soils are slightly on the low side while P and K contents are on the high side compared to the other selected countries. Fe and Mn contents are usually normal, while shortages of Cu, Zn, and Mo are to be expected in certain areas of Tanzania (Sillanpaa, 1982).

STATUS OF CADMIUM, LEAD, AND SELENIUM IN THE SOILS OF SELECTED AFRICAN COUNTRIES

The information and data presented below on the status of cadmium, lead, and selenium in soils (and plants) of selected African countries are based on the results of a longer term research study in the 1980s and early 1990s on the *Status of Cd, Co, Pb and Se in the Soils of Thirty Countries*, a collaborative effort between the FAO (Food and Agriculture Organization of the United Nations) and the Government of Finland through FINNIDA, and carried out by the Agricultural Research Centre of Finland, culminating in the publication of the obtained results in the FAO Soils Bulletin No. 65 in 1992. This publication was edited by M. Sillanpaa and H. Jansson, both of the FAO.

In carrying out the study, analytical data for these elements were based on two separate analyses; extraction of the elements by ammonium acetate acetic acid—ethylene diamine tetra acetate (AAAc-EDTA) from soils and analyses of two indicator crops, wheat and maize. In order to be able to compare the results of wheat analyses to those of maize, the analytical data on plants were pooled. For Se, the AAAC-EDTA extractable soil Se values were corrected for soil OC content.

The analytical data were presented in graphical form, the areas of the graphs being limited so that both the minimum and the maximum soil and respective plant values just remain inside the graph. Further, the graphs were divided into three plant/soil content zones (I–III) (Table 14.1), of which zone I represents the lowest international plant/soil decile, zone II "normal values," and zone III the highest decile. The international plant/soil regression line is also given as background information in each group. The national regression lines are presented, provided that they are statistically significant.

To facilitate the comparison of the relative abundance of the microelements in different countries, the frequency distributions of the analytical data in the three plant/soil content zones are summarized. Below, only study results of Cd, Pb, and Se are presented.

Table 14.1. Relative Frequencies (Percentages) of Four Elements in Three Plant/Soil Content Zones (I–III) in Various Countries[a,b]

	Cd			Pb			Co			Se		
Country	**I**	**II**	**III**	**I**	**II**	**III**	**I**	**II**	**III**	**I**	**II**	**III**
Belgium	–	12	88	–	5	95	27	73	–	–	100	–
Finland	–	93	7	20	80	–	53	47	–	88	12	–
Hungary	–	57	43	–	80	20	4	87	9	–	100	–
Italy	–	67	33	–	53	47	10	83	7	10	89	1
Malta	–	92	8	–	8	92	88	12	–	4	96	–
New Zealand	–	47	53	45	50	5	55	40	5	42	58	–
Argentina	–	97	3	12	72	16	–	100	–	9	91	–
Brazil	33	65	2	10	90	–	14	55	31	13	84	3
Mexico	–	78	22	2	81	17	7	84	9	12	81	7
Peru	1	86	13	1	92	7	12	87	1	41	49	10
India	14	86	–	11	89	–	9	90	1	–	64	36
Korea, Rep.	1	92	7	–	91	9	19	81	–	4	96	–
Nepal	16	84	–	8	92	–	31	69	–	12	76	12
Pakistan	2	94	4	–	95	5	0	98	2	–	48	52
Philippines	9	87	4	37	60	3	5	70	25	1	91	8
Sri Lanka	5	95	–	14	86	–	19	67	14	–	100	–
Thailand	19	75	6	27	73	–	5	80	15	3	93	4
Egypt	12	88	–	1	87	12	0	67	33	23	76	1
Iraq	–	96	4	9	90	1	3	96	1	–	63	37
Lebanon	–	75	25	6	94	–	6	50	44	12	88	–
Syria	3	89	8	16	84	–	8	84	8	11	78	11
Turkey	3	94	3	2	97	1	3	87	10	11	88	1
Ethiopia	11	88	1	26	74	–	5	82	13	8	91	1
Ghana	28	72	–	29	71	–	15	83	2	1	98	1
Malawi	46	54	–	45	54	1	6	70	24	17	83	–
Nigeria	31	68	1	4	83	13	9	87	4	8	89	3
Sierra Leone[c]				19	79	2	98	2	–	–	98	2
Tanzania	43	57	–	5	92	3	2	65	33	5	89	6
Zambia	18	82	–	11	82	7	11	85	4	33	67	–
Whole international material	10	80	10	10	80	10	10	80	10	10	80	10

[a] Sillanpaa and Jansson, 1992.
[b] Ecuador excluded, as soil samples were not available.
[c] Sierra Leone samples discarded due to contamination.

STATUS OF CADMIUM

Ethiopia

According to recent estimates, the national mean and median Cd values for Ethiopia are clearly lower than the respective international parameters, and thus Ethiopia's position in the "international Cd field" (Figure 14.1) is somewhat on the low side. About 11% of the Ethiopian plant/soil Cd values are in the lowest international decile and 88% in zone II. Only a single Cd point falls in zone III. The zone I Cd values originate from five areas of Ethiopia (Figure 14.2). In general, the Ethiopian data indicate no serious Cd problem.

Ghana

None of the plant Cd or soil Cd contents recorded from the Ghana sample material exceeds the respective international mean values (Figure 14.3) and Ghana occupies one of the lowest positions in the "international Cd field." More than one-fourth (28%) of the Ghana plant/soil Cd values fall in the lowest international decile (zone I), with the rest of them low in zone II. The relative frequency of the lowest (zone I) Cd values is about equal in the two main areas, the Ashanti and the Central Regions, sampled. According to the present data, the agricultural products of Ghana are among the lowest with regard to the Cd.

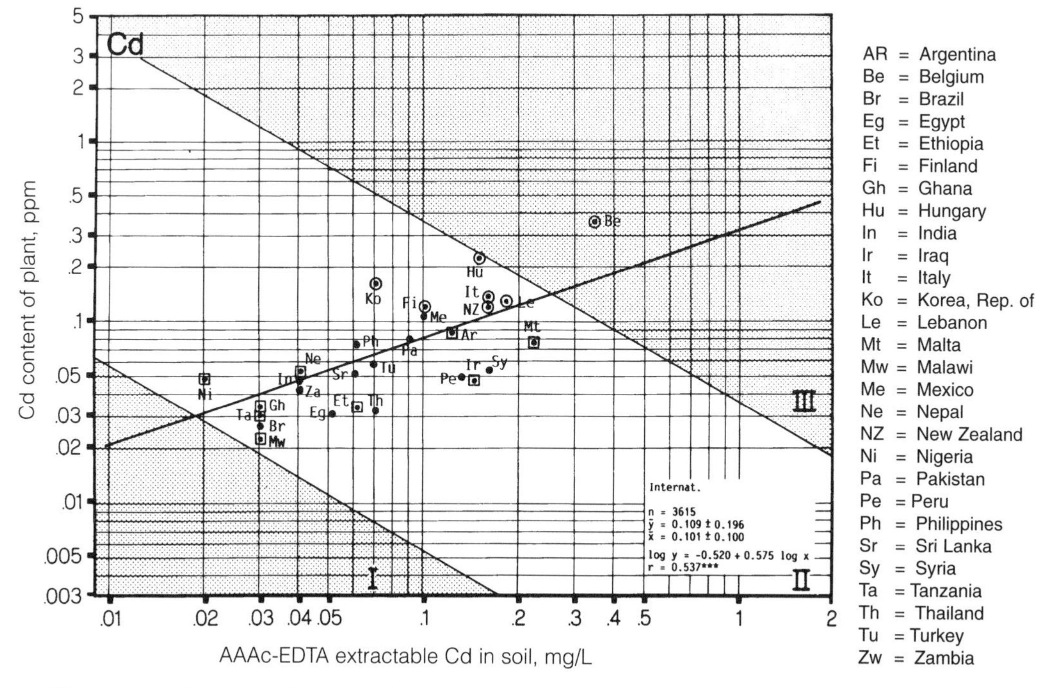

Figure 14.1. Locations of national plant Cd and soil Cd medians in the "international Cd field" where the Cd regression and three Cd content zones [the lowest and the highest plant Cd (soil Cd decile, screened)] are indicated. In addition, countries with high phosphorus fertilizer use [> 12 kg P/ha(yr)] are indicated by circles and those with low use (< 1 kg/ha) by squares (From Sillanpaa, M. and H. Jansson. Status of Cadmium, Lead, Cobalt and Selenium in Soils and Plants in Thirty Countries, United Nations FAO Soil Bulletin No. 65, Rome, 1992).

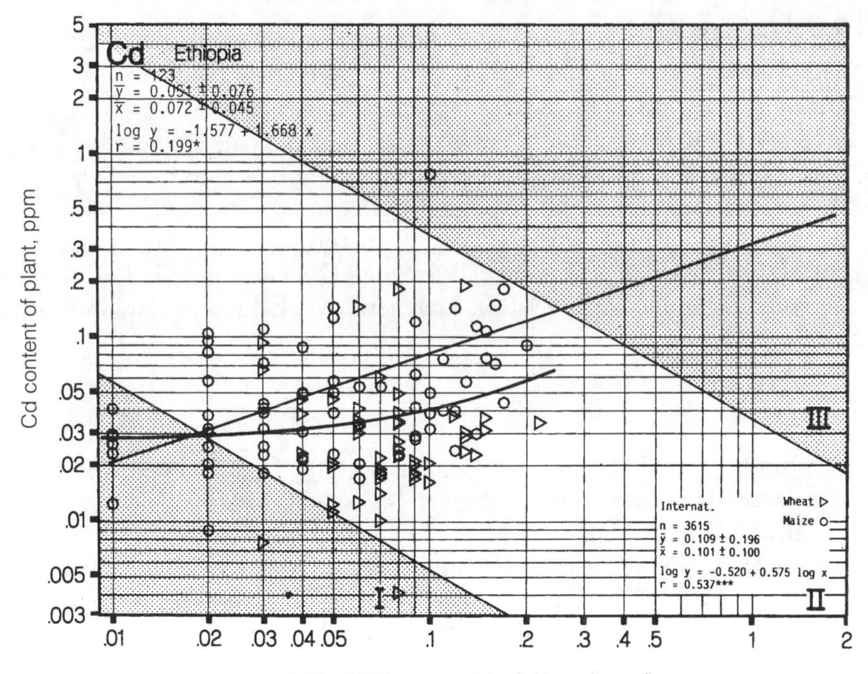

Figure 14.2. Cadmium data for Ethiopia (From Sillanpaa, M. and H. Jansson. Status of Cadmium, Lead, Cobalt and Selenium in Soils and Plants in Thirty Countries, United Nations FAO Soil Bulletin No. 65, Rome, 1992).

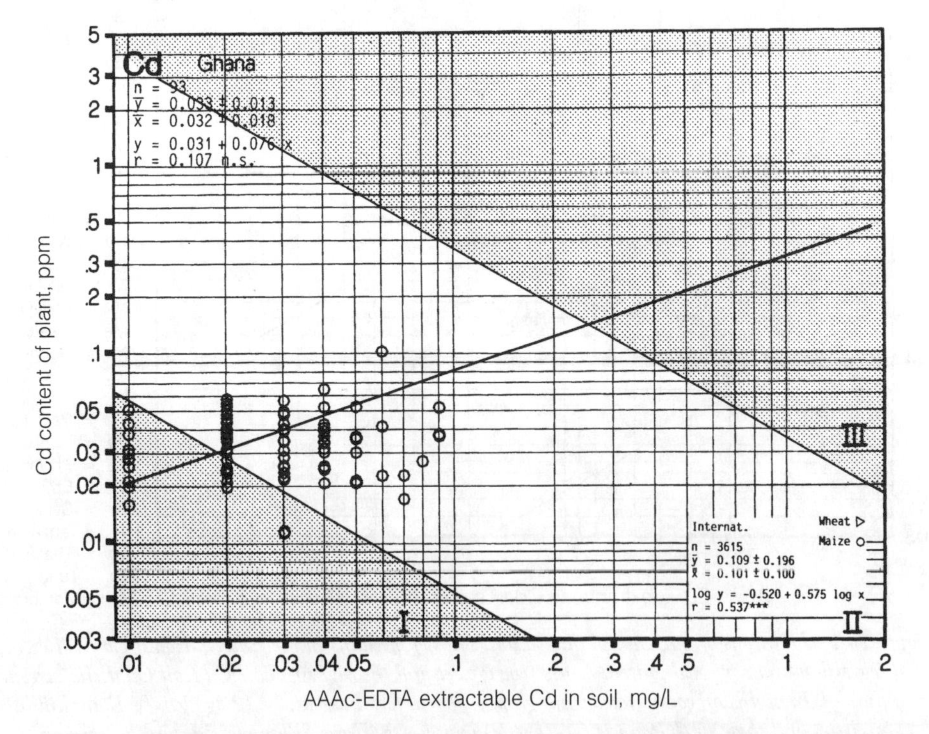

Figure 14.3. Cadmium data for Ghana (From Sillanpaa, M. and H. Jansson. Status of Cadmium, Lead, Cobalt and Selenium in Soils and Plants in Thirty Countries, United Nations FAO Soil Bulletin No. 65, Rome, 1992).

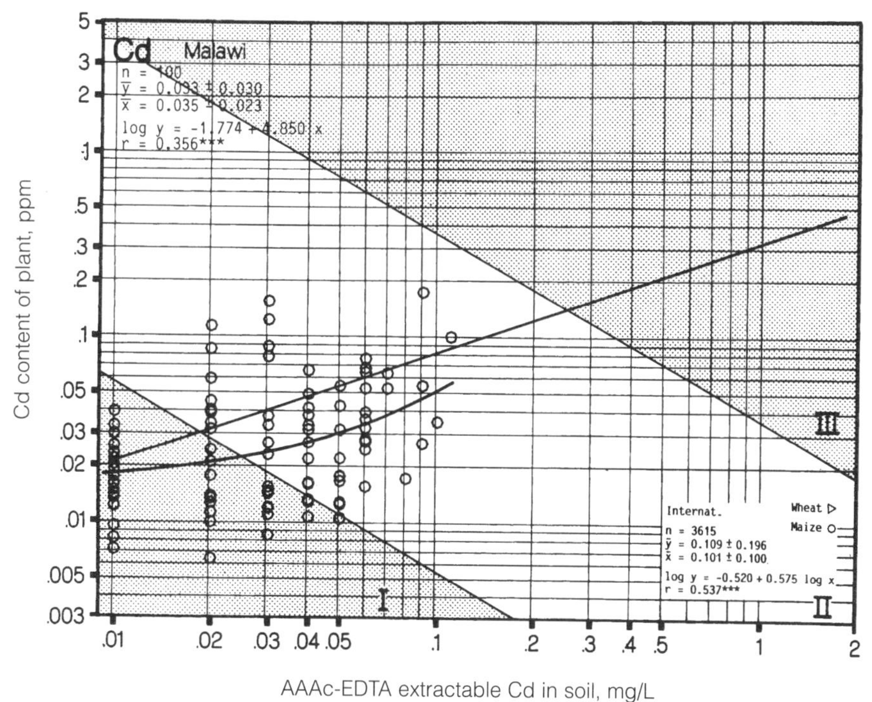

Figure 14.4. Cadmium data for Malawi (From Sillanpaa, M. and H. Jansson. Status of Cadmium, Lead, Cobalt and Selenium in Soils and Plants in Thirty Countries, United Nations FAO Soil Bulletin No. 65, Rome, 1992).

Malawi

Almost every second plant/soil Cd value of Malawi (46%) falls into the lowest international decile (zone I in Fig. 14.4) and even the rest are quite low in zone II. This means that Malawi has one of the lowest positions in the "international Cd field" (Fig. 14.1), and thus, with respect to Cd, uncontaminated agricultural products can be obtained in Malawi. Low Cd values are not typical for any distinct geographical area, but seem to be more frequent in the southern region of Malawi than elsewhere in the country.

Sierra Leone

Because of obvious Cd contamination of the Sierra Leone plant sample material, the results of plant Cd analyses were discarded. According to soil analyses, the Cd status of Sierra Leone was found to be lower than that of any other country.

Tanzania

With regard to cadmium, Tanzania's Cd status is favorable and the country occupies one of the lowest positions in the "international Cd field" (see Fig. 14.1). The percentage of zone I values (43%) is exceeded only by Malawi. Since no high Cd values were recorded in the Tanzanian data, its agricultural products can, in general, be considered low in Cd (Fig. 14.5).

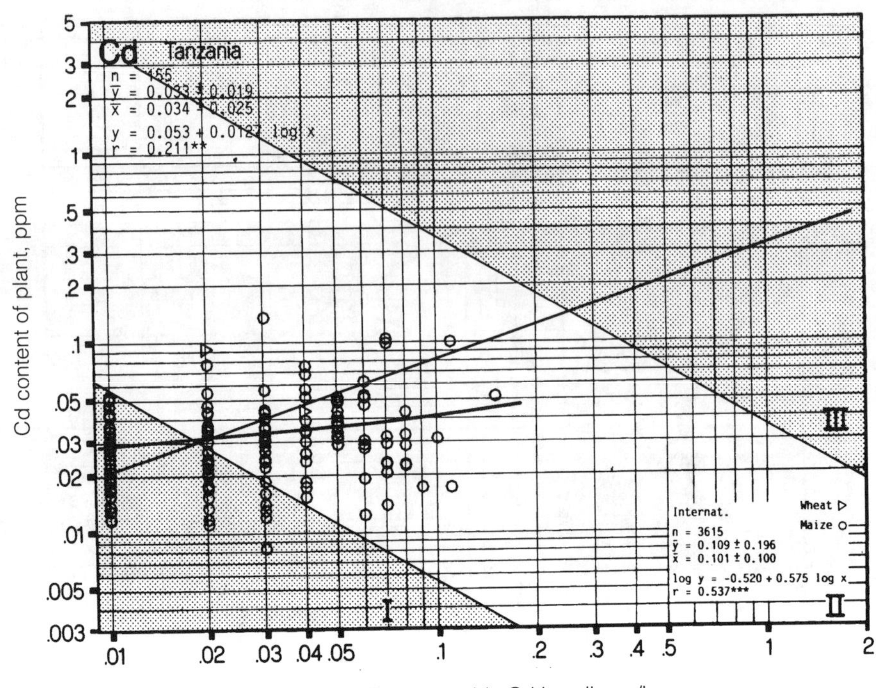

Figure 14.5. Cadmium data for Tanzania (From Sillanpaa, M. and H. Jansson. Status of Cadmium, Lead, Cobalt and Selenium in Soils and Plants in Thirty Countries, United Nations FAO Soil Bulletin No. 65, Rome, 1992).

STATUS OF LEAD

Ethiopia

The Pb level of Ethiopia is among the lowest in the international comparison (Fig. 14.6). The national median content of plant Pb is lower than that of any other country and more than one-quarter (26%) of the individual plant/soil Pb values are within the lowest international decile (zone I in Fig. 14.7). Most of the remaining Pb values are also low, falling in the lower region of zone II. A great majority of Ethiopian Pb data is below the international regression line, suggesting that pollution from airborne Pb may contribute only a relatively small part of the total Pb in Ethiopian crops. This fits well with a country of relatively sparse population and traffic.

Ghana

The national plant/soil Pb median of Ghana is one of the lowest in the international comparison (Fig. 14.6). As for Cd, more than one-fourth (29%) of the plant/soil Pb values belong to the lowest Pb range (zone I in Fig. 14.8), and most of the rest also occupy low positions in zone II. Almost all of the Pb data for Ghana are below the international regression line, i.e., the plant Pb contents are lower than could be anticipated on the basis of the soil data. This gives reason to assume that airborne Pb pollution plays only a minor role in Ghana compared to most other countries.

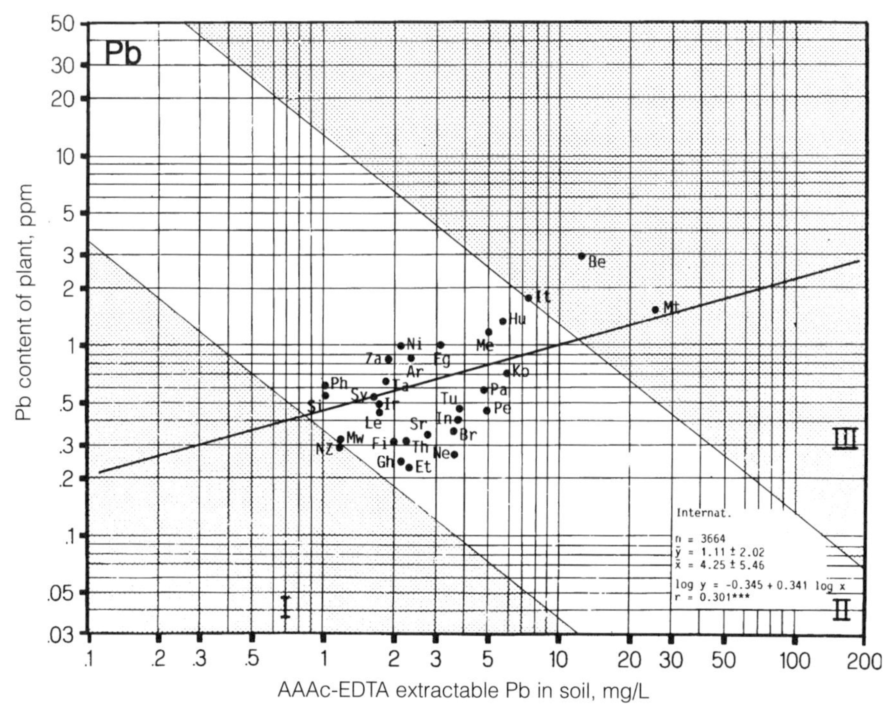

Figure 14.6. Locations of national plant Pb and soil Pb medians in the "international Pb field" where the plant/soil Pb regression and the three plant Pb × soil Pb content zones are indicated. For country abbreviations, see Figure 14.1. (From Sillanpaa, M. and H. Jansson. Status of Cadmium, Lead, Cobalt and Selenium in Soils and Plants in Thirty Countries.)

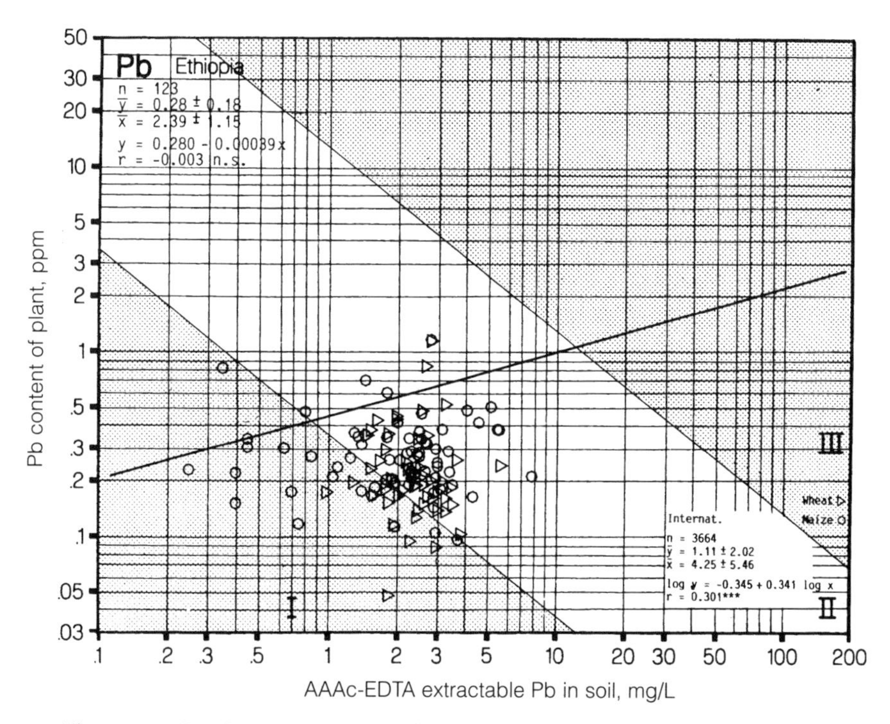

Figure 14.7. Lead data for Ethiopia. (From Sillanpaa, M. and H. Jansson. Status of Cadmium, Lead, Cobalt and Selenium in Soils and Plants in Thirty Countries.)

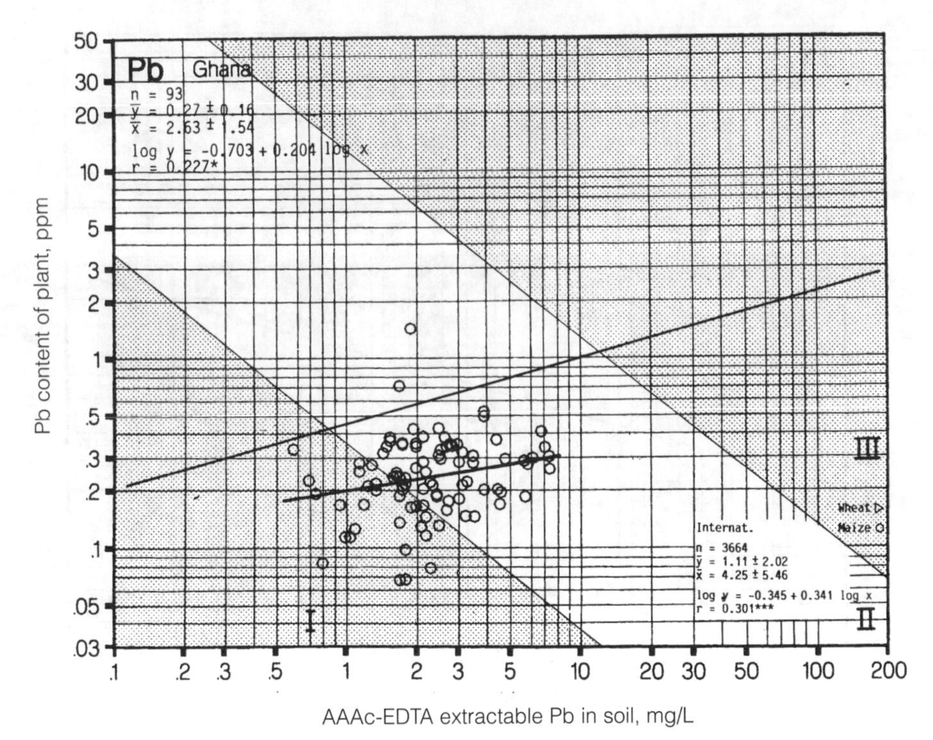

Figure 14.8. Lead data for Ghana. (From Sillanpaa, M. and H. Jansson. Status of Cadmium, Lead, Cobalt and Selenium in Soils and Plants in Thirty Countries.)

Malawi

As in the case of Cd, almost every second (45%) plant/soil Pb value of Malawi belongs to zone I (Fig. 14.9) and most of the rest are low in zone II. Consequently, Malawi occupies an almost equally low position in the "international Pb field" (Figure 14.6) as it does in the case of Cd. All the zone I Pb values come from the Southern and Central Regions while in the Northern Region, the Pb status is slightly higher, and only one of the 10 highest plant/soil Pb values originates outside the northern region. On the whole, Pb pollution is not a problem in agricultural production in Malawi.

Sierra Leone

Almost all Sierra Leone soils and plants have Pb contents lower than the respective international averages (Fig. 14.10) and the country occupies one of the lowest positions in the "international Pb field" (Figure 14.6). Only one relatively high, zone III Pb value was recorded, the rest being either within zone I (19%) or low in zonc II, indicating uncontaminated production with respect to Pb.

Tanzania

Most Tanzanian Pb data (92%) are within the normal range (zone II in Fig. 14.11) and only 5% fall in Zone I and 3% in zone III originating from Arusha and Ruvuma areas. In the international Pb comparison (Fig. 14.11), Tanzania is at a low medium Pb level, and

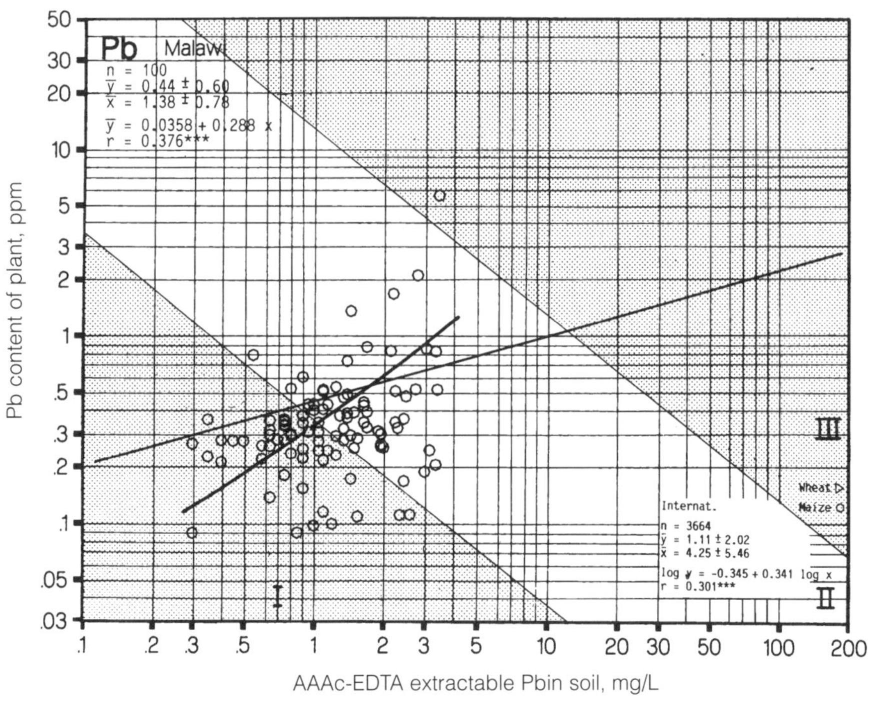

Figure 14.9. Lead data for Malawi. (From Sillanpaa, M. and H. Jansson. Status of Cadmium, Lead, Cobalt and Selenium in Soils and Plants in Thirty Countries.)

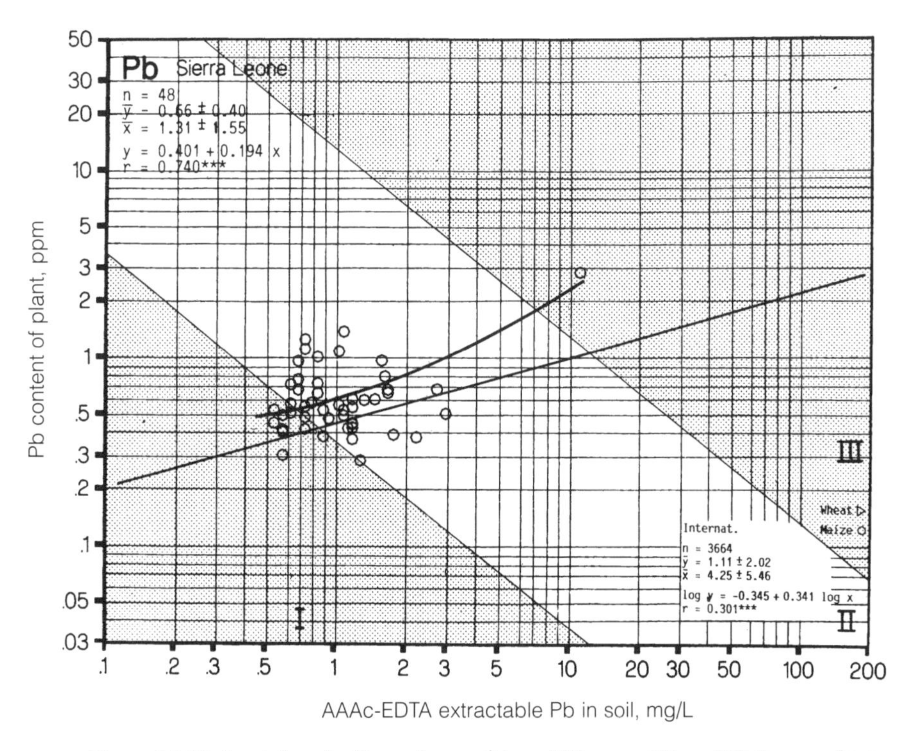

Figure 14.10. Lead data for Sierra Leone. (From Sillanpaa, M. and H. Jansson. Status of Cadmium, Lead, Cobalt and Selenium in Soils and Plants in Thirty Countries.)

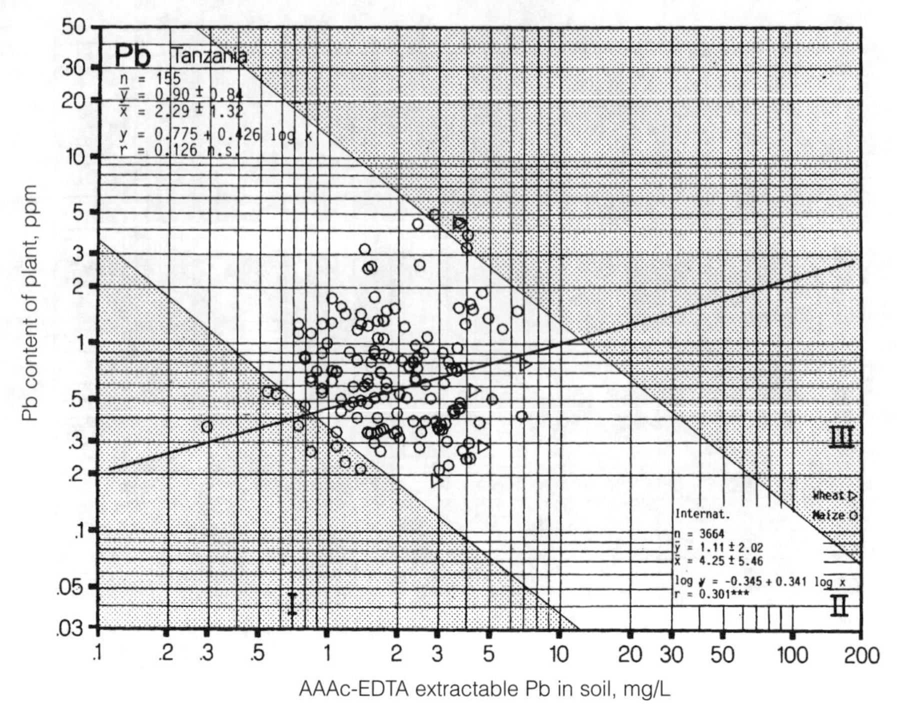

Figure 14.11. Lead data for Tanzania. (From Sillanpaa, M. and H. Jansson. Status of Cadmium, Lead, Cobalt and Selenium in Soils and Plants in Thirty Countries.)

since none of the Pb values recorded from this country are very high, the Pb status of Tanzania gives no cause for concern.

STATUS OF SELENIUM

Ethiopia

Most Ethiopian Se data are within the normal international range (zone II in Fig. 14.13) and Ethiopia is among the medium group of countries in the international comparison (Fig. 14.12). Due to relatively widely varying Se contents of plants, however, Ethiopia also has its share (8%) of zone I Se values. The 10 zone I Se values come from different areas of Ethiopia. Even though the Se status of Ethiopia in general seems to be satisfactory, Se deficiency of a local nature may exist and measures to correct it may be needed.

Ghana

Ghana's position in the "international Se field" is very central (Fig. 14.12) and its Se data fall almost totally within the normal Se range (zone II, Fig. 14.14). Thus, Se problems due to deficiency or toxicity are unlikely.

Malawi

In the international Se comparison, Malawi is clearly on the low side (Fig. 14.12), with 17% of its plant/soil Se values within the lowest international decile (zone I in Fig. 14.15) and most of the rest being in the low part of zone II.

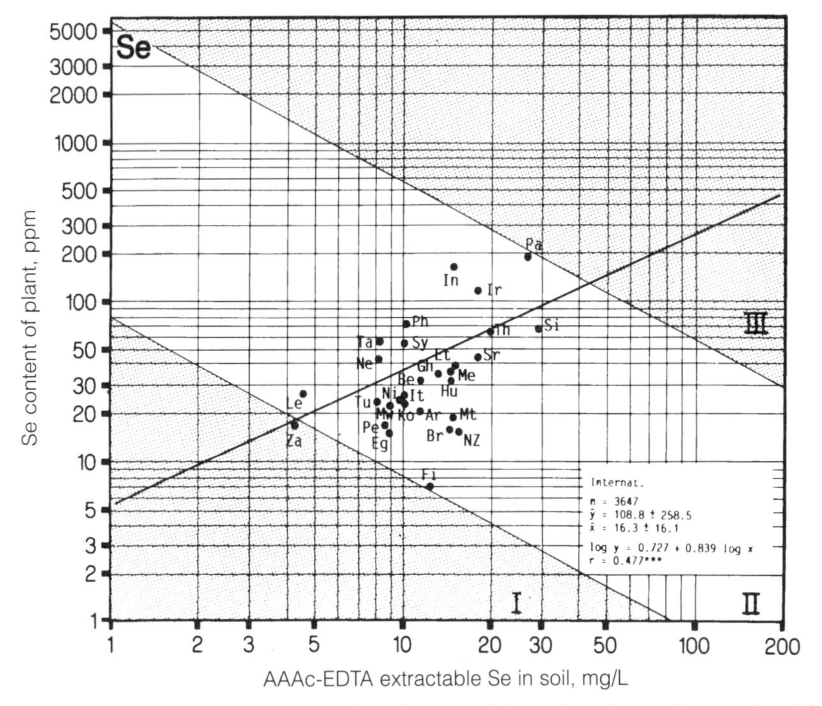

Figure 14.12. Locations of national plant Se and soil Se medians in the "international Se field" where the plant/soil Se regression and the three plant Se × soil Se content zones are given. For country abbreviations, see Figure 14.1. (From Sillanpaa, M. and H. Jansson. Status of Cadmium, Lead, Cobalt and Selenium in Soils and Plants in Thirty Countries.)

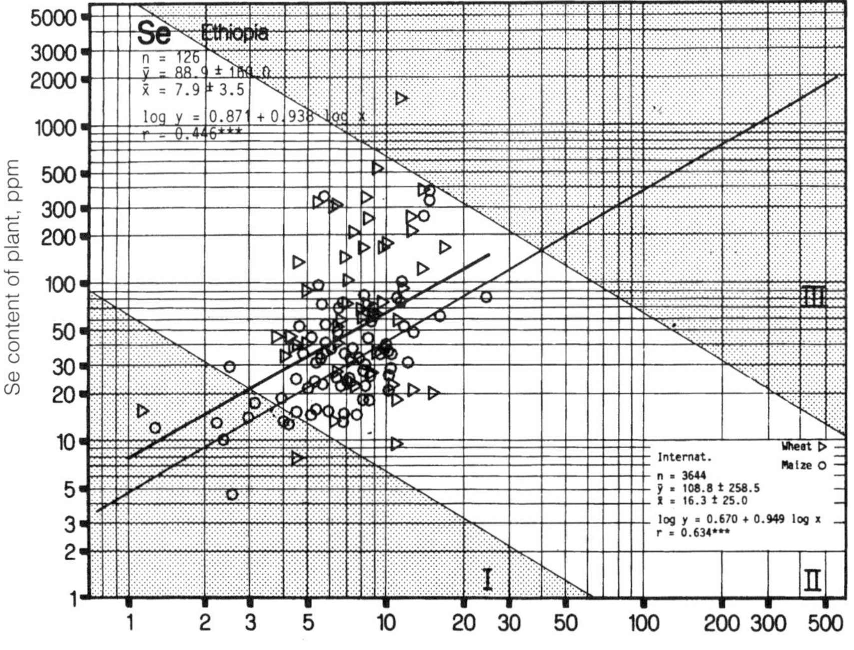

Figure 14.13. Selenium data for Ethiopia. (From Sillanpaa, M. and H. Jansson. Status of Cadmium, Lead, Cobalt and Selenium in Soils and Plants in Thirty Countries, United Nations FAO Soil Bulletin no. 65, Rome, 1992.)

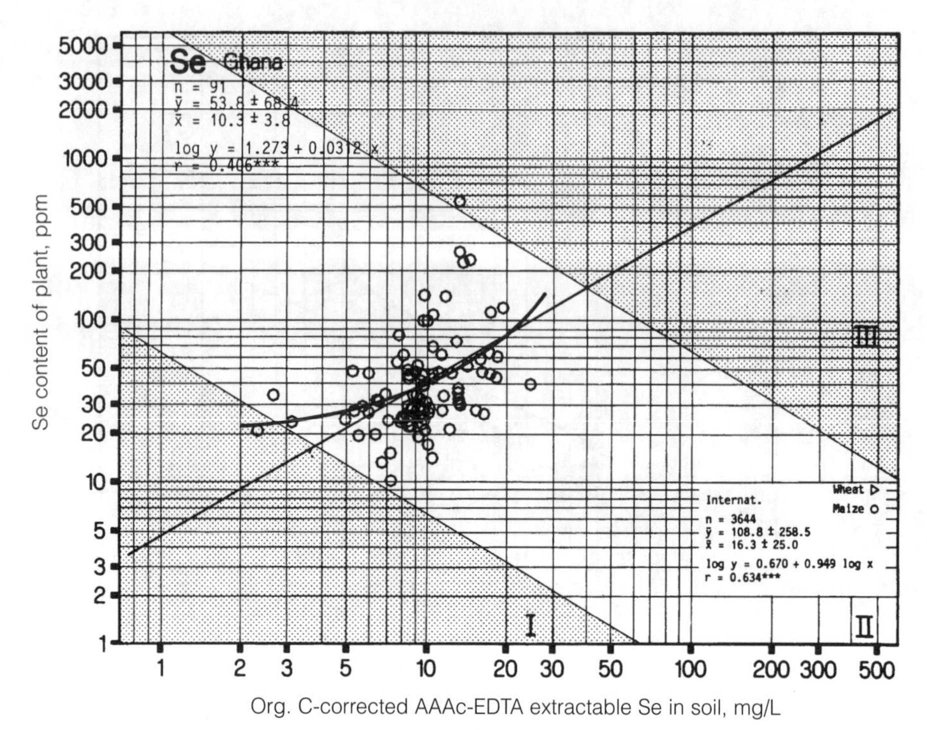

Figure 14.14. Selenium data for Ghana (From Sillanpaa, M. and H. Jansson. Status of Cadmium, Lead, Cobalt and Selenium in Soils and Plants in Thirty Countries, United Nations FAO Soil Bulletin No. 65, Rome, 1992.)

It seems that in a number of locations in Malawi, the Se status is too low to guarantee adequate nutritional availability of Se either for livestock or humans. Low-Se (zone I) sites seem to be relatively more frequent in the Northern Region (about every third site).

Sierra Leone

Because of relatively high OC contents in Sierra Leone soils, the availability of Se to plants is reduced and, consequently, the Se contents of plants are lower than would be expected on the basis of the AAAc-EDTA-extractable Se contents of soils (Fig. 14.12). Accordingly, the soil Se values decrease to about one-half when corrected for OC. After the correction, Sierra Leone occupies the sixth-highest position with respect to soil Se. The relatively uniform Se data from Sierra Leone fall mainly on the high side of zone 11 (Fig. 14.16), suggesting neither excess nor serious shortage of this micronutrient.

Tanzania

Very wide (over 400-fold) variations in plant Se contents with little corresponding variation in soil Se contents characterize the Tanzanian Se data (Figure 14.17). Even though 89% of the plant/soil Se values are within the normal range (zone II), 5–6% of the data fall in each of the extreme international deciles (zones I and III). The low, zone I values come mainly from southern Tanzania, while the zone III values originate in northern and central regions of the country. The highest plant Se contents values from Tanzania, although high, still do not quite reach the level considered toxic for animals, but the lowest

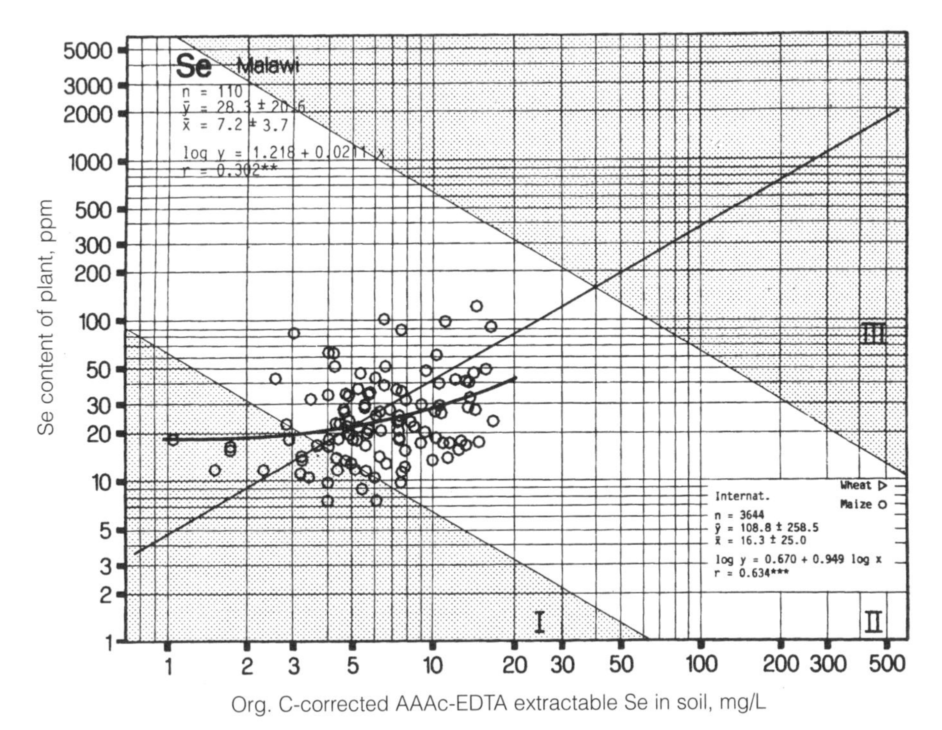

Figure 14.15. Selenium data for Malawi (From Sillanpaa, M. and H. Jansson. Status of Cadmium, Lead, Cobalt and Selenium in Soils and Plants in Thirty Countries, United Nations FAO Soil Bulletin No. 65, Rome, 1992.)

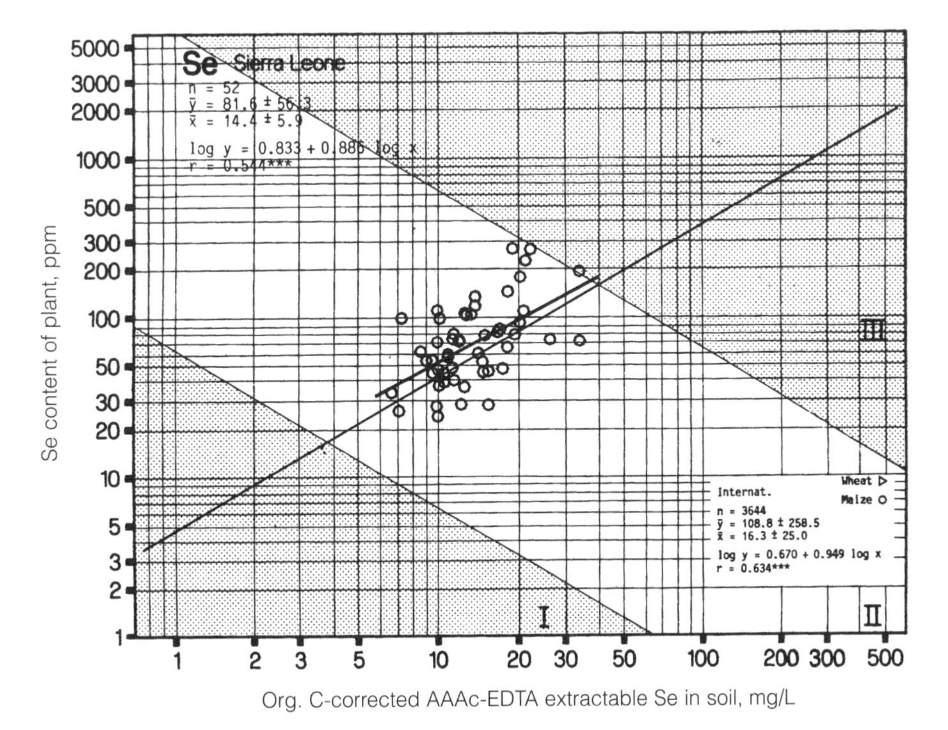

Figure 14.16. Selenium data for Sierra Leone (From Sillanpaa, M. and H. Jansson. Status of Cadmium, Lead, Cobalt and Selenium in Soils and Plants in Thirty Countries, United Nations FAO Soil Bulletin No. 65, Rome, 1992.)

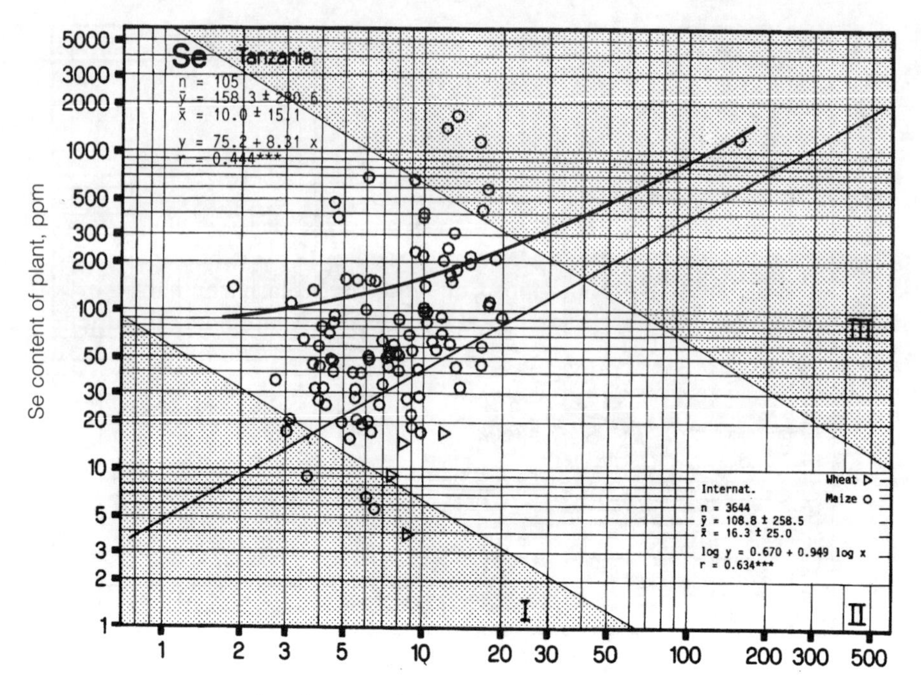

Figure 14.17. Selenium data for Tanzania (From Sillanpaa, M. and H. Jansson. Status of Cadmium, Lead, Cobalt and Selenium in Soils and Plants in Thirty Countries, United Nations FAO Soil Bulletin No. 65, Rome, 1992.)

values definitely indicate the possibility of Se shortage. In the international Se comparison, Tanzania places in the midrange (Fig. 14.12).

SOURCES, FORMS, AND DISTRIBUTION OF CADMIUM AND HUMAN EXPOSURE TO IT

Together with Zn and Hg, Cd, a metallic element, belongs to group IIb in the periodic table. Lead, Cu and to a greater extent Zn ores are the main sources of Cd. The ratio of cadmium to zinc is usually between 1:100 and 1:1000, depending on the source (Furkerson and Goetter, 1973). Cadmium is present in natural resources such as soils, water, and rocks as well as in petroleum and coal products. It is rarely found in a pure state. In the earth's crust, Cd is distributed at an average concentration of 0.1 mg/kg.

Cadmium's mobility in the environment and its effects on the ecosystem largely depend on the nature of the salts that the metal is able to form with organic acids, mainly as a result of its being bound to organic molecules such as proteins. In these forms, Cd is regarded as inorganic. The Cd which has a relatively high vapor pressure is quickly oxidized in air to form Cd oxide which, if in contact with carbon dioxide, water vapor, sulfur oxide or trioxide or hydrogen chloride, react in turn to produce carbonate, hydroxide, sulfite, sulfate or chloride, salts with Cd that may be emitted to the environment in stacks. Under the influence of acids and oxygen in the natural environment, some of the salts that are virtually insoluble in water include the sulfides, the carbonates and the oxides. The other salts—the sulfates, the nitrates, and the halogenates—are soluble in water.

Cadmium is released to the air, land, and water through mining, production, and consumption of Cd and other nonferrous metals or through the disposal of wastes containing Cd.

A third category consists of inadvertent sources where Cd is a natural constituent of the material being processed. Other sources include the application of phosphate fertilizers, atmospheric deposition, and sewage sludge (Page et al., 1981;UNEP/ILO/WHO, 1991).

The release of Cd to the soil results in increased uptake of the metal by plants, thus making the pathway of human exposure from agricultural crops become susceptible to increases in soil Cd. The soil pH, composition, and, particularly the nature of soil clays and organic matter content as well as the soil Cd level itself, influence the uptake of the metal into plants. Low soil pH creates conditions for higher cadmium uptake by crops and other plants. Acid rain and other processes that cause increased soil acidity, produce higher Cd levels in crops, and through the food chain, increased levels in the human body as well.

Crops growing near to atmospheric sources of Cd may contain higher levels of Cd. While many industrial sources of Cd possess tall stacks that bring about the wide dispersion and dilution of particular emissions, Cd deposition rates around smelter facilities are usually only distributed locally (Hirata, 1981). Rivers that are contaminated with Cd can pollute surrounding areas through irrigation practices, flooding, or dumping of dredged sediments. Another major source of Cd distribution by way of freshwater resources is industry.

A number of Cd effects on humans have been recorded. These include acute poisoning at very high Cd concentration levels, or even death among workers shortly after exposure to fumes, when Cd-containing materials have been heated to high temperatures. The main symptom in both fatal and nonfatal cases is respiratory distress due to pneumonitis and edema.

Humans can ingest cadmium through using Cd-plated cookware. This can rapidly cause severe nausea, vomiting and abdominal pain. Similar symptoms are observed when drinks containing cadmium are consumed.

Chronic effects usually occur when humans are exposed to low Cd concentrations for longer periods of time and result in renal dysfunction and emphysema. The kidney (and to a lower extent and at short-term peak exposures, the lung) is the critical organ. Renal function may become depressed and may result, at the advanced stage, in a combination of tubular and glomerular effects, coupled with impaired reabsorption of proteins, amino acids, and glucose, and an increase in blood creatinine in certain cases. In its early stage, renal dysfunction is known to cause mild proteinuria. From a point of view of preventive medicine, the detection of early effects on the kidneys is of particular importance in order to prevent more serious renal effects, as well as effects on the lungs.

Cadmium workers may also develop chronic injury to the respiratory system, a symptom that is apparent only after years of exposure. High inhalation exposure can cause acute pneumonitis with pulmonary edema, while high ingestion of Cd salts may cause acute gastroenteritis. Chronic inflammation of the nose, the pharynx, and the larynx have been reported and insomnia is a frequent symptom in Cd workers. Dyspnea, impaired lung function and increased residual volume, and reduced working capacity have been reported in Cd workers. Cadmium workers also suffer from anosmia after prolonged exposure. Other effects include the prevalence of high blood pressure among the workers in battery-producing factories.

In workers exposed to Cd dust from nickel-cadmium battery factories, several prostate cancers have been reported in England, Sweden, and Wales. Lung cancer has also been reported in the United Kingdom (Holden, 1980). These and other effects serve to demonstrate the carcinogenic effect of short- and long-term exposure to Cd.

Cadmium can be widely distributed in the human liver and eventually redistributed in the kidney. In these situations, the functions of both organs are very adversely affected. Low Cd concentrations are also usually found in the brain, bone, and fat. While there are no differences between indoor and outdoor Cd concentration levels, higher levels are evident in the dwellings of tobacco smokers since the tobacco plant carries Cd. Humans are also prone to Cd ingestion by way of feeding on freshwater fish.

Man is further exposed to varying Cd concentrations depending on whether he lives in a rural setting, in an urban area, or near an industrial complex.

ENVIRONMENTAL ASPECTS OF CADMIUM

Uptake by Selected Aquatic and Terrestrial Organisms

Cadmium is taken up from water by aquatic organisms mostly in the form of the free metal ion Cd (Sprague, 1985).

Fish uptake of Cd in water depends upon the size of the fish, temperature, and the availability of food (Pantreath, 1977). Aquatic mollusks such as oysters have demonstrated the capability for higher Cd uptakes from seawater, particularly when an uncontaminated phytoplankton food source was available.

With regard to other terrestrial organisms, earthworms have been demonstrated to concentrate Cd from soils amended with sewage sludge containing CdO. Addition of calcium carbonate to soils decreases the Cd uptake of worms only slightly, while high soil Zn levels decrease the Cd uptake considerably. Certain microorganisms also take up varying levels of Cd at higher pH levels.

Toxicity in a Few Selected Aquatic and Terrestrial Organisms

In two aquatic floating weeds, duckweed *(Lemna minor*) and floating fern (*Salvinia natans*) exposed to Cd concentrations of between 0.01 and 1.0 mg/L for up to 3 weeks, chlorosis was an observed symptom of Cd toxicity (Hutchinson and Czyrska, 1975). Under similar concentration levels, reduction of both the wet biomass and chlorophyll content of leaves occurred in the water hyacinth (Nir et al., 1990).

With regard to fish, increasing dissolved salt concentrations in water decreased Cd toxicity, whereas increasing temperatures increased it. Higher oxygen levels in water decreased the toxicity of Cd to freshwater fish (UNEP/ILO/WHO, 1992).

The effect of Cd on plants grown in soil and plants grown hydroponically is similar, though less pronounced in soil-grown plants. Interveinal chlorosis, stunting of leaves, eventual wilting with increased concentrations, and other symptoms are common in plants. At even higher doses, the number of roots is reduced (Mitchell and Fretz, 1977). Yields are also governed by the tissue concentration of Cd. For each plant type, concentrations below a certain level results in yield reduction, the amount depending on the type of soil. Stunted growth and toxic signs on leaves is another symptom of high or very high Cd concentrations in certain plants (Alloway et al., 1990).

SOURCES, FORMS, AND DISTRIBUTION OF LEAD AND HUMAN EXPOSURE TO IT

Lead is a soft metal, bluish or silver-gray in color. Lead occurs naturally in the earth's crust in concentrations of about 13 mg/kg, although this concentration may be higher in some areas (UNEP/ILO/WHO, 1977). The most important natural sources of Pb are igneous and metamorphic rocks, carbonaceous shales, sandstone (Davidson and Lakin, 1962), unconsolidated sediments in bodies of freshwater, shallow marine areas, and deep marine sediments (Riley and Skirrow, 1965).

Apart from the above-mentioned sources, soils, water, and plants are other Pb sources. Soils acquire lead either from natural sources or through pollution by man. Acidic salts generally have a lower Pb content than alkaline soils. The soil Pb content is also influenced by the nature of soil organic matter. All these factors play a major role in determining the Pb content of specific soils (Swaine, 1955).

Water (either sea water, natural surface water, or groundwater) is a poor source of Pb with concentrations of 0.08 to 0.5 μg/L in sea water, 1–10 μg/L in natural surface waters and 1–60 μg/L in groundwater (UNEP/ILO/WHO, 1977).

In plants, Pb occurs naturally. Extremely variable concentrations of Pb have been reported, but it is generally taken that normal concentrations are in the range of 2.5 mg/kg on dry water basis, in leaves and twigs of woody plants, and 0. 1 to 1.0 mg/kg in vegetables. Lead salts have a low solubility in water, except for such salts as the nitrates and the chloride. Tetraethyl lead, and tetramethyl lead, both volatile and poorly soluble in water, are used extensively as vehicle fuel additives and are regarded as a major source of Pb in the environment. When tetrakyl leads break down, they form triakyl lead compounds in the environment, which are less volatile and more readily soluble in water. Lead is mined usually in its sulfide form. Lead mining, smelting, and refining is one of the major sources of Pb and an important cause of environmental pollution, while burning of petroleum fuels is another major source and cause of air pollution. Burning of coal and oil causes only limited environmental pollution. During warfare, Pb deriving from shotgun cartridges may pollute the immediate environment.

The transport and distribution of lead from stationary or mobile sources into the environment is mainly through the atmosphere. Although large discharges may occur directly into natural waters, lead tends then to localize near the points of discharge owing to the very low solubility of the compounds that are formed in soil and water. Rain is a most efficient Pb clearing agent. Lead is quickly removed from rainwater and absorbed by soil surfaces. The transfer of air Pb to plants may be direct via the aboveground parts or indirectly through the soil. However, the concentration of Pb in the soil does not correlate well with the concentration in the plant.

Man is exposed to Pb through drinking water, food and beverages, and inhalation, as well as other miscellaneous pathways of exposure. Perhaps the most important exposure pathway occurs in the workers who come into contact with Pb in the workplace—during mining, Pb smelting, and various manufacturing processes which use Pb. Inhalation is the major pathway of exposure to Pb. Air Pb in the working environment of smelters and storage battery factories often exceeds 1000 μg/m^3 Although the concentration of Pb in drinking water is generally less than 10 μg/L, this concentration may steeply rise where Pb pipes and Pb-lined water tanks are used, resulting in elevated values of Pb in the blood of those who drink that water.

The contribution of food and beverages to man's exposure to Pb is highly variable, and no specific category of these items has been identified as being especially high in Pb content other than foods that are stored in Pb-soldered cans or lead glazed pottery. Among the miscellaneous exposure pathways of Pb are lead-glazed ceramics used for beverage storage, Pb-based paints in old houses and in the soil surrounding such homes, and atmospheric fallout (UNEP/ILO/WHO, 1977).

Two types of study characterizing the effects of Pb on man have been reported. These are retrospective studies of causes of mortality and morbidity in Pb-exposed populations compared to unexposed populations, and studies on the effects of Pb on specific human organs and systems. In both cases, the major objective of each study is to establish as far as possible the dose of, or exposure to Pb, which is associated to specific effects and the frequency of such effects (UNEP/ILO/WHO, 1977). Normally the dose for the general population is estimated from inhalation of air, ingestion of food, water, and other beverages, and various other contacts, including drugs and consumer products and, in children, ingestion of soil, settled dust, and paint chips. A more direct way of estimating the dose, however, is through taking measurements in body tissues and fluids such as blood, feces, sweat, or hair. To date, it has not been possible to establish precise doses of Pb responsible for biological effects of lead on men, and the use of Pb-B (blood lead level) remains the vital link between exposure and a biological effect, although, while some effects of Pb bear a close relationship to concurrent Pb-B levels, others do not. Renal effects of Pb may be caused by exposure to Pb long ago, which is reflected in the Pb-B level at the time the effect is first observed clinically. Because other variables such as age or the presence of disease may modify susceptibility to Pb exposure, Pb-B level indications are best used on communities rather than on individuals (UNEP/ILO/WHO, 1977).

Lead accumulation in workers in lead smelting or battery producing industries was found in studies carried out in the United Kingdom and the United States for over 10 years and was found not to have any correlation to mortality rates among workers when these were compared to nonexposed individuals. An exception was those workers who were suffering from chronic renal diseases, whereby a significant number of deaths was reported (UNEP/ILO/WHO, 1977). With regard to the effect of Pb on specific organs or systems, evidence for disturbances in heme synthesis has been shown in man, since Pb interferes with biosynthesis of heme at several enzymatic points. Furthermore, anemia caused by Pb poisoning may cause increased rates of erythrocyte breakdown (decreased erythrocyte life). Erythrocyte survival time has been known to have been reduced by up to 20% by occupationally exposed workers.

Inorganic Pb compounds variously affect the central nervous system, the nature of the effect depending on the duration and intensity of exposure. Chronic Pb exposure results in Pb encephalopathy, whose major features are dullness, restlessness, irritability, headaches, hallucinations, loss of memory and ability to concentrate, and muscular tremor both in adults and small children.

Brain lesions in fatal cases of Pb poisoning are symptomized by cerebral edema and changes in cerebral blood vessels. When inorganic lead affects the peripheral nervous system, weakness of the extensor muscles, particularly those used most, is heavily manifested (UNEP/ILO/WHO, 1977).

The effects of Pb on the kidney occur in two general types. The first occurs in the form of renal tubular damage characterized by such symptoms as hypophosphatemia manifested in decreased tubular reabsorption of glucose and alpha-amino acids. Lead's effect on the gastrointestinal tract appears in the form of colic, an early warning of potentially more serious effects likely to occur with prolonged periods of exposure, particularly industrial exposure.

There is no definite evidence for the effects of Pb on the liver. The findings to date are, however, inconclusive because no data on Pb-B levels or on indices of disturbed porphyrin metabolism were provided.

ENVIRONMENTAL ASPECTS OF LEAD

Uptake by Selected Aquatic and Terrestrial Organisms

In the environment, Pb is strongly adsorbed to soil and sediment, and only limited amounts are available to organisms. Lead also tends to precipitate because of the low solubility of its salts.

Temperature, water salinity, and pH as well as humic and alginic acid contents are some of the factors that influence Pb uptake by aquatic animals. In fish, Pb is accumulated mostly in gill, liver, kidney, and bone tissue. Lead taken up by fish reaches equilibrium only after weeks of exposure. Fish eggs show increased Pb levels with increased exposure concentrations, but those are only indications that Pb is present on the egg surface but not necessarily accumulated in the embryo.

Lead concentrations are higher in soils and plants and other organisms close to roads with high traffic density. These concentrations, inorganic in nature, derive almost entirely from alkylead compounds that are used as motor vehicle fuel additives. It goes without saying that soil and plant Pb levels decrease exponentially as one moves away from the road, to a background level at a distance of 500 m. Within this distance, soil Pb is found in the upper few centimeters of soil.

Lead Toxicity to Selected Aquatic and Terrestrial Organisms

Lead is acutely toxic to freshwater invertebrates at concentrations between 0.1 mg/L and 40 mg/L, when the metal is in the form of simple salts. For marine organisms, acute toxicity ranges between 2.5 and 500 mg/L, depending on the aquatic invertebrate communities. Young fish are more susceptible to Pb than adults. Spinal deformity and blackening of the caudal region are characteristic symptoms of fish Pb poisoning (UNEP/ILO/WHO, 1977).

Because of the capability of Pb to form highly insoluble salts together with its strong binding to soils, availability of metal to terrestrial plants is limited, and most bound Pb stays at the root or leaf surface. Furthermore, ingestion of Pb-contaminated bacteria and fungi by soil nematodes leads to impaired reproduction, while wood lice are tolerant to Pb as has been demonstrated by their prolonged exposure to soil containing externally added Pb (UNEP/ILO/WHO, 1977).

SOURCES, FORMS, AND DISTRIBUTION OF SELENIUM AND HUMAN EXPOSURE TO IT

Selenium exists naturally in several oxidation states and some of its chemical forms are volatile; there are also many analogies to organic sulfur compounds existing in nature.

Because Se is present in natural materials, its occurrence in any substance should not be always assumed to be as a consequence of human activities. Therefore, one needs to understand its occurrence in natural resources such as soils and rocks before evaluating its role as a pollutant.

Selenium exists in igneous rocks and metamorphic rocks in low concentrations of less than 1 mg/kg. Sedimentary rocks (sandstone, limestone, shales, and phosphorite), however, contain varying levels of Se ranging from less than 1 mg/kg to over 100 mg/kg. The parent material from which a soil type has been formed determines the concentration of Se levels in a soil. For example, sedimentary rocks, usually shales and chalks, are the parent material for the soils found in arid and semiarid areas. These are alkaline soils that favor the formation of selenate, which plants readily take up. Because of the limited rainfall in arid and semiarid areas, this selenate is redeposited in the surface soil, where it is still available to plants. This phenomenon renders surface soil analyses unsuitable measures of the potential of a soil type to produce plants containing toxic concentration levels of Se.

Coal contains an unusually high Se content, the highest ever identified being over 80,000 mg/kg in the People's Republic of China. The average is, however, 300 mg/kg globally. Coal selenium passes on to the soil through weathering, and leaching, and it is possible that it becomes available to crops through some biological action. Thus coal is an important source of Se. Lime and fertilizers are also another source of soil Se.

Several other factors influence the amount of Se plants are able to absorb. Apart from the amount of the element available in soil, the stage of growth of plant, and the plant species itself, the soil iron content, the soil pH, and other biochemical factors impact upon the availability of Se in the soil. All the tissues of the same plant species do not necessarily contain the same Se levels as these usually follow the tissue protein content. When that content is high, and this reflects the amount of selenium in animal feeds, the corresponding Se level in a plant tissue may become toxic for animals (Kovalskij and Ermakov, 1975). Such animal Se ultimately reaches the human body through the food chain. Thus, the amount of Se in the human diet is largely determined by the amount of Se available for absorption by plants. Deliberate addition of Se in animal feeds of poultry and livestock is yet another pathway for increased Se concentration levels in the human body (UNEP/ILO/WHO, 1987).

Generally, Se water concentration levels are in the range of 2–3 μg/L (Kovalskij and Ermakov, 1975). The highest natural concentration reported to date is 9000 μg/L. While surface waters are not very likely to contain high levels of Se, a small percentage of man-made wells were known to contain over 100 μg/L. Among the contributors to atmospheric selenium are soils, plants, animals, volcanoes, and microorganisms, all of which produce volatile forms of Se. Some anthropogenic activities such as mining, milling, smelting, and refining of Cn, Pb, Zn, Po_4, and U also contribute toward the increase of Se concentration levels in the atmosphere, as do such processes as the burning of fossil fuels and the use of Se in the manufacture of certain industrial products and the recovery and purification of Se itself (UNEP/ILO/WHO, 1987).

A few human activities are known to have caused Se deficiency in the food chain. These include sulfate fertilization, which results in decreases in the Se content in forage. The use of phosphate fertilizers has been known to cause Se deficiency in livestock (Judson and Obst, 1975). Heavy metal pollution, for instance by Ag, can also result in Se deficiency.

Much research has been conducted over the years to determine the human health effects of high or low Se intakes, through the food chain. Although farming populations living in seleniferous regions exhibited symptoms of ill health and symptoms of damage to the liver and kidney, icteroid discoloration of the skin, bad teeth, dermatitis, pathological nails, and rheumatoid arthritis, none of these could be regarded as specific for Se poisoning. Other findings have indicated that a high Se intake can cause dizziness, extreme lassitude accompanied by depression, fatigue, moderate emotional instability, and reduced capacity to concentrate. Other effects of high Se intake subsequently reported include loss of hair, abnormalities of the nervous system including peripheral anesthesia, and pain in the extremities. Hyperreflexia of the tendon, numbness, paralysis, and motor disturbance have been reported in the Hubei province of China due to Se toxicity.

Although none of these studies was conclusive, attempts to associate high Se intake with human diseases, dental caries, birth of malformed babies, amyotrophic lateral sclerosis in male farmers, and other symptoms indicate that Se might be an environmental factor predisposing people to the disease.

Workers who are occupationally exposed to industrial processes such as the manufacture of glass suffer from acute local, systematic, and possible long-term effects of selenium dioxide, the only oxide of Se found in the industrial environment. Acute local effects of exposure are seen in the lungs, gastric mucosa, skin, nails and eyes. Apart from these, sudden inhalation of large amounts of Se is known to cause edema due to a local irritation effect on lung alveoli. Allergy, increased indigestion, a metallic taste on the tongue are other symptoms that have been noticed in workers exposed to Se fumes and dusts for long periods of time (Glover, 1976). Other symptoms include nasal bleeding, headaches, and loss of weight.

Hydrogen selenite is also a known occupational poison. In the short term, hydrogen selenite has been reported by several researchers (UNEP/ILO/WHO, 1987) to cause such symptoms as weeping eyes, a running nose, lung edema, dyspnea, conjunctivitis, coughing, blood-stained sputum, bronchitis, difficulty in breathing, and many other symptoms.

The identification of any human disease due to low Se intake is difficult, as no clear-cut pathological condition attributable to Se deficiency alone has as yet been demonstrated. Attempts to associate low Se intake with human diseases have been made. These diseases include the Keshan disease, an endemic cardiomyopathy that primarily affects children in Keshan county, China, and the Kashin-Beck disease, an endemic chronic and disabling steathriopathy occurring in China and Siberia.

There exists, on the other hand, certain evidence to suggest that Se may be essential for man. For example, red blood cells contain glutathione peroxidase in quantities similar to those found in the enzyme isolated from animals (Ashwathi et al., 1975), while blood Se levels were found to be depressed in children suffering from kwashiakor (Burk, 1976). All in all, on the basis of the distribution of Se in various human tissues, it can be concluded that Se must be an essential element for man.

SUMMARY AND CONCLUSIONS/RECOMMENDATIONS

Cadmium contents were found to be higher in the industrialized than in developing countries of Europe and Oceania. In these countries the frequency of high (zone III) Cd values varied from 7 in Finland to 88 in Belgium. Among the developing countries, relatively high Cd values were recorded for Lebanon, Mexico, and Peru.

The lowest Cd levels were generally found in the African countries included in the study where the frequency of low-Cd (zone I) values varied from Ethiopia 15% to Malawi's 46%. In Africa, high-Cd (zone III) values were recorded only for one site in Ethiopia and one in Nigeria. Egypt had low Cd levels.

Cadmium is released to the air, land, and water by human activities. In general, the two major sources of contamination are the production and consumption of Cd and other nonferrous metals and the disposal of wastes containing Cd. The metal may also be introduced into the environment during processing of Cd-containing materials.

Increases in soil Cd content result in increases in the uptake of Cd by plants. The pathway of human exposure from agricultural crops depends upon increases in soil Cd. Other factors such as the soil pH and the application of phosphate fertilizers lead to increased soil Cd and subsequent higher crop and other plant uptake levels. Acid rain and other processes that acidify soil may increase the average Cd levels in foodstuffs. Organisms such as oysters, fish, earthworms, and microorganisms are natural accumulators of Cd.

High inhalation exposure to Cd oxide fumes and high ingestion exposure of soluble salts cause acute pneumonitis with pulmonary edema, and acute gastroenteritis, respectively. Long-term exposure to Cd is known to cause severe effects on the lungs and kidneys.

Public health measures for protection from Cd exposures would be improved by:

(1) Providing technical assistance to developing countries for the training of staff, particularly for Cd analysis;
(2) Monitoring of Cd levels in populations exposed to sources of high Cd level concentrations such as smelters and mines or exposed to high Cd levels in foodstuffs, and assembling data on comparisons with levels in nonexposed population groups both in the short and in long-term, and collecting more data from more countries on Cd levels in foodstuffs and the environment;
(3) Taking appropriate measures to improve the working conditions at the national level and establishing or strengthening national, regional and interregional networks for information dissemination on factors influencing Cd levels in humans and the environment. Also information exchange on measures necessary to minimize Cd waste discharge, particularly in freshwater systems and promoting practical measures to increase recycling of cadmium;
(4) Initiating national programs on safe and efficient use of chemical fertilizers, especially for women farmers;
(5) There is a need for improved analytical techniques for measuring Cd species and biological indicators of Cd exposure/toxicity, such as B_2—macroglobulin, in various matrices, and for establishing or strengthening national and international centers for quality assurance and training;
(6) International collaborative efforts should be encouraged to examine further the role of Cd in the development of human cancer in both the general population and industrial workers. Studies identifying all those workers who have shown evidence of an effect of Cd on the kidney should be conducted and data for renal dysfunction and cancer analyzed. Furthermore, more research on the effect of smoking should be pursued. In such studies, both morbidity and mortality data should be collected. The outcome would be studied not only for cancer but also for sequelae to renal dysfunction.

Lead contents in different countries follow the same distribution pattern as Cd, with industrialized countries having higher Pb levels than developing countries. However, in two developed countries, Finland and New Zealand, the lowest Pb levels were recorded among those studied, a fact that was attributed to their low population densities and low traffic per unit area. In addition to New Zealand and Finland, high frequencies of low Pb (zone I) values were measured in Malawi (45%), Ghana (29%), Ethiopia (26%), and Sierra Leone (19%)

Lead in the environment is strongly adsorbed onto sediment and soil particles, which reduces its availability to organisms. Because of the low Pb solubility in most soils, Pb tends to precipitate.

In aquatic and aquatic/terrestrial model ecosystems, uptake by primary producers and consumers seems to be determined by the bioavailability of the Pb. Bioavailability is generally much lower whenever organic material, sediment, or mineral (e.g., clay) are present. The uptake and accumulation of Pb by aquatic organisms from water and sediment are influenced by various environmental factors such as temperature, salinity, and pH, as well as humic and alginic acid content. Lead uptake by fish reaches equilibrium only after a number of weeks of exposure. Lead is accumulated in fish mostly in the gill, liver, kidney, and bone tissues.

Young stages of fish are more susceptible to Pb than adults or eggs. Typical symptoms of Pb toxicity in fish include spinal deformity and blackening of the caudal region. The acute toxicity of Pb is highly dependent on the presence of other ions in solution, and the measurement of dissolved lead in toxicity tests is essential for a realistic result. Organic compounds are more toxic to fish than inorganic lead salts. In the form of simple salts, Pb is acutely toxic to aquatic invertebrates at varying concentrations.

Lead concentrations are highest in soils and organisms close to roads where traffic density is high. The Pb measured is inorganic and derives almost exclusively from alkylead compounds added to petrol. The Pb in the soil and in the vegetation decreases exponentially with the distance from the road.

The tendency of inorganic Pb to form highly insoluble salts and complexes with various anions, together with its strong binding to soils, drastically reduces its availability to terrestrial plants via the roots. Even at very high concentrations, lead is not likely to be taken up by plant roots and translocated, but remains on the root or leaf surfaces.

Transport and distribution of Pb from stationary or mobile sources is mainly via air. Although large amounts are probably also discharged into soil and water, Pb tends to localize near the points of such discharge. The contribution of food to man's exposure to lead is highly variable, and no specific category of food has been identified as being especially high in Pb content other than wine and food that are stored in Pb-soldered cans or lead-glazed pottery. Various miscellaneous sources of Pb have been identified as being hazardous. These include Pb-glazed ceramics used for beverage storage. Lead-based paint in the houses and in the soil surrounding such homes, the soil surrounding Pb smelters, lead in street dust, miscellaneous lead-containing objects chewed or eaten by children are other possible sources of exposure, but their relative importance is not clear. The highest exposure occurs in workers who come into contact with Pb during mining, smelting, and various manufacturing processes where lead is used, the major pathway of exposure being inhalation.

With regard to research needs, it is recommended that

1. Future studies on the estimation of Pb in the human diet should include specifications concerning the characteristics of individuals for whom Pb consumption data are being reported, including sex, age, weight and physical activity, Pb-B levels and other indices such as beta-aminovulinic acid in urine specimens;
2. Estimates of the Pb concentrations of various components of the total diet be made as a means of controlling Pb in food;
3. More precise information is sought from both Pb workers and the general public regarding the contribution of airborne Pb to the total Pb concentration in the blood (Pb-B);
4. The sources of Pb most affecting infants and young children including the contribution of food, milk and other beverages, and air should be established. Studies could also be extended to cover paints, soil, dust and other miscellaneous sources of Pb;
5. Health effect studies of both inorganic and organic Pb compounds are conducted with particular attention being paid to Pb-B levels as compared with exposure levels;
6. Studies on the relationship between factors such as other pollutants, and nutritional status on man's response to Pb should be carried out.

Selenium contents of plants and soils of the industrialized countries are at low to medium level. Only one zone III Se value was found at an Italian site. The highest percentages of low-Se (zone I) values were recorded for Finland and New Zealand at 88% and 42%, respectively. African countries of low-Se status are Zambia (33%), Egypt (23%) and Malawi (17%). The highest Se contents, some of which approach toxic levels, were recorded from Pakistan (52% zone III Se values), Iraq (37%), and India (36%).

Selenium appears to be ubiquitous. However, its uneven distribution over the surface of the earth results in regions with very low or very high natural levels of the metal in the environment. Geographical, biological, and industrial processes are involved in the distribution and transport of the Se and its cycling, but the relative importance of these processes is not well established. However, the natural geophysical and biological processes are probably almost entirely responsible for the present status of Se in the general environment and this must be given primary consideration in any evaluation of the superimposed effects of man's activity on Se in the environment and food chains.

Some human activities are responsible for the redistribution of Se in the environment. Industrial sources of Se stem initially from Cu, Pb, Zn, PO_4, and U refining processes. During this refining and purification of the Se, there can be some loss of the element into the environment. There is concern in some countries with regard to the possible health effects of low and/or decreasing levels of Se in the soil. The use of fertilizers containing Se compounds in some countries is a remarkable example of intentional human intervention in the environmental distribution of Se.

Food constitutes the main route of exposure to Se for the general population. Because of geochemical differences, there exists a wide range of the estimates of adult human exposure to Se via the diet in different parts of the world; however, dietary intake more usually falls within the range of 20–300 μg/day and may be influenced by certain activities of man, such as the direct addition of Se to the food supply.

In view of the foregoing, the following actions should be taken/carried out at the national, regional and international levels:

1. Carry out further studies on the well-identified population segments overexposed to Se to clarify the signs and symptoms of Se overexposure in man and establish the relevant dose-response relationships.
2. Undertake further research on diseases attributed to exposure to Se, not only to improve prevention of such diseases, but also to provide basic information on the effects of low Se intake in man.
3. Conduct further research on the relationship between the level of Se intake and the incidence of cancer.
4. Establish networks for information exchange on all aspects of Se toxicity at all levels.

ACKNOWLEDGMENTS

I wish to express my heartfelt gratitude to Dr. W.G. Sombroek, Director, Land and Water Division, Food and Agriculture Organization of the United Nations, Rome, for granting UNEP permission to use, with full quotations of authorship and publishers, the relevant data from the FAO Soils Bulletin No. 65 "Status of Cadmium, Lead, Cobalt and Selenium in Soils and Plants of Thirty Countries."

I also wish to express my appreciation to Professor M. Mercier, Manager, the UNEP/ILO/WHO International Programme on Chemical Safety (IPCS), World Health Organization, Geneva for agreeing to the use of the documents of the Environmental Health Criteria series (EHC), particularly those on cadmium and on its environmental aspects, lead and on its environmental aspects, and selenium, as well as EHC documents on manganese, inorganic mercury, and nickel.

Appreciation is extended to Drs. A. Ayoub and T. Darnhofer, who helped during the finalization of the chapter after the death of Dr. B. Waiyaki.

Finally, I am indebted to Dr. C.K. Gachene of the Soils Department, Faculty of Agriculture, University of Nairobi, for his kind assistance in the use of the Faculty Library.

To all, I am very grateful for their assistance and cooperation.

REFERENCES

Alloway, B.J., A.P. Jackson, and H. Morgan. The Accumulation of Cadmium by Vegetables Grown on Soils Contaminated from a Variety of Sources. *Sci. Total Environ.*, 91, pp. 223–236, 1990.

Ashwathi, Y.C., E. Bentler, and S.K. Spivastava. Purification and Properties of Human Erythrocyte Glutathione Peroxides. J. Biol. Chem. 250, 1975.

Burk, R.F. Selenium in Man, in *Trace Elements in Human Health and Disease. 11 Essential and Toxic Elements*, Prasad, A.S., Ed., New York, Academic Press, 1976.

Davidson, E.E. and A.W. Lakin. Metal Content of Some Black Shales of the Western United States. U.S. Geological Survey Professional Paper 450C, 1962.

Furkerson, W. and H.E. Goetter. Cadmium, the Dissipated Element. Report ORNL-NSF-EP21 Oak Ridge National Laboratory, Oak Ridge, TN, 1973.

Glover, J.R. Environmental Health Aspects of Selenium and Tellurium, in *Proceedings of the Symposium on Selenium-Tellurium in the Environment*. Industrial Health Foundation, Pittsburgh, PA, 1976.

Hirata, H. Annaka: Land Polluted Mainly by Fumes and Dust From Zinc Smelter, in *Heavy Metal Pollution in the Soils of Japan*. Kitagishi. K. and Yamare, I., Eds., Japan Scientific Societies Press, Tokyo, 1981.

Holden, H. Health Status of European Cadmium Workers Exposed to Cadmium Fumes, in *Proceedings of the Seminar on Occupational Exposure to Cadmium*, London, 20 March 1980. Cadmium Association, 1980.

Hutchinson, T.C. and H. Czyrska. Heavy Metal Toxicity and Synergism to Floating Acquatic Weeds. *Verb 1. T. Ver. Limnol.*, 19, pp. 2102–2111, 1975.

Judson G.J. and J.M. Obst. Diagnosis and Treatment of Selenium Inadequacies in the Grazing Ruminant, in *Trace Elements in Soil-Plant-Animal Systems*, Nicholas, D.J.D. and Egan, A.R., Eds., New York, Academic Press, 1975.

Kovalskij, V.V. and V.V. Ermakov. Problem of Experimental Geochemical Ecology of Animals Under the Conditions of Selenium Regions and Some Approaches to the Study of Biological Role of Selenium, in: *Vitamins. VII. Biochemistry of Vitamin E and Selenium*. Nakova Dunka Publishing House, Kier, 1975, (in Russian).

Kovalskij, V.V. *Geochemical Ecology*, Nauka Publishing House, Moscow, 1974, (in Russian).

Mitchell, C.D. and T.A. Fretz. Cadmium and Zinc Toxicity in White Pine, Red Maple and Norway Spruce. *J. Am. Soc. Hortic. Sci.*, 102, pp. 81–84, 1977.

Nir, R., A. Gasth, and A.S. Perry. Cadmium Uptake and Toxicity to Water Hyacinth. Effect of Repeated Exposure under Controlled Conditions. *Bull. Environ. Contam. Toxicol.*, pp. 149–157, 1990.

Page, A.L., F.T. Bingham, and A.C. Chang, Cadmium, in Effect of Heavy Metal Pollution in Plants. *Applied Science*, Vol. 1, Lepp, N.W., Ed., London, 1981, pp. 77–109.

Pantreath, R.J. The Accumulation of Cadmium by the Plaice *Pleuronectes platessa* L. and the Thornbeck Ray *Rata cravats. J. Exp. Man. Biol. Ecol.*, 30, pp. 223–232, 1977.

Riley, J.P. and C.C. Skirrow. *Chemical Oceanography*, Vol. 2, Academic Press, New York, 1965.

Sillanpaa, M. Micronutrients and the Nutrient Status of Soils: A Global Study, United Nations FAO Soils Bulletin No. 48, Rome, 1982.

Sillanpaa, M. and H. Jansson. Status of Cadmium, Lead, Cobalt and Selenium in Soils and Plants of Thirty Countries, United Nations FAO Soils Bulletin No. 65, Rome, 1992.

Sprague, J. Factors That Modify Toxicity, in *Fundamentals of Aquatic Toxicology*. G.M. Rand and S.R. Petrocelli, Eds., Hemisphere Publishing Corp., New York, 1985.

Swaine, D.J. The Trace Element Content of Soils. Commonwealth Bur. Soil. Sci. Technol Comm. No. 48, 1955.

UNEP/ILO/WHO. Environmental Health Criteria 3—Lead, in *Environmental Health Criteria Nos. 1–5 (1976–78)*. World Health Organization, Geneva, 1977.

UNEP/ILO/WHO. Environmental Health Criteria 58—Selenium. World Health Organization, Geneva, 1987.

UNEP/ILO/WHO. Environmental Health Criteria 134—Cadmium. World Health Organization, Geneva, 1991.

UNEP/ILO/WHO. Environmental Health Criteria 135—Cadmium: Environmental Aspects. World Health Organization, Geneva, 1992.

Index

A

Acephate 302–304, 328
Acidification. *See also* Acid rain
  adsorption and 213
  control of 221–222
  of groundwaters 219–221
  leaching and 213–214
  plant growth 215–216
  processes 209–211
  sensitivity to 208, 216–218
  transport and 213–215
Acid rain
  in China 201, 202–204
  composition of 204
  consequences 211–216
  formation of 201–202
  prediction of soil pH 216–219
  sulfate-type 210–211
Actinomycetes 301–302
Active carbone 161
Adsorption
  acidification and 213
  kinetics 54, 56–57, 174
  mechanisms 235
  metals 279, 336
  pesticides 50–55, 231, 235–236
  soil properties 50–52
Africa 347–372. *See also individual countries*
African capeweed *(Arctotheta calendula)* 15, 26
Air pollution 60–61, 83, 85, 278. *See also* Atmospheric fallout
Akar 297, 328
Alachlor
  chemical name 251, 328
  degradation 311–313
  mobility 237, 308
  residues 300
  soil organisms and 342
Aldicarb
  adsorption 236
  chemical name 251
  degradation 230
  in groundwater 245–246
  mobility 237
Aldrin
  chemical name 251, 328
  degradation 232
  residues 61–62, 297–299, 316, 341
  in water 300
Algae 111
Aliette 251
Allolbophore caligionsa 242
Alopecurus myosuriodes 229
Aluminum 215
Aluminum phosphide 318
Americium 108, 119
Amitrole 231, 251
Ammoniacal-nitrogen 18, 258, 264–265, 304–305
Ammonium nitrate 18
Animal products 17, 26, 190
Animal wastes 4, 259–260
Anions 174, 213
Antimony 159
Aquifer, definition 65
Arctotheta calendula 15, 26
Arsenic
  in agricultural sprays 5, 10
  background values 2–3, 30, 151, 168–170, 272–273
  in contaminated soils 84, 277–278
  in groundwater 284
  kinetics and 174
  microbiology of 173
  in pastures 16
  plant uptake 173, 272–274, 280
  redox effects 157, 174, 191
  regulatory levels 29, 81, 276
  from smelters 7
  sodium arsenite 342
Asozine 251
Asumlam 328
Atmospheric fallout 4–6, 99–100, 102–103, 105, 111–118, 121
Atrazine
  adsorption 52–53, 236
  chemical name 251
  degradation 57, 231–232
  residues 63, 66
Australia
  animal products 17
  background values 2–3, 28, 30
  cadmium 4–5
  human intake 34, 94
  lead 5–7
  pesticide residues 59–60, 62–66, 70
  pesticide usage 43–45, 47
  regulatory levels 28–29
  sewage sludge 7–8, 33
  vulnerability assessments 67

Australian and New Zealand Environmental and Conservation Council (ANZECC) 28–29, 48–49
Automobile exhaust 5–7, 266
Avirosan 301, 305, 328
Azinophos-methyl 232, 251
Azocyclotin 234, 251
Azospirillum brasilense 244

B

Bacillus 310
Bacillus cerius 301, 303
Bacillus sphaericus 301, 303
Bacillus subtillis 301, 303
Background values
 in Australia 2–3, 28, 30
 in China 151, 168–171, 182–184
 concept of 1–2
 in Japan 169
 in Netherlands 30
 in New Zealand 2–3
 rare earth elements 169–171
 in United States 30, 169
Base-saturation percentage 217
Bauhina variegata 266
Baycid 304, 328. *See also* Fenthion
Behavior Assessment Model (BAM) 58, 67
Bensulfuron 251
Bensulfuron methyl 328
Bentazon 302, 307, 312, 317, 328
Bentazone. *See* Bentazon
Benthiocarb 328
Benzene 129–130
Benzoximate 234, 251
BHC
 chemical name 251, 328
 in China 26, 236
 degradation 232, 234, 309
 formulations 292
 in greenhouses 298–299
 in Korea 290, 294–298
 in Malaysia 333
 plant uptake 239–240, 307
 residues 59, 228–229, 298, 316
 soil nitrogen and 304
 soil organisms and 241–242, 244, 301, 303
 in water 244, 300
Bifenthrin 234, 251, 310, 312, 315, 328
Bioaccumulation 174
Biodegradation 55–57, 229–234, 309–313, 315, 318
Biological oxygen demand (BOD) 257–258, 264–265, 283
Bioremediation 69, 145
Blasticydin 304, 328
Boron 3, 30
BPMC 251, 296, 298, 300, 307–308, 315, 328
Brain lesions 366
Brassica alboglabra 34
Brassica campestris 280
Buffer strips 68
Buprofezin 234, 251
Butachlor
 chemical name 251, 328
 degradation 309–310, 315
 residues 300–303, 305
 soil organisms and 243
Bux 305, 328

C

Cabbage *(Brassica campestris)* 280
Cadmium
 in African soils 349–354, 369
 background values 2–3, 30, 151, 168–170, 272–273
 chloride and 19, 24–25
 in contaminated soils 277–278
 crop rotation and 21
 in crops 12–15, 25–28, 272–274
 ecological cycling 189–191, 362, 364
 in fertilizers 31, 86
 global input 3–4
 in groundwater 284–285
 historical input 10
 in humans 90–94, 363–364, 370
 loading from fertilizers 6–9, 13–16, 26–27, 159, 340, 351
 major ions and 279
 microbiology of 173
 organic matter and 192
 in pastures 15–17
 phosphates and 159
 plant uptake 119–120, 161–162, 173, 280, 363–364
 redox effects 157
 regulatory levels 29, 81–83, 155, 276
 in sewage sludge 8
 silicon fertilizer and 159–160
 soil loading capacity 172
 soil pH and 22–24
 sources 4–6, 10, 362–363
 speciation 279
 zinc and 19–21, 24, 27
CALculates Flow (CALF) 58–59
Captafol 296, 316
Captan 302
Captapol 328
Carbaryl 231–232, 234, 251, 317, 328

Carbendazim 233, 241, 243, 251
Carbofuran
  adsorption 236
  chemical name 251, 328
  degradation 232–234, 309–311
  effect on soils 305
  in Malaysia 333
  mobility 237, 307–309
  in paddy soils 302
  residues 300, 315, 317
  soil organisms and 241–243, 302–303
  volatilization 238
Carbosulfan 234, 251
Cartap 234, 251
Catfish *(Claria batrachu)* 342
Cation exchange 209–211
Cesium
  human intake 109–111
  migration 101–108
  plant uptake 111, 113, 115
  soil deposition 99–101
  transfer factors 119
Chemical oxygen demand (COD) 275, 283
Chemicals, Runoff, and Erosion from Agricultural Management Systems (CREAMS) 58–59
Chernobyl accident 100–101, 110, 122
China
  acid groundwater in 220–221
  acid rain in 201–204, 222
  background values 151, 168–171
  metal pollution in 150–163, 193–197
  pesticides in 225–228, 244–246
  soil sensitivity in 218–220
Chlorbenzilate 296
Chlordane 59, 61–62
Chlordimeforn 231, 251
Chlorfenvinphos 238, 251, 305, 314–315, 318–319
Chlorfluazuron 234, 251
Chloride 19, 24–25, 213
Chlorinated solvents 127–129, 132–133
Chlorothalonil 234, 251, 315, 328
Chlorpyrifos
  chemical name 251, 328
  degradation 231–232, 310–311, 315
  in Malaysia 333
  regulations 318–319
  residues 295
  soil organisms and enzymes 303–304
Chlorpyriphosmethyl (EC) 307
Chlorsulfuron 240, 251
Chlortoluron 229, 234, 238, 251
Chromium
  background values 30, 151–152, 168–169, 184, 272
  chemical forms 181
  enrichment coefficient 182–183
  in groundwater 126–127, 284–285
  organic matter and 159, 192
  plant uptake 272–273, 280
  redox effects 157, 191
  regulatory levels 29, 276
  in soils 84
Cinclozolin 329
Claria batrachu 342
Clays 51, 53–55
Clofentezine 234, 251
Clomazone 234
Clover *(Trifolium subterraneum)* 15, 27
CNP 301, 302, 304–306, 328
Cobalt
  in African soils 350
  background values 168–169, 182, 184, 339
  chemical forms 180–181
  distribution coefficients 108
  enrichment coefficient 182–183
  transfer factors 119
COD (chemical oxygen demand) 275, 283
Composting (bioremediation) 69, 145
Computer models 58, 67, 279
Cook Islands 45
Copper
  background values 30, 151, 168–169, 184, 272–273, 339–340
  chemical forms 180
  in contaminated soils 277–278
  enrichment coefficient 182–183
  in groundwater 284–285
  major ions and 279
  microbiology of 173
  molybdenum and 158, 189
  organic matter and 182
  plant uptake 173, 272–274, 280
  regulatory levels 29, 81, 155, 276
  in sewage sludge 8
Copper oxychloride 10, 333
Copper sulfate 251
Corn *(Zea mays)* 280
Costelytra zealandica 60
Cotton 64
CREAMS (Chemicals, Runoff, and Erosion from Agricultural Management Systems) 58–59
Crop rotations 21
Curium 119
Cyfluthrin 234, 251
Cyhalothrin 310, 315, 328
Cynodon dactylon 266
Cypermethrin 232, 233, 251, 317, 333

D

2,4–D 231, 251, 303, 306, 328, 333, 340
Dachtal 64
Dalapon 231, 251
Danjiami 241, 243, 253
2,4–DB 231, 251
DDT
  adsorption 236, 238
  banning of 43
  chemical names 251, 328
  in China 26
  degradation 55–56, 61, 231–232, 234
  formulations 292
  in greenhouses 298–299
  human breast milk 261
  in humans 342
  in Korea 290, 295–297
  plant uptake 239–240
  residues 10, 59–62, 229, 316
  soil organisms and 301, 303
  transport 59
Degradation and transformation
  of organochlorines 135–138
  of pesticides 55–57, 229–234, 307–313, 315, 318
Deltamethrin 232–234, 238, 244, 251
Deposition velocity 112
Desorption 54
Diazinon
  chemical name 251, 328
  degradation 231, 309–312, 315
  effect on soils 305
  organic matter 308
  residues 295–296, 298, 307, 316, 341
  soil enzymes and 304
  soil nitrogen and 304
  soil organisms and 303
  in water 300
Dibromochloropropane 236, 251
Dibromoethane 236
Dicamba 302
Dichlofluanid 298, 328
Dichloroethylenes 127–128, 130, 135–138
Dichloropropane 236
Dichlorovos 231–232, 251, 298, 328
Diclofop-methyl 234, 251
Dicofol 232, 251
Dieldrin
  chemical name 251, 328
  degradation 57, 232, 234
  in greenhouses 298–299
  human breast milk 261
  residues 59–62, 64, 261, 297–298, 316
  in water 300
Diethofencarb 315
Diflubenzuron 234, 251, 317
Dilution of soils 90
Dimehypo 241, 251
Dimepiperate 328
Dimetametryne 303, 328
Dimethazon 252
Dimethoate 232, 234, 252, 295
Dinoseb 231, 252, 328
1,4–Dioxane 129
Diquat 53, 55, 231, 252
Dissolved oxygen (DO) 257–258, 264–265
Distribution coefficients 96, 107–108
Disulfoton 230, 252
Diuron 252, 333
DNOC 230, 252
Dodine 315
DO (dissolved oxygen) 257–258, 264–265
DRASTIC 66–68
Drinking water 48–49, 194–195
Duckweed *(Lemna minor)* 364
Dutch soils 30
Dynamic loading capacity 172

E

E. coli 257, 264–265
Earthworms 240–241, 342
EC (chlorpyriphosmethyl) 307
EDTA extraction 11
Eisenia foetiela 242
Elensine indica 266
Encephalopathy 366
Endosulfan 63–64, 68, 296–297, 299–300, 328, 333
Endotrophic mycorrhiza (VAM) 244
Endrin 300
Enrichment coefficient/enrichment capacity 181–183
EPN 231, 252, 295
EPTC 236, 252
Escherichia coli 257, 264–265
Esfenvalerate 234, 252
Esprocarb 310, 315, 328
Ethalfluralin 328
Ethion 232, 252
Ethiopia 348, 351–352, 354–355, 358–359
Ethofenprox 234, 252
Ethoprophos 307–309, 315, 328
Ethylene dibromide 252
Ethylenethiourea (ETU) 231, 232, 252
Ethylmercury chloride 252
Eucalyptus 64
Eucalyptus nitens 68
Euparean 300
Europe 9

European Community 32
Exchangeable acidity 207
Extraction methods
  available metals 155
  EDTA 11
  sequential 154–155, 179–182, 184–186

F

Fenamiphos 56, 244, 252
Fenitrothion
  chemical name 252, 328
  degradation 231–232, 234, 308, 310, 315
  effects on soils 305
  residues 298
  in water 300
Fenothiocarb 234, 252
Fenoxaprop 342
Fenpropathrin 252
Fenpyroximate 234, 252
Fensulfothion 318
Fenthion 232, 234, 252, 328
Fenuron 252
Fenvalerate 232–233, 241, 244, 252
Ferbam 304, 328
Fertilizers. *See also* Phosphate fertilizers
  cadmium in 31, 86
  cadmium loading from 6–9, 13–16, 26–27, 159, 340, 351
  copper in 87
  experimental artifacts 18
  global metal input from 4
  lead from 6–7
  in Malaysia 331–335
  quality of 18, 29–32
  silicon 159–160
  zinc in 87
Fiji, pesticides in 45, 47, 60–62
Fires, lead from 5
Fish 342, 364, 367, 370–371
Floating fern *(Salvinia natans)* 364
Flooding 68, 91. *See also* Paddy fields
Fluazifop-butyl 234, 252
Flufenoxuron 234, 238, 252, 315, 329
Fluorine 30, 153, 168, 195–197, 253
Fluvalinate 234, 252
Fly ash 161
Fodders 188
Folpet 252
Fonofos 231, 238, 252
Food quality. *See also* Plant uptake; Regulations
  pesticides 48
  radionuclide uptake 108–118
  regulatory limits for metals 34, 83
  soil pH and 4
Food wastes 4
Forestry 46
Freundlich equation 53
Fthalide 305
Fuel combustion 5–7, 266
Fulvic acids 158–159, 239, 308
Fungicides
  in China 227
  in forestry 46
  in Korea 291–293, 320
  in Malaysia 333, 335
  metals in 10
Fungi in soil 301–302, 310
Furnace dust 161, 192

G

Gas emission 151–153
Gasoline 129–130, 266
Geographical information systems (GIS) 58
Ghana
  cadmium in 351–352
  lead in 354, 356
  selenium in 358, 360
  soil characteristics 349
Gingival bleeding 189
GIS (geographical information systems) 58
Glyphosate 306, 329
Goiter 195–197
Grass grub *(Costelytra zealandica)* 60
Greenhouse cultivation 298
Groundwater extraction 143–145, 147
Groundwaters
  acidification of 219–221
  definition 65
  in Hong Kong 263–265
  in Japan 126–132
  in Korea 284–285
  metals in 86, 126–127, 193–197, 284–285, 340
  nitrate in 194–195, 264–265
  pesticide residues 65–68, 244–246
  regulatory limits 82
  soil composition and 219–220
  volatile organochlorines in 127–132, 135
Groundwater Ubiquity Score (GUS) 58
Growth retardation index 52
Guam 45, 66

H

Haloxyfop 234, 252
Hand tremors 189
Harrisia 10
Heavy metals. *See also* Adsorption; Background values; *individual metals;* Transport
  computer modeling 58, 67, 279

definition 347
ecological cycling 189–191, 362, 364
interactions between 279
sequential extraction of 154–155, 179–182
soil loading capacity 171–179
soil organisms and 172–173
Heptachlor 59–62, 295–297, 299–301, 306–307, 329
Heptachlor epoxide 57, 61–62, 295, 297–300, 306–307
Herbicides. *See also* Pesticides
in Australia 46
in China 227
degradation 57
in Malaysia 333, 335
in New Zealand 46
Hexazinon 315
Hexythiazox 234, 252
Hong Kong
groundwater pollution 263–265
overview of 254–257
pollution sources in 256–263, 267–268
remediation 265–267
Human breast milk 261
Human intake
of basic food groups 109
of metals 90–94, 187, 279–281, 363–366, 368–372
of radionuclides 108–111
Humic acids 158–159, 192, 238–239, 308
Humin 239, 308
Hydrogen selenite 369
Hydrolysis 55, 230
Hydroponics 364
Hysteresis 54

I

IBP
chemical name 252
degradation 310, 315
in paddy soils 302, 305
residues 295–296, 298, 308
soil nitrogen and 304
in water 300
Insecticides. *See also* Pesticides
in China 227
in Korea 291–293, 320
in Malaysia 333, 335
In situ burial 68
Integrated pest management (IPM) 69–70, 225, 334
Iodine
distribution coefficients 108
in groundwater 195–197
plant uptake 111–118
radioisotopes 100
removal of 118–121
Ionic strength 55
IPM (integrated pest management) 69–70, 225, 334
Iprobenfos 309, 329
Iprothiolane 329
Iron 119
Iron oxides 186, 337, 339
Iron sulfate 191
Irrigation water
metals in 19, 272
from mining wastes 273–274, 278
water quality standards 283
Isazophos 234, 252
Isocarbofos 244, 252
Iso-ionic point 213
Isoprocarb 302
Isoprothiolane 301–302, 307–308, 310, 315
Isothiolane 303
Itai-itai disease 80

J

Japan
background values 169
daily intake of metals 90–94
groundwater pollution 126–132
metal pollution 80–87
meteorological data 97–98
radionuclides in 96–122
regulations 81–83, 133
remediation 87–90, 141–148
sewage sludge 91–95
volatile organochlorines 127–148
Jiajiyiliulin 241, 243, 253

K

Kaolinite 209, 337
Kaschin-Beck disease 170–171, 369
Kashin-Beck disease. *See* Kaschin-Beck disease
Kasugamin 304
Kasugamycin 329
Kecaoan 241, 243, 253
Kelthane 329
Keshan disease 170–171, 369
Kethane 296–297
Kidneys 366
Killfish *(Oryzias latipes)* 312
Kinetics 54, 56–57, 174
Kitazine 252
Korea
management strategies 281–284
metal contamination 272–278, 284–285

metal sources 270–272
pesticide effects 300–312
pesticide regulations 313–314, 316–323
pesticides in 290–300
Kwashiakor 369

L

Land disposal of sewage sludge 7–8, 11, 32–33, 82, 153–154, 174–178
Landfills 68, 261–262, 267
Langmuir equation 53, 336–337
Leaching 58, 88–89, 213–214, 244–245
Leaching Estimation And CHemistry Model (LEACHM) 58
Lead
in African soils 349–350, 354–358, 370
arsenate 5, 10, 252
from automobile exhaust 266
background values 2–3, 30, 151, 168–170, 184, 272–273
in contaminated soils 84, 277–278
in fertilizers 9, 29–30
in fish 370–371
global input 3–4
in humans 365–366, 371
microbiology of 173
in pastures 15–16
pH effects 174
plant uptake 272–274, 280, 355, 365, 367
pollution index grades 175–177
redox effects 157
regulatory levels 29, 155, 276
in sewage sludge 8
sources 4, 5–7, 10, 189, 365
transfer factors 119, 365
transport 365
in water 284–285, 367
Lead arsenate 5, 10, 252
Legumes 21, 57
Lemna minor 364
Leptophos 234, 252
Lime. *See also* Soil pH
acid rain and 221
cadmium uptake and 6, 23–24, 88, 157–158, 191–192
lead from 6
pesticide degradation 309
Lindane 231–232, 236, 238, 241, 243, 252
Linuron 231, 252, 302–304, 329
Livestock waste 4, 259–260
Loaches *(Misgurnus an guillicaudatus)* 312
Locusts *(Oxya chinensis)* 312
Logging 4
Lorox 304

M

Malathion
chemical name 252, 329
degradation 231–232, 318
in Malaysia 333
regulations 318–319
residues 295, 300, 314, 316
Malawi 349, 353, 356–358, 361
Malaysia
fertilizer and agrochemical usage 330–335
metals and agrochemicals pollution 335–344
Maleic hydrazide 231, 252
Mamet 301, 305
Management options. *See* Remediation
Mancozeb 333
Maneb 329, 333
Manganese
background values 3, 30, 168, 184, 339–340
distribution coefficients 108
in groundwater 284–285
plant growth 215–216
in sewage sludge 8
soil pH and 217
toxicity of 348
transfer factors 119
Manures 4, 259–260
Marathion 232
MCP 329
MCPA 231, 252, 333
MEP 295
Mercury
background values 30, 151–152, 168–170, 272
in contaminated soils 84, 277–278
plant uptake 119, 272–273
redox conditions 191
regulatory levels 29, 155, 276
sources 4, 189, 347
in water 194, 284–285, 340
Metalaxyl 252
Methabenzthiazuron 329
Methamidophos 341–342
Methomyl 232, 234, 252
Methothrin 252
Methoxychlor 61, 231, 252
Methsulfuron-methyl 252
Methyliodide 111–113
Methyl-parathion 231–232, 236, 238, 241, 243, 253
Metobromoron. *See* Metobromuron
Metobromuron 231, 252, 315
Metolachlor 234, 252, 329
Metribuzin 252, 329
Metsulfuron-methyl 240
Mevinphos 232, 252

Microorganisms. *See* Soil organisms
Midinyanglin 241, 243, 253
Mieyouniao 253
Migration. *See* Transport
Minamata disease 80
Mining 188, 196, 274, 285. *See also* Smelters
MIPC 296, 302, 329
Misgurnus an guillicaudatus 312
Mobility. *See* Transport
Models 58, 67, 279
Molinate 243, 252, 310, 315, 329
Molybdenum 2–3, 30, 158, 187–189, 196
Monitoring of soils 65–66, 281–282
Monocrotophos 241, 252
Monofluoroacetate 43
Monosodium methane arsonate (MSMA) 333
Monuron 231, 236, 252
Mosana shirak *(Oxya velox)* 313
MPP 295–296, 329
MTMC 252
Municipal waste 261–262, 278

N
NAC 296
Napropamide 306, 329
Neburon 253
Neoasozine 305
Neptunium 101–102, 105–106, 119
Netalaxyl 315
New Zealand 1–35
  animal products 17
  background metal concentrations 2–3
  lead 5–7
  metals in foods 34
  pesticide residues 50, 60–61, 65–66, 70
  pesticide usage 44–48
  regulatory levels 29
  sewage sludge 7–8, 33
  vulnerability assessments 68
Nickel
  background values 30, 151, 168–169, 182, 184, 272, 339
  chemical forms 180–181
  enrichment coefficient 182–183
  from mines and smelters 153
  plant uptake 280
  regulatory levels 29, 155
  in sewage sludge 8
  toxicity of 348
Nigeria 350
Nitralin 306, 329
Nitrate 194–195, 211, 264–265, 302–303
Nitrofen 301, 305, 309, 315, 329
Nitrogen
  ammonia 18, 258, 264–265, 304–305
  fixation 244
  nitrate 194–195, 211, 264–265, 302–303
Niue 45
Nonequilibrium index 54
Nongle 193
Nuclear weapons testing 99–100, 121. *See also* Atmospheric fallout

O
Ocean disposal ban 7
Oceania 42–70. *See also individual countries*
Octanol-water partition coefficients 52–53
Odontoseisis 189
Omethoate 253
Ophthalmospas 189
Organic matter
  cadmium and 192
  chromium and 159, 192
  copper and 182
  heavy metals toxicity and 158–159
  pesticide sorption 50–54, 235, 308
  zinc and 337
Organochorines (OCs) 59–60, 127–129
Oryza sativa. See Rice
Oryzias latipes 312
Oxadiazon 303, 329
Oxamyl 232, 253
Oxya chinensis 312
Oxya velox 313

P
Pacific Island countries 45, 47–49. *See also individual countries*
Paddy fields. *See also* Rice
  heavy metals in 191, 272, 275
  monitoring 282
  pesticides in 293–294, 297–300, 302
  soil pH 207–208, 221
  water budget 97–98
PAP 295
Papua New Guinea 45, 47
Paraquat
  chemical name 253, 329
  degradation 231, 315
  in Malaysia 333, 340–341
  soil organisms 243, 342
  sorption 308
Parathion
  chemical name 253, 329
  degradation 231–232, 234, 310
  regulations 318–319
  residues 295, 317

sorption 238
in water 300
Park soils 85
Partition coefficient 52–53. *See also* Adsorption; Distribution coefficient
Partitioning of pesticides 50–54
Pastures 15–17, 26–28, 46
PCBs (polychlorinated biphenyls) 261
PCE. *See* Tetrachloroethylene
PCNB 253, 297, 299, 329
PCP 230–231, 329
PCP-Na 243, 253
Pendimethalin 234, 253, 329
Perfluidone 329
Permethrin 232, 244, 253
PESTAN program 58
Pesticide Mobility Index (PMI) 58
Pesticide Root Zone Model (PRZM) 58
Pesticides
benefits of 42–43, 290
chemical names 251–253, 328–329
in China 225–253
degradation and transformation 55–57, 229–234, 307–313, 315, 318
effects on soils 304–306
formulation changes 292–293
in groundwater 65–68, 244–246
in Hong Kong 261
in Korea 290–312
lead and cadmium from 6, 10
in Malaysia 333–335, 340–341
in Oceania 42–70
persistence 228–229, 233–234, 237–244
photodecomposition 230–233
plant uptake 239–240, 306–307, 316–318
production of 226–228
prohibition of 320–321
public health guidelines 48–50
registration 47–48, 226
remediation 246–247
residue guidelines 48–50, 59–68, 342–343
soil organisms and 240–244
sorption 50–55, 231, 235–236
transport 53, 58–59, 235–237, 306, 322
usage 43–50, 293–294
volatilization 57–58, 236–238
pH 202–204, 217–218. *See also* Acidification; Soil pH
Phenazin 305
Phenothrin 253
Phenthoate 300
Pheretima guillelmi 242
Phorate 230, 231, 234, 238, 253
Phosalone 232, 253
Phosmet 232, 253, 318–319
Phosphate fertilizers
cadmium in 31, 86
cadmium loading from 6–9, 13–16, 26–27, 159, 340, 351
cadmium:phosphorus ratio 11
DDT degradation 69
heavy metal solubility and 159–160
in Malaysia 334–335
quality and type of 17–19
rock phosphate 8–9, 221, 335
Phosphate fixation 213–214
Phosphorus 8–9, 11, 13–15
Phosvel 304, 329
Photodecomposition of pesticides 230–233
Photolysis 55
Phoxim 234, 253
Phthalates 129–130
Phthalide 329
Phytosiderophores 19
Phytotoxicity 30, 52, 278–280
Phytotoxicity-based Investigation Levels (PILS) 30
Pigs, cadmium in 190
Pirimicarb 253
Plant growth, acidification and 215–216, 221–222
Plant growth regulators 320
Plant uptake
of arsenic 173, 280
of cadmium 12–15, 119–120, 161–162, 173, 280, 363–364
of copper 173, 280
of lead 280, 355, 365, 367
location in plant 119
of mercury 119
of nickel 280
of pesticides 239–240, 306–307, 316–318
of radionuclides 111–118
as remediation 118–121
of selenium 359, 368
temperature and 174
of zinc 280
Plutonium
human intake 111
migration 101–103, 105–106, 108
plant uptake 113
soil deposition 99–100
transfer factors 119
PMI (Pesticide Mobility Index) 58
Pollution, concept of 1–2
Pollution index grades 175
Polychlorinated biphenyls (PCBs) 261
Polyoxins 305, 329
Populus canadensis 162
Populus pekinensis 162

Populus xenrama robusta 162
Porina caterpillar *(Wiseana)* 60
Port Pirie smelter (Australia) 4–7, 10, 14
Potassium adsorption 213
Potassium fertilizers 18
Potatoes 14–15, 18, 20–21, 23–24, 26
Practical capacity 179
Pretilachlor 308, 329
Procymidone 298, 300, 308–309, 329
Profenophos 341
Prometryne 231, 253, 329
Propanil 231, 253, 329
Propazine 232, 253
Propoxur 253, 302
Propylene dichloride 253
Prothiocarb 253
PRZM (Pesticide Root Zone Model) 58
Public health guidelines 48–50. *See also* Regulations
Pursuit 234, 253
Pyrazosulfuron-ethyl 234, 253
Pyridaphenthion 233, 253

Q

Quinalphos 232, 234, 253
Quinclorac 234, 253, 329
Quizalofop-ethyl 234, 253

R

Radionuclides 96–122
  decontamination 118–121
  fallout of 99–100, 102–103, 105, 111–118, 121
  human intake 108–111
  migration 101–108
  overview 96
  plant uptake 108, 111–118
Radish *(Raphanus sativus)* 280
Radium 119
Raphanus sativus 280
Rare earth elements (REE) 169–171, 184–186, 193, 197
Redox reactions
  metals 156–157, 174, 191
  pesticides 55, 230, 305
  pH and 207–208
REE (rare earth elements) 169–171, 184–186, 193, 197
Regulations
  Australia and New Zealand 28–34
  China 155
  Japan 133
  Korea 276, 281, 313–314, 316–318
  Malaysia 342–343
Remediation
  bioremediation 69, 145
  classification of 142
  groundwater extraction 143–145, 147
  of metals 87–90, 156–163
  of organochlorines 141–147
  of pesticides 68–70, 246–247, 309–310, 314–320
  plant uptake 118–121
  redox control 156–157
  sequence of events 146–147
  soil dressing 156
  soil excavation 143
  soil vapor extraction 144–145, 147
Renal diseases 366
Resmethrin 253
Rice. *See also* Paddy fields
  cadmium in 87–91, 159–160, 173, 190, 192, 274–275
  human intake 280
  lead in 173, 274–275
  metals in 274–275, 280
  pesticides 240, 294, 319
  radionuclides 105–106, 111–118
  redox reactions 156–157
  water budget 97–98
Roads, metals from 85
Rock phosphate 8–9, 221, 335. *See also* Phosphate fertilizers
Rodenticides 335
Ruthenium 100

S

S-5439 232, 253
Salvinia natans 364
Sanggamma-C 329
Scandium 185. *See also* Rare earth elements
Seaweed 111
Sebufos 234, 253
Selenium
  in African soils 349–350, 358–362, 371
  background values 30, 168–171, 371
  deficiencies and disease 170–171, 368
  in humans 368–369, 372
  plant uptake 359, 368
  sources 367–368, 372
  volatilization of 163
Sepat siam *(Trichogaster pectoralis)* 342
Sequential extractions 154–155, 179–182, 184–186
Sethoxydim 253
Sevin 304
Sewage irrigation 175–177, 186–187, 190
Sewage sludge
  cadmium from 6–7
  in China 153–154
  global metal input from 4
  in Hong Kong 256–259